The Biochemistry of Methylotrophs

The Biochemistry of Methylotrophs

C. ANTHONY

Department of Biochemistry
University of Southampton
England

1982

ACADEMIC PRESS
A Subsidiary of Harcourt Brace Jovanovich, Publishers

London · New York
Paris · San Diego · San Francisco
São Paulo · Sydney · Tokyo · Toronto

ACADEMIC PRESS INC. (LONDON) LTD.
24—28 Oval Road
London NW1 7DX

United States Edition published by
ACADEMIC PRESS INC
111 Fifth Avenue
New York, New York 10003

Copyright © 1982 by
ACADEMIC PRESS INC. (LONDON) LTD.

All Rights Reserved

No part of this book may be reproduced in any form by photostat, microfilm, or any other means, without written permission from the publishers

British Library Cataloguing in Publication Data

Anthony, C.
   The biochemistry of methylotrophs.
   1. Bacteria   2. Microbiological chemistry
   I. Title
   576'.19'2   QR88

ISBN 0-12-058820-X

LCCCN 81-69589

Typeset by Bath Typesetting Ltd., Bath
and printed in Great Britain by
St Edmundsbury Press, Bury St Edmunds, Suffolk

Foreword

There are two kinds of organism which stand out as the biosynthetic virtuosi of the living world—the autotrophs and the methylotrophs. The autotrophs are organisms which can synthesize all their cell constituents from carbon dioxide as sole carbon source; methylotrophs can perform such total biosynthesis from reduced carbon compounds containing one or more carbon atoms but containing no carbon–carbon bonds. Autotrophs have long been the subject of intense study, if for no other reason than that they include the green plants as well as microorganisms. Although the first methylotroph, *Bacillus methanicus*, was discovered 76 years ago growing aerobically on methane, the next 40 to 50 years saw the discovery of only a few other methylotrophs growing on reduced C_1 compounds such as methane, methanol, methylamine and formate. They were regarded as somewhat exotic organisms and received scant biochemical attention; it was often speculated that they might merely couple the energy of oxidation of the C_1 substrate to the reduction of carbon dioxide and thus in effect be chemosynthetic autotrophs. If this had turned out to be the case, the subject of methylotrophy would have remained but a minor variant on a familiar theme. However, it was recognised in the 1950s that although some methylotrophs were indeed autotrophs in thin disguise, many other methylotrophic organisms assimilated C_1 units at levels more reduced than carbon dioxide by unknown assimilatory processes. As a result an explosive development of this whole field then took place. Biochemists espied virgin areas of intermediary metabolism and energy transduction, the range of methylotrophic microorganisms was greatly extended by development of new enrichment procedures and the new organisms were ripe for genetic analysis and manipulation. At the same time biotechnologists realized that methane and methanol offered entirely new industrial possibilities as cheap and readily available fermentation feedstocks, thus providing an additional stimulus for research and development.

A wealth of publications resulted from all this activity and for 20 years this burgeoning literature has been reviewed at frequent intervals by various investigators. With many of the major biochemical problems now solved, and the remaining ones clearly defined, it has been clear for some time that there was an urgent need for a definitive and wide-ranging textbook on the subject. The present book has been written to meet such a demand and it is hoped that it will serve as a standard reference book for all *aficionados* of this field, whether they be newcomers or veterans. The newcomers can be assured that it has been written by a biochemist who, since 1960, has himself been a highly active investigator and an authority on methylotrophy; the veterans will not need any such prefatory assurance from me.

Sheffield 1982   *J. R. Quayle*

Preface

In 1960, when I started working with methylotrophs I had half a dozen reprints (the total available) on these organisms. The growth of the literature since then has been exponential, and the decision to write this book is based on the well-established practice of harvesting in the mid-log phase; this is in order to avoid being completely submerged beneath the methylotrophic culture.

A preface usually defines the limits of the author's intentions; the ground covered and the likely readership. I shall do likewise; this book covers everything to do with the biochemistry of methylotrophs and it is written to suit everyone.

Perhaps I had better attempt to justify this bold claim. My first aim was to cover all aspects of methylotrophy in the detail required by anyone active in research in this field, whether they be microbiologists, microbial physiologists, "metabolic" biochemists or enzymologists. Such an aim, however, might tend to diminish the usefulness of a book to those using it as an introduction to any aspect of the subject; this is because more fundamental aspects can be obscured by detail, and submerged beneath lines of references to the original sources. I have tried to avoid this in two ways. When possible I have arranged references in separate tables so as to leave the text clear. More important, each chapter starts with a descriptive section which I have tried to make comprehensible to a senior undergraduate. The evidence for conclusions initially presented is then discussed in separate sections. When in doubt I have erred, I hope, on the side of too much explanation rather than too little.

As the subject of the book is biochemistry, the first chapter was to have been a very brief introductory guide to the methylotrophs themselves. This intention was abandoned when I realised that the biochemists for whom it was written could probably cope with any biochemistry they might come across in the literature, but might find the microbes themselves bewildering beyond belief. I hope that the first chapter now provides a complete account of the methylotrophic bacteria in a way that allows a ready evaluation of which are likely to be biochemically similar and which are probably different.

Chapters 2, 3, 4 and 5 cover the carbon assimilation pathways of methylotrophic bacteria. The pathways, and evidence for them, are first described; this is followed by summaries of the distribution of the pathways amongst the methylotrophs; the individual enzymes are then described and regulation of the assimilation pathways discussed.

Chapters 6 and 7 deal with the enzymes (many of them novel) involved in the oxidation of methane, methanol and methylated amines.

Chapter 8 is devoted to electron transport and energy transduction; and Chapter 9 attempts to interrelate all the biochemistry discussed in Chapters 2 to 8 in a discussion of growth yields and bioenergetics in methylotrophs.

Because the biochemistry of oxidation and assimilation of methanol in yeasts differs in most important respects from that in methylotrophic bacteria, Chapter 10 is exclusively concerned with metabolism in methylotrophic yeasts.

Chapter 11 is a self-contained chapter on methanogens and methanogenesis. Although few methanogens are actually methylotrophs, the recent rapid advances in our knowledge of the unique biochemistry of these "archaebacteria" made the temptation to include this complete review irresistible.

Chapter 12 deals with the commercial exploitation of methylotrophs. The biochemistry of microbes growing on methane or methanol makes them especially suitable for commercial exploitation; this includes their use as single cell protein (SCP); their use for the production of carboxylic acids, amino acids and vitamins; their use as biocatalysts; and their use in electroenzymology and biofuel cells. Although not dealing with detailed practical aspects of such processes, this chapter aims to provide a foundation for understanding this rapidly-developing area of biotechnology.

February 1982   *C. Anthony*

Acknowledgements

My first acknowledgement must be an expansion of the Dedication of this book, which is to K. G. Wiles, L. J. Zatman and J. R. Quayle, FRS.

"Willy" Wiles was my biology teacher in the sixth form at Watford Boys' Grammar School; his enthusiastic eccentric insistence that Life is "interesting", and worth some effort in understanding, was both exceptional and captivating. Len Zatman was first my tutor then my Ph.D. supervisor in the Microbiology Department at Reading University, and in spite of this he has remained a good friend. Rod Quayle is, of course, the "Godfather" of methylotrophy, honoured above all men by the international community of methylophiles. To him I owe a second thanks for generously agreeing to write a Foreword to "The Biochemistry of Methylotrophs".

I have received a great deal of help in the preparation and writing of this book, not least from the many scientists who have kindly sent me reprints and manuscripts prior to publication. Pat Goodwin (née Dunstan), my first research student, deserves a special thanks for her encouragement and valuable discussions in the early stages of preparation, and I must also thank my present research students and assistants who have probably suffered from both my absence and from my presence during the last two years (Matthew Beardmore-Gray, Steve Ford, Steve Froud, Ashley Lawton, David O'Keeffe and Dudley Page). David O'Keeffe has worked with me for seven years and to him I owe a special debt of gratitude — both for this, and for his very generous help in keeping the lab running and for accepting many extra burdens during the final stages of production of this book.

A final, most important word of thanks I owe my wife, Elizabeth, for her support and encouragement, without which "The Biochemistry of Methylotrophs" would not have been written.

To K. G. Wiles, L. J. Zatman
and J. R. Quayle, FRS

Contents

Foreword	v
Preface	vii
Acknowledgements	ix

CHAPTER 1. METHYLOTROPHIC BACTERIA 1
I. Introduction 2
II. Definition of methylotrophy and the taxonomy of methylotrophs . . 2
III. Microbes growing on methane (methanotrophs) 4
   A. General description of the methanotrophic bacteria 5
   B. Structural features of methanotrophic bacteria 13
   C. Biochemical and physiological features of methanotrophs used in their classification 20
IV. Methylotrophs unable to grow on methane 22
   A. Obligate methylotrophs able to use methanol or methylamine but not methane 23
   B. Facultative methylotrophs unable to grow on methane . . . 27
V. The place of methylotrophs in nature 38
   A. The microbial production and utilisation of methane . . . 38
   B. The microbial production and utilisation of methanol and formate . 39
   C. The microbial production and utilisation of methylated amines . . 40

CHAPTER 2. THE RIBULOSE BISPHOSPHATE PATHWAY OF CO_2 ASSIMILATION 42
I. Introduction: the definition of autotrophy 42
II. Description of the ribulose bisphosphate pathway 43
III. The occurrence and distribution of the ribulose bisphosphate pathway . 46
IV. The reactions of the ribulose bisphosphate pathway 51
V. Regulation of the ribulose bisphosphate pathway 58

CHAPTER 3. THE RIBULOSE MONOPHOSPHATE CYCLE OF FORMALDEHYDE FIXATION 60
I. Introduction 60
II. Description of the ribulose monophosphate pathways . . . 62
   A. The biosynthesis of a C_3 compound from formaldehyde . . . 62
   B. The biosynthesis of carbohydrates 68
   C. The oxidation of formaldehyde by a dissimilatory RuMP cycle . . 68
   D. The interconversion of pyruvate, phosphoenolpyruvate and oxaloacetate in bacteria with the RuMP cycle 70
III. The occurrence and distribution of the RuMP pathways . . . 74
   A. The RuMP pathways of formaldehyde assimilation 74
   B. The dissimilatory cycle of formaldehyde oxidation 76
IV. Reactions of the RuMP pathways 77
   A. Enzymes involved in the fixation phase (all variants) . . . 77
   B. Enzymes involved in the cleavage phase 81
   C. Enzymes involved in the rearrangement phase 85

xii   Contents

    D. The key enzyme of the RuMP dissimilatory cycle; 6-phosphogluconate dehydrogenase 89
    E. The interconversion of phosphoenolpyruvate with triose phosphates . 90
    F. The interconversion of pyruvate, phosphoenolpyruvate and oxaloacetate 91
  V. Regulation of the RuMP pathway of formaldehyde assimilation . . 93

CHAPTER 4. THE SERINE PATHWAY OF FORMALDEHYDE ASSIMILATION 95
I. Description of the serine pathway 96
II. Evidence for the serine pathway 98
    A. Evidence obtained by incubating whole cells with radioactive isotopes 98
    B. Evidence for the serine pathway from enzyme studies . . . 101
    C. Evidence for operation of the serine pathway from mutant studies . 105
III. The occurrence and distribution of the serine pathway . . . 109
    A. Methylotrophs using the pathway as their main assimilation pathway (Table 26) 109
    B. Distribution of icl$^+$ and icl$^-$ variants of the serine pathway (Table 27) . 113
IV. The reactions of the serine pathway 115
    A. Enzymes involved in the formation of oxaloacetate from glyoxylate, formaldehyde and CO_2 115
    B. The formation of glyoxylate and acetyl-CoA from oxaloacetate . 121
    C. The oxidation of acetyl-CoA to glyoxylate 124
V. Regulation of the serine pathway 126
    A. Regulation at the genetic level 126
    B. Regulation at the level of enzyme activity 132

CHAPTER 5. THE TCA CYCLE AND GROWTH OF METHYLOTROPHIC BACTERIA ON MULTICARBON COMPOUNDS . . 137
I. The operation of the TCA cycle in methylotrophs 137
II. The basis of obligate methylotrophy 142
III. The enzymes of the TCA cycle and their regulation 144
    A. Citrate synthase 144
    B. Isocitrate and malate dehydrogenases 145
    C. Pyruvate and 2-oxoglutarate dehydrogenases 146
IV. The growth of *Pseudomonas* AM1 on ethanol, lactate and 3-hydroxybutyrate 147
V. The growth of *Pseudomonas* AM1 on propane 1,2-diol and 4-hydroxybutyrate 150
VI. The metabolism of trimethylsulphonium salts by *Pseudomonas* MS . 151

CHAPTER 6. THE BACTERIAL OXIDATION OF METHANE, METHANOL, FORMALDEHYDE AND FORMATE 152
I. Introduction 153
II. The oxidation of methane to methanol 153
    A. Introduction 153
    B. The methane monooxygenase from *Methylococcus capsulatus* (Bath) . 155
    C. The methane monooxygenase from *Methylosinus trichosporium* (OB3b) 158
    D. Is there more than one type of methane monooxygenase? . . . 159
    E. Inhibitors of methane monooxygenase 161
    F. The substrate specificity of methane monooxygenases . . . 163
    G. The anaerobic oxidation of methane 167

III. Methanol dehydrogenase	167
A. Introduction	167
B. Substrate specificity of methanol dehydrogenase	171
C. Activators and inhibitors	173
D. The primary electron acceptor	174
E. Molecular weight, isoelectric point and amino acid composition	175
F. Serological relationships between methanol dehydrogenases	176
G. Localisation of the methanol dehydrogenase	176
H. The prosthetic group and mechanism of methanol dehydrogenase	177
I. The interaction of methanol dehydrogenase with the cytochrome system	186
IV. Formaldehyde oxidation	187
A. Introduction	187
B. NAD^+-dependent formaldehyde dehydrogenase	187
C. Dye-linked aldehyde dehydrogenases	190
D. N^5, N^{10}-methylenetetrahydrofolate dehydrogenase	191
E. Methanol dehydrogenase	192
F. A cyclic route for the oxidation of formaldehyde	192
G. Oxidation of formaldehyde by way of the serine pathway	193
V. Formate oxidation	194

CHAPTER 7. THE OXIDATION OF METHYLATED AMINES

I. Introduction	195
II. The oxidation of tetramethylammonium salts	197
III. The oxidation of trimethylamine to dimethylamine and formaldehyde	197
A. Trimethylamine dehydrogenase	197
B. The indirect route for trimethylamine oxidation	200
IV. The oxidation of dimethylamine to methylamine and formaldehyde	201
A. Dimethylamine monooxygenase	201
B. Dimethylamine dehydrogenase	203
V. The oxidation of methylamine to formaldehyde	204
A. Methylamine dehydrogenase	204
B. Methylamine oxidase	208
C. The oxidation of methylamine by way of methylated amino acids	210
VI. The distribution of metabolic routes for the oxidation of methylated amines	217
A. The oxidation of trimethylamine	217
B. The oxidation of dimethylamine	217
C. The oxidation of methylamine	217

CHAPTER 8. ELECTRON TRANSPORT AND ENERGY TRANSDUCTION IN METHYLOTROPHIC BACTERIA

	219
I. Introduction	219
II. The cytochromes c of methylotrophs	224
A. Introduction	224
B. Reaction of cytochromes c with CO	226
C. The autoreduction of cytochrome c in methylotrophs	227
III. The role of cytochrome c in the oxidation of methanol	229
A. Introduction	229
B. The reaction between methanol dehydrogenase and cytochrome c	230

xiv   Contents

IV. Electron transport and proton-translocating systems in methylotrophs . 232
   A. Introduction 232
   B. Electron transport and proton translocation in *Pseudomonas* AM1 . 233
   C. Electron transport and proton translocation in *Methylophilus methylotrophus* 236
   D. Electron transport and proton translocation in *Paracoccus denitrificans* 239
   E. Electron transport and proton translocation in *Methylosinus trichosporium* and other methanotrophs 240
V. The Coupling of methanol oxidation to ATP Synthesis . . . 242

CHAPTER 9. GROWTH YIELDS AND BIOENERGETICS OF METHYLOTROPHS 245
I. The concept of Y_{ATP} 245
II. Assumptions and methods for developing theoretical assimilation equations 249
III. The limitation of cell yields by ATP, NADH and carbon supply . . 251
   A. ATP-limited bacteria 257
   B. Carbon-limited bacteria 258
   C. NADH-limited bacteria 258
IV. The prediction of growth yields in methylotrophs 260
   A. Growth of non-photosynthetic organisms on methanol . . 260
   B. Growth of *Rhodopseudomonas acidophila* on methanol . . 262
   C. Growth on methane 264
V. The effect on electron transport systems of the oxidative and assimilatory pathways of methylotrophs 266

CHAPTER 10. METABOLISM IN THE METHYLOTROPHIC YEASTS 269
I. The methylotrophic yeasts 269
II. Oxidative metabolism in yeasts 276
   A. The oxidation of methanol to formaldehyde 276
   B. The formaldehyde dehydrogenase of yeasts 283
   C. Hydrolysis of *S*-formylglutathione, and the oxidation of formate . 284
III. The dihydroxyacetone (DHA) cycle of formaldehyde assimilation in yeasts 285
   A. Description of the DHA cycle 285
   B. Oxidation of formaldehyde by a dissimilatory DHA cycle . . 287
   C. Evidence for the DHA cycle of formaldehyde assimilation . . 288
   D. The reactions of the DHA cycle 290
IV. Regulation of methanol metabolism in yeasts 292
   A. Regulation of synthesis of methanol-metabolising enzymes . 292
   B. Regulation of enzyme activity 294

CHAPTER 11. METHANOGENS AND METHANOGENESIS . . 296
I. Introduction 296
II. The methanogens 300
III. Structure, cell wall structure and lipid composition of methanogens . 301
IV. The RNAs and DNA of methanogens 304
V. Biosynthesis in methanogenic bacteria 304
VI. Energy coupling in methanogens 306
   A. Introduction 306

B. Novel coenzymes from methanogens 308
C. ATP synthesis in methanogens 315
D. The reduction of CO_2 to CH_4 319
E. Methanogenesis from acetate, methanol, methylated amines and formate 323

CHAPTER 12. THE COMMERCIAL EXPLOITATION OF METHYLOTROPHS 328

I. The use of methylotrophs for the production of single cell protein (SCP). 328
  A. Introduction to SCP 328
  B. Single cell protein from methylotrophs 331
II. The use of methylotrophs for the overproduction of metabolites ("fermentation products") 338
  A. Introduction 338
  B. The production of vitamin B_{12} and riboflavin derivatives . . 339
  C. The production of carboxylic acids and amino acids by methylotrophs 339
  D. Commercial methanogenesis 342
III. The use of methylotrophs and their enzymes as biocatalysts. . . 342
  A. Enzymic and biological assay systems 342
  B. The oxidation of alcohols 343
  C. The oxidation of hydrocarbons and their derivatives by methane monooxygenase 344
  D. The use of methylotrophs and their enzymes in electroenzymology and in biofuel cells 348

References 351

List of figures 379
List of tables 381
Index 383

1
Methylotrophic bacteria

I. Introduction	2
II. Definition of methylotrophy and the taxonomy of methylotrophs	2
III. Microbes growing on methane (methanotrophs)	4
A. General description of the methanotrophic bacteria	5
1. The isolates of Whittenbury and his colleagues	5
2. Other obligate methanotrophs	9
3. The facultative methanotrophs (*Methylobacterium* spp.)	10
B. Structural features of methanotrophic bacteria	13
1. Resting stages	13
(a) Exospores	13
(b) Lipid cysts	14
(c) "*Azotobacter*-type" cysts	15
(d) "Immature *Azotobacter*-type" cysts	15
2. Internal membranes	15
3. Rosette formation, flagella and pigmentation	19
(a) Rosette formation	19
(b) Motility and flagella	19
(c) Pigmentation	19
C. Biochemical and physiological features of methanotrophs used in their classification	20
1. Major carbon assimilation pathways	20
2. The dehydrogenases for 2-oxoglutarate, malate, isocitrate, glucose 6-phosphate and 6-phosphogluconate	20
3. Growth temperatures, growth on methanol and nitrogen fixation	21
(a) Growth temperatures	21
(b) Growth on methanol	21
(c) Nitrogen fixation	21
4. DNA base ratios and predominant fatty acids	22
(a) DNA base ratio (percentage G + C)	22
(b) Predominant fatty acids	22
IV. Methylotrophs unable to grow on methane	22
A. Obligate methylotrophs able to use methanol or methylamine but not methane	23
B. Facultative methylotrophs unable to grow on methane	27
1. The pink facultative methylotrophs (Table 11)	27
2. The non-pigmented "pseudomonads" (Table 12)	29
3. Gram-negative (or variable), non-motile rods and coccal rods (Table 13)	29
4. Gram-positive facultative methylotrophs (Table 14)	33
5. Facultative autotrophs or phototrophs growing on methanol or formate (Table 15)	33
6. The *Hyphomicrobia*	35
7. Marine bacteria able to grow on methanol or methylated amines	37
V. The place of methylotrophs in nature	38
A. The microbial production and utilisation of methane	38
B. The microbial production and utilisation of methanol and formate	39
C. The microbial production and utilisation of methylated amines	40

I. Introduction

Once upon a time there was a lonely methylotroph called *Bacillus methanicus*, sole representative in the literature of a metabolic type of microorganism capable of growth on methane or methanol as sole source of carbon and energy (Söhngen, 1906). Relatively few additions were made to the list of known bacteria with these growth characteristics in the next half century and J.R. Quayle was able to quote only seven references to bacteria growing on methane or methanol in his first paper on *Pseudomonas* AM1 (Peel and Quayle, 1961). In the next ten years many new bacteria were described and much of their basic biochemistry elucidated. Since then an appreciation of the potential of these bacteria for production of single cell protein from methane or methanol has stimulated the isolation of many new species of bacteria and yeasts able to grow on these substrates. A measure of the growth of interest in these microbes and their metabolism is indicated by the holding of the International Symposia on Microbial Growth on C_1-compounds; the first informal symposium was in Edinburgh in 1973 and since then more formal symposia have been held in Tokyo and Putschino-on-Oka, the last being in Sheffield in the summer of 1980 (see Dalton, 1981).

The explosive development of this subject as outlined here has resulted in a vast literature presenting a daunting prospect to anyone approaching it for the first time. It is for such workers that this book has been written; it is also hoped that it will provide a useful guide for those who, like the author, have helped produce this literature and are now finding themselves becoming submerged beneath it.

II. Definition of methylotrophy and the taxonomy of methylotrophs

Microbes growing on C_1-compounds as their sole source of carbon and energy must make every carbon–carbon bond "de novo", and this problem is shared with microbes growing on compounds having more than one carbon atom but having no carbon–carbon bonds

These microbes are called methylotrophs by analogy with those autotrophs that are able to use CO_2 as their sole source of carbon.

Colby and Zatman (1972) have provided a generally accepted definition of methylotrophs which I use here in a slightly altered form to allow the inclusion of bacteria assimilating reduced C_1 compounds by way of CO_2 and the ribulose bisphosphate pathway. *Methylotrophs* are those microorganisms able to grow at the expense of reduced carbon compounds containing one or more carbon atoms but containing no carbon–carbon bonds. Obligate methylotrophs grow only on such compounds whereas facultative methylotrophs are also able to grow on a variety of other organic multicarbon compounds.

Table 1 lists those substrates known to support the growth of methylotrophs. Bacteria able to grow on carbon monoxide are, strictly speaking, methylotrophs, obtaining their energy by oxidising CO to CO_2 and assimilating the CO_2 by the "autotrophic" ribulose bisphosphate pathway. They are not discussed further in this book and the reader is referred to the following references for useful reviews of these organisms (Zavarzin and Nozhevnikova, 1976; Colby *et al.*, 1979; Schlegel and Meyer, 1981; Cypionka *et al.*, 1980; Meyer and Schlegel, 1980).

TABLE 1
Substrates used for methylotrophic growth

Compounds containing one carbon atom		Compounds containing more than one carbon atom	
Methane	CH_4	Dimethyl ether	$(CH_3)_2O$
Methanol	CH_3OH	Dimethylamine	$(CH_3)_2NH$
Methylamine	CH_2NH_2	Trimethylamine	$(CH_3)_3N$
Formaldehyde	$HCHO$	Tetramethylammonium	$(CH_3)_4N^+$
Formate	$HCOOH$	Trimethylamine N-oxide	$(CH_3)_3NO$
Formamide	$HCONH_2$	Trimethylsulphonium	$(CH_3)_3S^+$
Carbon monoxide	CO		

Ideally this first chapter would include a taxonomic survey of methylotrophs with some generally acceptable suggestions on nomenclature. However, this has proved impossible for a number of reasons. The first is that the property of methylotrophy is a very special property; very few bacteria in culture collections grow on methane or methanol unless they were first isolated on these substrates. Thus methylotrophs do not always fit neatly into previously characterised groups. A second difficulty is that the obligate methylotrophs have relatively few features that can be used in conventional taxonomic studies. A third problem is that many methylotrophs were initially isolated with the aim of studying their peculiar metabolism, and while this has been achieved, the original description may be inadequate or no longer correspond to the characteristics of the organism which is now available in culture collections and used for extensive biochemical investigations, or for production of millions of tons of single cell protein.

The best that this chapter can hope to achieve is to place the methylotrophs into groups having similar growth properties and, sometimes, structural features. These groups will be for convenient storage of information and they are not necessarily taxonomic groups with genetic relationships.

Methylotrophic yeasts (discussed separately in Chapter 10) do not pose such a problem as the bacteria, because they are typical yeasts which can be classified with previously described non-methylotrophic yeasts. The extra genetic information required for the oxidative and assimilatory pathways during growth on methanol may be sufficiently great, however, to warrant separate species for them.

The nomenclature of methylotrophs is confusing and likely to remain so. It has been stated by Cowan (1970) in his heretical taxonomy for bacteriologists that "names are only labels and have no significance except as a means of communication". Although this is a sound principle aimed at avoiding wrangles over nomenclature, it still leaves the problem of what is actually communicated in the name chosen. Many methylotrophs were first isolated by biochemists with little interest, expertise or experience in taxonomy or nomenclature; some chose to give completely new names, some gave the name of the most similar non-methylotrophic bacteria while others, in desperation I suppose, produced a prosaic series of numbers. The result of this is that the information communicated by the name may be wrong or non-existent. A further problem is that two methylotrophs may have the same generic name but be very different, while very similar bacteria may have completely different names.

In this book I have used the names in most common use and I have tried to indicate where differently named bacteria are very similar when this is particularly relevant.

III. Microbes growing on methane (methanotrophs)

A major landmark in the development of our knowledge of methanotrophs was the description in 1970 of more than 100 new isolates by Whittenbury and his colleagues at Edinburgh (Whittenbury *et al.*, 1970a). This was a landmark because previously only four methanotrophs had been described. The first was called *Bacillus methanicus* (Söhngen, 1906), reisolated as *Pseudomonas methanica* (Söhngen) by Dworkin and Foster (1956) and now renamed *Methylomonas methanica*. In the half century between the work of Söhngen and that of the Edinburgh group only three further methanotrophs were described; these were *Pseudomonas methanitrificans* (Davis *et al.*, 1964), *Methanomonas methano-oxidans* (Brown *et al.*, 1964) and *Methylococcus capsulatus* (Foster and Davis, 1966). All of these were strictly aerobic, obligate methylotrophs able to grow on methane, methanol or dimethylether but not on multicarbon compounds.

It was only in 1974 that Hanson and his colleagues at Wisconsin first described facultative methanotrophs (Patt *et al.*, 1974), and in 1979 that they first described methanotrophic yeasts (Chapter 10). It is now clear that methanotrophs are common in Nature and that there is a great biological and biochemical diversity amongst these organisms. The very important role of methanotrophs and methanogens (Chapter 11) in the global carbon cycle has been discussed fully elsewhere (Schlegel *et al.*, 1976; Vogels, 1979; Hanson, 1980).

A. General description of the methanotrophic bacteria

The main subject of this book is the biochemistry of methylotrophs and the following descriptions are primarily for the convenience of biochemists interested in their metabolism. It does not aim to evaluate taxonomic proposals but merely to collate the available information into tables and to make a few comments where the picture is rather confused. Whittenbury and his colleagues originally categorised their new isolates into groups and subgroups and they avoided genus and species designations. In the subsequent literature, however, they have referred to them as genera and species and I shall do likewise.

The descriptions of Whittenbury's 100 new isolates provides a convenient introduction to the methanotrophs. Because they only grow on C_1 compounds, the characteristics available for division into genera and species are necessarily structural and biochemical and these characteristics are discussed in later sections of this chapter.

1. The isolates of Whittenbury and his colleagues

Table 2 is a list of properties of the obligate methanotrophs in the collection of R. Whittenbury and these properties are shared by most other obligate methanotrophs. Table 4 is based on the main characteristics of the five genera initially described, and it includes only those features that might be expected to remain important in any classification scheme (Whittenbury *et al.*, 1970a,b; Davies and Whittenbury, 1970; Davey *et al.*, 1972). This table also includes the *Methylobacterium* species which are the facultative methanotrophs whose properties are more fully described in Table 7.

Table 3, based on that of Colby *et al.* (1979), includes some of the biochemical generalisations that can be made about the main groups of methanotrophs. The use of these biochemical properties in the classification of methanotrophs is discussed below.

Table 5 is based on the table of properties of subgroups or species of methanotrophs isolated by Whittenbury *et al.* (1970a) and is included as an illustration of the relative paucity of physiological properties available for classification of methanotrophs.

TABLE 2
Properties of the obligate methanotrophs in Professor Whittenbury's collection

Most of these properties are also shared by the majority of obligate methanotrophs isolated by other workers (see text).

All are Gram-negative, strictly aerobic, and are rods, vibrios or cocci.

All form a differentiated resting body (exospore or cyst).

All possess a complex internal arrangement of paired membranes.

All are catalase-positive, oxidase-positive and have typical cytochromes a, b and c.

All are (to some degree) sensitive to normal oxygen tension in air. Some do not grow in shaking flasks. During nitrogen fixation they become extremely oxygen-sensitive.

All use methane, methanol, methyl formate, dimethylcarbonate and (those tested) formaldehyde as sole carbon and energy sources.

All use ammonia as nitrogen source; most use nitrate and nitrite. Some use urea, amino acids and yeast extract. All those with a serine pathway (and some with the RuMP-pathway) fix molecular nitrogen.

All reduce nitrate to nitrite, but do not grow anaerobically on methane.

All oxidise (or co-oxidise) methane, methanol, ammonia, carbon monoxide, dimethylether, propane, ether, ethanol, propanol, butanol, formaldehyde and formate.

TABLE 3
Further properties of methanotrophic bacteria

	Type I bacteria (RuMP pathway) (Methylococcus, Methylomonas and Methylobacter)		Type II bacteria (serine pathway) (Methylosinus, Methylocystis and the facultative Methylobacterium)
TCA cycle	Incomplete (no oxoglutarate dehydrogenase)		Complete
Nitrogenase	Present in some		Present in all tested
Chain length of main fatty acid in lipid	16 carbon atoms		18 carbon atoms
Glucose-6-phosphate dehydrogenase	Present (NADP$^+$-specific)		Absent during growth on methane
6-phosphogluconate dehydrogenase	Present (NADP$^+$-specific)		Absent
Pentose phosphate isomerase	High level		Low level
	Methylococcus	*Methylomonas Methylobacter*	*All Type II bacteria*
Isocitrate dehydrogenase	NAD$^+$-specific	NAD$^+$ and NADP$^+$	NADP$^+$-specific
Malate dehydrogenase	Low activity	High activity	High activity
Ribulose bisphosphate carboxylase	+	−	−
Phosphoribulokinase	+	−	−
Growth at 45°C	+	some +	−

Not all strains have been tested for all characteristics and discrepancies in these generalisations may well arise. Most enzyme measurements are taken from Davey et al. (1972).

TABLE 4

Properties characterising the main groups (Genera) of methanotrophs

	Morphology	Resting stage[b]	Rosette formation	DNA base ratio (G + C)	Major carbon pathway	Motility[c]
Type I[a] (obligate)						
Methylococcus	Coccus	Azotobacter-type cyst (immature)	−	62–64%	RuMP pathway	−
Methylomonas	Rod	Azotobacter-type cyst (immature)	−	50–54%	RuMP pathway	+P
Methylobacter	Rod	Azotobacter-type cyst	−	50–54%	RuMP pathway	±P
Type II[a] (obligate)						
Methylosinus	Rod or pear-shaped	Exospore	+	62·5%	Serine pathway	+PT
Methylocystis	Rod or vibroid	'Lipid cyst'	+	62·5%	Serine pathway	−
Type II[a] (facultative)						
Methylobacterium	Rod	Exospore	+	58–66%	Serine pathway	+P

[a] Type I bacteria have internal membranes arranged as bundles of vesicular discs; Type II bacteria have membranes arranged in pairs around the cell periphery.
[b] Not all organisms form an identifiable resting stage.
[c] P indicates a single polar flagellum; PT indicates polar tufts of flagella; ± indicates that some strains are motile (see Table 5); all *Methylomonas* strains were motile except *M. streptobacterium*.

TABLE 5
Properties of species (sub-groups) of methanotrophs (taken from Whittenbury et al., 1970a)

Organism	Number of strains	Growth at 37°C	Growth at 45°C	Growth on methanol	Motility and flagellation	Capsule formed	Growth on methane enhanced by yeast extract, malate, acetate, or succinate (0·1%)
Type I bacteria							
Methylomonas methanica	30	−	−	+	+(P)	+	+
Methylomonas albus	3	+	−	+	+(P)	−	+
Methylomonas streptobacterium	5	−	−	−	−	+	−
Methylomonas agile	4	+	−	++	+(P)	−	+
Methylomonas rubrum	7	−	−	++	+(P)	++	−
Methylomonas rosaceous	2	−	−	−	+(P)	++	−
Methylobacter chroococcum	9	+	−	−	−	++	−
Methylobacter bovis	5	−	−	−	−	−	+
Methylobacter capsulatus	4	−	−	−	+(P)	+++	−
Methylobacter vinelandii	5	++	++	−	+(P)	++	−
Methylococcus capsulatus	3	++	−	−	−	++	−
Methylococcus minimus	5	−	−	−	−	++	+
Type II bacteria							
Methylosinus trichosporium	9	+	−	+	+(PT)	++	−
Methylosinus sporium	12	++	−	−	+(PT)	++	+
Methylocystis parvus	1	+	−	−	−	+	−

It has been suggested that *Methylococcus minimus* should be changed to *Methylomonas minimus* because it is rod-shaped (Whittenbury, personal communication). Slight variations from the original have been included after discussion with Professor Whittenbury. The capsules of Type II bacteria consisted of short fibres radiating from the cell wall; no structure was seen in capsules of Type I bacteria. Motility was by way of a polar flagellum (*P*) or by polar tufts of flagella (*PT*).

2. Other obligate methanotrophs

Methanomonas methano-oxidans (Stocks and McCleskey, 1964b; Brown *et al.*, 1964; Smith and Ribbons, 1970) is most similar to the *Methylosinus* strains. It is rod- or pear-shaped; it grows in rosettes, has Type II (peripheral) membranes; it assimilates its cell carbon by the serine pathway (Lawrence and Quayle, 1970) and it has a complete TCA cycle (Wadzinski and Ribbons, 1975b). It may differ from *Methylosinus* sp. in having only a single polar flagellum rather than a tuft of polar flagella and in forming microscopic colonies.

Pseudomonas methanitrificans was suggested by Whittenbury *et al.* (1970a) to be a strain of *Methylosinus*. Its original name was given because it was isolated using atmospheric nitrogen as the sole nitrogen source. As mentioned in Table 4 nitrogen fixation is a characteristic common to all Type II bacteria so far tested.

Pseudomonas methanica (Söhngen, 1906; Dworkin and Foster, 1956; Leadbetter and Foster, 1958) conforms to the description of the most commonly isolated methanotroph, *Methylomonas methanica*.

Methylococcus capsulatus was first isolated by Foster and Davis (1966), and the name adopted by Whittenbury *et al.* (1970a, b) for all their similar isolates.

The *Methylobacter vinelandii* strains of Whittenbury *et al.* (1970a, b) resemble the methanotrophs isolated by Leadbetter and Foster (1958).

Methylovibrio sohngenii strains (Hazeu and Steenis, 1970) were later identified as strains of *Methylosinus sporium* and *Methylocystis parvus* (Hazeu, 1975).

Methylococcus thermophilus. Malashenko (1976), in a study of thermotolerant and thermophilic methanotrophs described an organism very similar to *Methylobacter vinelandii* (Tables 4 and 5) but having an unusually high percentage (G + C) ratio (59%). In the same study he described a new species of thermophile which he named *Methylococcus thermophilus*. This differed from typical *Methylococcus* strains (Tables 2 and 3) in being motile.

Methylococcus mobilis. Hazeu *et al.* (1980) have recently published a thorough description of this novel species of obligate methanotroph. This non-pigmented coccus appears singly or in pairs, chains or tetrads that are occasionally motile and have one or two flagella. In some conditions multiple-bodied cysts like those of *Methylobacter* are formed. Carbon is fixed by the ribulose monophosphate pathway and electron micrographs reveal the presence of Type I membranes as well as reserve material; this is probably polysaccharide because these bacteria do not form poly 3-hydroxybutyrate. Growth only occurs on methane or methanol but peptone increases the growth rate on methane and in these conditions unusual tubular structures become visible in electron micrographs of cross-sections of bacteria. No fixation of atmospheric nitrogen was observed. The temperature optimum is 30°C but some growth occurs at 37°C. The percentage (G + C) ratio of 56·3 differs from values recorded for both *Methylobacter* and *Methylococcus* species (Table 4). The cysts are more like those of *Methylobacter* species but the new isolate is a coccus. This poses a problem of taxonomy which the authors have avoided by including this interesting organism in the more widely-defined *Methylococcus* genus as defined by Romanovskaya et al. (1978) (see below).

Many strains of obligate methanotrophs similar to those described above and in Tables 2, 3, 4 and 5 have been isolated in recent years (e.g. Malashenko, 1976; Trotsenko, 1976; Galchenko et al., 1978; Romanovskaya et al., 1978). A problem that has arisen during these studies has been due to the isolation of motile cocci and of thermophilic methanotrophs. Romanovskaya et al. (1978) in their proposed scheme for the classification of obligate methanotrophs have included these species together with all *Methylobacter* species (Tables 3 and 5) into the genus *Methylococcus;* the species proposed are given with a brief list of their properties in Table 6.

Besides the Gram-negative methanotrophs described above a number of Gram-positive bacteria (*Mycobacterium* sp. and *Nocardia* sp.) are said to be able to grow on methane but no full descriptions of these are available and little biochemical work on them has been reported (see Seto et al., 1975).

3. The facultative methanotrophs (Methylobacterium spp.)

Eleven strains of facultative methanotrophs have been described, four of these in some detail (Table 7). All are Gram-negative rods having the Type II membrane arrangement and assimilating C_1 compounds by the serine pathway. All were originally isolated on methane and many were isolated from freshwater lakes at depths where O_2 concentrations are relatively low; and some isolates (e.g. *M. organophilum* XX) grow best at low O_2 concentrations. Most are motile by polar flagella and have a (G + C) ratio of 66%; the exception is *Methylobacterium ethanolicum* which is non-motile and has a (G + C) ratio of only 58% (Lynch et al., 1980). Some isolates are resistant to high temperatures (Patt et al., 1974) and exospores were observed in the R6 strain (Patel et al., 1978). The facultative methanotroph described first and used for many biochemical investigations is *Methylobacterium organophilum* XX. This strain usually contains poly-3-hydroxybutyrate and is catalase- and oxidase-positive. It is similar to *Methylococcus capsulatus* in having phosphatidyl choline, squalene and sterols in its membranes (Patt et al., 1976; Bird et al., 1971).

TABLE 6
Summary of some properties of *Methylococcus* species as defined by Romanovskaya et al. (1978)

Organism	Shape	Motile in some growth phase	%(G + C) ratio	Temperature range and optimum
Methylococcus capsulatus	coccal	−	62·5%	20–50°C (37°C)
Methylococcus thermophilus	coccal or rod	+(P)	63·3%	37–62°C (55°C)
Methylococcus minimus	coccal	−	(62·5%)	20–30°C
Methylococcus luteus	oval	−	53·0%	20–37°C (33°C)
Methylococcus bovis (*Methylobacter bovis*)	coccal or rod	−	(50–54%)	20–37°C (30°C)
Methylococcus chroococcus (*Methylobacter chroococcum*)	oval	−	(50–54%)	20–30°C
Methylococcus ucrainicus	oval rods	+(P)	52·7%	20–37°C (30°C)
Methylococcus whittenburii (*Methylobacter capsulatus*)	oval rods	+(P)	(50–54%)	20–30°C
Methylococcus vinelandii (*Methylobacter vinelandii*)	oval rods	+(P)	(50–54%)	20–45°C

A complete description of the proposals is given in the paper on "A corrected diagnosis of the genera and species of methane-utilizing bacteria" by Romanovskaya et al. (1978). The names in parentheses are those given by Whittenbury et al. (1970a, b) (Table 5) when these differ from those given in this table. Percentage (G + C) ratios given in parentheses are those stated to be typical of the genus by Whittenbury et al. (1975).

TABLE 7
The facultative methanotrophs (*Methylobacterium* sp.)

	M. organophilum XX (ATCC 27886)	Seven other Methylobacterium sp.	Methylobacterium R6	M. hypolimneticum	M. ethanolicum
Colour on solid media	pink (white in first description)	white, yellow or orange	light-brown	white	yellow
Size	1 by 1·5 μm	Various	0·6 by 2·0 μm	0·4 by 0·9 μm	0·5 by 2·0 μm
Motility	+	n.d.	+	+	−
% (G + C) ratio	66%	n.d.	n.d.	66%	58%
Vitamin requirement	−	n.d.	−	+	−
Fixation of molecular N_2	+	n.d.	n.d.	n.d.	+
Spores or heat-stable forms	+	some	+	n.d.	n.d.
Alternative carbon sources					
nutrient broth	+	+	+	+	+
methanol	+	n.d.	+	−	−
methylamine	−	n.d.	−	n.d.	n.d.
formate	n.d.	n.d.	−	n.d.	n.d.
ethanol	−	+	+	−	+
acetate	+	+	+	±	±
succinate	+	+	+	±	+
malate	+	+	n.d.	±	+
fumarate	+	+	n.d.	n.d.	n.d.
glucose	−	+	+	+	+
fructose	n.d.	n.d.	n.d.	±	±
galactose	+	+	−	+	−
xylose	n.d.	n.d.	n.d.	−	+
sucrose	+	+	−	−	±
lactose	+	+	−	±	±

All strains (tested) have Type II membranes, the serine pathway and a complete TCA cycle. Some contain poly-3-hydroxybutyrate and are pleomorphic in shape. *Methylobacterium organophilum* XX is described in Patt *et al.* (1974, 1976) and Lynch *et al.* (1980). Seven other *Methylobacterium* sp. are described in Patt *et al.* (1974). *Methylobacterium* R6 is described in Patel *et al.* (1978b) and Lynch *et al.* (1980). *M. ethanolicum* and *hypolimneticum* are described in Lynch *et al.* (1980).

n.d., not determined; +, growth (not necessarily good growth); −, poor growth

In the original descriptions all facultative methanotrophs were described as white, yellow or orange and *M. organophilum* XX was said to be white, turning orange on ageing. However, subsequent descriptions of this organism have stated that it is pink on solid media and a culture of this strain that has lost the ability to grow on methane is said to be similar to the typical pink facultative methylotrophs that grow on methanol but not on methane (Green and Bousefield, 1981).

The main criterion by which the organisms in Table 7 are grouped into the single genus *Methylobacterium* is that of ability to grow on methane. It must therefore be emphasised that these facultative methanotrophs do not grow rapidly on this substrate (no information is available on cell yields on methane). Indeed, the species name "organophilum" was chosen to indicate the preference of these bacteria for multicarbon substrates. It has been suggested that the genetic information for growth on methane may be encoded on a plasmid and this would certainly explain the ease with which this characteristic is lost by some facultative methanotrophs (see Hanson, 1980; O'Connor, 1981).

B. Structural features of methanotrophic bacteria

1. Resting stages

In the natural environment most methanotrophs use only methane as energy source, even methanol being toxic to many of them. This apparently precarious existence is relieved by their ability to form resting stages (exospores and cysts) having survival properties not found in the vegetative organisms; the following description of these resting stages is taken from the first report of them by Whittenbury *et al.* (1970b) which includes many excellent electron micrographs as illustrations.

(*a*) *Exospores.* The only obligate methylotrophs to form exospores are *Methylosinus* sp. Exospore formation is also indicated in an electron micrograph of the facultative methanotroph *Methylobacterium* R6 (Patel *et al.*, 1978b). The exospores confer resistance to desiccation, surviving in the dried state for at least 18 months, and to heat, surviving 85°C for 15 min; they are also resistant to ultrasonication for 10 min.

(i) Spore formation. As cultures enter the stationary phase 5–95% of organisms (depending on the strain) elongate and taper to a pear-shape (*Methylosinus trichosporium*) or comma shape (*Methylosinus sporium*) at the opposite end from the flagella, the round spore being finally budded at this tapered end. In *Methylosinus trichosporium*, but not in *Methylosinus sporium* a second, ill-defined capsule forms around the developing spore and remains attached to it after budding. This spore capsule is more finely fibrous than that of the vegetative cell; neither capsule is stainable by conventional capsule stains and neither is removable by ultrasonication. After ageing for one to seven days many spores become refractile and acid-fast; for convenient observation they may be stained with malachite green spore stain. The immature exospores of *Methylosinus trichosporium* are cup-shaped and, in favourable conditions, are able to germinate before formation of the mature spherical exospore (Reed *et al.*, 1980).

(ii) Spore structure. The structure of the exospores bears some resemblance to the endospores of *Bacillus* species in having no well-defined cortex, respiratory activity or dipicolinic acid. The exospore has an outer coat derived from the vegetative organism, a laminated inner coat, an area corresponding to the cortex of endospores, a protoplasmic membrane and an inner protoplast. The exospores of *Methylosinus* are similar in many respects to those of *Rhodomicrobium vannielii* (Reed *et al.*, 1980; Dow and Whittenbury, 1980).

(iii) Spore germination. Methane is essential for spore germination which occurs within 2-15 days, the germination time being affected by the age of the spores and by treatments such as drying or heating. Before germination the capsular coat, only present on spores of *Methylosinus trichosporium*, is usually shed and appears to play no part in conferring the properties of heat and desiccation resistance. Spores do not swell before germination; the first observable change is that they lose refractility and the property of acid-fastness. This is followed by the emergence of a rod from the spore which increases in length and either divides or, rarely, forms another spore. The inner spore coat, but not the outer wall, remains an integral part of the germination organism for at least two divisions. The germination process, from loss of refractility to the first division, lasts 6-7 h.

On isolation all strains spore profusely but many strains tend to lose this property after subculturing a few times; this can sometimes be overcome by transferring to a medium lacking a nitrogen source.

(*b*) *Lipid cysts.* These have a complex wall structure and contain large lipid inclusions; they bear little resemblance to other bacterial cysts apart from their ability to resist desiccation. They are formed by only one strain, *Methylocystis parvus*, a Type II organism, very similar to *Methylosinus* sp. and which might therefore have been expected to form exospores.

In the exponential phase the bacteria are small rods, free of inclusions and stainable by conventional stains. As the stationary phase is entered a proportion (5–90%) of bacteria increase in size and form large lipid inclusions consisting mainly of poly-3-hydroxybutyrate which, after 7–14 days, can only be stained with acid-fast stain. The only difference in fine structure of these "lipid organisms" compared with the vegetative organisms is the increased complexity in their walls and their lack of a complex internal membrane system.

The "lipid organisms" fail to resist 65°C for 30 min but are resistant to desiccation, 80% surviving after a week. On transfer to a fresh medium the lipid organisms become sensitive to desiccation, losing their lipid inclusions and the property of acid-fastness.

(c) "Azotobacter-*type*" *cysts*. These are formed only by *Methylobacter* sp. "Immature *Azotobacter*-type" cysts are formed by most other Type I bacteria (see below). The cysts are single, double or multiple-bodied and resemble in all ways the cysts formed by *Azotobacter* species. They are produced as cultures enter the stationary phase, the rod or oval-shaped bacteria becoming larger, round and refractile; this process is often accompanied by a change from creamy white pigmentation to various shades of yellow to brown. The mature cysts have a central body, an inner wall and an outer coat or exine. They are not heat-resistant but are resistant to desiccation.

(d) "*Immature* Azotobacter-*type*" *cysts*. These are produced by those Type I bacteria not forming mature cysts, *Methylomonas* sp. and *Methylococcus* sp. These bacteria produce rounded bodies which appear to be an immature form of the *Azotobacter*-type cysts produced by *Methylobacter* sp. (see above); they possess a wall structure intermediate between that of a vegetative cell and a mature cyst and survive the absence of methane for 5–6 weeks in contrast to the vegetative bacteria which survive only 2–3 days.

2. *Internal membranes*

All methanotrophs have a complex internal membrane structure (Procter *et al.*, 1969; Davies and Whittenbury, 1970; de Boer and Hazeu, 1972; Patt *et al.*, 1974; Hazeu *et al.*, 1980) resulting in a surface area 4–8 times that of the cytoplasmic membrane. The Type I bacteria (*Methylococcus*, *Methylomonas* and *Methylobacter*) have bundles of disc-shaped vesicles which appear to be formed by invagination of the cytoplasmic membrane (Fig. 1) while Type II bacteria *Methylosinus*, *Methylocystis* and *Methylobacterium* have a system of paired peripheral membranes (Fig. 2). Although all methanotrophs are able to form these membranes the extent of formation depends on the growth conditions. In the Type I *Methylococcus capsulatus* the membranous development observed in methane-grown bacteria persisted when they were grown on methanol (Linton and Vokes, 1978) but internal membranes were absent from the facultative Type II organism *Methylobacterium organophilum*

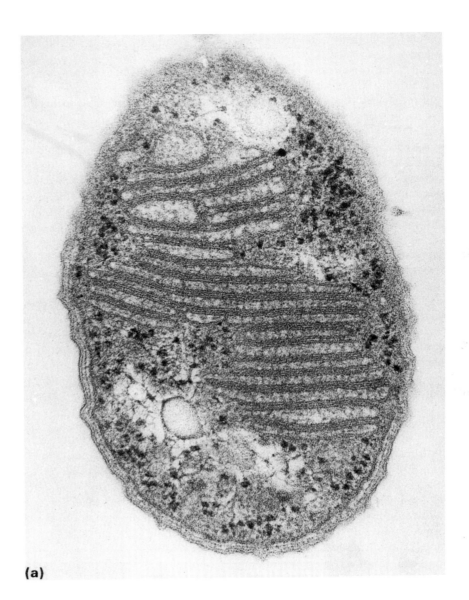

(a)

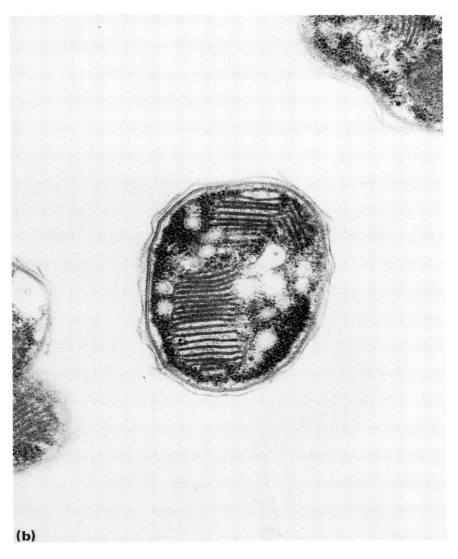

(b)

Fig. 1. *Methylomonas methanica* (a) and *Methylococcus capsulatus* (b) showing Type I membranes. This electron micrograph was kindly provided by Professor D. W. Ribbons. Type I membranes are typically arranged as bundles of disc shaped vesicles.

18   The biochemistry of methylotrophs

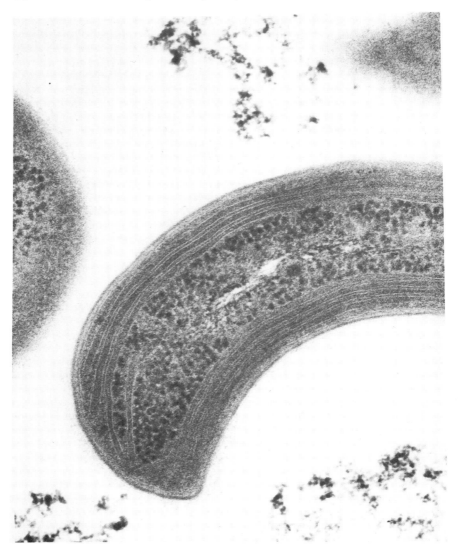

Fig. 2. *Methanomonas methano-oxidans* showing Type II membranes. This electron micrograph was kindly provided by Professor D. W. Ribbons. Type II membranes are typically arranged as a system of paired peripheral membranes.

when grown on methanol or glucose (Patt *et al.*, 1974; Patt and Hanson, 1978; O'Connor, 1981). In this organism the internal membranes increased in extent under low oxygen tension during growth on methane. Low oxygen tension also appears to stimulate membrane production in the obligate

Type II bacterium *Methylosinus trichosporium;* a review by Higgins *et al.*, (1981a) suggests that membranes are only produced under conditions of oxygen-limitation and that cultures growing rapidly under conditions of rapid gas transfer do not contain membranes (see also Best and Higgins, 1981; Scott *et al.*, 1981).

The role of the internal membranes is a matter for speculation (see Procter *et al.*, 1969; Davies and Whittenbury, 1970; Higgins, 1979a; Higgins *et al.*, 1981a) but it is probable that they are associated with peculiar oxidative properties of methanotrophs because similar membrane systems are found in photosynthetic bacteria, ammonia and nitrite oxidisers, cyanobacteria and some higher hydrocarbon utilisers all of which have specialised electron transport requirements. As methylotrophs able to grow on methanol but not methane do not possess internal membranes, then it seems reasonable to conclude that they perform some special function in the initial step in methane hydroxylation. The membranes may merely act by anchoring the oxygenase components or they may be involved as a solvent for methane or oxygen as proposed by Higgins (1979a).

3. Rosette formation, flagella and pigmentation

(*a*) *Rosette formation.* The majority of Type II, but not Type I, bacteria are able to form rosettes of bacteria held together and anchored at their non-flagellated poles by polysaccharide "holdfast" material which is visible under the light microscope (Whittenbury *et al.*, 1970a).

(*b*) *Motility and flagella.* Many methanotrophs are motile at some stage during growth; this is by way of polar flagella which are single or (in *Methylosinus*) arranged in tufts. Motility is often lost during laboratory culture and may thus be a confusing characteristic in classification schemes.

(*c*) *Pigmentation.* The most commonly isolated methanotrophs are the pink or red species very similar to those described by Leadbetter and Foster (1958). These are now named *Methylomonas methanica;* the majority of methanotrophs able to form pink-red carotenoid pigments are included in *Methylomanas methanica, Methylomonas rubrum* or *Methylomonas rosaceous.* When pigmentation is recorded it is usually that of colonies grown on solid media; pigmentation recorded in pellets of bacteria centrifuged from aerobically-grown cultures is more difficult to use in classification because it may often be due to cytochromes (particularly cytochrome *c*) rather than to carotenoid pigments. Water-soluble pigments are rarely formed by methanotrophs; *Methylobacter* produces a yellow pigment and, on iron-deficient medium, *Methylosinus sporium* produces a brown-black pigment and *Methylo-*

monas methanica a green to sapphire pigment. Many cyst-forming species produce a brown pigment whenever the cysts are present.

C. Biochemical and physiological features of methanotrophs used in their classification (see Table 3)

1. Major carbon assimilation pathways

Two major carbon assimilation pathways occur in methanotrophs; the serine pathway in Type II bacteria and the ribulose monophosphate (RuMP) pathway in Type I bacteria (Table 8). Because methanotrophs usually have high levels of enzymes of only one of the pathways, some of the enzymes peculiar to each pathway have been used as indicator enzymes. That for the serine pathway has been hydroxypyruvate reductase and that for the RuMP pathway has been hexulosephosphate synthase. Reports of low levels of hydroxypyruvate reductase in Type I methanotrophs in which the RuMP pathway is predominant emphasise that the mere presence of an indicator enzyme is insufficient evidence for the presence of one pathway or the other. Furthermore, the low levels of ribulose bisphosphate carboxylase and phosphoribulokinase in *Methylococcus capsulatus* (but not other methanotrophs) indicates that CO_2 can be fixed by the RuBP (autotrophic) pathway to some extent (very slight) even in the presence of a major alternative assimilation pathway (Taylor *et al.*, 1980).

2. The dehydrogenases for 2-oxoglutarate, malate, isocitrate, glucose-6-phosphate and 6-phosphogluconate

The levels of most TCA cycle enzymes are generally lower in Type I than in Type II methanotrophs (Chapter 5). In particular, Type I bacteria (*Methylococcus* and *Methylomonas*) contain no 2-oxoglutarate dehydrogenase whereas the levels of this enzyme in Type II bacteria (*Methylosinus*, *Methylocystis* and *Methylobacterium*) are similar to those measured in non-methanotrophic bacteria growing on methanol. Thus it appears that those bacteria in which the major assimilation pathway is the RuMP pathway (Type I bacteria) have an incomplete TCA cycle whereas those with the serine pathway have a complete TCA cycle.

It has been suggested (Whittenbury *et al.*, 1976) that the function of the TCA cycle is to provide NADH for the serine pathway and for the hydroxylation step in methane oxidation. This is unlikely to be the case because the production of acetyl-CoA from methane by the serine pathway requires the same amount of NADH as is produced by its subsequent oxidation by the TCA cycle. Perhaps a more likely function of the cycle is the occasional provision of NADH for methane hydroxylation by oxidation of stored poly 3-hydroxybutyrate; this might be particularly important for the initial hydroxylation of methane after germination of resting stages or after a period in the absence of methane.

The analysis of Davey et al. (1972) has indicated that dehydrogenases for glucose-6-phosphate and 6-phosphogluconate only occur in Type I organisms and are $NADP^+$-specific. The facultative Type II methanotroph, *Methylobacterium*, is typical in lacking these enzymes during growth on methane but they are induced during growth on methanol and glucose (when they are NAD^+-specific). The isocitrate dehydrogenases of Type II bacteria are $NADP^+$ specific, but use NAD^+ or $NADP^+$ in the Type I bacteria with the exception of *Methylococcus capsulatus* in which the enzyme uses only NAD^+. This important organism is also the exception in its low activity of malate dehydrogenase compared with all other methanotrophs. It is also unusual in having the enzymes ribulose bisphosphate carboxylase and phosphoribulokinase, albeit at very low levels compared with those in bacteria whose major path of carbon assimilation is the ribulose bisphosphate pathway (Taylor et al., 1980).

3. Growth temperatures, growth on methanol and nitrogen fixation

(*a*) *Growth temperature.* Ability to grow at 45°C was initially used to distinguish between various subgroups ("species") by Whittenbury et al. (1970a). It is of interest that no Type II species is able to grow at this temperature. *Methylococcus minimus* was originally separated from *Methylococcus capsulatus* by its inability to grow at 37°C or 45°C but it has now been suggested (Whittenbury, personal communication) that this should be included with the *Methylomonas* species, because recent observations have shown it to be rod-shaped, and not a coccus. The thermophilic methanotrophs described by Malashenko (1976) are all Type I *Methylococcus* or *Methylobacter* species.

(*b*) *Growth on methanol.* Ability or otherwise to grow on methanol is a difficult characteristic to determine with any degree of certainty. Many methanotrophs grow on methanol only if this substrate is provided in the vapour phase or when it is the rate-limiting nutrient (and hence at negligible concentration) in continuous culture.

(*c*) *Nitrogen fixation.* Many methanotrophs fix atmospheric nitrogen, using this as their sole nitrogen source. The discrepancy between earlier reports (Whittenbury et al., 1970a) and summaries included in recent reviews (Colby et al., 1979) is because false negative results were originally obtained when using the acetylene reduction test for nitrogenase. Nitrogenase requires an added reductant and this was supplied as methane. This has been shown to be unsuitable because acetylene inhibits the oxidation of methane; a better test for the presence of nitrogenase thus uses methanol as the reductant in the acetylene reduction test (Dalton and Whittenbury, 1976; Chapter 6).

4. DNA base ratios and predominant fatty acids

(a) *DNA base ratio* (*percentage G + C*). This is usually assumed to be a fundamental characteristic and is thus given particular importance in taxonomy. This is one reason why the merging of *Methylobacter*, having a percentage (G + C) ratio of less than 56%, into the *Methylococcus* group (percentage G + C, 62·5%), as suggested by Romanovskaya *et al.* (1978), has not been followed here.

(b) *Predominant fatty acids*. Type I and Type II bacteria differ in the predominant fatty acids in their membranes. As shown first by Ribbons and co-workers, in Type I bacteria the predominant fatty acid is a saturated fatty acid with 16 carbon atoms (16:0) while in Type II bacteria the predominant fatty acid is a mono-saturated fatty acid with 18 carbon atoms (18:1) (see Colby *et al.*, 1979 and Higgins, 1979a for references).

IV. Methylotrophs unable to grow on methane

I start this section by emphasising that although there may be some taxomomic validity in the groupings used here, my approach when devising these groups has been essentially pragmatic. A taxonomist would no doubt abhor the division of methylotrophs into two major groups on the apparently single characteristic of growth on methane, but this division is convenient and to some extent justifiable. Ability to grow on methane is correlated with possession of methane mono-oxygenase, a complex membrane system and the ability to form spores or resting stages. The organisms in the following section have none of these characteristics. The existence of the facultative methanotrophs (previous section) does, of course, confuse this neat division to some extent and the possibility of an organism transferring from the previous section into this one merely by losing a plasmid is rather irritating. This is especially so if the rumour is justified that some methanotrophs can be cured of "methane plasmids" by means of flight across the Atlantic ocean!

It is not always obvious to which group individual organisms should be allocated in this section. When in doubt, I have tended to group together those organisms with biochemical similarities and this has sometimes involved placing bacteria of the same (apparent) genus into different groups.

This reflects my experience that it is probably easier to distinguish between biochemical pathways than to allocate to a genus, with any certainty, an organism having, for example, some of the characteristics of a *Pseudomonas* sp. and some of an *Arthrobacter* sp.

Because of the importance I have placed on their biochemistry, I have included in this section a summary of the distribution of the main carbon assimilation pathways operating in the methylotrophs (Table 8).

TABLE 8
Distribution of carbon assimilation pathways in methylotrophic bacteria

The RuBP pathway (Chapter 2)
This occurs only in facultative autotrophs and *Pseudomonas oxalaticus* (Table 15)

The RuMP pathway (Chapter 3)
(a) Type I obligate methanotrophs (KDPG/TA variant) (Tables 3 and 4)
(b) All obligate methylotrophs unable to use methane (KDPG/TA variant) (Tables 9 and 10)
(c) Some facultative methylotrophs unable to use methane; mainly non-motile species; most *Arthrobacter* and *Bacillus* sp. (FBP aldolase/SBPase variant) (Tables 13 and 14)
(d) *Pseudomonas oleovorans* (a Gram-negative, motile, facultative methylotroph unable to use methane) (KDPG/TA variant). This organism is an exception to many generalisations in this book.

The serine pathway (Chapter 4)
(a) Type II obligate methanotrophs (Tables 3 and 4)
(b) Type II facultative methanotrophs (icl⁻) (Tables 3, 4 and 7)
(c) Pink facultative methylotrophs (icl⁻) (Table 11)
(d) The *Hyphomicrobia* (icl⁻)
(e) Most non-pigmented pseudomonads (icl⁺) (Table 12)
(f) A few non-motile, Gram-negative facultative methylotrophs (Table 13)

This is a rough guide; full distribution tables are presented as follows: RuBP pathway, Table 16; RuMP pathway, Tables 18 and 19; serine pathway, Tables 26 and 27.

A. Obligate methylotrophs able to use methanol or methylamine but not methane

Under this heading I am including the more restricted facultative methylotrophs. The distinction between obligate and facultative methylotrophs among the methanol- and methylamine-utilisers is not as well-defined as might at first appear; this is because some facultative methylotrophs are almost obligate in nature. Some grow (very slowly) on only one or two multicarbon substrates (usually glucose or fructose) (e.g. organism W3A1 and W6A) and these have been called the (Type *M*) restricted facultative methylotrophs (Colby and Zatman, 1975a). Other organisms grow rapidly on a larger but still very limited range of multicarbon compounds and these have been called the less (Type *L*) restricted facultative methylotrophs; these include the *Hyphomicrobia* (Harder and Attwood, 1978) and the *Bacillus* species S2A1 and PM6 (Colby and Zatman, 1975a).

Under the heading of obligate methylotrophs, I have included the more restricted facultative organisms, whereas the less restricted organisms have been included in the next section with the facultative methylotrophs.

The non-methanotrophic obligate methylotrophs are listed in Table 9 and a summary of their properties presented in Table 10. This summary shows that all such obligate methylotrophs are very similar to one another and that they could perhaps be included in a single genus (perhaps *Methylophilus*). A taxonomic study of these organisms with many similar new isolates by Urakami and Komogata (1979) appears to confirm this. Byrom (1981), however, has suggested (on the basis of DNA/DNA hybridisation experiments) that while they do form a separate group, this group itself might be divisible into two. In some features these bacteria are similar to the obligate methane-utilisers whose properties are summarised in Table 2. The main differences are that none grow on methane, none have a complex internal arrangement of membranes, none form a differentiated resting body, none fix atmospheric nitrogen and all grow well on methanol. The obligate methanol-utilisers are most similar to the methanotrophs *Methylomonas* and *Methylobacter* (Table 4), being rod-shaped, motile, having the RuMP assimilation pathway for carbon assimilation, having a (G + C) ratio of 52–56% and having fatty acids with 16 carbon atoms as the predominant fatty acids (Urakami and Komogata, 1979). It is unfortunate that some workers, on the "authority" of Bergey's Manual, have given the name *Methylomonas* to obligate methanol-utilisers because this name was originally used for obligate methane-utilisers.

An unusual characteristic of some of the non-methanotrophic obligate methylotrophs is that they have a rather thick cell envelope and a wavy outline. This is clearly seen in electron micrographs of *Methylophilus methylotrophus*, *Pseudomonas* C, *Pseudomonas methanolica* (Rokem et al., 1978b) and *Pseudomonas* W1 (Dahl et al., 1972), but not in facultative methylotrophs. In some cases the wavy outlines may be so extreme that they constitute clear protrusions of the outer membrane. The thick cell envelope is correlated with relatively high concentrations of lipid in these obligate methylotrophs and it has been calculated that the amounts of lipid per unit of cell surface is between two and five times that in facultative methylotrophs.

TABLE 9
Obligate methylotrophs unable to use methane but able to use methanol or methylamine

Organism	References
Pseudomonas C	Chalfan and Mateles (1972); Goldberg and Mateles (1975); Rokem *et al.* (1978b); Urakami and Komogata (1979)
Pseudomonas W1 (Organism W1)	Dahl *et al.* (1972)
Organism C2A1	Colby and Zatman (1973); Urakami and Komogata (1979); Byrom (1981)
Organism 4B6 (non-motile and grows only on methylated amines)	Colby and Zatman (1973)
Pseudomonas RJ3	Mehta (1973a)
Methylomonas methylovora (ATCC 21850) (about 50 strains; 3 were non-motile)	Kouno *et al.* (1973); Kouno and Ozaki (1975); Mehta (1977); Patel *et al.* (1979b); Urakami and Komogata (1979); Byrom (1981)
Pseudomonas W6	Babel and Miethe (1974)
Methylophilus methylotrophus (*Pseudomonas methylotropha*, NC1B 10592; 10515)	Maclennan *et al.* (1974); Byrom and Ousby (1975); Taylor (1977); Brooks and Meers (1973); Rokem *et al.* (1978b); Urakami and Komogata (1979); Byrom (1981); Large and Haywood (1981)
Methylomonas M15	Sahm and Wagner (1975)
Organisms W3A1 and W6A (also grow slowly on glucose)	Colby and Zatman (1975a); Byrom (1981)
Methylomonas aminofaciens 77a	Ogata *et al.* (1977)
Methylomonas P11	Drabikowska (1977); Michalik and Raczynska-Bojanowska (1976); Budohoski *et al.* (1978); Michalik *et al.* (1979)
Organism BC3	Chen *et al.* (1977); Urakami and Komogata (1979)
Methylomonas methanolica (*Pseudomonas methanolica* ATCC 21704)	Rokem *et al.* (1978b); Amano *et al.* (1975); Urakami and Komogata (1979); Byrom (1981)
Pseudomonas J	Matsumoto (1978)
Organism L3	Hirt *et al.* (1978)
Methylomonas clara	Drozd and Linton (1981); Byrom (1981)

The properties of these bacteria are summarised in Table 10; in spite of their different names most of these bacteria are very similar to one another. Names in parentheses are previously used names for the same organism. The name *Methylomonas* was used initially for bacteria using methane (Tables 3, 4 and 5) and not for the bacteria described here. *Methylophilus methylotrophus* and *Pseudomonas* C were originally thought to be facultative methylotrophs but they are known to be obligate methylotrophs growing only on C$_1$ compounds and perhaps also fructose (Taylor, 1977; Goldberg, personal communication; Urakami and Komogata, 1979; Byrom, 1981)

TABLE 10

Properties common to the obligate methylotrophs unable to use methane

All are strictly aerobic, Gram-negative, non-pigmented, non-sporing, slender rods, motile by a single polar flagellum and occurring singly or in pairs.

All (except organism 4B6) grow well on methanol; none grow on methane or formate but a few (e.g. organism L3) grow on formaldehyde. Methylamine(s) support the growth of *Pseudomonas* W1, J and RJ3, *Methylophilus methylotrophus*, organisms C2A1, W3A1, W6A, some strains of *Methylomonas methylovora*, *Methylomonas methanolica* and nine out of 52 of Kouno's isolates (3 of these 9 were non-motile). None grow on typical laboratory nutrient media and the majority grow on no other carbon source tested (except fructose in a few cases).

All (except organisms W3A1, W6A and 4B6) have a percentage (G + C) ratio between 52% and 56%, the majority being between 52% and 54% (exceptions are given below).

All tested grow by way of the Ribulose monophosphate pathway of carbon assimilation (KDPG aldolase/transaldolase variant).

All use ammonia, and the majority use nitrate, as nitrogen source. Some grow (poorly) with organic nitrogen sources but none have been reported to fix atmospheric nitrogen.

All tested have coenzyme Q_8 and have fatty acids with 16 carbon atoms as the predominant fatty acids (Urakami and Komogata, 1979).

The optimum temperature for growth is usually about 30°C but the majority grow fairly well at 37°C; none grow above 42°C.

The majority are oxidase-positive and catalase-positive.

A few require a vitamin for growth (Kouno et al., 1973; Colby and Zatman, 1973).

Some (e.g. *Methylophilus methylotrophus*, organism BC3 and *Pseudomonas* C) are particularly sensitive to high concentrations of phosphate in growth media.

Many grow at very high rates and to high growth yields on methanol; this may reflect the purpose for which they were isolated (production of single cell protein).

This Table lists those properties common to the majority of obligate methylotrophs unable to use methane but able to use methanol as sole source of carbon and energy (Table 9).

B. Facultative methylotrophs unable to grow on methane

Most of these bacteria (listed in Tables 11 to 15) were isolated by elective culture on methanol or methylated amines; very few methylotrophic bacteria have been isolated by other means and then "discovered" in culture collections. Although the majority of facultative methylotrophs isolated on methanol are able to grow on methylamine, the opposite does not appear to be true. Apart from the pink facultative methylotrophs and the *Hyphomicrobia*, bacteria isolated on methylated amines do not usually grow on methanol (Levering *et al.*, 1981a).

1. The pink facultative methylotrophs (Table 11)

These Gram-negative bacteria are the facultative methylotrophs most readily isolated on methanol which they assimilate by way of the serine pathway. They are usually motile by a single polar flagellum and form red carotenoid pigments so that colonies often appear scarlet on nutrient agar (Peel and Quayle, 1961). They grow rather slowly on nutrient media (such as nutrient broth) and on solid media, seven days being required for full development of colonies. They are large, non-sporing rods containing inclusions of poly-3-hydroxybutyrate and are strictly aerobic, non-fermentative and are catalase- and oxidase-positive. None grow autotrophically, none grow anaerobically with nitrate as electron acceptor, none fix molecular nitrogen and none grow on methane (but see comment on *M. organophilum*, below).

Most strains grow on methanol, formate, methylamine, ethylamine, ethanol and/or acetate, lactate, pyruvate, 3-hydroxybutyrate, C_4 carboxylic acids and propanediol; a few also grow on oxalate. Only a few grow on dimethylamine or trimethylamine (Large, 1981).

Their DNA base ratios are between 60% and 70% (G + C) (Green and Bousefield, 1981); they all form coenzyme Q_{10} and they all have an unsaturated fatty acid with 18 carbon atoms (18:1) as their predominant fatty acid (Natori *et al.*, 1978; Urakami and Komogata, 1979).

It should be noted that neither these pink facultative methylotrophs nor the non-pigmented *Pseudomonas* sp. listed in Table 12 are typical *Pseudomonas* sp.; of the 267 typical Pseudomonads selected and classified by Stanier *et al.* (1966) none were able to grow on methanol, methylamine or oxalate as sole carbon and energy source.

In their valuable taxonomic survey of 150 strains of pink facultative methylotrophs, Green and Bousefield (1981) also emphasise that the pseudomonad properties of these bacteria are not very convincing: "the Gram reaction is often equivocal, flagellation is not always polar and the cellular size, pleomorphism and branching shown by most strains is not typical of the genus *Pseudomonas*". They suggest that all these bacteria should be placed in

TABLE 11
The pink facultative methylotrophs

Organism	References
Pseudomonas extorquens NC1B 9399 (*Bacillus extorquens*, *Vibrio extorquens*)	Bassalik (1913); Bhat and Barker (1948) (both cited in Peel and Quayle, 1961); Stocks and McCleskey (1964a, b); Tonge et al. (1974); Urakami and Komogata (1979)
Protaminobacter ruber	den Dooren de Jong (1926) (cited by Peel and Quayle (1961); Sato et al. (1976, 1977); Sato (1978); Urakami and Komogata (1979)
Pseudomonas PRL-W4	Kaneda and Roxburgh (1959a, b, c)
Pseudomonas AM1 (NC1B 9133)	Peel and Quayle (1961); Urakami and Komogata (1979)
Pseudomonas AM2	Blackmore and Quayle (1970)
Pseudomonas M27	Anthony and Zatman (1964a); Urakami and Komogata (1979)
Pseudomonas PP	Ladner and Zatman (1969)
Pseudomonas 3A2	Colby and Zatman (1972, 1973); Urakami and Komogata (1979)
Pseudomonas RJ1	Mehta (1973a, b)
Pseudomonas TP1	Sperl et al. (1974); Bellion and Spain (1976)
Organism FM02T	Toraya et al. (1975)
'Pink *Pseudomonas*' (59 strains)	Kouno and Ozaki (1975)
Pseudomonas sp. 1 and 135	Rock et al. (1976); Urakami and Komogata (1979)
Pseudomonas sp. YR, JB1 and PCTN	Bellion and Spain (1976)
Pseudomonas methylica sp. 2 and 15	Kirikova (1970); Netrusov et al. (1977); Loginova and Trotsenko (1979a)
Pseudomonas rosea (NC1B 10597 to 10612)	Urakami and Komogata (1979)
Pseudomonas 2941	Yamanaka and Matsumoto (1977b, 1979)
Pseudomonas AT2	Boulton et al. (1980)
Pseudomonas 80	Kortstee (1980)
'Pink chromogens' (C₁ compounds not tested) [*Methylobacterium organophilum* XX]	Austin et al. (1978); Austin and Goodfellow (1979) [Patt et al., 1974, 1976]

These rod-shaped bacteria form red carotenoid pigments and are usually motile by single polar flagella. They grow on methanol, formate and usually methylated amines which they assimilate by the serine pathway. Besides the strains referred to below many more similar strains are discussed in taxonomic surveys by Urakami and Komogata (1979) and Green and Bousefield (1981). *Methylobacterium organophilum* XX is not typical, being able to grow on methane and being able to fix atmospheric nitrogen (see text). Oxalate is also assimilated by the serine pathway (Blackmore and Quayle, 1970).

a separate genus having two species, one of which corresponds to the description of *Pseudomonas mesophilica* (Austin *et al.*, 1978; Austin and Goodfellow, 1979). They also suggest that the genus *Methylobacterium* might be a suitable generic name. When first described, *Methylobacterium organophilum* XX was said to be a white facultative methanotroph (Patt *et al.*, 1974) and this obviously obscured any similarity to the typical pink facultative methylotrophs, none of which grows on methane. A later description stated that this organism is pink but its ability to fix atmospheric nitrogen and its ability to resist high temperatures appears to set it apart from all other organisms listed in Table II. Green and Bousefield (1981) conclude that *Methylobacterium organophilum* XX is a typical pink facultative methylotroph but their strain had lost the ability to grow on methane. Other strains of *Methylobacterium* which are not pink are said to be more stable with respect to this growth characteristic (O'Connor, 1981). This is obviously an interesting problem awaiting more work for its resolution and I think it best to avoid the confusion that would arise if the name of these unusual organisms were to be used as the generic name for the typical pink facultative methylotrophs.

2. The non-pigmented "pseudomonads" (Table 12)

As mentioned above, no typical *Pseudomonas* sp. grows on methanol or methylamine. The term pseudomonad is used here because these bacteria were originally designated *Pseudomonas* sp. They are Gram-negative, non-pigmented, non-sporing rods, motile by a single polar flagellum and they are catalase- and oxidase- positive. Most of them were isolated on methylated amines; this type of methylotroph is not usually isolated on methanol although some do grow on this substrate. Most grow well on methylated amines, formate and a wide range of multicarbon compounds and on nutrient agar. Most of them assimilate C_1 compounds by the serine pathway (icl+ variant). The exception is *Ps. oleovorans* which uses the RuMP pathway of formaldehyde assimilation. Although a considerable amount is known about its biochemistry I have failed to find any published description of this unusual methylotroph.

3. Gram-negative (or variable), non-motile rods and coccal rods (Table 13)

The bacteria listed in Table 13 comprise a diverse group. They are all Gram-negative at some stage of growth but the *Arthrobacter* species are usually Gram-positive at some stage and are often pleomorphic. The bacteria listed here are white, yellow, orange or (rarely) red. Most are catalase-positive and said to be oxidase-negative or variable. Many of the Gram-negative, non-motile rods are difficult to allocate to a genus and they have been assigned tentatively to *Alcaligenes*, *Achromobacter* or *Acinetobacter* groups.

TABLE 12
The non-pigmented "pseudomonads"

Organism	Reference
Bacteria able to grow on methylated amines but unable to grow on methanol	
Pseudomonas aminovorans (NC1B 9039)	Peel and Quayle (1961); Eady *et al.* (1971)
Pseudomonas MA (Shaw strain)	Shaw *et al.* (1966)
Pseudomonas MS (ATCC 25262)	Kung and Wagner (1970a)
Pseudomonas 2A3	Colby and Zatman (1973)
Bacteria able to grow on methanol and (usually) methylated amines	
Pseudomonas aminovorans (some strains only)	den Dooren de Jong (1926, 1927) (cited in Peel and Quayle, 1961)
Pseudomonas sp. 1A3, 1B1, 7B1 and 8B1 (isolated on trimethylamine)	Colby and Zatman (1973)
Pseudomonas S25*	Yamanaka and Matsumoto (1979)
Pseudomonas (*methylica*) 20*	Loginova and Trotsenko (1977b, 1979a)
[**Pseudomonas oleovorans***]	[Loginova and Trotsenko, 1977a, b]

All these bacteria are Gram-negative rods, motile by a single polar flagellum. The majority grow on methylated amines, formate and a wide range of conventional growth substrates. All those tested assimilate formaldehyde by the icl⁺ variant of the serine pathway except for *Ps. oleovorans* which has the RuMP pathway (no full description of this methylotroph is available). The facultative autotrophs are listed separately in Table 13. An asterisk indicates that no full description is available.

TABLE 13
Gram-negative (or variable), non-motile rods

Organism	References
Bacteria able to grow on methanol and (usually) methylated amines	
Arthrobacter rufescens (red)	Akiba et al. (1970)
Arthrobacter 1A1 and 1A2	Colby and Zatman (1973)
Arthrobacter 2B2 (yellow)	Colby and Zatman (1973)
Arthrobacter (yellow/orange, 10 strains)	Kouno and Ozaki (1975)
Arthrobacter globiformis SK-200	Tani et al. (1978b)
Organism 5B1 (requires thiamine) (*Alcaligenes*/*Achromobacter* group)	Colby and Zatman (1972, 1973)
Klebsiella 101	Nishio et al. (1975a)
Alcaligenes/*Acinetobacter* (28 strains)	Kouno and Ozaki (1975)
Strain S50	Yamanaka and Matsumoto (1979)
Bacteria unable to grow on methanol but able to grow on methylated amines	
"*Diplococcus*" PAR	Leadbetter and Gottlieb (1967); Bellion and Spain (1976)
Organism 5H2	Hampton and Zatman (1973)
Arthrobacter globiformis B-175, B-126 and B-53	Loginova and Trotsenko (1976a)
Arthrobacter P1 (NCIB 11625)	Levering et al. (1981a, b)

This is a diverse (and rather arbitrary) group of white, yellow, orange or (rarely) red bacteria. They are all Gram-negative except the *Arthrobacter* which are Gram-variable, with the response to the Gram stain varying with the stage of growth. The majority grow on methylated amines and on a wide range of multicarbon substrates. The majority are catalase-positive and oxidase-negative (or variable) (organisms 5B1 and S50 are oxidase-positive). The facultative autotrophs are listed separately in Table 15.

TABLE 14
Gram-positive facultative methylotrophs

Organism	References
Bacillus cereus M-33-1 (grows on methanol)	Akiba *et al.* (1970)
Bacillus PM6 (grows on methylamine, not methanol)	Myers and Zatman (1971); Colby and Zatman (1975a)
Bacillus S2A1 (grows on methylamine, not methanol)	Colby and Zatman (1975a)
Streptomyces 239 (grows on methanol)	Kato *et al.* (1974d)
Mycobacterium vaccae (grows on methanol and methylamine)	Loginova and Trotsenko (1977b)
Brevibacterium fuscum 24 (grows on methylamine)	Loginova and Trotsenko (1977b)

The *Bacillus* species are restricted facultative methylotrophs (Type L). The *Arthrobacter* sp. are often Gram-positive during some stage of growth but they are arbitrarily included with the Gram-negative bacteria in Table 11 because many were originally described as Gram-negative. Facultative autotrophs are listed separately in Table 15.

Some of the bacteria in Table 13 use the serine pathway and others the RuMP pathway. In the recent detailed study of *Arthrobacter* P1 it was suggested that a high proportion of rapidly-growing facultative methylotrophs isolated on methylamine are *Arthrobacter* sp. assimilating methylated amines by way of the RuMP pathway (Levering *et al.*, 1981a). These authors were presumably referring to the very rapidly-growing bacteria and excluding the pink facultative methylotrophs and the *Hyphomicrobia* from this generalisation.

4. Gram-positive facultative methylotrophs (Table 14)

Besides the Gram-variable *Arthrobacter* sp. (Table 13), very few methylotrophic bacteria stain positively with the Gram stain. Two of the *Bacillus* sp. (PM6 and S2A1) are restricted facultative methylotrophs (Type *L*). They are unable to grow on methanol but grow on methylated amines, glucose, gluconate, citrate, alanine, betaine and nutrient agar (no growth was recorded on about 50 alternative carbon sources). All of the bacteria in Table 14 have the RuMP pathway of formaldehyde assimilation but the specific activities of key enzymes were low in *Streptomyces* sp. 239 and low activities of some serine pathway enzymes were also present.

5. Facultative autotrophs or phototrophs growing on methanol or formate (Table 15)

Table 15 is a list of miscellaneous bacteria assimilating methanol or formate by way of their oxidation product CO_2 and the RuBP cycle of CO_2 fixation. The *Rhodopseudomonas* sp. are red, motile, prosthecate, phototrophic bacteria that reproduce by budding. They grow anaerobically in the light using methanol or formate (not methylamine) as reductant, and Seifert and Pfennig (1979) have now shown that in the right conditions they can grow aerobically in the dark using methanol or formate as the sole source of carbon and energy.

Paracoccus denitrificans is a Gram-negative, oxidase-positive, non-motile coccus or coccal rod; it is a facultative chemoautotroph, able to grow aerobically on $H_2 + CO_2$ and it was originally called *Micrococcus denitrificans*. It is able to grow anaerobically on methanol with nitrate as terminal electron acceptor instead of oxygen. Like *Rhodopseudomonas* sp. it grows better on methanol or methylamine when the growth medium contains a trace of yeast extract.

Many bacteria that grow on methanol or methylamine also grow on formate and they assimilate all three substrates by the same route. However, some bacteria that are able to grow on formate cannot use other C_1 substrates

TABLE 15
Facultative autotrophs or phototrophs growing on methanol or formate

Organism	References
Rhodopseudomonas palustris	Quadri and Hoare (1969); Stokes and Hoare (1969)
Rhodopseudomonas acidophila and some other Rhodospirillaceae	Quayle and Pfennig (1975); Douthit and Pfennig (1976); Sahm et al. (1976a); Seifert and Pfennig (1979)
Thiobacillus A2 (formate only)	Kelly et al. (1979); Gottschal and Kuenen (1981); Wood and Kelly (1981)
Thiobacillus novellus	Chandra and Shethna (1977)
Nitrobacter agilis	Ida and Alexander (1965)
Alkaligenes eutrophus H-16 (formate only)	Friedrich et al. (1979)
Paracoccus denitrificans (methanol, methylamine)	Cox and Quayle (1975)
Pseudomonas oxalaticus (formate only)	Khambata and Bhat (1953); Quayle (1961); Dijkhuizen and Harder (1979a, b)
Pseudomonas gazotropha[a]	Romanova and Nozhevnikova (1977)
Pseudomonas 8[a]	Loginova and Trotsenko (1979c)
Achromobacter 1L[a]	Loginova and Trotsenko (1979c)
Mycrocyclus aquaticus (e.g. ATCC 25396)	Namsarev et al. (1971); Loginova et al. (1978); Urakami and Komogata (1979)
Mycrocyclus ebrunous (ATCC 21373)	Kouno and Ozaki (1975); Urakami and Komogata (1979)
Blastobacter viscosus	Loginova and Trotsenko (1979c)
Mycobacterium 50[a]	Loginova and Trotsenko (1979c)

Except for *Mycobacterium* 50, all are Gram-negative. Some grow only very slowly on methanol. The *Rhodopseudomonas* species are facultative photoorganotrophs. Except for *Ps. oxalaticus* the others are facultative autotrophs, oxidising H_2 as energy source. All use the RuBP pathway of CO_2 assimilation.

[a] Denotes an organism for which there is no full description.

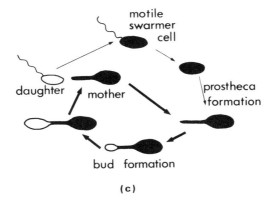

Fig. 3. Photomicrographs and electron micrographs of *Hyphomicrobium* sp. The pictures were kindly provided by Dr. C. S. Dow. (a) Phase contrast photomicrograph of a *Hyphomicrobium* species. (b) Gold/palladium shadowed electron micrograph of a *Hyphomicrobium* species. (c) Diagrammatic representation of the *Hyphomicrobium* cell cycle. The motile swarmer cell undergoes obligate morphogenesis and differentiation to become a reproductive, prosthecate cell. Subsequent growth is by budding to generate a flagellate cell at the end of the prostheca. Division is asymetric yielding a motile swarmer (daughter) and a prosthecate (mother) cell. Two cell cycles are therefore apparent—the reproductive (mother) cycle and the swarmer cell cycle.

7. Marine bacteria able to grow on methanol and methylated amines

Remarkably few marine methylotrophs have been fully described although the isolation of methane-utilisers from a marine environment has been reported (Hutton and Zobell, 1949) and marine *Hyphomicrobia* have been reported (Attwood and Harder, 1972; Hirsch, 1974). More recently, 65 new marine isolates have been described by Yamamoto et al. (1978a, b; 1980). These organisms are presented in this separate section because there is insufficient physiological or biochemical information available to allocate them to the various groups used in this chapter. All 65 strains were isolated on methanol in media containing 2% NaCl. All are Gram-negative, non-sporing white or yellow rods, motile by polar flagella. All require NaCl (1–4%) and vitamin B_{12} and all grow at 42°C (temperature optima are 30–37°C). The percentage (G + C) ratios of representative strains are in the range 43·8% to 47·6%. All isolates grow on methanol but none grow on methane. The majority (63 out of 65) grow on methylamine and 40 out of 65 grow on dimethylamine and trimethylamine. Although identical in all other respects the 63 isolates able to grow on methylamine were divided into two completely new species on the basis of their ability to grow on fructose (no other carbon source appears to have been tested); those growing on fructose (39 strains) were called *Alteromonas thalassomethanolica* and the remaining 24 isolates were called *Methylomonas thalassica* (Yamamoto *et al.*, 1980).

It was concluded that those strains growing on fructose were facultative methylotrophs whereas those unable to grow on this substrate were obligate methylotrophs. This conclusion, however, is clearly suspect; some typical facultative methylotrophs do not grow on fructose and the one substrate most often supporting growth of otherwise obligate methylotrophs is fructose. The failure of any of these marine isolates to grow on nutrient media (containing NaCl) is more typical of the obligate methylotrophs than of any facultative methylotrophs and it remains to be seen whether or not these new organisms are really different from the non-marine obligate (and near-obligate) methylotrophs listed in Table 9.

V. The place of methylotrophs in nature

I have insufficient expertise in ecology to match the grandeur of this title and I can do little more than to make a few obvious generalisations about this subject.

There are two main questions to consider: where are methylotrophs found in nature and what is their function in relation to biological carbon and nitrogen cycles? Over the last 20 years it has become obvious that methylotrophs are ubiquitous, reflecting the physiological diversity of this vast group of microorganisms. Which methylotrophs are found in which environments depends on the predominant C_1 substrate and its concentration; the presence or absence of O_2 and alternative terminal electron acceptors (nitrate and sulphate); the available nitrogen source (amino acids, amines, ammonia, nitrate, nitrite and N_2); and the prevailing conditions of temperature and pH.

A. The microbial production and utilisation of methane

The ecology of methane-utilising bacteria has been reviewed extensively by Quayle (1972), Whittenbury *et al.* (1970a, 1976) and Hanson (1980). Methane is produced anaerobically by methanogenic bacteria (see Chapter 11). When this occurs at depth in marine sediments it diffuses upwards and most of it is oxidised anaerobically by bacteria at the base of the sulphate-reducing zone using sulphate as the terminal electron acceptor (Reeburgh, 1976, 1980, 1981; Reeburgh and Heggie, 1977). Most methane is produced and utilised within 1 m of the surface of the earth (above or below) and an idea of the proportions produced in different environments may be obtained from a consideration of the amounts released into the atmosphere. The total released annually is about 10^{15}g; 45% of this is from swamps, 20% from ruminants,

25% from paddy fields and most of the remainder from river and lake mud (Schlegel et al., 1976; Vogels, 1979; Hanson, 1980).

Once formed, the methane remains for 1·5–7 years in the atmosphere where interaction with ozone and its radicals produces H_2, CO and CO_2.

Many methanotrophs fix atmospheric nitrogen, most form drought-resistant spores or cysts and many are able to grow at high temperatures; as a group they are thus able to take advantage of the widespread availability of their carbon substrate. Although most methanotrophs grow well at low oxygen tensions, one niche not available to most methanotrophs is where strictly anaerobic conditions prevail and nitrate must be used as an alternative electron acceptor to O_2. Besides their importance in fixing atmospheric nitrogen, the methanotrophs also play a role in the nitrogen cycle by oxidising ammonium ions (by way of the methane mono-oxygenase).

B. The microbial production and utilisation of methanol and formate

Methanol arises in nature by the oxidation of methane and by the hydrolysis of methyl ethers and esters present in pectin and lignin which are major structural components of plants. So, like methanotrophs, methanol-utilisers abound in nature and they are readily isolated from almost any sample of soil, water or sewage. There appears to be no correlation between the sites from which they have been isolated and whether or not they are facultative or obligate methylotrophs. Yeasts are likely to predominate at low pH whereas bacteria are more likely to be found in neutral conditions. Very few form spores, few, if any, fix atmospheric nitrogen and relatively few are able to grow anaerobically using nitrate as terminal electron acceptor. The important exceptions are the *Hyphomicrobia* which can be specifically isolated by anaerobic elective culture on methanol or methylated amines with nitrate (Sperl and Hoare, 1971; Attwood and Harder, 1972). These organisms are widespread in nature and, in part, this is related to their ability to grow on very low concentrations of carbon substrate. This ability to grow in oligotrophic environments may be related to their prosthecae which are used to fasten them to solid substrates and which increase their relative surface area, enabling them to transport low concentrations of carbon substrate more effectively (Harder and Attwood, 1978; Meiberg, 1979; Dow and Whittenbury, 1980). The attachment to solid surfaces may also help in this respect because it has been suggested that limiting nutrients may be concentrated at the surface-water interface.

Hyphomicrobia are often found to be associated with methanotrophs in enrichment cultures on methane and it has been suggested that *Hyphomi-*

crobia predominate over other methanol-utilisers in such mixed cultures because of their relatively high affinity for methanol (Harrison, 1973; Wilkinson and Harrison, 1973; Wilkinson *et al.*, 1974). However, this may not be the only explanation because some other methanol utilisers have equally low Ks values for growth on methanol (Harder and Attwood, 1978; Meiberg, 1979). Some methanotrophs are inhibited by methanol and are able to grow in laboratory culture on methane only in the presence of *Hyphomicrobia* which remove the methanol (Wilkinson and Harrison, 1973, and see Harrison, 1978, for a useful review of growth of bacteria in mixed populations in the laboratory).

Formate is produced by oxidation of C_1 compounds and, more important in producing formate for bacterial growth, it is a major end-product of bacterial fermentation (see Chapter 11). It is used by many typical facultative methylotrophs, by some facultative autotrophs, by *Pseudomonas oxalaticus*, and also by many photosynthetic bacteria as an electron donor (see Table 15).

C. The microbial production and utilisation of methylated amines

The methylated amines used by methylotrophs are listed in Table 1 and their origin in nature reviewed by Meiberg (1979). Trimethylamine is produced by microbial degradation (usually anaerobic) of carnitine and of lecithin and other choline derivatives (Hayward and Stadtman, 1959; Baker *et al.*, 1962; Neill *et al.*, 1978). Trimethylamine *N*-oxide is an important constituent of fish (possibly having a role in osmoregulation) and during their decomposition it acts as a terminal electron acceptor for some bacteria, the product being trimethylamine which leads to the smell of rotting fish (Ishimoto and Shimokawa, 1978; Kim and Chang, 1974; Madigan and Gest, 1978). Dimethylamine is produced during the degradation of some pesticides (Tate and Alexander, 1976) and is a product of trimethylamine oxidation. Methylamine is a constituent of some plant material and is produced during the oxidation of all other methylated amines.

No yeasts can use methylated amines as the sole source of carbon and energy although many can use them as the sole nitrogen source. Probably more methylotrophic bacteria are known to be able to grow on methylated amines than on methanol and these bacteria have been isolated from most natural sources. As with the methanol-utilisers, obligate and facultative utilisers of methylated amines have been isolated from similar environments. Very few, if any, fix atmospheric nitrogen and very few (some *Bacillus* species) form spores.

Most *Hyphomicrobia* are able to grow on low concentrations of methylated amines and they are the only methylotrophs able to grow anaerobically with nitrate as terminal electron acceptor (except for *Paracoccus denitrificans*). An important conclusion of Meiberg *et al.* (1980) is that some *Hyphomicrobia* growing on dimethylamine (and probably other methylated amines) may play a role in the denitrification process (conversion of nitrate to N_2) even in aerobic conditions provided the growth rate is sufficiently low.

2
The ribulose bisphosphate pathway of CO_2 assimilation

I. Introduction: the definition of autotrophy 42
II. Description of the ribulose bisphosphate pathway 43
III. The occurrence and distribution of the ribulose bisphosphate pathway . 46
IV. The reactions of the ribulose bisphosphate pathway . . . 51
V. Regulation of the ribulose bisphosphate pathway 58

I. Introduction: the definition of autotrophy

The diversity of microbes able to grow on C_1 compounds as the sole source of carbon and energy (Chapter 1) is impressive, but even more remarkable is the diversity of their assimilation pathways. There are four different pathways by which aerobic methylotrophs assimilate carbon substrates into cell material and each has at least two potential variants; one route operates in yeasts (Chapter 10) and the other three in bacteria.

The ribulose bisphosphate pathway or Calvin cycle was first elucidated in photosynthetic autotrophs, and because the designations autotroph and methylotroph will be used frequently in this chapter, a brief digression on terminology is appropriate.

An autotroph was originally an organism able to grow with CO_2 as its major carbon source, heterotrophs by contrast requiring a reduced source of carbon. Later definitions emphasised the inorganic nature of the energy source—light or oxidation of inorganic compounds—as well as the inorganic carbon source. When it was found that most autotrophs use the ribulose bisphosphate pathway for CO_2 assimilation, the operation of this pathway became loosely synonymous with autotrophy, thereby changing from a nutritional to a metabolic basis for the definition. This sowed the seeds of later confusion when it was found that some bacteria grow on reduced C_1 compounds by way of the ribulose bisphosphate pathway. Three main problems arise when designating organisms as autotrophs or methylotrophs on a metabolic, rather than on a nutritional basis; the first is that the metabolism of many organisms may be insufficiently investigated to determine the nature of the assimilation pathway; the second is that some organisms (such as *Methylococcus capsulatus*) may use more than one pathway; the third is that some photosynthetic bacteria (e.g. *Chlorobium thiosulphatophilum*) possibly have no ribulose bisphosphate pathway, but may instead have a reductive carboxylic acid cycle (Evans *et al.*, 1966; Quayle, 1972; McFadden, 1973; Cole, 1976). Such problems would to some extent be avoided by adopting the suggestion of Whittenbury and Kelly (1977) of a return to a nutritional basis of designation, but these authors have, in my view, also unnecessarily included methylotrophs under the heading of autotrophs, thereby losing the advantages of grouping the organisms at all. An alternative solution (Zatman, 1981), of introducing the term pseudo-methylotroph for those bacteria growing on reduced C_1 compounds by the ribulose bisphosphate pathway, can be confusing because this implies (etymologically speaking) that the organism does not *really* grow on reduced C_1 compounds.

A further argument against using the operation of the ribulose bisphosphate pathway as the basis of sub-dividing the methylotrophs into two groups is its arbitrary nature. After all, the ribulose monophosphate and ribulose bisphosphate pathways are very similar, so bacteria using these pathways might just as well be separated from those using the serine pathway for assimilation of C_1 compounds; or, on another arbitrary basis; the serine pathway bacteria might be grouped with the ribulose bisphosphate pathway bacteria because more than one third of the cell carbon arises from carbon dioxide.

Clearly this sort of debate can be entertaining but in order to achieve clarity of presentation I shall use the nutritional definitions, and metabolic types will be indicated by naming their assimilation pathway when necessary. For this purpose autotrophs are considered as those organisms able to grow on CO_2 as their major carbon source, and methylotrophs are those able to grow on reduced carbon compounds containing one or more carbon atoms but no carbon–carbon bonds. Some bacteria, of course, are able to grow autotrophically, methylotrophically and heterotrophically according to the carbon source present in the growth medium. It is appreciated that organisms such as *Pseudomonas oxalaticus*, growing by the ribulose bisphosphate pathway will be unhappy wherever we arbitrarily place them, because the name (but not the definition of) methylotroph implies that they can grow on "methyl" compounds whereas the only C_1 substrate supporting growth of *Ps. oxalaticus* is formate.

II. Description of the ribulose bisphosphate pathway

In the ribulose bisphosphate (RuBP) pathway or Calvin cycle (Figs 4 and 5), all the cell's carbon is assimilated at the oxidation level of CO_2, some or all of

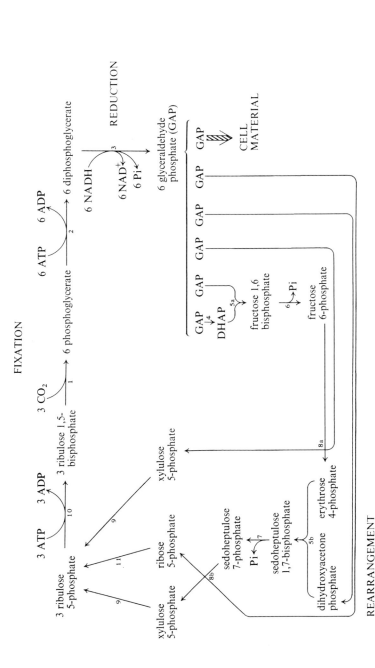

Fig. 4. The ribulose bisphosphate cycle of CO_2 fixation (sedoheptulose bisphosphatase variant). The enzyme numbers refer to those listed in the text: (1) ribulose bisphosphate carboxylase; (2) phosphoglycerate kinase; (3) glyceraldehyde phosphate dehydrogenase; (4) triose phosphate isomerase; (5a, b) aldolase; (6) fructose bisphosphatase; (7) sedoheptulose bisphosphatase; (8) transketolase; (9) pentose phosphate epimerase; (10) phosphoribulokinase; (11) pentose phosphate isomerase.

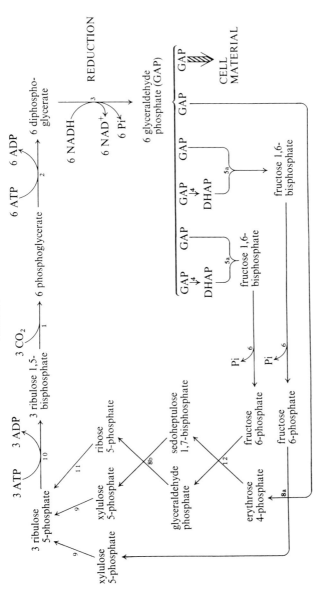

Fig. 5. The ribulose bisphosphate cycle of CO_2 assimilation (transaldolase variant). The enzyme numbers refer to those in the text: (1) ribulose bisphosphate carboxylase; (2) phosphoglycerate kinase; (3) glyceraldehyde phosphate dehydrogenase; (4) triose phosphate isomerase; (5a, b) aldolase; (6) fructose bisphosphatase; (8) transketolase; (9) pentose phosphate epimerase; (10) phosphoribulokinase; (11) pentose phosphate isomerase; (12) transaldolase.

which arises from oxidation of reduced C_1 compounds during methylotrophic growth. The overall reaction cycle synthesises one molecule of glyceraldehyde 3-phosphate from 3 molecules of CO_2:

$$3\ CO_2 + 6\ NAD(P)H + 6\ H^+ + 9\ ATP \rightarrow \text{glyceraldehyde}$$
$$\text{3-phosphate} + 6\ NAD(P)^+ + 9\ ADP + 8\ Pi$$

The first part of the cyclic sequence (*fixation*) is the carboxylation of 3 molecules of ribulose 1,5-bisphosphate (RuBP), yielding 6 molecules of 3-phosphoglycerate (PGA) which is then phosphorylated to 1,3-diphosphoglycerate.

The second part of the cycle is the *reduction* of the diphosphoglycerate to 6 molecules of glyceraldehyde 3-phosphate (GAP), one of which constitutes the final product of the cycle.

The third part of the cycle (*rearrangement*) regenerates the 3 molecules of ribulose bisphosphate from 5 molecules of glyceraldehyde 3-phosphate. At least two variants of the cycle may occur; they are the same in the fixation and reduction reactions but differ in the reactions involved in the rearrangement part of the cycle. The sedoheptulose bisphosphatase (SBP) variant (Fig. 4) involves this enzyme but not a transaldolase, whereas the transaldolase (TAL) variant (Fig. 5) involves transaldolase but not the sedoheptulose bisphosphatase.

The following section discusses the evidence for the operation of the pathway in methylotrophs and this is followed by a description of the individual enzymes of the pathway.

III. The occurrence and distribution of the ribulose bisphosphate pathway

This pathway (the RuBP pathway) operates in relatively few methylotrophic bacteria (Table 16). In all essential features it is the same as that first proposed by Calvin, Bassham and their colleagues (Bassham *et al.*, 1954; Bassham, 1971a, b) for the fixation of CO_2 during plant photosynthesis and which subsequently has been shown to operate in those autotrophs obtaining their energy by oxidation of inorganic compounds (the chemolithotrophs) (for reviews see Vishniac *et al.*, 1957; Quayle, 1961; Kelly, 1971; McFadden, 1973; Walker, 1977; Whittenbury and Kelly, 1977; Quayle and Ferenci, 1978). It is not known which variant (Figs 4 and 5) of the pathway operates in autotrophic or methylotrophic bacteria.

2. The ribulose bisphosphate pathway

In the RuBP pathway the cell carbon is assimilated at the oxidation level of CO_2, some or all of which, in the methylotrophs, arises from initial oxidation of the reduced C_1 substrates. With the exception of *Pseudomonas oxalaticus*, the methylotrophs using this pathway are facultative autotrophs using light energy or the oxidation of H_2 as the ATP source; and hence acquisition of the ability merely to oxidise C_1 compounds would, without the further elaboration of biosynthetic enzymes, confer upon the organisms the capability of growth on such compounds. It has been suggested (Sahm et al., 1976a) that "the gain in nutritional versatility, at such little cost in fashioning of extra enzymes, probably outweighs the disadvantage of using this more energetically costly pathway instead of a formaldehyde fixation sequence such as the RuMP pathway or serine pathway".

To establish the occurrence of the RuBP pathway with the same degree of thoroughness as in green plants it is necessary to satisfy the following requirements: phosphoglycerate must be the first labelled compound during assimilation of $^{14}CO_2$, and other phosphorylated intermediates must also be labelled at early times; the labelling pattern in intermediates must conform to those predicted by consideration of the proposed pathway; all the enzymes necessary for the operation of the cycle must be present in the organism. In practice, the presence of two key enzymes ribulose bisphosphate carboxylase (carboxydismutase) and phosphoribulokinase is often taken as indication of the presence of the pathway. A summary of the evidence for its occurrence in a range of autotrophs is given in the review by Quayle (1961).

Whilst the occurrence of the RuBP pathway in most of the methylotrophs listed in Table 16 has usually not been established as rigorously as for autotrophic bacteria it is certain that many methylotrophs do use this pathway and what evidence is available is described below.

The first methylotroph shown to assimilate its carbon substrate by the RuBP pathway was *Pseudomonas oxalaticus*, an otherwise typical heterotroph, which grows on formate as sole carbon and energy source but not on other C_1 compounds (Khambata and Bhat, 1953; Quayle and Keech, 1958, 1959a, b, c, 1960). It was thoroughly established that the RuBP pathway is involved during growth on formate by showing that 94% of the carbon of formate is converted to CO_2 prior to assimilation into cell material; isotope from ^{14}C-formate or ^{14}C-bicarbonate was incorporated most rapidly into phosphoglycerate and then into phosphorylated sugars; and that ribulose bisphosphate carboxylase and phosphoribulokinase were present during growth on formate but not on other substrates. It is of interest that oxalate, which might be expected to be assimilated by the RuBP pathway after oxidation by way of formate to CO_2, is assimilated after reduction to glyoxylate (Quayle et al., 1961); hence although oxalate is at a slightly higher oxidation level than formate its carbon–carbon bond is preserved for biosynthetic purposes.

TABLE 16
Methylotrophs able to use the ribulose bisphosphate pathway

Organism	References
PHOTOSYNTHETIC BACTERIA	
Rhodopseudomonas acidophila (10050) (uses methanol or formate)	Quayle and Pfennig (1975); Douthit and Pfennig (1976); Sahm *et al.* (1976a)
Rhodopseudomonas gelatinosa, Rhodomicrobium vannielii, Rhodospirillum rubrum, Rhodospirillum tenue (all use methanol)	Quayle and Pfennig (1975)
Rhodopseudomonas palustris (formate)	Yoch and Lindstrom (1967); Stokes and Hoare (1969); Quadri and Hoare (1969)
Spirulina platensis (CO)	Cited in Colby *et al.* (1979)
NON-PHOTOSYNTHETIC BACTERIA (and potential photo-synthetics growing aerobically in the dark)	
Pseudomonas oxalaticus (uses formate)	Quayle and Keech (1958, 1959a, b, c, 1960); Dijkhuizen *et al.* (1978); Dijkhuizen and Harder (1979a, b)
Paracoccus denitrificans (*Micrococcus denitrificans*) (uses methanol, methylamine or formate); this bacterium also grows on methanol and methylamine anaerobically with nitrate	Cox and Quayle (1975); Shively *et al.* (1978)
(*Rhodopseudomonas acidophila*) (methanol or formate)	Seifert and Pfennig (1979)
Thiobacillus novellus (methanol, formate or formamide)	Chandra and Shethna (1977)

Thiobacillus A2 (methylamine or formate)	Kelly *et al.* (1979)
Hydrogenomonas eutropha, *Alcaligenes* FOR$_1$, *Bacterium formoxidans* (all use formate); *Pseudomonas gazotropha* (uses CO); *Bacterium* sp. 7d (methylamine)	All cited in Colby *et al.* (1979)
Microcyclus aquaticus (methanol or formate); *Achromobacter* 1L, *Pseudomonas* 8, *Mycobacterium* 50 (all use methanol)	Loginova *et al.* (1978), Loginova and Trotsenko (1979c)
Blastobacter viscosus (methanol)	Loginova and Trotsenko (1979a)

With the exception of *Ps. oxalaticus*, methylotrophs using this pathway are facultative autotrophs using light or the oxidation of hydrogen as the source of ATP. No methanotroph uses this pathway. The distribution of other pathways is described in Tables 8, 16 and 18.

The evidence for inclusion of some of the bacteria in this table is not extensive; it is usually based on the presence of ribulose bisphosphate carboxylase and the absence of key enzymes of alternative pathways. A number of photosynthetic bacteria whose similarity to *Rhodopseudomonas acidophila* on methanol which is 10·6 h). The anaerobic photosynthetic bacteria use light as the source of ATP for biosynthesis, and the C_1 substrate as reductant and source of CO_2. During aerobic growth in the dark the reduced C_1 substrate provides ATP (by oxidative phosphorylation), reductant and CO_2.

The first methylotroph shown to grow on methanol by the RuBP pathway was *Paracoccus denitrificans* (*Micrococcus denitrificans*) which also grows autotrophically on $H_2 + CO_2 + O_2$ and on a range of conventional substrates (Cox and Quayle, 1975). During growth on methanol, or on H_2 plus CO_2, ribulose bisphosphate carboxylase is induced but (except for constant low levels of hydroxypyruvate reductase) enzymes for the alternative pathways for assimilation of reduced C_1 compounds are absent. As all carboxylases have oxygenase activity it is possible that the product of this activity (phosphoglycollate) might act as a precursor for glycine and serine biosynthesis and it has been suggested (Colby *et al.*, 1979) that this unexplained presence of hydroxypyruvate reductase in *Paracoccus denitrificans* growing on methanol (Bamforth and Quayle, 1977) might be involved in "recycling" the serine produced in this way.

The anaerobic growth of *Rhodopseudomonas acidophila* on methanol (Sahm *et al.*, 1976a) in the light is the second example of growth of a methylotroph by way of the RuBP pathway. The pattern of labelling of metabolites from ^{14}C methanol and bicarbonate was consistent with this pathway as was the high level of ribulose bisphosphate carboxylase compared with that present during growth on succinate. Dehydrogenases for oxidation of methanol, formaldehyde and formate were also induced during anaerobic growth on methanol in the light. Growth yield measurements indicate that *Rhodopseudomonas acidophila* assimilates methanol by a process which approximates to:

$$2 CH_3OH + CO_2 \rightarrow 3 (HCHO) \text{ (cell material)} + H_2O$$

In this overall process all carbon is assimilated at the oxidation level of CO_2, the NAD(P)H requirement for the reductive steps of the RuBP pathway being provided by the oxidation of methanol, the first steps of which must be coupled to reduction of $NAD(P)^+$ by reversed electron transport from methanol dehydrogenase (see Chapter 9). It has also been shown that *Rhodopseudomonas acidophila* can grow aerobically on methanol in the dark, presumably also by way of the RuBP pathway. The low growth yield suggests that the first step in aerobic methanol oxidation, which is presumably by way of methanol dehydrogenase plus an electron transport system, cannot be coupled to ATP synthesis.

Although it is probable that methylotrophs assimilate their cell carbon predominantly from only one of the four main pathways, the distribution of key enzymes of these pathways is not as exclusive as earlier results first indicated. The presence of the serine pathway enzyme, hydroxypyruvate reductase in *P. denitrificans*, has already been mentioned (above) and a second example is probably even more significant. *Methylococcus capsulatus*

(Bath) assimilates most of its cell carbon by way of the RuMP pathway of formaldehyde fixation and the first observation of ribulose bisphosphate carboxylase and phosphoribulokinase in this methanotroph was both important and surprising (Taylor, 1977). Recent experiments in which incorporation of $^{14}CO_2$ into this organism was measured have confirmed that the ribulose bisphosphate carboxylase does operate *in vivo*, the label appearing in phosphoglycerate at early times. At these early times, label also appeared in aspartate, malate and citrate indicating that the conventional heterotrophic carboxylation reactions were occurring (Taylor *et al.*, 1981). The presence of phosphoglycollate phosphatase and low levels of hydroxypyruvate reductase suggests that any phosphoglycollate produced by the oxygenase activity of the RuBP carboxylase is metabolised by way of glycollate, glyoxylate, serine, hydroxypyruvate and glycerate. This perhaps accounts for the presence of some serine pathway enzymes in this methanotroph in which the predominant assimilation pathway is the RuMP pathway (Taylor *et al.*, 1981). It was concluded that only about 2·5% of cell carbon arose from CO_2 during growth on methane as carbon substrate. As the short-term labelling experiments suggested that about 75% (very approximately) of this was by way of conventional carboxylation reactions it is apparent that the presence of RuBP carboxylase in *Methylococcus capsulatus* is of little physiological significance. Even so, if the RuBP pathway of CO_2 fixation evolved from the RuMP pathway of formaldehyde assimilation as suggested by Quayle and Ferenci (1978) then *Methylococcus capsulatus*, having the possibility of both assimilation pathways, represents an important transition organism.

IV. The reactions of the ribulose bisphosphate pathway

The following section describes the individual reactions of the pathway and the enzymes catalysing them. Descriptions of enzymes isolated from methylotrophic bacteria are relatively few and the reader is referred to the following more general discussions for more information on these enzymes (Vishniac *et al.*, 1957; Bassham, 1963, 1971a, b; Wood, 1966, 1975a, b). The numbers in parentheses refer to the number of the reaction as designated in Figs 4 and 5.

Ribulose bisphosphate carboxylase (carboxydismutase) (1)

This enzyme, first described in 1954 (Quayle *et al.*, 1954; Weissbach *et al.*, 1954), catalyses the carboxylation of ribulose 1,5-bisphosphate in a reaction requiring Mg^{++} ions for activation. The reaction yields 2 molecules of

52   The biochemistry of methylotrophs

3-phosphoglycerate and proceeds by way of a number of (probable) enzyme-bound intermediates (written in parentheses) (Walsh, 1979):

$$\begin{array}{c} CH_2O\text{-}P \\ | \\ C=O \\ | \\ H\text{-}C\text{-}OH \\ | \\ H\text{-}C\text{-}OH \\ | \\ CH_2O\text{-}P \end{array} \longrightarrow \left\{ \begin{array}{c} CH_2O\text{-}P \\ | \\ C\text{-}OH \\ || \\ C\text{-}OH \\ | \\ H\text{-}C\text{-}OH \\ | \\ CH_2O\text{-}P \end{array} \xrightarrow{CO_2} \begin{array}{c} CH_2O\text{-}P \\ | \\ HOOC\text{-}C\text{-}OH \\ | \\ C=O \\ | \\ H\text{-}C\text{-}OH \\ | \\ CH_2O\text{-}P \end{array} \xrightarrow{H_2O} \begin{array}{c} CH_2O\text{-}P \\ | \\ HOOC\text{-}C\text{-}OH \\ | \\ HO\text{-}C\text{-}OH \\ | \\ H\text{-}C\text{-}OH \\ | \\ CH_2O\text{-}P \end{array} \right\} \longrightarrow \begin{array}{c} CH_2O\text{-}P \\ | \\ H\text{-}C\text{-}OH \\ | \\ COOH \\ \\ COOH \\ | \\ H\text{-}C\text{-}OH \\ | \\ CH_2O\text{-}P \end{array}$$

ribulose 1,5-bisphosphate   3-phosphoglycerate

The carboxylase also catalyses the oxygenation of ribulose bisphosphate, the product of this reaction being one molecule of 3-phosphoglycerate plus a molecule of phosphoglycollate; this oxygenation reaction appears to be an inevitable consequence of the reaction mechanism and hence the enzyme is also named ribulose bisphosphate carboxylase/oxygenase. The oxygenase activity of the enzyme is the basis for the photorespiration occurring in some plants (see review by Andrews and Lorimer, 1978). The first intermediate in this oxygenase reaction is probably the same as in the carboxylation reaction but this intermediate is oxygenated instead of carboxylated:

$$\begin{array}{c} CH_2O\text{-}P \\ | \\ C=O \\ | \\ H\text{-}C\text{-}OH \\ | \\ H\text{-}C\text{-}OH \\ | \\ CH_2O\text{-}P \end{array} \longrightarrow \left\{ \begin{array}{c} CH_2O\text{-}P \\ | \\ C\text{-}OH \\ || \\ C\text{-}OH \\ | \\ H\text{-}C\text{-}OH \\ | \\ CH_2O\text{-}P \end{array} \xrightarrow{O_2} \begin{array}{c} CH_2O\text{-}P \\ | \\ HOO\text{-}C\text{-}OH \\ | \\ C=O \\ | \\ H\text{-}C\text{-}OH \\ | \\ CH_2O\text{-}P \end{array} \xrightarrow{H_2O} \begin{array}{c} CH_2O\text{-}P \\ | \\ HOO\text{-}C\text{-}OH \\ | \\ HO\text{-}C\text{-}OH \\ | \\ H\text{-}C\text{-}OH \\ | \\ CH_2O\text{-}P \end{array} \right\} \xrightarrow{H_2O} \begin{array}{c} CH_2O\text{-}P \\ | \\ COOH \\ \text{phosphoglycollate} \\ COOH \\ | \\ H\text{-}C\text{-}OH \\ | \\ CH_2O\text{-}P \end{array}$$

ribulose 1,5-bisphosphate   3-phosphoglycerate

Ribulose bisphosphate carboxylases have been isolated, purified and characterised from a wide variety of sources (for reviews see Kawashima and Wildman, 1970; Seigel *et al.*, 1972; McFadden, 1973, 1978; Jensen and Bahr, 1977; Lawlis *et al.*, 1979; Taylor and Dow, 1980). On the basis of their quaternary structure the RuBP carboxylases have been divided into two categories. Those termed the O enzymes consist entirely of 2, 6 or 8 large subunits of about 55 000 daltons and are only found in some "less evolved" procaryotes (including *Rhodospirillum rubrum*, *Chlorobium thio-*

sulphatophilum and *Thiobacillus intermedius*). Enzymes in the second category, the T enzymes, are large (about 500 000 daltons) and consist of similar large subunits to those found in the O enzymes and in addition an equal number of small subunits (about 15 000 daltons). Catalytic activity is usually associated with the larger subunits while the smaller subunits have a binding site for the usual inhibitor 6-phosphogluconate. Enzymes of this type (T enzymes) have been described in *Hydrogenomonas eutropha*, *Ectothiorhodospira*, *Halophila*, *Chromatium*, *Euglena gracilis* and in higher plants and they usually have 8 large and 8 small subunits.

The ribulose bisphosphate carboxylase from the facultative methylotroph and autotroph *Paracoccus denitrificans* has been shown to be a typical large (T) type of enzyme having 8 large, and 8 small subunits and being inhibited (94%) by 1 mM 6-phosphogluconate; the K_m values for ribulose bisphosphate and CO_2 are 0·166 mM and 0·051 mM respectively (Shively et al., 1978).

The carboxylase from *Pseudomonas oxalaticus* (grown on formate) also has a relatively high K_m for ribulose bisphosphate (0·22 mM) and is competitively inhibited by 6-phosphogluconate (K_i, 0·27 mM) (Lawlis et al., 1979). The subunit structure of this carboxylase is unusual, having only 6 large and 6 small subunits and in this feature it is similar to the enzymes from *Rhodomicrobium vannielii* (grown on pyruvate and malate) (Taylor and Dow, 1980) and *Methylococcus capsulatus* (Bath) (Taylor et al., 1980). It should be noted, however, that Andrews et al. (1981) have recently suggested that assignment of such hexameric structures might be erroneous if based on gel filtration experiments alone.

As would be expected for an enzyme functioning uniquely in the ribulose bisphosphate pathway, the RuBP carboxylase is usually induced during autotrophic and methylotrophic growth, being present in facultative heterotrophs at much lower levels during growth on typical multicarbon substrates (Quayle, 1961; Kelly, 1971; McFadden, 1973, 1978; Cox and Quayle, 1975; Sahm et al., 1976a; Dijkhuizen and Harder, 1979a, b).

Phosphoglycerate kinase (2)

This enzyme catalyses the phosphorylation of the first stable product of CO_2 fixation, 3-phosphoglycerate to 1,3-diphosphoglycerate using ATP as phosphate donor:

$$\begin{array}{c} \text{COOH} \\ | \\ \text{H-C-OH} \\ | \\ \text{CH}_2\text{O-P} \end{array} + \text{ATP} \leftrightarrow \begin{array}{c} \text{COO-P} \\ | \\ \text{H-C-OH} \\ | \\ \text{CH}_2\text{O-P} \end{array} + \text{ADP}$$

3-phosphoglycerate   1,3-diphosphoglycerate

54   The biochemistry of methylotrophs

Glyceraldehyde phosphate dehydrogenase (3)

This enzyme catalyses the reduction of 1,3-diphosphoglycerate to glyceraldehyde 3-phosphate using NAD(P)H as reductant. In plants this enzyme is specific for NADPH but this is not necessarily the case in all bacteria (McFadden, 1978):

$$\begin{array}{c} \text{COO-P} \\ | \\ \text{H-C-OH} \\ | \\ \text{CH}_2\text{O-P} \end{array} + \text{H}^+ + \text{NAD(P)H} \leftrightarrow \begin{array}{c} \text{CHO} \\ | \\ \text{H-C-OH} \\ | \\ \text{CH}_2\text{O-P} \end{array} + \text{NAD(P)}^+ + \text{Pi} + \text{H}_2\text{O}$$

1,3-diphosphoglycerate   glyceraldehyde 3-phosphate

Triose phosphate isomerase (4)

This enzyme catalyses the isomerisation of glyceraldehyde 3-phosphate and dihydroxyacetone phosphate:

$$\begin{array}{c} \text{CHO} \\ | \\ \text{H-C-OH} \\ | \\ \text{CH}_2\text{O-P} \end{array} \longleftrightarrow \begin{array}{c} \text{CH}_2\text{OH} \\ | \\ \text{C=O} \\ | \\ \text{CH}_2\text{O-P} \end{array}$$

glyceraldehyde   dihydroxyacetone
3-phosphate   phosphate

Aldolase

Aldolase catalyses the condensation of an aldose with a ketose giving a larger ketose sugar. The first aldolase-catalysed reaction in the RuBP pathway (5a) is between dihydroxyacetone phosphate and glyceraldehyde 3-phosphate yielding fructose 1,6 bisphosphate:

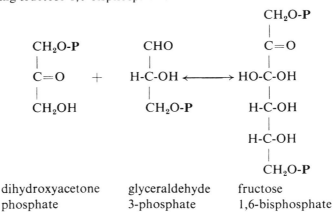

dihydroxyacetone   glyceraldehyde   fructose
phosphate   3-phosphate   1,6-bisphosphate

In the sedoheptulose bisphosphatase variant of the pathway (Fig. 4), an aldolase also catalyses the condensation (reaction 5b) of a second molecule of dihydroxyacetone phosphate with erythrose 4-phosphate to give sedoheptulose 1,7-bisphosphate (this second reaction does not occur in the alternative transaldolase variant):

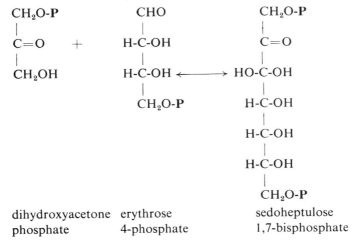

dihydroxyacetone phosphate   erythrose 4-phosphate   sedoheptulose 1,7-bisphosphate

Fructose bisphosphatase (6)

This enzyme catalyses the hydrolysis of fructose 1,6-bisphosphate giving the monophosphate:

$$\begin{array}{c} CH_2O\text{-}P \\ | \\ C=O \\ | \\ HO\text{-}C\text{-}H \\ | \\ H\text{-}C\text{-}OH \\ | \\ H\text{-}C\text{-}OH \\ | \\ CH_2O\text{-}P \end{array} + H_2O \longrightarrow \begin{array}{c} CH_2OH \\ | \\ C=O \\ | \\ HO\text{-}C\text{-}H \\ | \\ H\text{-}C\text{-}OH \\ | \\ H\text{-}C\text{-}OH \\ | \\ CH_2O\text{-}P \end{array} + P_i$$

fructose 1,6-bisphosphate   fructose 6-phosphate

Sedoheptulose bisphosphatase (7)

In the sedoheptulose bisphosphate variant of the pathway this enzyme catalyses hydrolysis of the bisphosphate to sedoheptulose 7-phosphate. In plants this phosphatase is not the same as that hydrolysing fructose bisphosphate; it is usually a separate enzyme although in bacteria this may not be the case (McFadden, 1973).

Transketolase (8)

Transketolase catalyses the reversible transfer of a glycolaldehyde moiety from a donor ketose to an acceptor aldose, the reaction typically requiring thiamine pyrophosphate and a divalent cation as cofactors. Two transketolase-catalysed reactions occur in the RuBP pathway, both using glyceraldehyde 3-phosphate as the aldose acceptor and both producing xylulose 5-phosphate. In the first reaction (8a) the glycolaldehyde donor is fructose 6-phosphate and the product is erythrose 4-phosphate:

```
   CH₂OH
    |
    C=O                           CH₂OH
    |                              |
   HO-C-H                          C=O              CHO
    |                              |                 |
   H-C-OH      +     CHO   ⇌     HO-C-H    +       H-C-OH
    |                 |            |                 |
   H-C-OH           H-C-OH        H-C-OH            H-C-OH
    |                 |            |                 |
   CH₂O-P           CH₂O-P        CH₂O-P            CH₂O-P
  fructose       glyceraldehyde   xylulose         erythrose
 6-phosphate      3-phosphate    5-phosphate      4-phosphate
```

In the second transketolase-catalysed reaction (8b) the glycolaldehyde donor is sedoheptulose 7-phosphate and the product is ribose 5-phosphate. Usually the same transketolase is able to catalyse both of these reactions.

```
   CH₂OH
    |
    C=O
    |
   HO-C-H                         CH₂OH             CHO
    |                              |                 |
   H-C-OH                          C=O              H-C-OH
    |                              |                 |
   H-C-OH            CHO          HO-C-H            H-C-OH
    |                 |            |                 |
   H-C-OH    +      H-C-OH  ⇌    H-C-OH    +       H-C-OH
    |                 |            |                 |
   CH₂O-P           CH₂O-P        CH₂O-P            CH₂O-P
 sedoheptulose   glyceraldehyde  xylulose          ribose
  7-phosphate     3-phosphate   5-phosphate      5-phosphate
```

2. The ribulose bisphosphate pathway

Pentose phosphate epimerase (9)

This enzyme catalyses the epimerisation of xylulose 5-phosphate to ribulose 5-phosphate:

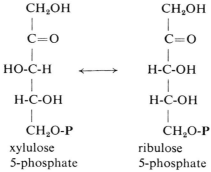

```
    CH₂OH                    CH₂OH
    |                        |
    C=O                      C=O
    |                        |
    HO-C-H      ⇌            H-C-OH
    |                        |
    H-C-OH                   H-C-OH
    |                        |
    CH₂O-P                   CH₂O-P
    xylulose                 ribulose
    5-phosphate              5-phosphate
```

Phosphoribulokinase (10)

Phosphoribulokinase catalyses the regeneration of the CO_2-acceptor molecule, ribulose 1,5-bisphosphate from ribulose 5-phosphate in an irreversible reaction requiring ATP:

```
    CH₂OH                              CH₂O-P
    |                                  |
    C=O                                C=O
    |                                  |
    H-C-OH    +   ATP  ⟶              H-C-OH    +   ADP
    |                                  |
    H-C-OH                             H-C-OH
    |                                  |
    CH₂O-P                             CH₂O-P
 ribulose 5-phosphate           ribulose 1,5-bisphosphate
```

Pentose phosphate isomerase (11)

This enzyme catalyses the isomerisation of the ribose 5-phosphate produced in the second transketolase reaction (8b) to ribulose 5-phosphate:

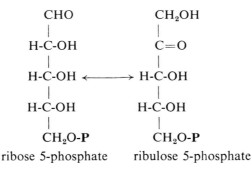

```
    CHO                      CH₂OH
    |                        |
    H-C-OH                   C=O
    |                        |
    H-C-OH      ⇌            H-C-OH
    |                        |
    H-C-OH                   H-C-OH
    |                        |
    CH₂O-P                   CH₂O-P
    ribose 5-phosphate       ribulose 5-phosphate
```

Transaldolase (12)

This enzyme is only involved in the transaldolase variant of the pathway (Fig. 5). It catalyses the reversible transfer of a dihydroxyacetone moiety from the donor ketose, fructose 6-phosphate to the acceptor aldose erythrose 4-phosphate, the products being glyceraldehyde 3-phosphate and sedoheptulose 7-phosphate:

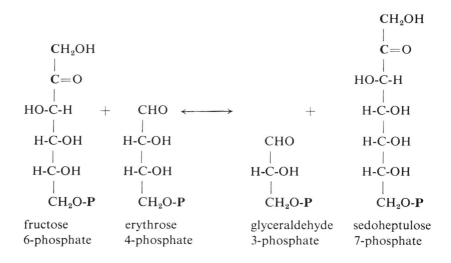

V. Regulation of the ribulose bisphosphate pathway

A more extensive discussion of general aspects of regulation in methylotrophs is presented in Chapter 4 (Section V).

The key enzymes of the RuBP pathway (RuBP carboxylase and phosphoribulokinase) are both regulated at the genetic level in facultative autotrophs (McFadden, 1978), and in those facultative methylotrophs able to grow on reduced C_1 compounds by the RuBP pathway (references in Table 16). This has been particularly studied in *Pseudomonas oxalaticus* which has this pathway during growth on formate but not on oxalate or other multicarbon compounds. Regulation of synthesis of the RuBP pathway enzymes in this organism is especially suitable for study because the change to this pathway from typical heterotrophic pathways of assimilation involves only a change in carbon metabolism and not the change in energy metabolism which must occur in most typical facultative autotrophs (Blackmore and Quayle, 1968; Dijkhuizen *et al.*, 1978; Dijkhuizen and Harder, 1979a, b).

Although a considerable amount is known about regulation of the RuBP pathway at the enzymic level in autotrophic bacteria there is little information on this subject in methylotrophs. Regulation in autotrophs has been recently reviewed by McFadden (1978) and this is summarised below.

The phosphoribulokinase of a number of chemolithotrophs is inhibited by AMP which is an indicator of low "energy charge" and hence low potential for biosynthesis. Other enzymes similarly inhibited in some autotrophs are phosphoglycerate kinase and fructose bisphosphatase. In the non-photosynthetic autotrophs and perhaps also the methylotrophs, NADH, instead of NADPH is the reductant in assimilation of CO_2 and related to this fact is the demonstration that NADH (not NADPH) is a potent allosteric activator of phosphoribulokinase in these bacteria. Thus it appears that the CO_2 assimilation cycle is regulated not only by the supply of ATP and NADH but also by specific effects on regulatory enzymes which ensure that the operation of the cycle is increased when there is a sufficient supply of ATP and NADH to support it.

A further regulation of the cycle may be achieved by feedback inhibition by "products" of the pathway. For example, PEP inhibits phosphoribulokinase and/or fructose bisphosphatase in some autotrophs and in others this enzyme is inhibited by glyceraldehyde 3-phosphate. Of particular interest is the inhibition by low concentrations of 6-phosphogluconate of the RuBP carboxylase. Inhibition of this enzyme in photosynthetic organisms has been taken to indicate a regulatory role for 6-phosphogluconate in changing from light to dark metabolism during which this metabolite becomes an intermediate in oxidative carbon metabolism. A similar inhibition of RuBP carboxylase occurs in the hydrogen bacteria, this inhibition becoming important when the 6-phosphogluconate concentration increases during a change to heterotrophic metabolism. This is also presumably the significance of the inhibition by 6-phosphogluconate of the RuBP carboxylase of the facultative methylotrophs *Paracoccus denitrificans* and *Pseudomonas oxalaticus* (Shively et al., 1978).

3
The ribulose monophosphate cycle of formaldehyde fixation

I. Introduction	60
II. Description of the ribulose monophosphate pathways	62
A. The biosynthesis of a C_3 compound from formaldehyde	62
B. The biosynthesis of carbohydrates	68
C. The oxidation of formaldehyde by a dissimilatory RuMP cycle	68
D. The interconversion of pyruvate, phosphoenolpyruvate and oxaloacetate in bacteria with the RuMP cycle	70
III. The occurrence and distribution of the RuMP pathways	74
A. The RuMP pathways of formaldehyde assimilation	74
B. The dissimilatory cycle of formaldehyde oxidation	76
IV. Reactions of the RuMP pathways	77
A. Enzymes involved in the fixation phase (all variants)	77
B. Enzymes involved in the cleavage phase	81
1. KDPG aldolase variants	81
2. Fructose bisphosphate aldolase variants	84
C. Enzymes involved in the rearrangement phase	85
1. All variants	85
2. Transaldolase variants	87
3. Sedoheptulose bisphosphatase variants	88
D. The key enzyme of the RuMP dissimilatory cycle; 6-phosphogluconate dehydrogenase	89
E. The interconversion of phosphoenolpyruvate with triose phosphates	90
F. The interconversion of pyruvate, phosphoenolpyruvate and oxaloacetate	91
V. Regulation of the RuMP pathway of formaldehyde assimilation	93

I. Introduction

This cycle operates only during methylotrophic growth of bacteria (not yeasts), fixing carbon at the oxidation level of formaldehyde. The first intimation that carbon is not assimilated at the level of CO_2 was obtained during studies of $^{14}CO_2$ assimilation in *Methylomonas methanica* during growth on methane by Leadbetter and Foster (1958), and by Johnson and Quayle (1965) who showed that ribulose bisphosphate carboxylase, the key enzyme of the ribulose bisphosphate pathway of CO_2 fixation is absent from this organism. That a novel pathway was operating was confirmed (Johnson and Quayle, 1965), by showing that the ^{14}C-labelled compounds accumulating

3. The ribulose monophosphate cycle 61

at early times during incubation with $^{14}CH_4$ or $^{14}CH_3OH$ were mainly glucose and fructose phosphates, and not the first product of the ribulose bisphosphate pathway of CO_2 fixation. A formaldehyde-condensing enzyme was then demonstrated in crude extracts of *Methylomonas methanica* by Kemp and Quayle (1965, 1966), who suggested that this enzyme was catalysing the hydroxymethylation of ribose 5-phosphate by formaldehyde to give allulose phosphate, and that it was thus likely to be a key enzyme in a novel pentose phosphate cycle for C_1 assimilation. Such a cycle was subsequently proposed, in which a novel hydroxymethylation of ribose phosphate to allulose phosphate is followed by cleavage of fructose 1,6-bisphosphate, catalysed by aldolase, and typical pentose phosphate rearrangements catalysed by transketolase and transaldolase (Kemp and Quayle, 1967). This type of cycle was shown to be consistent with the labelling in radioactive hexose phosphates obtained from whole-cell experiments, and consistent with the demonstration that the label from $^{14}CH_3OH$ was found predominantly in the C-1 of hexoses at the earliest times (Kemp and Quayle, 1967).

The same, or a similar cycle was then shown to operate in *Methylococcus capsulatus*, by demonstrating that $^{14}CH_3OH$ was rapidly incorporated into hexose phosphates, and that an enzyme system is present which is able to condense formaldehyde with pentose phosphate to give a mixture of hexose phosphates, consisting predominantly of fructose phosphate and what appeared to be allulose phosphate (Lawrence *et al.*, 1970). It was by using extracts of this organism that Kemp (1972, 1974) was able to show that the true substrate for the first enzyme in the cycle is ribulose monophosphate rather than ribose phosphate, and that the product is not allulose phosphate but the very similar novel sugar D-*erythro*-L-*glycero* 3-hexulose 6-phosphate (shorter names, D-arabino-hexulose 6-phosphate or hexulose 6-phosphate). The ribulose monophosphate cycle of formaldehyde fixation was eventually completed by the purification and characterisation of the hexulose phosphate synthase and isomerase from *Methylococcus capsulatus* (Ferenci *et al.*, 1974); and by demonstrating the presence of essential cleavage and rearrangement enzymes in *Methylomonas methanica* and *Methylococcus capsulatus* (Strom *et al.*, 1974), and in the obligate and facultative methylotrophs which do not use methane (organisms, 4B6, C2A1, W3A1, W6A and *Bacillus* sp. S2A1 and PM6) (Colby and Zatman, 1975c).

In a highly stimulating review, Quayle and Ferenci (1978) have argued the case that the RuMP cycle of formaldehyde assimilation might have been an evolutionary precursor of the RuBP cycle of CO_2 assimilation.

II. Description of the ribulose monophosphate pathways

A. The biosynthesis of a C_3 compound from formaldehyde

In the ribulose monophosphate pathways (RuMP pathways) (Figs 6 to 9) all the cell carbon is assimilated at the oxidation level of formaldehyde, produced by oxidation of methane, methanol or methylated amines. The overall reaction cycle synthesises one molecule of a C_3 compound from three molecules of formaldehyde, this C_3 compound being either pyruvate or dihydroxyacetone phosphate.

The first part of the cyclic sequence (*fixation*) is the aldol condensation of formaldehyde with three molecules of ribulose 5-phosphate to give 3-hexulose 6-phosphate (hexulose phosphate), which is then isomerised to fructose 6-phosphate (FMP) (three molecules).

In the second part of the cycle (*cleavage*) one of the molecules of FMP is converted to either fructose 1,6-bisphosphate (FBP) by phosphofructokinase, or to 2-keto 3-deoxy 6-phosphogluconate (KDPG) by the Entner/Doudoroff enzymes; these molecules are then cleaved by aldolases to glyceraldehyde 3-phosphate plus the "product" of the pathway, which is either pyruvate (from KDPG) or dihydroxyacetone phosphate (from FBP).

In the final part of the cycle (*rearrangement*) the remaining two molecules of FMP and the glyceraldehyde phosphate undergo a series of reactions, similar to those occurring in the RuBP cycle of CO_2 fixation, leading to regeneration of three molecules of ribulose 5-phosphate. As in the RuBP pathway there are two rearrangement variants; the sedoheptulose bisphosphatase (SBPase) variant involves this enzyme but not a transaldolase, whereas the transaldolase (TA) variant involves transaldolase but not the SBPase.

Because there are two possible cleavage enzymes and two rearrangement sequences there is a total of four potential variants of the RuMP pathway. Although not equally important, they are all written out in full for convenience and comparison (Figs 6 to 9).

As it is essential to produce both triose phosphate and pyruvate for biosynthesis, bacteria with the KDPG aldolase variant must contain enzymes for the synthesis of triose phosphate from pyruvate. Likewise, bacteria with the fructose bisphosphate aldolase variant must also have all the "glycolytic" enzymes for production of pyruvate from triose phosphate. The choice as to whether pyruvate or a phosphorylated C_3 compound is taken as the "endproduct" of the pathway is, to some extent arbitrary, but as the greater proportion of carbon flow must be by way of phosphoenol pyruvate (PEP), phosphoglycerate and triose phosphates, I have chosen to consider phosphoglycerate as the "end-product". This choice has the added advantage that the pathways can be more readily compared with respect to their energy requirements.

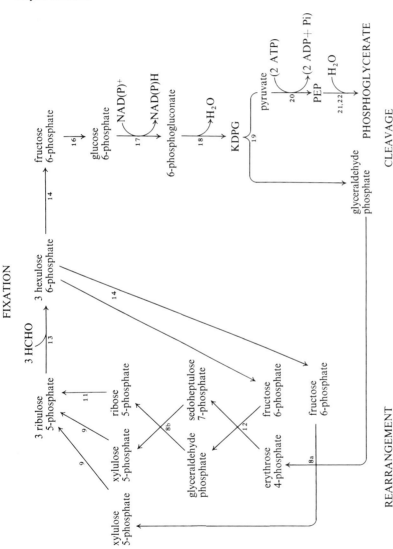

Fig. 6. The ribulose monophosphate (RuMP) cycle of formaldehyde assimilation (KDPG aldolase/transaldolase variant). This variant occurs predominantly in obligate methylotrophs. The enzyme numbers refer to those listed in the text: (8) transketolase; (9) pentose phosphate epimerase; (11) pentose phosphate isomerase; (12) transaldolase; (13) hexulose phosphate synthase; (14) hexulose phosphate isomerase; (16) glucose phosphate isomerase; (17) glucose phosphate dehydrogenase; (18) phosphogluconate dehydrase; (19) 2-keto, 3-deoxy, 6-phosphogluconate (KDPG) aldolase; (20) PEP synthetase or equivalent enzyme(s); (21) enolase; (22) phosphoglyceromutase.

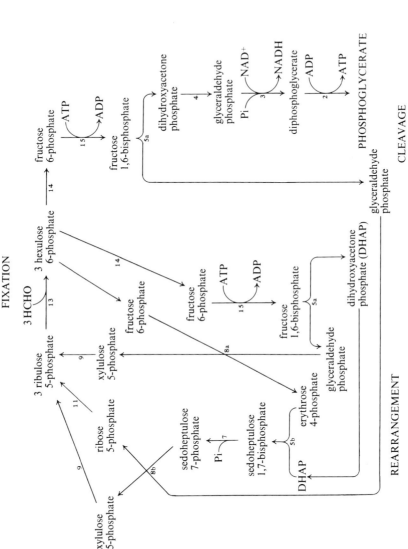

Fig. 7. The ribulose monophosphate (RuMP) cycle of formaldehyde assimilation (Fructose bisphosphate aldolase/sedoheptulose bisphosphatase variant). This variant occurs predominantly in facultative methylotrophs. The enzyme numbers refer to those listed in the text: (2) phosphoglycerate kinase; (3) glyceraldehyde phosphate dehydrogenase; (4) triose phosphate isomerase; (5) aldolase; (7) sedoheptulose bisphosphatase; (8) transketolase; (9) pentose phosphate epimerase; (11) pentose phosphate isomerase; (13) hexulose phosphate synthase; (14) hexulose phosphate isomerase; (15) phosphofructokinase.

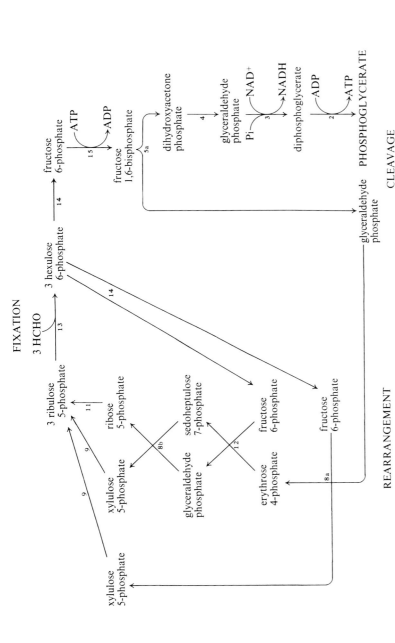

Fig. 8. The ribulose monophosphate (RuMP) cycle of formaldehyde assimilation (Fructose bisphosphate aldolase/transaldolase variant). This variant does not occur frequently in methylotrophs. The enzyme numbers refer to those listed in the text: (2) phosphoglycerate kinase; (3) glyceraldehyde phosphate dehydrogenase; (4) triose phosphate isomerase; (5) aldolase; (8) transketolase; (9) pentose phosphate epimerase; (11) pentose phosphate isomerase; (12) transaldolase; (13) hexulose phosphate synthase; (14) hexulose phosphate isomerase; (15) phosphofructokinase.

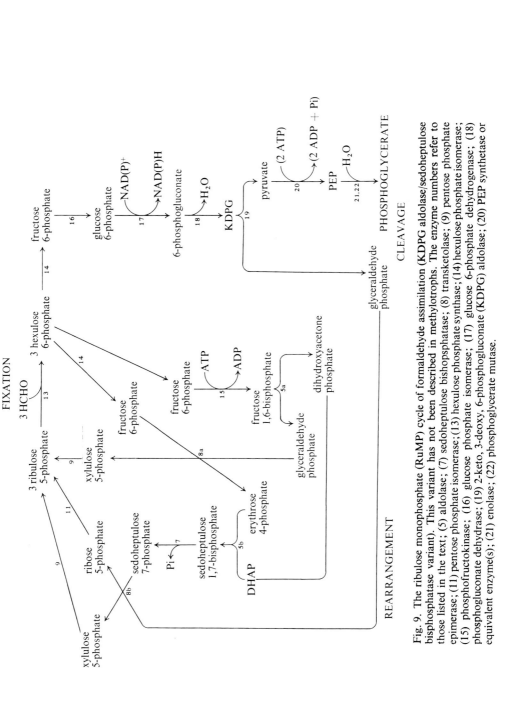

Fig. 9. The ribulose monophosphate (RuMP) cycle of formaldehyde assimilation (KDPG aldolase/sedoheptulose bisphosphatase variant). This variant has not been described in methylotrophs. The enzyme numbers refer to those listed in the text; (5) aldolase; (7) sedoheptulose bishopsphatase; (8) transketolase; (9) pentose phosphate epimerase; (11) pentose phosphate isomerase; (13) hexulose phosphate synthase; (14) hexulose phosphate isomerase; (15) phosphofructokinase; (16) glucose phosphate isomerase; (17) glucose 6-phosphate dehydrogenase; (18) phosphogluconate dehydrase; (19) 2-keto, 3-deoxy, 6-phosphogluconate (KDPG) aldolase; (20) PEP synthetase or equivalent enzyme(s); (21) enolase; (22) phosphoglycerate mutase.

TABLE 17
Summary of variants of the ribulose monophosphate cycle of formaldehyde fixation

	Cleavage phase	Rearrangement phase	Occurrence	Other key enzymes
(1) (Fig. 6)	KDPG aldolase	transaldolase	mainly in obligate methylotrophs	requires Entner/Doudoroff enzymes; enzymes for conversion of pyruvate to glyceraldehyde phosphate; NOT phosphofructokinase or fructose bisphosphate aldolase
(2) (Fig. 7)	FBP aldolase	SBPase	mainly in facultative methylotrophs	requires phosphofructokinase and other "glycolytic" enzymes
(3) (Fig. 8)	FBP aldolase	transaldolase	minor importance	requires phosphofructokinase and other "glycolytic" enzymes
(4) (Fig. 9)	KDPG aldolase	SBPase	not yet described	requires Entner/Doudoroff enzymes; enzymes for conversion of pyruvate to glyceraldehyde phosphate and phosphofructokinase; N.B. this variant would have both "cleavage" enzymes

Summary equations for production of phosphoglycerate from formaldehyde as in Figs 6 to 9

(1) (Fig. 6) KDPGA/TA $3\ HCHO + NAD^+ + 2\ ATP \longrightarrow$ phosphoglycerate $+ NADH + H^+ + 2\ ADP + Pi$
(2) (Fig. 7) FBPA/SBPase $3\ HCHO + NAD^+ + ATP \longrightarrow$ phosphoglycerate $+ NADH + H^+ + ADP$
(3) (Fig. 8) FBPA/TA $3\ HCHO + NAD^+ + Pi \longrightarrow$ phosphoglycerate $+ NADH + H^+$
(4) (Fig. 9) KDPGA/SBPase $3\ HCHO + NAD^+ + 3\ ATP \longrightarrow$ phosphoglycerate $+ NADH + H^+ + 3\ ADP + 2\ Pi$

68 The biochemistry of methylotrophs

Table 17 summarises the four potential variants of the RuMP pathway, the distribution of which will be discussed in more detail in Section III (below). The most commonly occurring variants are the KDPG aldolase/transaldolase variant and the FBP aldolase/SBPase variant, and these differ by only one ATP in their energy requirements. The less important FBP aldolase/transaldolase variant requires no ATP and the "non-existent" KDPG aldolase/SBPase variant requires the most ATP of the four potential variants. It should be noted that these differences are very slight compared with the great differences in ATP requirement between the major C_1 assimilation pathways.

B. The biosynthesis of carbohydrates

Pentose phosphate can be produced from triose phosphate, plus two molecules of formaldehyde by the enzymes of the complete cycle acting without the cleavage enzyme; hexose phosphate can be synthesised from this by fixation of another molecule of formaldehyde (Fig. 10).

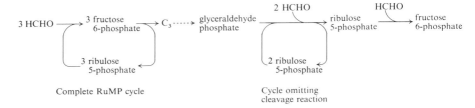

Fig. 10. The biosynthesis from formaldehyde of triose, pentose and hexose. This route can operate in bacteria having either variant of the RuMP pathway.

Erythrose monophosphate can be produced by way of the enzymes of the complete RuMP cycle operating to produce one molecule of glyceraldehyde phosphate and one of fructose 6-phosphate from four molecules of formaldehyde; the operation of the ubiquitous transketolase (reaction 8a) can then regenerate pentose phosphate and yield one molecule of erythrose phosphate (Fig. 11a). In bacteria containing transaldolase, an alternative route might operate which involves transaldolase, acting together with transketolase, to effect the rearrangement of two molecules of pentose phosphate (Fig. 11b) (Quayle and Ferenci, 1978).

C. The oxidation of formaldehyde by a dissimilatory RuMP cycle

Bacteria having the Entner/Doudoroff enzymes also have the potential for oxidising formaldehyde completely to CO_2 in a cyclic manner (Fig. 12),

3. The ribulose monophosphate cycle

Route (a)

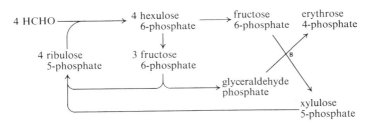

Route (b)

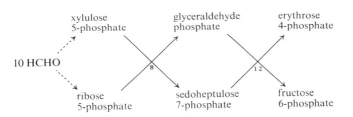

Fig. 11. The biosynthesis of erythrose 4-phosphate from formaldehyde. Reaction 8 is catalysed by transketolase and reaction 12 by transaldolase. Route (a) can operate in bacteria with either of the variants of the RuMP pathway. Route (b) can only operate in those bacteria having the transaldolase variant of the pathway.

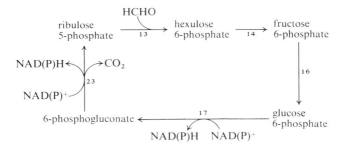

Fig. 12. The oxidation of formaldehyde to CO_2 by a dissimilatory RuMP cycle. The enzymes of this cycle are those of the assimilatory cycle plus the dissimilatory enzyme 6-phosphogluconate dehydrogenase (reaction 23). Other enzymes: (13) hexulose phosphate synthase; (14) hexulose phosphate isomerase; (16) glucose phosphate isomerase; (17) glucose 6-phosphate dehydrogenase.

provided that they also have an active 6-phosphogluconate dehydrogenase (Strom *et al.*, 1974; Colby and Zatman, 1975c).

The summary equation of the cycle is:

$$HCHO + 2\ NAD(P)^+ \rightarrow CO_2 + 2\ NAD(P)H + 2\ H^+$$

In the oxidative pentose phosphate cycle of higher organisms the glucose phosphate and phosphogluconate dehydrogenases are coupled to synthesis of NADPH, and the cycle is an important source of this reductant for biosynthesis. In methylotrophs this may also be the case but a main function of the cycle is the provision of energy, and the dehydrogenases are usually coupled to reduction of both NAD^+ and $NADP^+$ (see below); the electron transport chains of some of these bacteria are able to oxidise both NADH and NADPH (Colby and Zatman, 1975a; Cross and Anthony, 1980b).

D. The interconversion of pyruvate, phosphoenolpyruvate (PEP) and oxaloacetate in bacteria with the RuMP cycle (Fig. 13)

Bacteria having the fructose bisphosphate aldolase variant of the cycle produce glyceraldehyde 3-phosphate (GAP) and hence PEP as the "end-product". For biosynthesis of cell material, routes must also exist for formation from PEP of pyruvate, acetyl-CoA and oxaloacetate (OAA). These bacteria must therefore possess pyruvate kinase (giving pyruvate from PEP) and (usually) pyruvate dehydrogenase for production of acetyl-CoA. Alternatively, malyl-CoA lyase catalyses the formation of acetyl-CoA from malyl-CoA, and this might provide an alternative route for acetyl-CoA production in bacteria having enzymes for malyl-CoA synthesis and sufficiently high activities of malyl-CoA lyase. The most likely enzymes for carboxylation of a C_3 compound to a C_4 compound are PEP carboxylase or pyruvate carboxylase, both of which yield oxaloacetate (Fig. 13).

Bacteria having the KDPG aldolase variant produce pyruvate as their "end-product" and must, therefore, have routes for synthesis from pyruvate of acetyl-CoA, PEP and OAA. One possibility for PEP production involves its direct synthesis from pyruvate by way of PEP synthetase; the alternative route involves carboxylation of pyruvate to OAA, catalysed by pyruvate carboxylase followed by decarboxylation of the OAA to PEP, catalysed by PEP carboxykinase. If this second route operates then PEP carboxylase will not be necessary whereas bacteria having a PEP synthetase will not necessarily require pyruvate carboxylase.

There have been very few complete studies of these enzymes in bacteria having the RuMP pathway, and not all of the published work can be readily related to the analysis presented above. This may be to some extent because of the methods of enzyme assay used; these have sometimes been based on single determinations in crude extracts using radioactive substrates and this can lead to positive results in which the specific activity may be so low as to be physiologically insignificant.

One organism that can be related satisfactorily to the analysis presented above is *Methylophilus methylotrophus*, an obligate methanol-utiliser having

3. The ribulose monophosphate cycle

KDPG aldolase variant. This synthesises PEP and oxaloacetate from pyruvate by way of pyruvate carboxylase (which is not activated by acetyl-CoA) and PEP carboxykinase; the direct route by way of PEP synthetase is absent from this organism (Aperghis and Quayle, 1981). *Pseudomonas oleovorans* and sp. W6, which also have the KDPG aldolase variant also contain pyruvate carboxylase that does not require acetyl-CoA for activity (Loginova and Trotsenko, 1977a, 1979d; Babel and Loffhagen, 1977; Loffhagen et al., 1979). Unexpectedly, these bacteria also have low levels of PEP carboxylase during growth on methanol (Loginova and Trotsenko, 1977a, 1979d; Babel and Loffhagen, 1977; Loffhagen and Babel, 1978; Loffhagen et al., 1979). *Pseudomonas* W6 also contains pyruvate kinase (Loffhagen and Babel, 1978) but its function in this organism is not obvious. PEP carboxykinase, the second enzyme essential for synthesis of PEP from pyruvate is present at low levels compared with those on acetate in *Pseudomonas oleovorans* (Loffhagen et al., 1979), but its presence in *Pseudomonas* W6 has not been reported. The presence (or absence) of PEP synthetase has not been reported in these two bacteria.

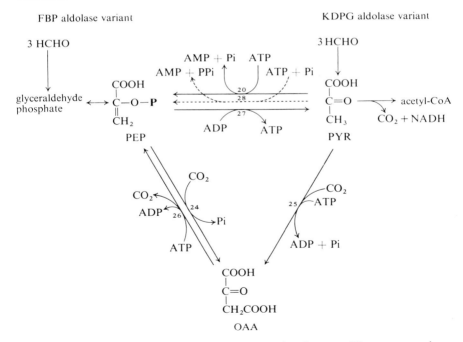

Fig. 13. The interconversion of phosphoenolpyruvate and oxaloacetate. The enzyme numbers refer to numbers listed in the text: (20) PEP synthetase; (24) PEP carboxylase; (25) pyruvate carboxylase; (26) PEP carboxykinase; (27) pyruvate kinase; (28) pyruvate phosphate dikinase.

TABLE 18
The occurrence and distribution of the RuMP pathway of formaldehyde fixation

OBLIGATE METHANOTROPHS
(all tested have type I membranes)

Methylomonas methanica[a,b] — Johnson and Quayle (1965); Kemp and Quayle (1965, 1966, 1967) Lawrence and Quayle (1970); Strom et al. (1974); Ferenci et al. (1974)

Methylomonas agile and *M. rosaceous*[c] — Lawrence & Quayle (1970)
Methylomonas rubrum[c] — Shishkina et al. (1976)
Methylomonas GB3 and GB8[a] — Babel and Mothes (1978)
Methylococcus capsulatus[a,b] — Lawrence et al. (1970); Lawrence and Quayle (1970); Kemp (1972, 1974); Strom et al. (1974); Ferenci et al. (1974); Whittenbury et al. (1975)

Methylococcus minimus[c] — Lawrence and Quayle (1970)
Methylococcus ucrainicus and *M. thermophilus*[c] — Shishkina et al. (1976)
Methylobacter capsulatus[c] — Lawrence and Quayle (1970)
Methylobacter bovis and *M. vinelandii*[c] — Shishkina et al. (1976); Trotsenko (1976)
Methylobacter chroococcum[c] — Trotsenko (1976)

OBLIGATE METHYLOTROPHS UNABLE TO USE METHANE

Pseudomonas W1[c] — Dahl et al. (1972)
Organism 4B6[a] — Colby and Zatman (1972)
Pseudomonas C[a] — Stieglitz and Mateles (1973); Goldberg and Mateles (1975); Ben-Bassat and Goldberg (1977); Ben-Bassat et al. (1980)

Pseudomonas W6[a] — Babel and Miethe (1974)
Organism C2A1[a] — Colby and Zatman (1975c)
Organisms W3A1 and W6A[a] — Colby and Zatman (1975c)
(these also grow slowly on glucose)
Methylomonas M15[a] — Sahm et al. (1976b); Steinbach et al. (1978)
Methylomonas aminofaciens 77a[a] — Kato et al. (1977b, 1978)
Methylophilus methylotrophus[a] — Beardsmore et al. (1982); Large and Haywood (1981)
Organism L3[c] — Hirt et al. (1978)

FACULTATIVE METHYLOTROPHS UNABLE TO USE METHANE

Arthrobacter[a]	Cox and Zatman (1974)
Bacillus sp. PM6 and S2A1[a]	Colby and Zatman (1975c)
Arthrobacter globiformis[a]	Loginova and Trotsenko (1976b)
Streptomyces sp. 239[c] (low specific activity)	Kato et al. (1977b)
Pseudomonas oleovorans[a]	Loginova and Trotsenko (1977a); Sokolov and Trotsenko (1978a, b); Muller and Sokolov (1979)
Organisms MB53, 55, 56, 57, 58, 59 and 60[a]	Babel and Mothes (1978)
Brevibacterium fuscum 24[a,b]	Loginova and Trotsenko (1979a)
Mycobacterium vaccae 10[a,b]	Loginova and Trotsenko (1979a)
Arthrobacter P1[a]	Levering et al. (1981)

The distribution of other pathways is described in Tables 8, 16 and 18. The distribution of variants is discussed in Table 19.

[a] Hexulose phosphate synthase plus some other enzymes of the pathway demonstrated.
[b] *In vivo* incorporation of ^{14}C substrates indicates operation of the pathway.
[c] Only evidence is a demonstration of hexulose phosphate synthase and this is sometimes at a low specific activity.

74   The biochemistry of methylotrophs

The facultative methylamine-utiliser *Arthrobacter globiformis* B-175 has the FBP aldolase cleavage variant and so must be able to synthesise oxaloacetate from C_3 compounds. This is probably achieved by way of PEP carboxylase, but low levels of pyruvate carboxylase and PEP carboxykinase are also present in this organism (Loginova and Trotsenko, 1979d).

III. The occurrence and distribution of the RuMP pathways

A. The RuMP pathways of formaldehyde assimilation

Because each species of bacteria has only a single predominant carbon assimilation pathway, the presence of high levels of key enzymes of a pathway can be used as a criterion for initially establishing the presence of that pathway in a given species. The absence of key enzymes of alternative pathways is an important confirmation and, ideally, an indication of ^{14}C labelling patterns is also obtained. On the basis of these criteria a number of useful generalisations can be made; but it must be emphasised that any such generalisations are merely guidelines, one of whose functions is to highlight exceptional cases as they may arise.

The first demonstration of the RuMP pathway was in obligate methane-utilising bacteria (Table 18), but the assay of key enzymes soon showed that this pathway is also important in obligate methylotrophs unable to use methane (Dahl *et al.*, 1972; Colby and Zatman, 1972; Stieglitz and Mateles, 1973; Babel and Miethe, 1974; Colby and Zatman, 1975c); and in facultative methylotrophs (Cox and Zatman, 1974; Colby and Zatman, 1975c). These observations eliminated the possibility that the RuMP pathway is uniquely associated with methanotrophy or with obligate methylotrophy; but it remains true to say that more facultative methylotrophs appear to use the serine pathway, and that all those obligate methylotrophs which are unable to use methane as substrate use the RuMP pathway for carbon assimilation.

The original RuMP cycle as proposed by Kemp and Quayle (1966, 1967) for *Methylomonas methanica* and *Methylococcus capsulatus* involved fructose bisphosphate (FBP) aldolase in the cleavage phase and sedoheptulose bisphosphate (SBPase) in the rearrangement phase (Fig. 7), but the first measurements of these enzymes (Strom *et al.*, 1974) showed that the alternative route involving KDPG aldolase and transaldolase (Fig. 6) is probably the major route operating in these bacteria. Although FBP aldolase is also present in *Methylomonas methanica*, the level of phosphofructokinase, the other essential enzyme for the FBP aldolase/transaldolase variant (Fig. 8), is so low as to indicate that this route can only be of minor importance (Zatman, 1981).

TABLE 19
Distribution of cleavage and rearrangement variants of the RuMP cycle

Bacteria with the KDPG aldolase/transaldolase variant (Fig. 6)

OBLIGATE METHYLOTROPHS USING METHANE
Methylococcus capsulatus (Texas)	Strom *et al.* (1974)
Methylomonas methanica[a]	Strom *et al.* (1974)

OBLIGATE METHYLOTROPHS NOT USING METHANE
Pseudomonas W6[b]	Babel and Miethe (1974)
Organisms 4B6 and C2A1	Colby and Zatman (1975c)
Organisms W3A1 and W6A (these also use glucose)	Colby and Zatman (1975c)
Methylophilus methylotrophus	Taylor (1977); Beardsmore *et al.* (1982)
Pseudomonas C	Ben-Bassat and Goldberg (1977); Ben-Bassat *et al.* (1980)

Facultative methylotrophs not using methane
Pseudomonas oleovorans	Loginova and Trotsenko (1977a)

Bacteria with the FBP aldolase/sedoheptulose bisphosphatase variant (Fig. 7)

FACULTATIVE METHYLOTROPHS NOT USING METHANE
Bacillus species PM6 and S2A1	Colby and Zatman (1975c)
Arthrobacter globiformis B-175[d]	Loginova and Trotsenko (1976a, b, 1977b)
Brevibacterium fuscum 24[c]	Loginova and Trotsenko (1977b, 1979a)
Mycobacterium vaccae 10[c]	Loginova and Trotsenko (1977b, 1979a)
Arthrobacter P1	Levering *et al.* (1981)

The variants are described and summarised in Figs 6 to 9 and in Table 17; their distribution is discussed further by Zatman (1981).
[a] Also contains fructose bisphosphate aldolase but has very low levels of phosphofructokinase.
[b] Measurements not available for the transaldolase.
[c] Sedoheptulose bisphosphatase not measured.
[d] No value quoted for the FBP aldolase.

It is now evident that the KDPG aldolase/transaldolase variant (Fig. 6) (Strom et al., 1974; Babel and Miethe, 1974; Colby and Zatman, 1975c, Zatman, 1981) is the major route in most obligate methylotrophs (which usually lack FBP aldolase and SBPase); whereas in most facultative methylotrophs the FBP aldolase/SBPase variant is the major route (Table 19). The one exception to this generalisation so far known is *Pseudomonas oleovorans*, a facultative methylotroph having the KDPG aldolase variant of the RuMP pathway.

B. The dissimilatory cycle of formaldehyde oxidation

Except for methane-utilisers, all bacteria having the RuMP pathway appear to have high levels of glucose 6-phosphate dehydrogenase and 6-phosphogluconate dehydrogenase, and to oxidise formaldehyde predominantly by the dissimilatory RuMP cycle (Fig. 12) (Zatman, 1981). This even includes the methylamine-utilising *Bacillus* species which do not have the KDPG aldolase (Colby and Zatman, 1975c). All methane-utilisers have relatively low levels of the dehydrogenases of the dissimilatory cycle, and they appear to oxidise formaldehyde predominantly by way of formaldehyde and formate dehydrogenases (Zatman, 1981).

The generalisation that all obligate methylotrophs (except methanotrophs) having the RuMP pathway oxidise formaldehyde by the cyclic route is an important one, and it is thus essential to examine critically the evidence against this generalisation presented by Ben-Bassat et al. (1980). This evidence is based on work with the obligate methanol-utiliser *Pseudomonas* C. Although the *in vitro* levels of formaldehyde and formate dehydrogenases are very low compared with the enzymes of the cyclic oxidation route in this organism (Ben-Bassat and Goldberg, 1977), it is claimed, on the basis of complex ^{14}C-labelling experiments, that more methanol is oxidised by the direct route, by way of formate, than by the cyclic route; and that the *in vivo* activities of the formaldehyde and formate dehydrogenases must be higher than those actually measured.

A critical part of this work is the evaluation of the amount of 6-phosphogluconate being assimilated, compared with that being decarboxylated by way of the cyclic route. The method depends on the use of uniformly-labelled ^{14}C-glucose as a tracer of the 6-phosphogluconate metabolism occurring during carbon-limited growth in continuous culture on methanol containing a low concentration of glucose.

However, this assumes that the use of uniformly-labelled glucose necessarily ensures the presence of uniformly-labelled 6-phosphogluconate, and this assumption is not valid. The 6-phosphogluconate will have relatively less label at the C-1 position, and hence measurement of $^{14}CO_2$ will give a

3. The ribulose monophosphate cycle

false low value for estimating the proportion of formaldehyde being oxidised by the cyclic route. This is because almost all the subsequent metabolism of uniformly-labelled ribulose 5-phosphate produced by decarboxylation of 6-phosphogluconate must involve condensation with non-radioactive formaldehyde. This produces hexulose monophosphate, and hence 6-phosphogluconate which is not labelled at the C-1 position, and this becomes CO_2 (unlabelled) in all subsequent decarboxylations.

Using the data of Ben-Bassat *et al.* (1980), and taking into account these considerations, it can be estimated that more than 90% of the oxidation of methanol is by way of the cyclic route in *Pseudomonas* C. This confirms the enzymic evidence and supports the suggestion that the RuMP dissimilatory cycle is the predominant route for formaldehyde oxidation in methanol-utilising methylotrophs having the RuMP assimilation pathway.

IV. Reactions of the ribulose monophosphate pathways

The following section describes the individual reactions of the pathway and the enzymes catalysing them. The numbers in parentheses refer to the numbers of the reactions as described in Figs 6 to 13. Some of these enzymes are the same as those involved in the ribulose bisphosphate pathway and these have retained the same numbers.

A. Enzymes involved in the fixation phase (all variants)

3-*Hexulose phosphate synthase* (D-arabino-3-hexulose 6-phosphate formaldehyde lyase) (13)

This first, key enzyme of the RuMP pathway catalyses the aldol condensation of formaldehyde with ribulose 5-phosphate to give a novel 3-hexulose 6-phosphate (Kemp and Quayle, 1965, 1966; Kemp, 1972, 1974; Ferenci *et al.*, 1974):

$$\text{HCHO} + \begin{array}{c} CH_2OH \\ | \\ C=O \\ | \\ H\text{-}C\text{-}OH \\ | \\ H\text{-}C\text{-}OH \\ | \\ CH_2O\text{-}P \\ \text{ribulose} \\ \text{5-phosphate} \end{array} \longrightarrow \begin{array}{c} CH_2OH \\ | \\ HO\text{-}C\text{-}H \\ | \\ C=O \\ | \\ H\text{-}C\text{-}OH \\ | \\ H\text{-}C\text{-}OH \\ | \\ CH_2O\text{-}P \\ \text{hexulose} \\ \text{6-phosphate} \end{array}$$

This enzyme was first purified and characterised by Ferenci et al. (1974) from the obligate methanotroph *Methylococcus capsulatus*. More than 80% was membrane-bound and was released by incubation in 1M-NaCl prior to a 40-fold purification. The synthase is probably a hexamer of identical subunits of about 49 000 daltons (total mol. wt 310 000); dissociation into monomers occurs at low pH (4·6) or in buffer of low ionic strength (less than 15 mM). Ions of Mg^{++} or Mn^{++} are absolutely essential for activity and also for stabilisation of the enzyme which thus resembles a Class 11 aldolase (Horecker et al., 1972).

The equilibrium constant of the reaction is in favour of synthesis (Keq. (M) is $2·5 \times 10^{-4}$ in the direction of synthesis). Low concentrations of reactants and product (less than 0·1 mM) will tend to favour cleavage rather than condensation but the reaction will be "pulled" in favour of synthesis by its coupling to the subsequent isomerisation reaction which gives fructose 6-phosphate. The synthase from all sources is specific for ribulose 5-monophosphate as the acceptor for condensation with formaldehyde, and for 3-hexulose phosphate in the direction of cleavage. No significant inhibition by ATP, ADP, AMP, NADH, fructose 6-phosphate or phosphoenolpyruvate occurs with synthase from any source and this has led to the suggestion (Ferenci et al., 1974) that regulation of the ribulose monophosphate cycle must lie elsewhere.

The hexulose phosphate synthases from methylotrophs not growing on methane are similar in most respects to that first described in *Methylococcus capsulatus* except that they appear to be small, soluble, dimeric proteins (see Table 20) and it has been suggested that their solubility reflects the lack of a complex internal membrane system in non-methanotrophs (Kato et al., 1978).

A feature of the synthases from some non-methanotrophs is that they may exhibit unusual kinetic characteristics showing intermediary plateau regions in plots of rate against substrate concentration; this has been observed with the synthases from obligate methanol-utilisers (*Pseudomonas* W6 and *Methylomonas methylovora*), and from facultative methanol-utilisers (bacterium MB58 and *Pseudomonas oleovorans*), but not from the methanotrophs "*Methylomonas*" GB3 and GB8. (For a summary in English of this work see Muller and Sokolov, 1979; Muller and Babel, 1980.) These kinetics have been interpreted in terms of multiple and interconvertible enzyme forms, each having slightly differing kinetics, and it has been suggested that "this mechanism exerts the main influence on the regulation of the hexulose phosphate synthases in all non-methane-utilising methylotrophic bacteria" (Muller and Sokolov, 1979). How such regulation might work, and whether or not such regulation of this enzyme is necessary for regulation of the ribulose monophosphate pathway, is not immediately obvious.

TABLE 20
Hexulose phosphate synthase

	Methylococcus capsulatus (Ferenci et al., 1974)	Methylomonas M15 (Sahm et al., 1976b)	Methylomonas amino-faciens (Kato et al., 1977a, 1978)	Pseudomonas oleovorans (Sokolov and Trotsenko, 1978a, b; Muller and Sokolov, 1979)	Methylophilus methylotrophus (Beardsmore et al., 1982)
Bacterial type	Obligate methanotroph	Obligate methanol-utiliser	Obligate methanol-utiliser	Facultative methanol-utiliser	Obligate methanol-utiliser
Location of enzyme	membrane fraction	soluble fraction	soluble fraction	soluble fraction	soluble fraction
pH optimum	7.0	7.5–8.0	8.0	7.0	7.2
Molecular weight	310 000	40 000–43 000	45 000–47 000	45 000	40 000
Subunit mol. wt	49 000	22 000	23 000	—	22 500
Activator (K_m)	Mg^{++} or Mn^{++}	Mg^{++} (0·25 mM) or Mn^{++}	Mg^{++} (0·17 mM) or Mn^{++}	Mg^{++} or Mn^{++}	Mg^{++} (0·25 mM) or Mn^{++}
K_m values: formaldehyde (HCHO)	0·49 mM (at 0·57 mM-RuMP)	1·1 mM	0·29 mM (at ∞ RuMP)	about 1 mM	0·53 mM (at 1·67 mM-RuMP)
ribulose 5-phosphate (RuMP)	0·083 mM (at 4·0 mM-HCHO)	1·6 mM	0·059 mM (at ∞ HCHO)	—	0·136 mM (at 5mM-HCHO)
hexulose 6-phosphate	0·075 mM		0·036 mM		0·041 mM
Specific activity of most pure preparation (μmol (min)$^{-1}$ (mg protein)$^{-1}$)	69	66·5	53	196	97·4

80   The biochemistry of methylotrophs

3-*Hexulose phosphate isomerase* (D-arabino-3-hexulose 6-phosphate 3,2-ketolisomerase) (14)
This second key enzyme of the RuMP cycle catalyses the isomerisation of the product of the first initial condensation reaction, 3-hexulose 6-phosphate and fructose 6-phosphate:

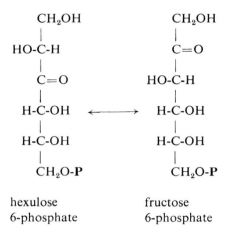

hexulose   fructose
6-phosphate   6-phosphate

As far as is known this represents the first example of an enzyme catalysing the isomerisation of a 3-ketulose to a 2-ketulose. This soluble isomerase is highly active in extracts of *Methylococcus capsulatus* and it has been purified about 150-fold from this source (specific activity of the 90% pure preparation, 1560 μmoles (min)$^{-1}$(mg protein)$^{-1}$). Characterisation of this isomerase preparation (Ferenci *et al.*, 1974) has shown it to have a pH optimum of 8·3, a molecular weight of 67 000 and to be inhibited by divalent cations. The apparent K_m (pH 8·3) for fructose 6-phosphate is 1·1 mM. The equilibrium of the isomerisation is in favour of fructose 6-phosphate formation, the equilibrium constant, calculated from K_m and V_{max} values, being $1·9 \times 10^2$. The isomerase appears to be specific with respect to the isomerisation of 3-hexulose 6-phosphate, no reaction being observed with ribulose 5-phosphate, xylulose 5-phosphate or allulose 6-phosphate. No appreciable inhibition was observed with various sugar phosphates nor with ATP, ADP, AMP, NAD$^+$ or NAD(P)H. A similar, highly active isomerase from *Methylophilus methylotrophus* has been purified and characterised (Beardsmore *et al.*, 1982).

The hexulose phosphate isomerase has also been partially purified from the obligate methanol-utiliser *Methylomonas aminofaciens* 77a (specific activity, 20 μmoles (min)$^{-1}$(mg protein)$^{-1}$) (Kato *et al.*, 1977a). It has a pH optimum of 7·5 and K_m values for hexulose phosphate and fructose 6-phosphate of 0·029 mM and 0·67 mM respectively.

B. Enzymes involved in the cleavage phase

1. KDPG aldolase variants (Figs 6 and 9)

Glucose phosphate isomerase (16)

This enzyme catalyses the formation of glucose 6-phosphate from fructose 6-phosphate:

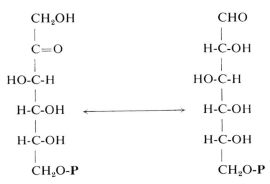

fructose 6-phosphate   glucose 6-phosphate

Glucose 6-phosphate dehydrogenase (17)

This enzyme catalyses the oxidation of glucose 6-phosphate to 6-phosphogluconate, by way of 6-phosphogluconolactone, with the concomitant formation of NADH or NADPH:

```
    CHO                                    COOH
     |                                      |
   H-C-OH                                 H-C-OH
     |                                      |
   HO-C-H   + H₂O + NAD(P)⁺  ⟶   HO-C-H   +  NAD(P) + H⁺
     |                                      |
   H-C-OH                                 H-C-OH
     |                                      |
   H-C-OH                                 H-C-OH
     |                                      |
   CH₂O-P                                 CH₂O-P
```

glucose 6-phosphate   6-phosphogluconate

TABLE 21
The glucose 6-phosphate dehydrogenases of methylotrophs

	Methylomonas M15	Pseudomonas C	Pseudomonas W6	Methylophilus methylotrophus	Candida boidinii
	(Steinbach et al., 1978)	(Ben-Bassat and Goldberg, 1980)	(Miethe and Babel, 1976)	(Beardsmore et al., 1982)	(Kato et al., 1979b)
Extent of purification	192-fold (homogenous)	220-fold (homogenous)	not stated ("partially pure")	190-fold (homogenous)	1714-fold (homogenous)
Specific activity of pure enzyme (μmol (min)$^{-1}$ (mg)$^{-1}$)	122.8	150	not stated	304	516
V_{max} NAD$^+$/V_{max} NADP$^+$	about 1.3	about 1.0	1.6	1.51	NADP$^+$-specific
Molecular weight (subunits)	108 000 (55 000)	150 000	—	110 000 (52 000)	118 000 (61 000)
pH optimum	about 9.0	8.5–9.5	8.6–9.2	9.0	8.5–9.0
K_m values					
glucose 6-phosphate	0.29 mM (with NAD$^+$)	0.16 mM (with NAD$^+$)	0.18 mM (with NAD$^+$)	1.03 mM (with NAD$^+$)	—
	0.11 mM (with NADP$^+$)	0.25 mM (with NADP$^+$)	0.13 mM (with NADP$^+$)	1.02 mM (with NADP$^+$)	0.86 mM (with NADP$^+$)
NAD$^+$	0.2 mM	0.29 mM	—	0.39 mM	—
NADP$^+$	0.014 mM	0.022 mM	—	0.018 mM	0.014 mM
Inhibitors with NAD$^+$ as acceptor	1 mM ATP (35%) NADH, NADPH	ATP (K_i 1.1 mM) NADH, NADPH	ATP and GTP (1 mM) NAD, (not NADPH) acetyl-CoA	—	—
with NADP$^+$ as acceptor	1 mM ATP (16%) NADH, NADPH	ATP ("less pronounced") NADH, NADPH	ATP and GTP (1 mM) not acetyl-CoA	ATP NADPH (not NADH)	1 mM ATP (90%) NADPH (not NADH) glyceraldehyde 3-phosphate

These dehydrogenases are all from obligate methanol-(or methylamine-) utilising bacteria or from yeast.

The glucose 6-phosphate dehydrogenases from a number of obligate methanol-utilisers (*Methylomonas* M15, *Pseudomonas* C and *Methylophilus methylotrophus*) have been extensively purified and shown to be similar in most important respects (Table 21). The enzyme is almost equally active with NAD^+ and $NADP^+$. Hyperbolic kinetics are usually followed and K_m values with $NADP^+$ are considerably lower than with NAD^+. None of a wide range of sugar phosphates or carboxylic acids tested (including acetyl-CoA) are activators or inhibitors (Steinbach *et al.*, 1978; Ben-Bassat and Goldberg, 1980) but all dehydrogenases studied are to some extent inhibited by ATP and by NADH or NADPH (or both). The glucose 6-phosphate dehydrogenase from yeast is included for comparison; it is specific for $NADP^+$ and is probably only involved in provision of NADPH for biosynthesis (Kato *et al.*, 1979b).

Assays of glucose 6-phosphate dehydrogenase in crude extracts of a number of obligate and restricted facultative methylotrophs having the KDPG aldolase variant of the RuMP cycle suggests that the glucose 6-phosphate dehydrogenase in these bacteria also uses both NAD^+ and $NADP^+$ as oxidant, the rates measured with $NADP^+$ usually being greater than with NAD^+. (In such experiments it is not shown that a single enzyme is responsible for both activities). These bacteria included organisms 4B6, C2A1, W3A1 and W6A (Colby and Zatman, 1975c) and *Pseudomonas oleovorans* (a facultative methylotroph) (Loginova and Trotsenko, 1977a; Sokolov *et al.*, 1980). The methylamine-utilising *Bacillus* species S2A1 and PM6 have no KDPG aldolase but they oxidise formaldehyde by the dissimilatory RuMP cycle; in these bacteria the glucose 6-phosphate dehydrogenase is $NADP^+$-specific (Colby and Zatman, 1975c).

6-phosphogluconate dehydrase (18)

By removal of water from 6-phosphogluconate this enzyme catalyses formation of 2-keto, 3-deoxy, 6-phosphogluconate (KDPG):

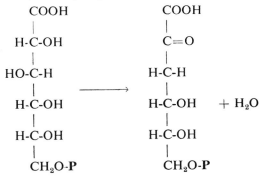

84 The biochemistry of methylotrophs

KDPG aldolase (2-keto, 3-deoxy, 6-phosphogluconate aldolase) (19)

This is the cleavage enzyme of KDPG aldolase variants, catalysing the formation of pyruvate and glyceraldehyde 3-phosphate from one molecule of KDPG:

```
    COOH
    |
    C=O
    |
    H C-H
    |
    H-C-OH           COOH              CHO
    |                |                 |
    H-C-OH   ──→     C=O        +      H-C-OH
    |                |                 |
    CH₂O-P           CH₃               CH₂O-P

    KDPG             pyruvate          glyceraldehyde
                                       3-phosphate
```

2. *Fructose bisphosphate aldolase variants* (Figs 7 and 8)

The following two enzymes are involved in the rearrangement phase of the sedoheptulose bisphosphatase variants; the only variant not requiring them is the KDPG aldolase/transaldolase variant (Fig. 6 and Table 17).

Phosphofructokinase (15)

This enzyme catalyses the irreversible phosphorylation by ATP of fructose 6-phosphate to fructose 1,6-bisphosphate:

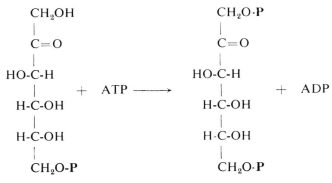

3. The ribulose monophosphate cycle

Fructose bisphosphate aldolase (5a)

This catalyses the typical "glycolytic" cleavage of fructose 1,6-bisphosphate to two triose phosphate molecules. Although the equilibrium constant (M) (about 10^{-4}) favours condensation, at low concentrations of reactants and products the cleavage reaction will be favoured.

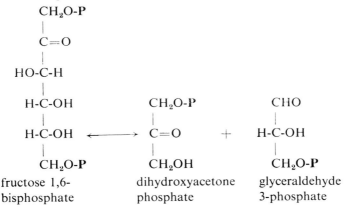

```
        CH₂O-P
         |
         C=O
         |
        HO-C-H
         |
        H-C-OH              CH₂O-P              CHO
         |                   |                   |
        H-C-OH    ⇌         C=O       +        H-C-OH
         |                   |                   |
        CH₂O-P              CH₂OH               CH₂O-P
      fructose 1,6-      dihydroxyacetone     glyceraldehyde
      bisphosphate         phosphate           3-phosphate
```

C. Enzymes involved in the rearrangement phase

1. All variants

Transketolase (8a and 8b)

Transketolase catalyses the reversible transfer of a glycolaldehyde moiety from a donor ketose to an acceptor aldose, the reaction typically requiring thiamine pyrophosphate and a divalent cation as cofactors. Two transketolase catalysed reactions occur in the RuMP pathway, both using glyceraldehyde 3-phosphate as the aldose acceptor and both producing xylulose 5-phosphate. In the first reaction (8a) the glycolaldehyde donor is fructose 6-phosphate and the product is erythrose 4-phosphate:

```
   CH₂OH
    |
    C=O                              CH₂OH
    |                                 |
   HO-C-H                             C=O                 CHO
    |                                  |                   |
   H-C-OH    +    CHO        ⇌       HO-C-H    +        H-C-OH
    |              |                   |                   |
   H-C-OH         H-C-OH              H-C-OH              H-C-OH
    |              |                   |                   |
   CH₂O-P         CH₂O-P              CH₂O-P              CH₂O-P
  fructose     glyceraldehyde        xylulose            erythrose
  6-phosphate   3-phosphate         5-phosphate         4-phosphate
```

86   The biochemistry of methylotrophs

In the second transketolase-catalysed reaction (8b) the glycolaldehyde donor is sedoheptulose 7-phosphate and the product is ribose 5-phosphate; usually the same transketolase is able to catalyse both of these reactions:

$$
\begin{array}{c}
\text{CH}_2\text{OH} \\
| \\
\text{C}=\text{O} \\
| \\
\text{HO-C-H} \\
| \\
\text{H-C-OH} \\
| \\
\text{H-C-OH} \\
| \\
\text{H-C-OH} \\
| \\
\text{CH}_2\text{O-P}
\end{array}
\;+\;
\begin{array}{c}
\text{CHO} \\
| \\
\text{H-C-OH} \\
| \\
\text{CH}_2\text{O-P}
\end{array}
\;\rightleftharpoons\;
\begin{array}{c}
\text{CH}_2\text{OH} \\
| \\
\text{C}=\text{O} \\
| \\
\text{HO-C-H} \\
| \\
\text{H-C-OH} \\
| \\
\text{CH}_2\text{O-P}
\end{array}
\;+\;
\begin{array}{c}
\text{CHO} \\
| \\
\text{H-C-OH} \\
| \\
\text{H-C-OH} \\
| \\
\text{H-C-OH} \\
| \\
\text{CH}_2\text{O-P}
\end{array}
$$

sedoheptulose 7-phosphate   glyceraldehyde 3-phosphate   xylulose 5-phosphate   ribose 5-phosphate

Pentose phosphate epimerase (9)

This enzyme catalyses the epimerisation of xylulose 5-phosphate to ribulose 5-phosphate:

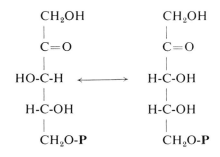

xylulose 5-phosphate   ribulose 5-phosphate

3. The ribulose monophosphate cycle

Pentose phosphate isomerase (11)

This enzyme catalyses the isomerisation of the ribose 5-phosphate produced in the second transketolase reaction (8b), giving ribulose 5-phosphate:

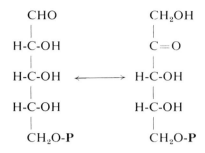

ribose 5-phosphate   ribulose 5-phosphate

2. Transaldolase variants (Figs 6 and 8)

Transaldolase (12)

This enzyme is only involved in the transaldolase variants of the pathway (Figs 6 and 8). It catalyses the reversible transfer of a dihydroxyacetone moiety from the donor ketose, fructose 6-phosphate to the acceptor aldose, erythrose 4-phosphate, the products being glyceraldehyde 3-phosphate and sedoheptulose 7-phosphate:

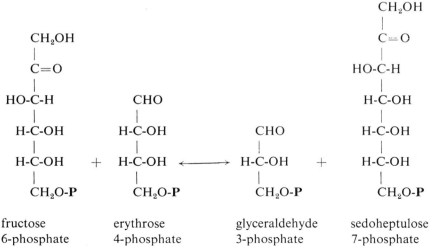

fructose 6-phosphate   erythrose 4-phosphate   glyceraldehyde 3-phosphate   sedoheptulose 7-phosphate

3. Sedoheptulose bisphosphatase variants (Figs 7 and 9)

Aldolase (5a and 5b)

Aldolase catalyses the reversible condensation of an aldose with a ketose to give a larger ketose sugar. In bacteria with the sedoheptulose bisphosphatase rearrangements (Figs 7 and 9) aldolase is involved twice in the rearrangement phase. In the first reaction (5a) fructose 1,6-bisphosphate is cleaved to give glyceraldehyde 3-phosphate and a molecule of dihydroxyacetone phosphate (DHAP) which then combines with erythrose monophosphate in a condensation reaction also catalysed by aldolase (5b) to give sedoheptulose 1,7-bisphosphate. The same aldolase probably catalyses both reactions:

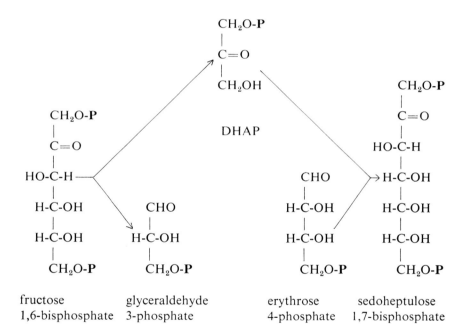

Sedoheptulose bisphosphatase (7)

In the sedoheptulose bisphosphatase variants of the pathway this enzyme catalyses the hydrolysis of sedoheptulose 1,7-bisphosphate to sedoheptulose 7-phosphate.

Phosphofructokinase (15)
(See above.)

D. The key enzyme of the RuMP dissimilatory cycle; 6-phosphogluconate dehydrogenase (23)

This enzyme catalyses the irreversible oxidative decarboxylation of 6-phosphogluconate to ribulose 5-phosphate:

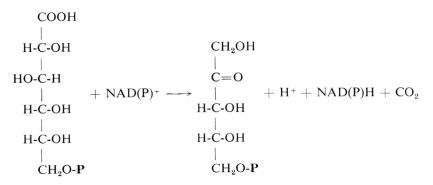

As shown in Fig. 12, possession of this enzyme provides a mechanism for the cyclic oxidation of formaldehyde in some bacteria.

The dehydrogenase has been purified 245-fold from the obligate methanol-utiliser *Pseudomonas* C, giving a homogeneous preparation with a specific activity of 56·4 IU/mg of protein (Ben-Bassat & Goldberg, 1980). It has a pH optimum between pH 8 and 9, it requires Mg^{++} ions for maximum activity and its molecular weight is 220 000 with a subunit mol. wt of 55 000. The enzyme shows normal hyperbolic kinetics. The maximum activity with NAD^+ is 60% of that with $NADP^+$ and the K_m for $NADP^+$ (0·02 mM) is lower than that for NAD^+ (0·5 mM). The K_m for 6-phosphogluconate is 0·04 mM (with NAD^+) and 0·1 mM (with $NADP^+$).

NADH and NADPH inhibit the dehydrogenase with either NAD^+ or $NADP^+$ as coenzyme, 50% inhibition being measured between 28 and 75 μM NAD(P)H. ATP is also a potent inhibitor when NAD^+ is the coenzyme, this inhibition being competitive with respect to NAD^+ (K_i, 0·6 mM). This inhibition by ATP is less pronounced with $NADP^+$ as coenzyme and is non-competitive with respect to 6-phosphogluconate.

The overall effect of the stronger inhibition by ATP of the NAD^+-linked oxidation of glucose 6-phosphate and 6-phosphogluconate will ensure that when a plentiful supply of ATP is available the overall activity of the dissimilatory cycle will decrease, and that the remaining activity will continue to provide NADPH for biosynthesis.

Although the metabolism of the obligate methanol-utiliser *Methylophilus methylotrophus* is often similar to that of *Pseudomonas* C, it appears to differ from it in having two separate 6-phosphogluconate dehydrogenases; one is specific for NAD^+ and the other for $NADP^+$. Both are inhibited by ATP but the $NADP^+$ enzyme is inhibited by NADH and NADPH whereas the NAD^+-linked enzyme is only inhibited by NADH and not NADPH. These enzymes and their regulation are fully discussed in Beardsmore et al. (1982).

Measurements of 6-phosphogluconate dehydrogenase activity in crude extracts of other methylotrophs show that all are able to use both NAD^+ and $NADP^+$, but whether this is due to a single enzyme or to 2 different enzymes with differing coenzyme specificity is not known (Babel and Miethe, 1974; Colby and Zatman, 1975c; Loginova and Trotsenko, 1977a, 1979a; Ben-Bassat and Goldberg, 1977; Sokolov et al., 1980).

In the methylamine-utilising *Bacillus* species S2A1 and PM6 which have only $NADP^+$-specific glucose 6-phosphate dehydrogenase, the 6-phosphogluconate dehydrogenase activity is much greater with $NADP^+$ than with NAD^+ and it is suggested that this might correlate with the high levels of NADPH oxidase in these bacteria (Colby and Zatman, 1975a, c).

E. The interconversion of phosphoenolpyruvate with triose phosphates

The enzymes catalysing this reversible interconversion are required by all RuMP pathway bacteria. In those with the FBP-aldolase variant they are needed for conversion of dihydroxyacetone phosphate to phosphoglycerate and phosphoenolpyruvate for biosynthesis; and in those with the KDP-aldolase variant they are essential for formation of triose phosphate from PEP. The direction of each reaction in a particular bacterium will thus depend on which cleavage variant of the pathway is operating.

Phosphoglycerate kinase (2)

$$\begin{array}{c} COOH \\ | \\ H\text{-}C\text{-}OH \\ | \\ CH_2O\text{-}P \end{array} + ATP \rightleftharpoons \begin{array}{c} COO\text{-}P \\ | \\ H\text{-}C\text{-}OH \\ | \\ CH_2O\text{-}P \end{array} + ADP$$

3-phosphoglycerate   1,3-diphosphoglycerate

Glyceraldehyde phosphate dehydrogenase (3)

$$\begin{array}{c} CHO \\ | \\ H\text{-}C\text{-}OH \\ | \\ CH_2O\text{-}P \end{array} + NAD^+ + Pi + H_2O \rightleftharpoons \begin{array}{c} COO\text{-}P \\ | \\ H\text{-}C\text{-}OH \\ | \\ CH_2O\text{-}P \end{array} + NADH + H^+$$

glyceraldehyde 3-phosphate   1,3-diphosphoglycerate

Triose phosphate isomerase (4)

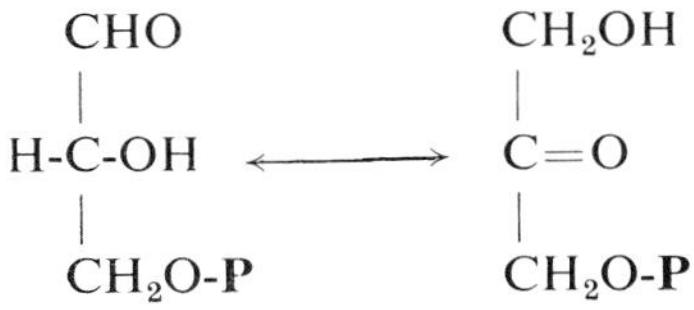

$$\begin{array}{c} \text{CHO} \\ | \\ \text{H-C-OH} \\ | \\ \text{CH}_2\text{O-P} \end{array} \longleftrightarrow \begin{array}{c} \text{CH}_2\text{OH} \\ | \\ \text{C=O} \\ | \\ \text{CH}_2\text{O-P} \end{array}$$

glyceraldehyde 3-phosphate   dihydroxyacetone phosphate

Enolase (21)

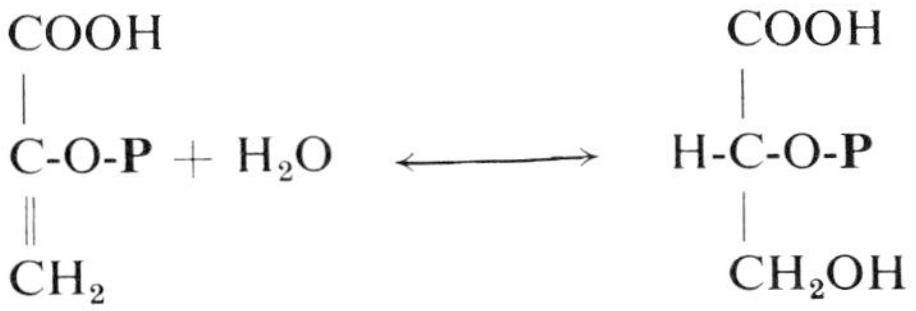

$$\begin{array}{c} \text{COOH} \\ | \\ \text{C-O-P} + \text{H}_2\text{O} \\ \parallel \\ \text{CH}_2 \end{array} \longleftrightarrow \begin{array}{c} \text{COOH} \\ | \\ \text{H-C-O-P} \\ | \\ \text{CH}_2\text{OH} \end{array}$$

phosphoenolpyruvate   2-phosphoglycerate

Phosphoglyceromutase (22)

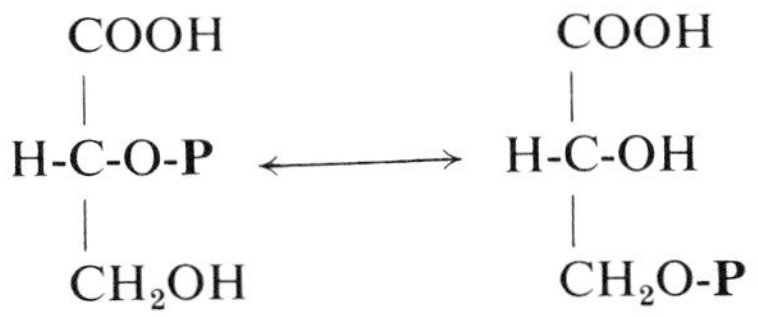

$$\begin{array}{c} \text{COOH} \\ | \\ \text{H-C-O-P} \\ | \\ \text{CH}_2\text{OH} \end{array} \longleftrightarrow \begin{array}{c} \text{COOH} \\ | \\ \text{H-C-OH} \\ | \\ \text{CH}_2\text{O-P} \end{array}$$

2-phosphoglycerate   3-phosphoglycerate

F. The interconversion of pyruvate, phosphoenolpyruvate (PEP) and oxaloacetate (Fig. 13)

Phosphoenolpyruvate synthetase (20)

This reversible enzyme catalyses the synthesis of PEP from pyruvate during growth of some bacteria (e.g. enteric bacteria) on C_3 compounds (Cooper and Kornberg, 1967). ATP is the phosphate donor and AMP and phosphate are produced; a second ATP is required to synthesise ADP from AMP and thus the reaction in effect uses two molecules of ATP for every PEP synthesised:

Pyruvate + ATP ⟶ phosphoenolpyruvate + AMP + Pi

Phosphoenolpyruvate carboxylase (24)

PEP carboxylase catalyses the irreversible carboxylation of PEP to oxaloacetate; it is the main route for formation of C_4 compounds from C_3 compounds in bacteria with the serine pathway for assimilation of C_1

compounds (Large et al., 1962b; Salem et al., 1973a; Loginova and Trotsenko, 1979d):

$$\text{Phosphoenolpyruvate} + CO_2 \longrightarrow \text{oxaloacetate}$$

Pyruvate carboxylase (25)

This biotin-containing enzyme is the first enzyme in the mammalian gluconeogenic route from pyruvate and it also functions as an anaplerotic enzyme "filling up" the TCA cycle when intermediates are being removed for biosynthesis. It is in this context that the mammalian (and some microbial) enzymes are activated by acetyl-CoA. It catalyses a reversible reaction but in all cases reported it functions in the direction of carboxylation:

$$\text{Pyruvate} + ATP + CO_2 \longrightarrow \text{oxaloacetate} + ADP + Pi$$

In methylotrophs this enzyme is not usually activated by acetyl-CoA (Loginova and Trotsenko, 1979d; Babel and Loffhagen, 1977; Loffhagen et al., 1979; Aperghis and Quayle, 1981).

Phosphoenolpyruvate carboxykinase (26)

This enzyme usually functions in the decarboxylation of oxaloacetate to PEP. In mammals GTP is usually required but in microorganisms ATP may replace it. It is the principal enzyme by which C_3 compounds are synthesised from C_4 compounds during growth on these and it is also involved in the synthesis of PEP from pyruvate in those bacteria lacking a direct (single enzyme) route.

$$\text{Oxaloacetate} + ATP \longrightarrow \text{phosphoenolpyruvate} + ADP + CO_2$$

Pyruvate kinase (27)

This enzyme catalyses the irreversible formation of pyruvate from PEP with the concomitant formation of ATP.

$$\text{Phosphoenolpyruvate} + ADP \longrightarrow \text{pyruvate} + ATP$$

Pyruvate phosphate dikinase (28)

This enzyme is an alternative enzyme to PEP synthetase, catalysing the synthesis of PEP from pyruvate in propionic acid bacteria (Evans and Wood, 1971) and in some tropical grasses (Hatch and Slack, 1968). (For a review see Wood et al., 1977.) The extra phosphate involved in the phosphorylation reaction allows the synthesis of pyrophosphate whose subsequent hydrolysis drives the reaction in the direction of phosphorylation:

$$\text{Pyruvate} + ATP + Pi \longrightarrow \text{phosphoenolpyruvate} + AMP + PPi$$

V. Regulation of the RuMP pathway for formaldehyde assimilation

The problem of balancing the flow of formaldehyde into assimilation and oxidation in bacteria with the RuMP pathways is different in one respect from that in bacteria having the serine pathway (see Chapter 4, Section V). In the serine pathway formaldehyde enters the assimilatory route by way of methylene tetrahydrofolate, whereas in bacteria with the RuMP pathway the substrate for oxidation and assimilation is free formaldehyde. Furthermore, in those bacteria having the cyclic route for formaldehyde oxidation (Fig. 12), the branch point in metabolism is not at the level of formaldehyde, but at 6-phosphogluconate. In these bacteria, the hexulose phosphate synthase and isomerase have both dissimilatory and assimilatory roles, and the properties of the purified enzymes (above) suggest that they are not involved in regulation. Regulation of the RuMP pathways is unlikely to be achieved by way of the freely-reversible transaldolases, transketolases, epimerases and isomerases whose function is to maintain appropriate proportions of phosphorylated sugars in the pathways. However, one "rearrangement" enzyme that might be involved in regulation is the irreversible sedoheptulose bisphosphatase (when present) because it is a regulatory enzyme in the RuBP pathway in some organisms (McFadden, 1978).

The "cleavage" phase is the most probable site of regulation of assimilation of formaldehyde by the RuMP pathway. In bacteria having the FBP aldolase variant (Fig. 7), a likely enzyme is the same as that regulated in glycolysis. This enzyme, phosphofructokinase, is usually inhibited by citrate and ATP, thus reflecting the function of the glycolytic enzymes as the first part of the energy-yielding metabolism of glucose. However, ATP would appear to be an unlikely inhibitor of phosphofructokinase in methylotrophs in which its function is assimilatory; to my knowledge this has not been investigated.

In bacteria having the KDPG aldolase variant (Fig. 6), the enzymes of gluconeogenesis are involved in triose phosphate formation from pyruvate. The most probable site for regulation of this route is the enzyme(s) responsible for the phosphorylation of pyruvate to phosphoenolpyruvate but no information is available on this point. The balance of formaldehyde oxidation and assimilation in these bacteria appears to be regulated by way of the dehydrogenases for glucose 6-phosphate and 6-phosphogluconate. The glucose 6-phosphate dehydrogenase is involved in both assimilation and oxidation, whereas 6-phosphogluconate dehydrogenase is involved only in the oxidative cycle. This cycle produces both NADH and NADPH and it is usually assumed that the NADH is used for ATP synthesis and the NADPH as a reductant in biosynthesis. The dehydrogenases involved have rather similar regulatory properties, both being multimeric enzymes showing sigmoidal kinetics and being inhibited by ATP and by NADH or NADPH (or both) (see pages 83 and 89). Regulation of the proportion of NADH to NADPH is probably achieved by inhibition of NAD^+-coupled dehydrogenase activities by NADH, and of $NADP^+$-coupled dehydrogenase activities by NADPH. It should be noted that the separation of the functions of NADH and NADPH may not be as well defined as at first appears, because in *Methylophilus methylotrophus* (and perhaps in some other methylotrophs) the electron transport chain oxidises both NADH and NADPH. There are slight variations from one organism to another in the details of the regulation of the two dehydrogenases but the principles appear to be the same in all bacteria having the cyclic route for formaldehyde oxidation. When there is sufficient ATP and NAD(P)H for biosynthesis, these metabolites diminish the activities of both dehydrogenases. However, the extent of inhibition needs to be different; the 6-phosphogluconate dehydrogenase will need to be increased to a greater extent than the glucose 6-phosphate dehydrogenase because some activity of this latter enzyme will still be required for the biosynthetic pathway. As emphasised by Ben-Bassat and Goldberg (1980) elucidation of the *in vivo* regulation of these enzymes requires a knowledge of the intracellular concentrations of substrates and inhibitors. A start has been made in this direction (Steinbach *et al.*, 1978) but there is insufficient information to build a complete picture at the present time.

4
The serine pathway of formaldehyde assimilation

I. Description of the serine pathway 96
II. Evidence for the serine pathway 98
   A. Evidence obtained by incubating whole cells with radioactive isotopes . 98
   B. Evidence for the serine pathway from enzyme studies 101
     1. The formation of oxaloacetate 101
     2. The cleavage reaction 101
     3. The oxidation of acetyl-CoA to glyoxylate 103
       (a) The icl$^+$-serine pathway 103
       (b) The icl$^-$-serine pathway 103
   C. Evidence for operation of the serine pathway from mutant studies . 105
III. The occurrence and distribution of the serine pathway . . . 109
   A. Methylotrophs using the pathway as their main assimilation pathway (Table 26) 109
   B. Distribution of icl$^+$ and icl$^-$ variants of the serine pathway (Table 27) . 113
IV. The reactions of the serine pathway 115
   A. Enzymes involved in the formation of oxaloacetate from glyoxylate, formaldehyde and CO_2 115
   B. The formation of glyoxylate and acetyl-CoA from oxaloacetate . 121
   C. The oxidation of acetyl-CoA to glyoxylate 124
     1. Oxidation of acetyl-CoA to glyoxylate in icl$^+$ bacteria . . 124
     2. Oxidation of acetyl-CoA to glyoxylate in icl$^-$ bacteria . . 125
V. Regulation of the serine pathway 126
   A. Regulation at the genetic level 126
     1. Investigations using predominantly biochemical techniques . 126
     2. Investigations using genetic techniques 130
       (a) Preliminary experiments on gene transfer in methylotrophs . 130
       (b) Gene mapping in the facultative methanotroph *Methylobacterium organophilum* XX 131
   B. Regulation at the level of enzyme activity 132
     1. Introduction 132
     2. Regulatory enzymes of the serine pathway 134

This bacterial pathway of formaldehyde assimilation was first proposed in outline by Quayle and his colleagues about 20 years ago as the result of experiments in which serine was found to be the first labelled compound to accumulate during incubation of the pink facultative methylotroph *Pseudomonas* AM1 with radioactive methanol. Although one of the first methylotrophic pathways to be proposed, after 20 years of research (both elegant and devious) the pathway in *Pseudomonas* is only now nearing something like completion.

The first section below is a brief description of the serine pathway and its variants; this is followed by a summary of the evidence for the pathway, its occurrence and distribution in methylotrophs, descriptions of the individual enzymes, and a discussion of its regulation.

I. Description of the serine pathway

The serine pathway differs from other formaldehyde assimilation pathways in the nature of its intermediates which are carboxylic acids and amino acids rather than carbohydrates, and in the key formaldehyde-assimilating enzyme; this is serine transhydroxymethylase which catalyses the addition of formaldehyde to glycine, thus giving the key intermediate in the pathway, serine. The serine pathway operates in bacteria during growth on a variety of C_1 compounds varying in reduction level between methane and formate. For ease of comparison with other assimilation pathways the serine pathway is drawn in Fig. 14 as it operates to synthesise phosphoglycerate from two molecules of formaldehyde plus one of CO_2. To achieve this, the first part of the pathway (the synthesis of phosphoglycerate from glyoxylate) is doubled up. Two molecules of formaldehyde plus two of glyoxylate give two molecules of 2-phosphoglycerate. One is assimilated into cell material by way of 3-phosphoglycerate, while the other is converted to phosphoenolpyruvate (PEP) whose carboxylation yields oxaloacetate and subsequently malyl-CoA. In the final part of the pathway this malyl-CoA is cleaved to glyoxylate plus acetyl-CoA, whose oxidation to glyoxylate completes the cycle. There are two variants of the serine pathway, differing in how the final oxidation of acetyl-CoA to glyoxylate is achieved. In the icl$^+$-serine pathway this involves a glyoxylate cycle in which isocitrate lyase is the key enzyme (Fig. 14). This enzyme is absent from bacteria, such as *Pseudomonas* AM1, which have no isocitrate lyase. In such bacteria having the icl$^-$-serine pathway, the route for oxidation of acetyl-CoA has not yet been completely elucidated, but it might involve homoisocitrate lyase and a homoisocitrate lyase-glyoxylate cycle in which all the intermediates are homologues of those of the glyoxylate cycle (see below).

The summary equation of the icl$^+$-serine pathway is as follows (FPH$_2$ is the reduced flavoprotein, succinate dehydrogenase):

$$CO_2 + 2\,HCHO + 2\,NADH + 2\,H^+ + 3\,ATP \rightarrow phosphoglycerate \\ + 2\,NAD^+ + 3\,ADP + 2\,Pi + FPH_2$$

The summary equation for the icl$^-$-serine pathway is likely to be the same except perhaps, for the nature of the two reducing equivalents produced during the oxidation of acetyl-CoA to glyoxylate.

4. The serine pathway   97

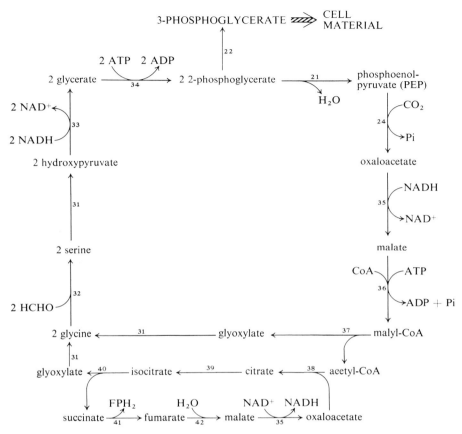

Fig. 14. The serine pathway of formaldehyde assimilation (icl⁺ variant). The icl⁻ variant differs from this in lacking a measurable malate thiokinase and in having an alternative route for oxidation of acetyl-CoA to glyoxylate not involving isocitrate lyase. Precursors can be removed from the cycle at the level of oxaloacetate or of succinate (see Fig. 15). The enzyme numbers refer to those in the text: (21) enolase; (22) phosphoglycerate mutase; (24) PEP carboxylase; (31) serine-glyoxylate aminotransferase; (32) serine transhydroxymethylase; (33) hydroxypyruvate reductase; (34) glycerate kinase; (35) malate dehydrogenase; (36) malate thiokinase; (37) malyl-CoA lyase; (38) citrate synthase; (39) aconitase; (40) isocitrate lyase; (41) succinate dehydrogenase; (42) fumarase.

The serine pathway must be able to provide precursors for biosynthesis of all cell components from the C_1 substrate. The pathway is redrawn in Fig. 15 to illustrate the biosynthesis of oxaloacetate and of succinate from two molecules of formaldehyde plus two of CO_2. For biosynthesis of carbohydrates from phosphoglycerate the usual gluconeogenic enzymes are also required and pyruvate will be produced from PEP by way of pyruvate kinase. The usual source of acetyl-CoA will presumably be the malyl-CoA lyase reaction; this will be the only route in those bacteria such as obligate methylotrophs and restricted facultative methylotrophs (e.g. *Hyphomicrobia*) which lack pyruvate dehydrogenase.

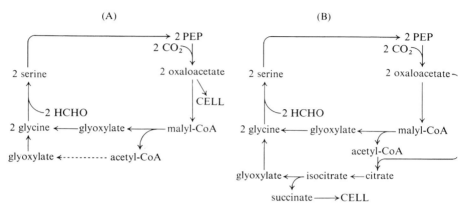

Fig. 15. The serine pathway operating to synthesise oxaloacetate (A) and succinate (B). Details of the pathway are as in Fig. 14.

II. Evidence for the serine pathway

The evidence is discussed in more or less chronological order and is considered under three main headings: isotope studies with whole cells; enzymological studies; studies with mutant bacteria.

A. Evidence obtained by incubating whole cells with radioactive isotopes

Much of the earlier work leading to the formulation of the serine pathway was done by Quayle and his colleagues using the facultative methanol- and formate-utilisers *Pseudomonas* AM1 and *Hyphomicrobium vulgare*. The first short-term incubation experiments with methanol, formate and bicarbonate gave essentially the same results in these two organisms (Large et al., 1961), and also in methanol-grown *Pseudomonas* sp. PRL-W4, methane-grown *Methanomonas methanooxidans* and methylamine-grown *Diplococcus* PAR (Table 22):
(a) radioactivity from all substrates was found at early times predominantly in serine, glycine, aspartate and malate;

(b) phosphorylated compounds were never important early intermediates, always appearing later in the incubations;
(c) there was more radioactivity at early times in serine than in glycine when ^{14}C methanol, methylamine or formate were used as tracer; but glycine contained more radioactivity than serine when ^{14}C bicarbonate was used as tracer.

TABLE 22

Evidence for the serine pathway from ^{14}C-labelling experiments

Pseudomonas PRL-W4	Kaneda and Roxburgh (1959b)
Pseudomonas AM1	Large *et al.* (1961, 1962a); Salem *et al.* (1972); Dunstan *et al.* (1972a, b); Dunstan and Anthony (1973)
Hyphomicrobium vulgare	Large *et al.* (1961)
Methanomonas methanooxidans	Lawrence *et al.* (1970)
Diplococcus PAR	Leadbetter and Gottlieb (1967)
Pseudomonas MS	Wagner and Quayle (1972)

These experiments clearly indicated that the reduced C_1 substrates were not being assimilated by way of the ribulose bisphosphate cycle after prior oxidation to CO_2. This was substantiated by the observation that the specific radioactivity of cellular material obtained from *Pseudomonas* AM1 grown on ^{14}C methanol was decreased by about 50% when CO_2 was bubbled through the growing culture (Large *et al.*, 1961). This indicated that only 50% of the carbon incorporated into cell material is from CO_2, rather than the 90–100% expected if the ribulose bisphosphate cycle of CO_2 assimilation were operating as in *Pseudomonas oxalaticus* (Quayle and Keech, 1959a). Before a metabolic pathway could be formulated from these results it was necessary to determine the distribution of isotope in the early-labelled intermediates, and this was done using *Pseudomonas* AM1 incubated with ^{14}C methanol and ^{14}C bicarbonate (Large *et al.*, 1962a). These experiments showed that:
(a) the carboxyl group of glycine is mainly derived from carbon dioxide, and the methylene carbon from methanol;
(b) the hydroxymethyl group of serine (C-3) is derived from methanol;
(c) the distribution of radioactivity between C-1 and C-2 of serine is the same under all conditions as that between C-1 and C-2 of glycine;
(d) the labelling pattern of serine and malate are consistent with formation of malate by carboxylation of a C_3 compound derived from serine.

These data were seen to be consistent with two possible routes for incorporation of methanol and CO_2 into serine, glycine and malate; both involving the hydroxymethylation of glycine by formaldehyde to give serine, and the

100   The biochemistry of methylotrophs

carboxylation of PEP (or pyruvate) to give oxaloacetate, which could give rise to the radioactive malate and aspartate measured in the experiments. In the direct route (Fig. 16a) glycine arises from an initial (hypothetical) condensation of CO_2 and a methanol derivative, whereas in the alternative, cyclic route the glycine comes from cleavage of a C_4 compound into two C_2 units, one of which (at least) acts as a precursor of glycine (Fig. 16b).

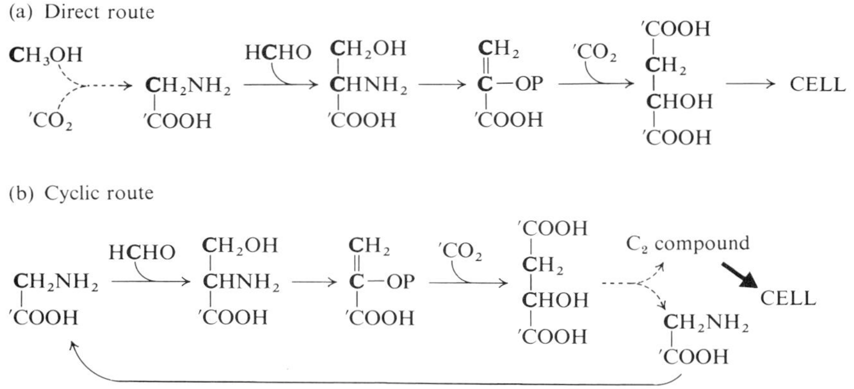

(a) Direct route

$$CH_3OH \atop {}^\prime CO_2 \longrightarrow {CH_2NH_2 \atop {}^\prime COOH} \xrightarrow{HCHO} {CH_2OH \atop \overset{|}{C}HNH_2 \atop {}^\prime COOH} \longrightarrow {CH_2 \atop \overset{\|}{C}-OP \atop {}^\prime COOH} \xrightarrow{{}^\prime CO_2} {{}^\prime COOH \atop \overset{|}{C}H_2 \atop \overset{|}{C}HOH \atop {}^\prime COOH} \longrightarrow CELL$$

(b) Cyclic route

$${CH_2NH_2 \atop {}^\prime COOH} \xrightarrow{HCHO} {CH_2OH \atop \overset{|}{C}HNH_2 \atop {}^\prime COOH} \longrightarrow {CH_2 \atop \overset{\|}{C}-OP \atop {}^\prime COOH} \xrightarrow{{}^\prime CO_2} {{}^\prime COOH \atop \overset{|}{C}H_2 \atop \overset{|}{C}HOH \atop {}^\prime COOH} \longrightarrow {C_2 \text{ compound} \atop \searrow \atop {CH_2NH_2 \atop {}^\prime COOH}} \longrightarrow CELL$$

Fig. 16. Potential routes for assimilation of radioactive (^{14}C) methanol and CO_2 into glycine, serine and malate in *Pseudomonas* AM1.

It was more than ten years after the first appreciation of the problem of alternative potential routes for the regeneration of glycine that the cyclic route was shown to operate in *Pseudomonas* AM1. This depended finally on the discovery of an enzyme able to catalyse the cleavage of a C_4 compound (malyl-CoA) into two C_2 units (see below). Before this, the operation of the cyclic route was demonstrated by measuring the incorporation of radioactivity into various metabolites during brief incubations of whole cells with 2,3-^{14}C-succinate (Salem *et al.*, 1972). Label was rapidly incorporated into glycine in methanol-grown cells, whereas the proportion in glycine was much lower when succinate-grown cells were used, thus indicating the operation of a cleavage mechanism having special importance during growth on methanol. By analysis of the distribution of label in glycine and serine it was shown that the succinate was cleaved symmetrically giving rise to glycine and thence serine labelled predominantly in the C-2 carbon atoms. In order to show that the glycine was not being produced from serine arising from succinate by way of phosphoserine, the experiments were repeated, with similar results, using a mutant of *Pseudomonas* AM1 (mutant 20S) lacking phosphoserine phosphatase.

The serine pathway, as now understood, proposes that acetyl-CoA, a product of malyl-CoA lyase, is oxidised to glyoxylate whose amination regenerates glycine (Fig. 14). Before the discovery of malyl-CoA lyase some indirect evidence from mutant studies had indicated that a cyclic route was operating, and that the oxidation of acetate or acetyl-CoA to glyoxylate (and thence glycine) must be an essential part of this pathway (Dunstan et al., 1972a, b). This was confirmed by showing that ^{14}C acetate was rapidly converted into glycine by whole cells of methanol-grown *Pseudomonas* AM1; this did not occur in ethanol-grown cells in which the serine:glyoxylate aminotransferase was not induced (Dunstan and Anthony, 1973).

B. Evidence for the serine pathway from enzyme studies

1. The formation of oxaloacetate

During the early work on the serine pathway, enzymes catalysing the formation of oxaloacetate from glyoxylate, formaldehyde and CO_2 were shown to be present in *Pseudomonas* AM1 (Fig. 14), and all of them have been shown to be induced to higher specific activities during growth on C_1 compounds (see Table 23 for references). The final step in oxaloacetate formation is catalysed by a highly active PEP carboxylase which is the only enzyme functioning in the carboxylation of C_3 compounds to C_4 compounds in *Pseudomonas* AM1 (Large et al., 1962b; Salem et al., 1973a). The serine-glyoxylate aminotransferase, the hydroxypyruvate reductase and the glycerate kinase were shown to be coordinately regulated suggesting that they may be encoded on an operon (Dunstan et al., 1972b), as has since been shown by gene mapping in the facultative methanotroph *Methylobacterium organophilum* (O'Connor and Hanson, 1978). A variant of this first part of the serine pathway was initially proposed in *Pseudomonas* MA (Bellion and Hersh, 1972), but these bacteria have since been shown to be similar with respect to oxaloacetate formation from glyoxylate (Newaz and Hersh, 1975). The next part of the serine pathway is the cleavage of a C_4 compound which, in all cases is the coenzyme A ester of the malate produced from oxaloacetate by way of malate dehydrogenase.

2. The cleavage reaction

The key cleavage enzyme is an inducible malyl-CoA lyase, first described in *Pseudomonas* MA by Hersh (1973, 1974a) and independently in a number of different methylotrophs including *Pseudomonas* AM1, *Hyphomicrobium*, *Pseudomonas* MS and *Methylosinus trichosporium* by Salem et al. (1973b). A malate thiokinase is also present in methylamine-grown *Pseudomonas* MA and this enzyme, acting together with malyl-CoA lyase constitutes the ATP malate lyase found in extracts of *Pseudomonas* MA (Hersh and Bellion, 1972;

TABLE 23
Evidence for the serine pathway from enzyme studies

Enzyme	References
Serine transhydroxymethylase	Large and Quayle (1963); Harder and Quayle (1971a)
Serine-glyoxylate aminotransferase	Large and Quayle (1963); Blackmore and Quayle (1970); Harder and Quayle (1971b); Dunstan et al. (1972b)
Hydroxypyruvate reductase	Large and Quayle (1963); Dunstan et al. (1972b); Newaz and Hersh (1975)*
Glycerate kinase	Large and Quayle (1963); Dunstan et al. (1972b); Hill and Attwood (1974); Newaz and Hersh (1975)*
PEP carboxylase	Large et al. (1962b); Salem et al. (1973a); Newaz and Hersh (1975)*
Malate thiokinase (Malyl-CoA synthetase)	Hersh (1973, 1974b)*
Malyl-CoA lyase	Salem et al. (1973b); Hersh (1973, 1974a)*
Isocitrate lyase	Bellion and Hersh (1972)*; Bellion and Woodson (1975)*
Homoisocitrate lyase	Kortstee (1980, 1981)
Methylene tetrahydrofolate dehydrogenase	Large and Quayle (1963)
Tetrahydrofolate formylase	Large and Quayle (1963)

These references are to the early observations on those enzymes and their regulation which have been particularly important in the elucidation of the serine pathway. Further references to these enzymes are in the section on the reactions of the serine pathway and in the tables of mutants (Tables 24 and 25). References are predominantly to work with *Pseudomonas* AM1 except for those marked with an asterisk which refer to work with *Pseudomonas* MA (Shaw strain).

Bellion and Hersh, 1972; Hersh, 1973, 1974b), *Rhodopseudomonas spheroides* (Tuboi and Kikuchi, 1962, 1963, 1965; Mue *et al.*, 1964) and *Hyphomicrobium* (Harder *et al.*, 1973; Salem *et al.*, 1973b). It is likely that the ATP malate lyase found in organism 5H2 (Cox and Zatman, 1973), *Pseudomonas aminovorans* and *Pseudomonas* MS (Large and Carter, 1973) is also due to the concerted action of malate thiokinase and malyl-CoA lyase. No enzyme has yet been described able to catalyse the conversion of malate to malyl-CoA in *Pseudomonas* AM1 or in other methylotrophs which also lack isocitrate lyase (Quayle, 1975).

3. The oxidation of acetyl-CoA to glyoxylate

(a) The icl^+-serine pathway. In bacteria with this pathway, isocitrate lyase (icl) operates together with some of the TCA cycle enzymes to achieve the cyclic oxidation of acetyl-CoA to glyoxylate (Fig. 14) exactly as occurs in bacteria growing on C_2 compounds by the glyoxylate cycle. The involvement of isocitrate lyase in growth on methylated amines was first demonstrated in *Pseudomonas* MA by Bellion and Hersh (1972) and subsequently in organism 5H2 (Cox and Zatman, 1973), *Pseudomonas aminovorans* and *Pseudomonas* MS (Large and Carter, 1973) and in a range of other methylotrophs isolated on methylamine (Bellion and Spain, 1976) (see page 113 and Table 27).

(b) The icl^--serine pathway. Although radioactive-labelling and mutant studies have provided strong evidence for a route for oxidation of acetyl-CoA to glyoxylate in *Pseudomonas* AM1, the enzymic basis for this remains uncertain (Dunstan *et al.*, 1972a, b; Dunstan and Anthony, 1973; Anthony, 1975a). This organism has no isocitrate lyase during growth on C_1 compounds or on C_2 compounds, and mutant evidence (next section) indicates that the same enzymes are involved in the oxidation of acetyl-CoA to glyoxylate during growth on both classes of compound.

The only sign of a solution to this long standing problem is due to the work of Kortstee (1980, 1981) who has proposed that acetyl-CoA is oxidised to glyoxylate by a pathway homologous to the isocitrate lyase route (Fig. 17).

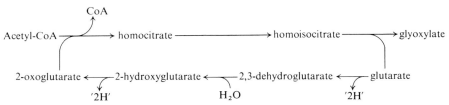

Fig. 17. The homoisocitrate glyoxylate cycle for oxidation of acetyl-CoA to glyoxylate in icl^--serine pathway bacteria as proposed by Kortstee (1980, 1981).

Instead of citrate synthase condensing acetyl-CoA with oxaloacetate to give citrate, a homocitrate synthase condenses acetyl-CoA with 2-oxoglutarate yielding homocitrate. The key enzyme of the proposed cycle is homoisocitrate lyase which cleaves homoisocitrate to glyoxylate plus glutarate, whose oxidation by way of 2,3-dehydroglutarate and 2-hydroxyglutarate regenerates 2-oxoglutarate and thus completes the cycle. The evidence for the operation of this cycle was obtained using two similar pink facultative methylotrophs *Pseudomonas* AM1 and *Pseudomonas* 80 (Kortstee, 1980, 1981):

(a) Crude extracts of *Pseudomonas* 80 catalysed the production of glyoxylate and glutarate from a compound assumed to be homoisocitrate. This substrate was prepared enzymically using extracts of *Pseudomonas* 80 and was characterised solely by chromatography with no physical data being reported for it. Clearly an enzymic characterisation of this key cleavage reaction is essential for this pathway to be generally accepted.

(b) Crude extracts of acetate-grown *Pseudomonas* 80 and of ethanol-grown *Pseudomonas* AM1 catalysed the formation of ^{14}C glyoxylate from 1-^{14}C or 2-^{14}C acetyl-CoA in the presence of 2-oxoglutarate (but not oxaloacetate). An unidentified dialysable cofactor appeared to be required for the reaction and NADH was included in the reaction mixture although its function (if any) was not discussed.

(c) The rate of formation (specific activity) of glyoxylate from acetyl-CoA plus 2-oxoglutarate was greater with extracts of *Pseudomonas* 80 grown on methanol, ethanol and acetate than on succinate; and it was greater in extracts of *Pseudomonas* AM1 derived from methanol- or pyruvate-grown cells than from succinate-grown cells.

(d) When acetate-grown *Pseudomonas* 80 was exposed for 20 s to 1-^{14}C acetate, radioactive malate, succinate, homocitrate and "homoisocitrate" accumulated (neither a time course of incorporation nor evidence of identification is available).

Considerable doubt now exists with respect to the homoisocitrate cycle because of the difficulty experienced by a number of workers in repeating these results with closely related facultative methylotrophs. An extensive study by Bellion *et al.* (1981) has failed to detect any enzyme in *Pseudomonas* AM1, "*Diplococcus*" PAR or *Pseudomonas* MA which is able to catalyse the cleavage of homoisocitrate to glyoxylate. In their investigation, all four stereoisomers (chemically prepared and characterised) of homoisocitrate were used but none was a suitable substrate. It also proved impossible to repeat the reported conversion of ^{14}C-acetyl-CoA to ^{14}C-glyoxylate using extracts of *Pseudomonas* AM1 incubated with 2-oxoglutarate. Likewise, no homocitrate synthase was detected in extracts of *Pseudomonas* AM1 or *Pseudomonas* MA.

4. The serine pathway   105

These studies suggest that the homoisocitrate cycle as written in Fig.17 does not operate in the oxidation of acetyl-CoA to glyoxylate in icl$^-$ methylotrophs. This proposed route is, however, an attractive one and it remains possible that a similar route does operate but perhaps with derivatives (e.g. CoA esters) of the intermediates shown in Fig. 17.

C. Evidence for operation of the serine pathway from mutant studies

The isolation and characterisation of mutants of *Pseudomonas* AM1 (pioneered by Heptinstall and Quayle, 1970), and later of *Methylobacterium organophilum*, has been a valuable approach in the study of the serine pathway, being used to confirm proposed metabolic sequences and to indicate the presence of novel ones (Tables 24 and 25). The interpretation of some of these studies is not always straightforward. The mutagen most often used is N-methyl,N'-nitro,N-nitrosoguanidine and this can lead to production of double or multiple lesions; although the secondary isolation of revertants can avoid some erroneous conclusions such revertants cannot always be isolated. There is also the problem that metabolic routes operating in mutants may not be "normal"; that is, they may not operate in the wild-type bacteria. Furthermore, some mutants may have a misleading phenotype because they may differ from the wild-type bacteria in the concentrations of metabolic intermediates which function as feedback inhibitors, or they may be more sensitive to inhibition by added growth supplements.

Mutants have been especially valuable for confirming suggested specific functions for particular enzymes in the serine pathway, loss of the enzymes by mutation being concomitant with loss of ability to grow on C_1 compounds; a list of such mutants and their lesions is presented in Table 24.

An example of the use of mutants to point the way to unpredicted features of metabolism has involved those mutants of *Pseudomonas* AM1 listed as being unable to oxidise acetate (or acetyl-CoA) to glyoxylate in Table 24. Mutant PCT 48 was isolated by Pat Dunstan as being unable to grow on methanol but able to grow on succinate. It was also unable to grow on ethanol or 3-hydroxybutyrate, both precursors of acetyl-CoA. Supplements of glycollate or glyoxylate were sufficient to allow growth on C_1 compounds or ethanol or 3-hydroxybutyrate. These results suggested that, unexpectedly, some reactions may be common to the pathways for assimilation of both C_1 and C_2 compounds and they led us to investigate the route for assimilation of C_2 compounds in *Pseudomonas* AM1. The conclusion that isocitrate lyase is not involved in this C_2 pathway, nor in the serine pathway, contrasted with the situation in *Pseudomonas* MA where isocitrate lyase is involved in growth on both methylamine and C_2 compounds. *Pseudomonas* MA thus has the icl$^+$-variant

TABLE 24
Mutants of *Pseudomonas* AM1

Mutant designation	Enzyme or reaction(s) missing in mutant	References
Mutants deficient in serine pathway enzymes		
82G^r	Serine transhydroxymethylase (N.B. the missing enzyme is that involved in growth on succinate)	Harder and Quayle (1971a)
2OST-1^r	Serine-glyoxylate aminotransferase (N.B. also lacks phosphoserine phosphatase)	Harder and Quayle (1971b)
82GT-1^r	Serine-glyoxylate aminotransferase (also lacks serine transhydroxymethylase)	Harder and Quayle (1971b)
20B-L^r	Hydroxypyruvate reductase	Heptinstall and Quayle (1970)
PCT 64^r	Glycerate kinase	Dunstan et al. (1972b); Dunstan and Anthony (1973)
PCT 57^r	Malyl-CoA lyase	Dunstan et al. (1972b); Dunstan and Anthony (1973); Salem et al. (1974); Cox and Quayle (1976a, b)
Mutants deficient in the oxidation of C_1 compounds		
M-15A^r	Methanol dehydrogenase	Heptinstall and Quayle (1970); Cox and Quayle (1976b); McNerney and O'Connor (1980)
PCT 29^r	Methanol dehydrogenase	Dunstan et al. (1972a); Dunstan and Anthony (1973); Anthony (1975b); Bolbot and Anthony (1980a)
PCT 76^r	Cytochrome c (also lacks methylamine dehydrogenase)	Dunstan and Anthony (1973); Anthony (1975b); Bolbot and Anthony (1980a)
PCT 761	Cytochrome c (derived from PCT 76 also lacks carotenoid pigments)	Anthony (1975b)

Mutants unable to oxidise acetyl-CoA to glyoxylate

PCT 48, G5, C5 and JAB 40	None of these 4 mutants grows as well as wild-type bacteria on C_1, C_2 or C_3 compounds unless supplemented with succinate, glycollate or glyoxylate	Dunstan et al. (1972b); Dunstan and Anthony (1973); Bolbot and Anthony (1980a, b); Salem et al. (1973a)

Mutants altered in pyruvate- or 2-oxoglutarate dehydrogenase

JAB 20–25	Pyruvate dehydrogenase (possibly E2 component)	Bolbot and Anthony (1980a, b)
JAB 30	Pyruvate and oxoglutarate dehydrogenases (possibly altered E2 components)	Bolbot and Anthony (1980a, b)
ICT 41[r]	2-Oxoglutarate dehydrogenase (E2?)	Taylor and Anthony (1976a)
JAB 10–13[r]	2-Oxoglutarate dehydrogenase (E2?)	Bolbot and Anthony (1980a, b)
JAB 14	2-Oxoglutarate dehydrogenase (E1)	Bolbot and Anthony (1980a)

Miscellaneous mutants

ICT 51[r]	Malyl-CoA hydrolase (part of "malate synthase")	Taylor and Anthony (1976b); Cox and Quayle (1976a, b)
ICT 54	Acetyl-CoA synthetase (acetate thiokinase)	Taylor and Anthony (1976b)
20S	Phosphoserine phosphatase	Harder and Quayle (1971a); Salem et al. (1972)
PCT 7	No carotenoid synthesis	Anthony (1975b)

Revertants have been isolated from mutants marked with a superscript [r].

TABLE 25
Mutants of *Methylobacterium organophilum*, *Pseudomonas aminovorans* and *Pseudomonas* MS

Mutants of M. organophilum (usually isolated as methanol$^-$, succinate$^+$)

82^r	Lacks glycerate kinase
24C	Lacks cytochrome *c*
22C	Has altered cytochrome *c*
17M	Lacks methanol dehydrogenase
3a^r	Lacks isocitrate dehydrogenase (a glutamate auxotroph; O'Connor *et al.*, 1977)
8A^r	Regulatory mutant; lacks methanol dehydrogenase, hydroxypyruvate reductase and serine-glyoxylate aminotransferase
7A	Probably deletion; lacks methanol dehydrogenase, hydroxypyruvate reductase and serine transhydroxymethylase
4M^r	Possibly operator; lacks methanol dehydrogenase, hydroxypyruvate reductase, serine-glyoxylate aminotransferase, glycerate kinase, serine-glyoxylate aminotransferase and glycerate kinase (cytochrome *c*$^+$)
17A	Regulatory mutant; lacks cytochrome *c*, methanol dehydrogenase, glycerate kinase, serine-glyoxylate aminotransferase, serine transhydroxymethylase, hydroxypyruvate reductase

Mutants of Ps. aminovorans (Bamforth and O'Connor, 1979) (isolated as methylamine$^-$, succinate$^+$)

These are all regulatory mutations and are pleiotropic.

M3	All the following are absent or present at very low specific activities; formaldehyde dehydrogenase (dye-linked); formate dehydrogenase, *N*-methylglutamate dehydrogenase, serine-glyoxylate aminotransferase; hydroxypyruvate reductase; PEP carboxylase; isocitrate lyase
M5	As for M3; it also lacks *N*-methylglutamate synthase and γ-glutamylmethylamide synthetase
M4	As for M3 but has low levels of PEP carboxylase
M2	As for M3 but has low levels of isocitrate lyase
M1	As for M3 but has low levels of hydroxypyruvate reductase, and both PEP carboxylase and isocitrate lyase are present

Mutants of Pseudomonas *MS* (Wagner and Levitch, 1975) (isolated as methylamine$^-$, pyruvate$^+$)

Type II	Lack ATP malate lyase and isocitrate lyase

The mutants of *M. organophilum*, a facultative methanotroph, were described and used for gene mapping by O'Connor and Hanson (1977, 1978) (Fig. 18). *Ps. aminovorans* is a facultative methylotroph able to grow on methylated amines but not methanol. The superscript ^{r} indicates that revertants were obtained. (N.B. mutants with low levels of enzyme are recorded as lacking it.)

of the serine pathway whereas *Pseudomonas* AM1 has the alternative icl⁻ variant. The available evidence indicates that mutant PCT 48 and similar mutants (Table 24) lack enzymes involved in oxidation of acetyl-CoA to glyoxylate in *Pseudomonas* AM1 and that this route is also involved in growth on other substrates assimilated by way of acetyl-CoA such as ethanol, malonate, 3-hydroxybutyrate, pyruvate and lactate (see Chapter 5).

III. The occurrence and distribution of the serine pathway

A. Methylotrophs using the pathway as their main assimilation pathway (Table 26)

A commonly-used indicator of the operation of the RuBP, RuMP and DHA pathways of C_1 assimilation is the presence of the first enzyme in which a carbon–carbon bond is made in each pathway; that is, ribulose bisphosphate carboxylase, hexulose phosphate synthase or dihydroxyacetone synthase. The equivalent enzyme in the serine pathway, serine transhydroxymethylase, cannot be used for this purpose because the same or similar enzyme also functions in the biosynthesis of glycine from serine during growth on multicarbon compounds and is probably present in all bacteria. Alternative enzymes must therefore be used as indicator enzymes for the serine pathway. Because of its high specific activity in methylotrophs, and because it is easy to assay, hydroxypyruvate reductase has been used extensively for this purpose. However, this alone is insufficient evidence for the operation of the pathway in a given organism and other key enzymes of the pathway should also be demonstrated: malyl-CoA lyase, serine-glyoxylate aminotransferase, glycerate kinase, PEP carboxylase and also isocitrate lyase (or its absence) in order to assess the likely route for oxidation of acetyl-CoA to glyoxylate.

Table 26 lists those methylotrophs in which the serine pathway is the main assimilation pathway. Most serine pathway bacteria are facultative methylotrophs and these include the facultative methanotrophs. The only obligate methylotrophs that use the serine pathway are the obligate methanotrophs having Type II membranes: no obligate methylotroph unable to use methane has the serine pathway.

Some bacteria whose main assimilation pathway is the RuMP- or RuBP-pathway contain low levels of hydroxypyruvate reductase and/or malyl-CoA lyase. The presence of these enzymes does not necessarily indicate the operation of the pathway as a complete pathway of carbon assimilation and their function will be discussed further in the individual descriptions of the enzymes (and see Salem *et al.*, 1973b; Shishkina *et al.*, 1976; Bamforth and Quayle, 1977; Taylor *et al.*, 1981).

TABLE 26

Methylotrophs using the serine pathway as their main assimilation pathway

Obligate methylotrophs able to use methane (all have Type II membranes)

Methylosinus trichosporium[a]	Lawrence and Quayle (1970); Salem *et al.* (1973b); Trotsenko (1976)
Methylosinus sporium[c]	Lawrence and Quayle (1970)
Methylocystis parvus[a]	Lawrence and Quayle (1970); Trotsenko (1976)
Methanomonas methanooxidans[ab]	Lawrence *et al.* (1970)

Facultative methylotrophs able to use methane (all have Type II membranes)

Methylobacterium organophilum XX[a]	O'Connor and Hanson (1977)
Methylobacterium R6[a]	Patel *et al.* (1978b)
Methylobacterium ethanolicum[a]	Lynch *et al.* (1980)
Methylobacterium hypolimneticum[a]	Lynch *et al.* (1980)

Facultative methylotrophs unable to use methane

Pink facultative methylotrophs (usually use methanol and methylated amines, Table 11)

Pseudomonas AM1[ab]	Large *et al.* (1961, 1962a, b); Large and Quayle (1963); Salem *et al.* (1973b)
Pseudomonas extorquens[a]	Quayle (1975)
Pseudomonas 3A2[a]	Colby and Zatman (1972); McNerney and O'Connor (1980)
Pseudomonas sp. 1, 135 and M27[a]	Rock *et al.* (1976)
Pseudomonas sp. TP1, YR, JB1 and PCTN[c]	Bellion and Spain (1976)
Pseudomonas methylica[ab]	Loginova and Trotsenko (1979a)
Pseudomonas AT2[c]	Boulton *et al.* (1980)

Facultative methylotrophs using methanol or methylamines (not pink bacteria) (Tables 12 and 13)

Organism 5B1[a]	Colby and Zatman (1972, 1973)
Arthrobacter[c]	Tani *et al.* (1978b)
Hyphomicrobium[ab]	Large *et al.* (1961); Harder *et al.* (1973); Salem *et al.* (1973b); Attwood and Harder (1978)
Streptomyces 239[a] (this has very low levels of enzymes of serine and RuMP pathways)	Kato *et al.* (1977b)

Facultative methylotrophs unable to use methanol but using methylamines (Tables 12 and 13)

Pseudomonas aminovorans[a]	Large and Carter (1973); Bamforth and O'Connor (1979)
Pseudomonas MA[a]	Bellion and Hersh (1972); (Hersh, 1973)
Pseudomonas MS[ab]	Wagner and Quayle (1972); Large and Carter (1973); Salem *et al.* (1973b); Wagner and Levitch (1975)
Organism (*Diplococcus*) PAR[bc]	Leadbetter and Gottlieb (1967); Bellion and Spain (1976)
Organism 5H2[a]	Cox and Zatman (1973)

Facultatively methylotrophic eucaryotes

Gliocladium deliquescens[ab]	Sakaguchi *et al.* (1975)
Paecilomyces varioti[ab]	Sakaguchi *et al.* (1975)

See Tables 22, 23 and 24 for more complete references to the evidence for the pathway. Most serine pathway bacteria are facultative methylotrophs; no obligate methylotrophs unable to use methane use the serine pathway. The distribution of other pathways is described in Tables 8, 16, and 18. The distribution of variants of this pathway is described in Table 27.

[a] Hydroxypyruvate reductase plus other enzymes of the pathway present.
[b] *In vivo* incorporation of ^{14}C substrates indicate operation of the pathway.
[c] Hydroxypyruvate reductase only enzyme demonstrated.

TABLE 27
Distribution of the icl$^+$ and icl$^-$ variants of the serine pathway

Methylotrophs having the icl$^+$ variant	
Pseudomonas MA	Bellion and Hersh (1972); Bellion and Kim (1978); Bellion *et al.* (1981)
Organism 5H2	Cox and Zatman (1973)
Pseudomonas MS	Large and Carter (1973); Wagner and Levitch (1975)
Pseudomonas aminovorans (low specific activity but glyoxylate identified as product)	Large and Carter (1973); Bellion and Spain (1976); Bamforth and O'Connor (1979)

All these are Gram-negative facultative methylotrophs using methylated amines but not able to use methanol. They all contain "ATP malate lyase" activity presumably due to malyl-CoA lyase. They all grow on C$_2$ compounds forming isocitrate lyase during this growth.

Methylotrophs having the icl$^-$ variant	
Pink facultative methylotrophs (*Pseudomonas* AM1, M27, YR, JB1, PCTN, 3A2, *Protaminobacter ruber*)	Large and Quayle (1963); Dunstan and Anthony (1972a); Colby and Zatman (1972); Quayle (1975); Bellion and Spain (1976); Ueda *et al.* (1981); Bellion *et al.* (1981)
Methylobacterium organophilum XX	Patt *et al.* (1974); Bellion and Spain (1976)
Methylobacterium ethanolicum	Lynch *et al.* (1980)
Methylobacterium hypolimneticum	Lynch *et al.* (1980)
Organism ("*Diplococcus*") PAR	Bellion and Kelley (1979); Bellion and Kim (1979); Bellion *et al.* (1981)
Organism 5B1 (produces icl on acetate but not C$_1$ compounds)	Colby and Zatman (1972)
Hyphomicrobium spp. (some strains appear to have low levels of icl)	Bellion and Spain (1976); Attwood and Harder (1977); Harder and Attwood (1978)
Streptomyces sp. 239	Kato *et al.* (1977b)

Some of these facultative methylotrophs grow on both methanol and methylated amines but some are unable to grow on methanol. Most of them contain no malyl-CoA synthetase and hence no "ATP malate lyase". Isocitrate lyase is usually absent on C$_1$ and on C$_2$ compounds.

B. Distribution of icl⁺ and icl⁻ variants of the serine pathway (Table 27)

In bacteria having the icl⁺ variant, isocitrate lyase is involved in the oxidation of acetyl-CoA to glyoxylate and is thus the key enzyme for this part of the pathway. Bacteria having the icl⁻ variant of the pathway lack isocitrate lyase but no alternative enzymes have been sufficiently described to form a positive basis for identification of the particular variant (see page 103). The assay for isocitrate lyase is the only criterion for distinguishing the variants and so the interpretation of such assays is exceptionally critical. It can be calculated that for bacteria with a doubling time of 5 h the specific activity of isocitrate lyase during growth on C_1 compounds must be about 115 nmoles (min)$^{-1}$ (mg protein)$^{-1}$, and half this value during growth on C_2 compounds (Quayle, 1975). Low specific activities may be due to genuine isocitrate lyase but they may also be due to isocitrate dehydrogenase whose product, 2-oxoglutarate, gives a similar colour in the usual phenylhydrazine assay for glyoxylate; the enzymic products must therefore be chromatographed to identify them with certainty (Quayle, 1975; Attwood and Harder, 1977).

Table 27a lists those bacteria in which the specific activity of isocitrate lyase is sufficiently high to consider that the icl⁺ serine pathway is operating and Table 15b lists those in which the activity is very low or absent. Those bacteria having isocitrate lyase during growth on C_1 compounds also have it during growth on C_2 compounds although it may not be an identical enzyme (Bellion and Woodson, 1975). These icl⁺ bacteria are all Gram-negative, facultative methylotrophs able to use methylated amines but not methanol or methane. All bacteria listed in Table 27a lack isocitrate lyase during growth on C_1 and C_2 compounds except for organism 5B1 which is unusual in having high levels of isocitrate lyase during growth on acetate but low levels during growth on trimethylamine.

An unexpected observation is that those bacteria lacking isocitrate lyase also lack malate thiokinase and hence "ATP malate lyase" activity. In relation to this it is of interest that Wagner and Levitch (1975) have shown that a mutant of *Pseudomonas* MS (an icl⁺ organism) which is unable to grow on methylamine or acetate has lost both "ATP malate lyase" and isocitrate lyase. The authors suggest that loss of the two activities in a single mutant may be because of the similarity in the cleavage reaction catalysed by the two enzymes. However, an alternative explanation (highly speculative) might be that the activation of malate to malyl-CoA is related in some way to the function of the isocitrate lyase (or alternative enzyme in icl⁻ bacteria).

Although sensitivity to growth inhibitors is rarely used for indicating the presence of particular assimilation pathways in bacteria it is possible that such methods might be developed for this purpose in methylotrophs. One example that has not been further developed is that of inhibition of growth of *Pseudomonas* AM1 on C_1 compounds by sulphanilamide at concentrations not affecting growth of the same organism on succinate and not affecting growth of methylotrophs assimilating C_1 compounds by routes other than the serine pathway (Hollinshead, 1966). Two further examples relate to the possibility of confirming (or not) the *in vivo* activity of isocitrate lyase in methylotrophs.

The first, described by Cox and Zatman (1976), depends on the inhibitor fluoroacetate which gives rise *in vivo* to fluorocitrate, a potent inhibitor of aconitate hydratase. This enzyme catalyses the conversion of citrate to isocitrate and so it is important during growth on substrates where the TCA cycle must function at a high rate and during growth on C_1 compounds if the oxidation of acetyl-CoA to glyoxylate involves isocitrate lyase. If the icl$^-$ variant does not involve aconitase, or a similar sensitive enzyme, then bacteria having this variant should be relatively less sensitive to fluoroacetate. These predictions were confirmed by experiments with the organism 5H2 (icl$^+$), *Pseudomonas* AM1 (icl$^-$) and organism 5B1 (icl$^-$ on C_1 compounds). Growth of organism 5H2 (icl$^+$) on methylamine or trimethylamine was very sensitive to inhibition (10 μM-fluoroacetate completely inhibited growth), whereas growth on succinate was relatively insensitive (some growth occurring even at 1 mM-fluoroacetate). By contrast, the sensitivity of growth of *Pseudomonas* AM1 (icl$^-$) and organism 5B1 to fluoroacetate was the same on all substrates and was similar to the low sensitivity of organism 5H2 (icl$^+$) during growth on succinate, when the activity of aconitate hydratase is relatively less important.

A second possible way of testing for the presence of an isocitrate lyase-type of reaction *in vivo* is to use itaconate, a known inhibitor of isocitrate lyase (Williams *et al.*, 1971), which inhibits growth of bacteria when isocitrate lyase is obligatory for such growth but not when it is not required (McFadden and Purohit, 1977). Bellion and Kelley (1979) have shown that growth of *Pseudomonas* MA (icl$^+$) is more sensitive to itaconate during growth on methylamine or acetate than during growth on glucose. By contrast organism ("*diplococcus*") PAR (icl$^-$) and *Pseudomonas* AM1 (icl$^-$) were hardly sensitive to this inhibitor. Unfortunately the use of this inhibitor does not provide unequivocal results because it has been found in similar experiments with *Pseudomonas* AM1 that when the concentration of itaconate is raised (from 10 mM to 20 mM or 50 mM), growth is markedly inhibited on methanol and 3-hydroxybutyrate but not on pyruvate (Bolbot, 1979).

Thus sensitivity to some inhibitors may be a useful guide to differences in the metabolism of various substrates in one organism but it is unlikely to be as useful for distinguishing between different types of metabolism in different bacteria.

IV. The reactions of the serine pathway

The following section describes the individual reactions of the pathway and the enzymes catalysing them. The numbers in parentheses refer to the numbers of the reactions as shown in Fig. 14 (page 97). References to early descriptions in methylotrophs are given in Table 23 and references to mutants lacking some of the enzymes are given in Tables 24 and 25.

A. Enzymes involved in the formation of oxaloacetate from glyoxylate, formaldehyde and CO_2

Serine transhydroxymethylase (32)

This enzyme catalyses a reversible aldol condensation between formaldehyde and glycine to give serine:

$$\text{HCHO} \; + \; \begin{array}{c} \text{CH}_2\text{NH}_2 \\ | \\ \text{COOH} \end{array} \longleftrightarrow \begin{array}{c} \text{CH}_2\text{OH} \\ | \\ \text{CHNH}_2 \\ | \\ \text{COOH} \end{array}$$

$$\qquad\qquad\qquad\text{glycine} \qquad\quad \text{serine}$$

Tetrahydrofolate is required for the reaction during which it acts as a "protective" carrier of formaldehyde to the active site (Jordan and Akhtar, 1970). There are two species of this enzyme in *Pseudomonas* AM1; one functions in the serine pathway and is induced during growth on C_1 compounds whereas the other provides for glycine synthesis from serine during growth on multicarbon compounds (Large and Quayle, 1963; Harder and Quayle, 1971a). The same applies to the facultative methanotroph *Methylobacterium organophilum* XX which produces two serine transhydroxymethylase isoenzymes, differing in their physical properties, their regulation *in vitro* and their response to additions to the growth medium (O'Connor and Hanson, 1975).

In *M. organophilum* one isoenzyme is produced predominantly during growth on methanol or methane and is activated by glyoxylate whereas the second isoenzyme, predominant during growth on succinate, is insensitive to glyoxylate. Although both enzymes have been purified, their properties have been determined with relatively impure preparations obtained by separating the isoenzymes by anion-exchange chromatography. The isoenzymes are similar in a number of respects; the pH optimum is between 8·5 and 8·8; the K_m value for serine is about 1 mM; activity is stimulated by Ca^{++}, K^+ and Na^+; and both show the expected product-inhibition by glycine when measured with serine as substrate. The isoenzymes also differ in a number of respects. The isoelectric point of the "methanol enzyme" is higher than that of the "succinate enzyme". Its molecular weight is also higher, being about 200 000 and it has four identical subunits (mol. wt, 50 000); the molecular weight of the "succinate enzyme", by contrast, is about 100 000 and it has no subunit structure. As well as the ions mentioned above, the "succinate enzyme" is also activated by Mg^{++}, Mn^{++}, and Zn^{++}.

The most significant difference in the two enzymes is the activation of the "methanol enzyme" by glyoxylate (4·5-fold activation by 5 mM glyoxylate) which is not observed with the "succinate enzyme". During growth on methanol, this activation by glyoxylate might be important in stimulating formaldehyde utilization for biosynthesis when the level of glyoxylate is sufficiently high to support biosynthesis of cell material by the serine pathway. Glyoxylate not only increases the activity of preformed enzyme from methanol-grown bacteria but it also increases its specific activity when included in the growth medium (0·02% w/v). This effect is also observed when the main carbon source is succinate; in this case the added glyoxylate causes a switch in serine transhydroxymethylase synthesis to the glyoxylate-activated ("methanol") form (O'Connor and Hanson, 1975).

Serine-glyoxylate aminotransferase (31)

This enzyme, which is induced during growth on C_1 compounds, catalyses two reactions in the serine pathway. It effects the formation of glycine from its precursor (glyoxylate) and in the same reaction it catalyses the next step after serine synthesis which is the conversion of serine to hydroxypyruvate:

$$\begin{array}{c} CH_2OH \\ | \\ CHNH_2 \\ | \\ COOH \end{array} + \begin{array}{c} \\ | \\ CHO \\ | \\ COOH \end{array} \longleftrightarrow \begin{array}{c} CH_2OH \\ | \\ C=O \\ | \\ COOH \end{array} + \begin{array}{c} \\ | \\ CH_2NH_2 \\ | \\ COOH \end{array}$$

serine   glyoxylate   hydroxypyruvate   glycine

This enzyme is usually assayed by measuring formation of hydroxypyruvate using NADH and hydroxypyruvate reductase (added or present in crude extracts) (Harder and Quayle, 1971b). The aminotransferase is probably specific for serine and glyoxylate. Neither aspartate nor glutamate is able to replace serine as amino-group donor when glyoxylate is the acceptor (Blackmore and Quayle, 1970); and the transamination of serine using pyruvate or 2-oxoglutarate as amino-group acceptor is much slower than with glyoxylate as acceptor (Large and Quayle, 1963).

4. The serine pathway

Hydroxypyruvate reductase (33)

This inducible enzyme, first described in *Pseudomonas* AM1 by Large and Quayle (1963), catalyses the reduction of hydroxypyruvate to glycerate and, at 10% of the rate, glyoxylate to glycollate:

$$\begin{array}{c} CH_2OH \\ | \\ C=O \\ | \\ COOH \end{array} + NADH + H^+ \longleftrightarrow \begin{array}{c} CH_2OH \\ | \\ CHOH \\ | \\ COOH \end{array} + NAD^+$$

hydroxypyruvate   glycerate

The pH optimum of the reductase in *Pseudomonas* AM1 is 4·5 but it is higher in *Pseudomonas* MS (about 6·5) (Wagner and Levitch, 1975) and in *Pseudomonas aminovorans* (pH 6·0) (Large and Carter, 1973). The reductase is inhibited (competitively with respect to hydroxypyruvate) by citrate (cited in Bamforth and Quayle, 1977).

As mentioned above, the presence of this enzyme has sometimes been used as the sole indicator of operation of the serine pathway in particular methylotrophs. The problems associated with this practice are discussed by Bamforth and Quayle (1977) who have reported the properties of a constitutive hydroxypyruvate reductase (NADP-dependent) in *Paracoccus denitrificans* which grows on methanol by way of the RuBP pathway and so does not appear to need the enzyme. A possible explanation (Taylor *et al.*, 1981) for the presence of this "transvestite" enzyme in *Paracoccus denitrificans* and in *Methylococcus capsulatus* is that it is involved in the production of phosphoglycerate from the phosphoglycollate arising from the (fortuitous) oxygenase activity of the ribulose bisphosphate carboxylase.

Glycerate kinase (34)

This enzyme catalyses the phosphorylation of glycerate:

$$\begin{array}{c} CH_2OH \\ | \\ H-C-OH \\ | \\ COOH \end{array} + ATP \longrightarrow \begin{array}{c} CH_2OH \\ | \\ H-C-OP \\ | \\ COOH \end{array} + ADP$$

glycerate   2-phosphoglycerate

The product of this reaction in *Pseudomonas* AM1, *Hyphomicrobium* X and *Pseudomonas* MA (and probably other methylotrophs) is 2-phosphoglycerate (Harder *et al.*, 1973; Hill and Attwood, 1974; Newaz and Hersh, 1975).

Phosphoglycerate mutase (22)

This enzyme catalyses the formation of 3-phosphoglycerate from 2-phosphoglycerate:

$$\begin{array}{ccc} CH_2OH & & CH_2OP \\ | & & | \\ H\text{-}C\text{-}OP & \longleftrightarrow & CHOH \\ | & & | \\ COOH & & COOH \end{array}$$

That the product of the glycerate kinase reaction in methylotrophs is 2-phosphoglycerate, rather than 3-phosphoglycerate (above), has drawn attention to the fact that phosphoglycerate mutase is the first enzyme specifically involved in gluconeogenesis during growth on C_1 compounds. This conclusion is supported by the observation that during growth of *Hyphomicrobium* X on methanol, the mutase is induced 17-fold and by the demonstration that the precursor for gluconeogenesis cannot be a C_4 compound because of the low specific activity of PEP carboxykinase (Harder *et al.*, 1973). The purified phosphoglyceromutases from *Pseudomonas* AM1 and *Hyphomicrobium* X are similar to one another (and to other phosphoglyceromutases) (Hill and Attwood, 1976a; Harder and Attwood, 1978). The molecular weight (both enzymes) is about 32 000, the pH optimum is about 7·3 and 2,3-diphosphoglycerate is required as activator (K_m for this is about 10^{-5}M). The K_m for 2-phosphoglycerate is $3·7$–$6·9 \times 10^{-4}$M and about 10 times this for 3-phosphoglycerate. The enzyme is freely reversible, the equilibrium constant in the direction of 3-phosphoglycerate formation being about 11·3.

Preliminary work using crude extracts of methanol-grown *Hyphomicrobium* X had suggested that ATP might be an activator of the mutase and AMP an inhibitor (Harder *et al.*, 1973). However, although this makes sense in terms of regulation, it has since been shown that the pure enzymes from *Hyphomicrobium* X and *Pseudomonas* AM1 are not regulated by energy charge and that the effects of nucleotides were due to interference in the assay procedure by other enzymes (Hill and Attwood, 1976b).

Enolase (PEP hydratase) (21)

Enolase catalyses the reversible dehydration of 2-phosphoglycerate to phosphoenolpyruvate (PEP):

$$\begin{array}{c} CH_2OH \\ | \\ H\text{-}C\text{-}OP \\ | \\ COOH \end{array} \longleftrightarrow \begin{array}{c} CH_2 \\ || \\ C\text{-}OP \\ | \\ COOH \end{array} + H_2O$$

2-phosphoglycerate   PEP

Phosphoenolpyruvate carboxylase (24)

This enzyme carboxylates PEP to oxaloacetate, liberating inorganic phosphate during the reaction which has an absolute requirement for divalent cations (for review see Utter and Kolenbrander, 1972):

$$\begin{array}{c} CH_2 \\ || \\ C\text{-}OP \\ | \\ COOH \end{array} + CO_2 \longrightarrow \begin{array}{c} CH_2COOH \\ | \\ C=O \\ | \\ COOH \end{array} + Pi$$

PEP   oxaloacetate

This carboxylase is induced during growth on C_1 compounds and is the only enzyme with appreciable activity in carboxylating C_3 compounds to C_4 compounds in the following methylotrophs: *Pseudomonas* AM1 (Large et al., 1962b; Salem et al., 1973a; Wagner and Quayle, 1972); *Hyphomicrobium* (Attwood and Harder, 1974; Loginova and Trotsenko, 1979d); *Pseudomonas* MS and *Ps. aminovorans* (Large and Carter, 1973); *Pseudomonas* MA (Newaz and Hersh, 1975); *Pseudomonas methylica* and *Methylosinus trichosporium* (Loginova and Trotsenko, 1979d).

The first PEP carboxylase isolated from a methylotroph was that from *Pseudomonas* AM1 (Large et al., 1962b). It has a sharp pH optimum at pH 8·5 and a K_m value for PEP of 0·4 mM. PEP carboxylases have been grouped into three categories by Utter and Kolenbrander (1972) according to their mode of regulation. Enzymes in the first of these groups are activated by acetyl-CoA, this activation being related to their function as anaplerotic enzymes, replenishing intermediates of the TCA cycle when they are removed for biosynthesis (Kornberg, 1966). During growth of *Pseudomonas* MA and *Pseudomonas* AM1 on succinate, an acetyl-CoA-dependent PEP carboxylase is induced which is not present during growth on C_1 compounds (Newaz and Hersh, 1975) and a similar observation has been made in the facultative methanotrophs *Methylobacterium ethanolicum* and *M. hypolimneticum* (Lynch et al., 1980). During methylotrophic growth, the enzymes of the TCA cycle are not usually involved in the oxidation of acetyl-CoA to CO_2 nor in the provision of C_4 precursors for biosynthesis, and so it is not surprising that the PEP carboxylases present during methylotrophic growth are not regulated in the same way. The enzyme from methanol-grown *Pseudomonas* AM1 is not regulated by any metabolite tested (except perhaps NADH) and this appears to be true for other PEP carboxylases described in the methylotrophs listed above. The PEP carboxylases present during methylotrophic growth thus appear to be in the third group (non-regulated) of Utter and Kolenbrander (1972).

One definite exception to this is the NADH-activated PEP carboxylase from *Pseudomonas* MA which has been described in detail by Hersh and his colleagues (Newaz and Hersh, 1975; Millay and Hersh, 1976; Millay *et al.*, 1978). Before describing this further it should be noted that most reaction mixtures for assaying PEP carboxylase referred to above contain NADH (including those used in radioactive assay methods) and that some of the first observations of the enzyme in *Pseudomonas* AM1 could be interpreted as indicating a requirement for NADH (Large *et al.*, 1962b). Having said this, Newaz and Hersh (1975) state that their experiments indicated no stimulatory effect of NADH on the PEP carboxylase of *Pseudomonas* AM1. Further investigations are obviously required before any generalisations can be made about the regulation by NADH of the PEP carboxylases of methylotrophs other than *Pseudomonas* MA.

The PEP carboxylase of *Pseudomonas* MA is a tetramer (molecular weight about 320 000) of four identical subunits (mol. wt about 88 000); its K_m value for bicarbonate is about 0·4 mM and Mg^{++} or Mn^{++} are required for activity. It is an allosteric enzyme, being activated by NADH and inhibited by ADP, and the interactions between enzyme, substrates, activators and inhibitors give rise to complex kinetics. The following discussion will be limited to those aspects which have the most obvious bearing on the *in vivo* regulation of the enzyme.

The carboxylase shows cooperative kinetics with respect to its substrate PEP; there is negative cooperativity below about 3 mM in the presence or absence of its activator NADH and positive cooperativity above 3 mM PEP, but only in the presence of NADH. NADH is a potent activator of the enzyme, 0·2 mM NADH giving a 50-fold activation in the presence of 1 mM PEP. NADPH is not as good an activator as NADH. The extent of activation depends on the Mg^{++} ion concentration, the K_a for NADH varying inversely with the concentration of Mg^{++} — indicating that the Mg^{++} facilitates interaction of NADH with the enzyme. The K_a for NADH is 0·1 mM at about 10 mM Mg^{++} (total concentration). NAD^+ is an inhibitor acting competitively with respect to NADH (the Ki for NAD^+ is about 1·3 mM). The activation by NADH is time-dependent, requiring 2–3 min for maximum activation and it involves conversion of the inactive tetramer to the active dimeric form of the carboxylase. ADP is an allosteric inhibitor of the enzyme, reversing the effect of NADH and thus producing the inactive tetramer from active dimers. Because of the number of interacting species, no single figure can summarise the effectiveness of ADP as inhibitor so an example must suffice: 0·1 mM ADP gives 86% inhibition when measured with 2 mM PEP, 0·2 mM NADH, and 5 mM Mg^{++}. Less ADP is required for a similar extent of inhibition when the NADH concentration is lower and, conversely, more ADP is required to inhibit at high NADH concentrations. It has been pointed out by Millay *et al.* (1978) that although there is a correlation between activation by NADH and conversion of tetramer to dimer, it is possible that the change is not required for enzyme activity but merely reflects a difference in association properties caused by the activation.

B. The formation of glyoxylate and acetyl-CoA from oxaloacetate

The first part of the serine pathway produced oxaloacetate. In this next part of the pathway, the oxaloacetate is reduced to malate by an NAD^+-dependent malate dehydrogenase which is usually a constitutive enzyme present (but not described in detail) in methylotrophs. Malate thiokinase then catalyses the formation of malyl-CoA, which is then cleaved to acetyl-CoA plus glyoxylate, in a reaction catalysed by malyl-CoA lyase. Not all bacteria with the serine pathway have a malate thiokinase, but no alternative enzyme (such as a Coenzyme A transferase to malate) has been described. In the first descriptions of malate thiokinase in methylotrophs, it was measured in crude extracts containing malyl-CoA lyase, and the concerted activities were described as ATP malate lyase.

Malate thiokinase (malyl-CoA synthetase) (36)

This enzyme catalyses the synthesis of L-malyl-CoA (3-hydroxysuccinyl-CoA) from L-malate, coenzyme A and ATP in a reaction analogous to the succinate thiokinase reaction:

HOCHCOOH   HOCHCOOH
| + Coenzyme A + ATP → | + ADP + Pi
CH_2COOH CH_2CO-CoA

The enzyme described here is that isolated and characterised from *Pseudomonas* MA by Hersh and his colleagues (Hersh and Bellion, 1972; Hersh, 1973, 1974b; Elwell and Hersh, 1979). It is similar, in those characteristics investigated, to that described for the purified enzyme isolated from *Rhodopseudomonas spheroides* by Mue *et al.* (1964).

The malate thiokinase of *Pseudomonas* MA is induced during growth on methylamine but not on glycerol, and is a separate enzyme from succinate thiokinase (mol. wt, 150 000) which is constitutive and has no malate thiokinase activity (Hersh, 1973). For kinetic studies, the malate thiokinase has been purified about 70-fold, the final preparation being 80% pure and lacking the constitutive succinate thiokinase. The K_m values for malate, ATP and coenzyme A were 0·3 mM, 0·36 mM and 0·037 mM respectively. The thiokinase is equally active with L-malate and succinate, and shows considerable activity (about 27%) with D-malate and DL-isocitrate (all substrates present at 5 mM). The heats of inactivation and sensitivity to sulphydryl reagents were almost identical for the malate- and succinate-thiokinase activities of the purified enzyme, and the K_m and V_{max} values were the same for the two substrates. A thorough kinetic investigation of the purified malate thiokinase has shown that its reaction mechanism is very similar in kinetic terms to that of succinate thiokinase from other organisms (Hersh, 1974b). Both react in a random or partially random mechanism involving quaternary complex formation, and in both cases, phosphoenzyme and phosphorylated substrates (succinyl-enzyme and presumably malyl-enzyme) are involved as reaction intermediates.

The malate thiokinase has now been purified to homogeneity by affinity chromatography using 3'5'-ADP-Sepharose, and characterised with respect to subunit structure (Elwell and Hersh, 1979). Its overall molecular weight is about 290 000 and it consists of two non-identical subunits (α subunit, mol. wt 34 000; β subunit mol. wt 42 000) yielding an $\alpha_4\beta_4$ structure for the native enzyme. This compares with an $\alpha_2\beta_2$ structure for the succinate thiokinase of *Escherichia coli* and an $\alpha\beta$ structure for mammalian succinate thiokinase. Incubation of malate thiokinase with ATP leads to phosphorylation of all the α subunits present but none of the β subunits. It is suggested that the acid lability and the base stability of the phosphorylated thiokinase are characteristic of phosphorylation at a histidine residue. Dephosphorylation occurs on incubation of phosphorylated enxyme with ADP or malate. Phosphorylation with ATP or binding of coenzyme A causes the enzyme to dissociate to an $\alpha_2\beta_2$ structure but whether or not such a dissociation occurs during the catalytic cycle has not been determined.

Malyl-CoA lyase (37)

Malyl-CoA lyase catalyses the reversible cleavage of 4-malyl-CoA (3-hydroxysuccinyl-CoA) to glyoxylate plus acetyl-CoA (Salem *et al.*, 1973; Hersh, 1973; Hacking and Quayle, 1974).

$$\begin{matrix} \text{HOCHCOOH} \\ | \\ \text{CH}_2\text{CO-CoA} \end{matrix} \longleftrightarrow \text{CHOCOOH} + \text{CH}_3\text{CO-CoA}$$

4-malyl-CoA   glyoxylate   acetyl-CoA

4. The serine pathway

This enzyme was first suggested to be part of the "ATP malate lyase" discovered in *Rhodopseudomonas spheroides* by Tuboi and Kikuchi (1963). Their suggestion was confirmed by the separation of "ATP malate lyase" of *Pseudomonas* MA and *Rhodopseudomonas spheroides* into malate thiokinase plus malyl-CoA lyase (Hersh, 1973; Tuboi and Kikuchi, 1965) and by characterisation of the malyl-CoA lyase from *Pseudomonas* MA (Hersh, 1974a) and from *Pseudomonas* AM1 which lacks the malate thiokinase (Salem *et al.*, 1973b Hacking and Quayle, 1974). This key cleavage enzyme of the serine pathway (malyl-CoA lyase) has been found (when tested) in all methylotrophs having this pathway. It is also present in some methylotrophs having alternative pathways (*Methylococcus capsulatus* and *Methylomonas methanica*), albeit at relatively low specific activities (Salem *et al.*, 1973b). In these cases it probably serves to supply glyoxylate and thence glycine for porphyrin and protein biosynthesis, as suggested for *Rhodopseudomonas spheroides* growing on malate plus glutamate (Tuboi and Kikuchi, 1963).

The enzyme from *Pseudomonas* AM1 has been purified (20-fold) to homogeneity and shown to have no malate synthase or citrate synthase activity. This was tested because it is known that, in the absence of substrate oxaloacetate, citrate synthase catalyses the hydrolysis of malyl-CoA. The enzyme shows a broad pH optimum between 7·2 and 8·2. It has a molecular weight of 190 000 and an absolute requirement for bivalent metal ions (Mg^{++}, Co^{++}, Mn^{++}, or Zn^{++}). It is insensitive to sulphydryl reagents. In the presence of 10 mM Mg^{++}, the K_m values were 66 μM for malyl-CoA, 15 μM for acetyl-CoA (with 10mM glyoxylate) and 1·7 mM for glyoxylate (with 0·4 mM acetyl-CoA). In the direction of cleavage the K_m for Mg^{++} was 1·2 mM (with 0·2 mM malyl-CoA).

The equilibrium constant (M) (at 30°C) in the direction of cleavage is $4·7 \times 10^{-4}$ ($\Delta G°$, + 19·2 kJ mol^{-1}). Thus the equilibrium position of the reaction is such that at low concentrations of reactants and products (mM or less) substantial reaction in either direction is feasible.

The lyase is specific for (2S)-4-malyl-CoA, the alternative optical isomer (2R)-4-malyl-CoA not being cleaved but being able to inhibit the reaction. Neither citryl-CoA, (DL)-3-hydroxybutyryl-CoA nor 3-hydroxy-3-methyl-glutaryl-CoA act as alternative substrates. In the reverse direction only propionyl-CoA can replace acetyl-CoA when measured with glyoxylate as second substrate (giving 10% of the rate); *n*-butyryl CoA, succinyl-CoA and oxalyl-CoA cannot replace acetyl-CoA; and glycollate, glycolaldehyde and pyruvate cannot substitute for glyoxylate.

It has been pointed out by Hacking and Quayle (1974) that malyl-CoA lyase is the first enzyme demonstrated to catalyse a readily reversible reaction of the type:

$$RR'C(OH)CH_2CO\text{-}CoA \longrightarrow RR'CO + CH_3CO\text{-}CoA$$

Citramalyl-CoA lyase and 3-hydroxy-3-methylglutaryl-CoA lyase appear to catalyse similar reactions, but their reactions differ from that catalysed by malyl-CoA lyase in being essentially irreversible. Malyl-CoA lyase is also different in requiring a divalent cation and in being insensitive to sulphydryl reagents.

The malyl-CoA lyase from *Pseudomonas* MA (Hersh, 1973) has not been as fully characterised as that from *Pseudomonas* AM1 but in those aspects studied the two enzymes are very similar. A comprehensive investigation of the kinetics of the purified enzyme from *Pseudomonas* MA by Elwell and Hersh (1979) has shown that, in the direction of malyl-CoA synthesis, glycollate and oxalate are inhibitors, being competitive with respect to glyoxylate and non-competitive with respect to acetyl-CoA. Propionyl-CoA, on the other hand, is a competitive inhibitor with respect to acetyl-CoA and an uncompetitive inhibitor with respect to glyoxylate. In the direction of malyl-CoA cleavage, glyoxylate, glycollate and oxalate are competitive inhibitors with respect to malyl-CoA, whereas acetyl-CoA and propionyl-CoA inhibit non-competitively. These results suggest that (in the direction of synthesis) malyl-CoA lyase catalyses a sequential ordered reaction in which glyoxylate binds to the enzyme prior to acetyl-CoA. In apparent contradiction to this, the enzyme catalyses proton abstraction from the methyl group of acetyl-CoA, showing that the latter can bind to free enzyme even in the absence of glyoxylate. The explanation for this is that the K_s for acetyl-CoA is 0·34 mM, which is considerably greater than the K_m value (0·04 mM), suggesting that the affinity of acetyl-CoA for free enzyme is lower than its affinity for the enzyme-glyoxylate complex.

C. The oxidation of acetyl-CoA to glyoxylate

This final part of the serine pathway involves different enzymes in the two variants (icl+ and icl−).

1. Oxidation of acetyl-CoA to glyoxylate in icl+ bacteria

As its name implies, this route for oxidation of acetyl-CoA to glyoxylate involves isocitrate lyase and some of the enzymes of the TCA cycle (citrate synthase, aconitate hydratase, succinate dehydrogenase, fumarate hydratase and malate dehydrogenase) (Fig. 13); and this must be taken into account when considering regulatory properties of these enzymes in relation to their function. In methylotrophs they usually only function in biosynthesis and not in the oxidation of acetyl-CoA to CO_2 (but see possible exception

4. The serine pathway   125

proposed by Newaz and Hersh discussed on p. 135). The TCA cycle enzymes are described in more detail in Chapter 5; this section will describe the key enzyme of the icl⁺ variant of the serine pathway, isocitrate lyase.

Isocitrate lyase (40)

This enzyme catalyses the cleavage of isocitrate to glyoxylate plus succinate:

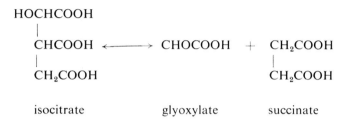

| isocitrate | glyoxylate | succinate |

The properties of isocitrate lyases from non-methylotrophs have been thoroughly described (Kornberg, 1966; McFadden, 1969; Spector, 1972; Johanson *et al.*, 1974a, b) but little is known about the enzyme in methylotrophs except for that in *Pseudomonas* MA (Bellion and Woodson, 1975).

Before the importance of this enzyme during growth of some methylotrophs on C_1 compounds (Bellion and Hersh, 1972) was understood, its sole function was thought to be in the glyoxylate cycle during growth on compounds assimilated exclusively by way of acetyl-CoA (Kornberg, 1966). Because the methylotrophs having the icl⁺ serine pathway are all able to grow on such compounds, it might be assumed that an identical isocitrate lyase would operate during growth on both C_1 and C_2 compounds. However, this is not the case in *Pseudomonas* MA which produces two different enzymes, one on methylamine and the other on acetate (Bellion and Woodson, 1975). They are very similar (but not identical) in molecular weight (about 160 000) and in K_m values for isocitrate (about 0·3 mM at pH 7·2), but they differ in isoelectric points, rates of thermal denaturation and in the effect of pH on the K_m value for isocitrate.

2. Oxidation of acetyl-CoA to glyoxylate in icl⁻ bacteria

None of the enzymes of the proposed homoisocitrate cycle (Fig. 17) has been characterised; all the available evidence relating to these enzymes (Kortstee, 1980, 1981) has been described in an earlier section (page 103).

V. Regulation of the serine pathway

During growth on C_1 compounds the enzymes of the serine pathway are essential, whereas most are unnecessary during growth on multicarbon compounds. To conserve energy and precursors the biosynthesis of these enzymes must be regulated at the genetic level. Once formed, their catalytic activity must be regulated, and the main requirement here is to balance the flow of substrate carbon into the serine pathway to provide carbon skeletons for biosynthesis, and into the oxidative route to provide the necessary NAD(P)H and ATP for biosynthesis. This brief review of regulation of the serine pathway will first consider regulation at the genetic level and then at the level of enzyme activity. Regulation in microorganisms having the RuBP, RuMP and DHA pathways of C_1 assimilation are discussed on pages 58, 93 and 292 respectively; and regulation of citrate synthase is described on page 144.

A. Regulation at the genetic level

The synthesis of enzymes involved in the oxidation and assimilation of C_1 compounds is regulated at the genetic level, as indicated by the higher specific activities measured during growth on these compounds (Section II.B1; Table 23). Because genetic techniques using methylotrophs have only recently been developed, little is known about this regulation. The main questions to be considered are:
(a) Are the genes for the C_1 enzymes grouped on the bacterial chromosome in operons?
(b) If operons exist is their expression regulated in a positive or negative mode?
(c) Is the synthesis of oxidative enzymes regulated together with that of the assimilatory enzymes?
(d) Is there any system of catabolite repression superimposed upon specific induction mechanisms?
(e) What is the metabolic "signal" regulating gene expression? Is it the same for all C_1 enzymes? For example, is methanol an inducer or is a product of methanol oxidation the actual inducer?

1. Investigations using predominantly biochemical techniques

Even in the absence of genetic techniques, it has been shown that the genes coding for some enzymes of the serine pathway in *Pseudomonas* AM1 might be encoded on an operon. This conclusion was based on the observation that the relative specific activities of three enzymes (serine-glyoxylate aminotransferase, hydroxypyruvate reductase and glycerate kinase) remain constant during growth in a variety of conditions (Dunstan et al., 1972b). By contrast, this coordinate regulation was not observed with the methanol and methylamine dehydrogenases. Measurements of enzymes in bacteria grown on a mixture of methanol and succinate have led to the conclusion that succinate or a product of its metabolism acts as a catabolite repressor of the serine pathway enzymes but not of the two dehydrogenases. These observations have been confirmed for *Pseudomonas* AM1 by McNerney and O'Connor (1980) who have also confirmed the suggestion of Salem et al. (1973b) that malyl-CoA lyase may be regulated coordinately with the three assimilatory enzymes discussed above. By contrast, the synthesis of the inducible serine transhydroxymethylase and PEP carboxylase is not repressed when succinate is present together with methanol during growth (McNerney and O'Connor, 1980; O'Connor, 1981).

A product of methanol metabolism rather than methanol itself may be responsible for induction of the serine pathway enzymes in *Pseudomonas* AM1. This is perhaps indicated by the observation that a mutant lacking cytochrome c and, therefore, unable to oxidise methanol, only produces low activities of the serine pathway enzymes when transferred from succinate to a methanol medium (Anthony, 1975b). This conclusion is supported by similar results obtained with a mutant lacking methanol dehydrogenase (McNerney and O'Connor, 1980; O'Connor, 1981).

In summary, in *Pseudomonas* AM1 synthesis of some of the C_1 assimilation enzymes (serine-glyoxylate aminotransferase; hydroxypyruvate reductase; glycerate kinase and malyl-CoA lyase) are probably regulated together by way of an operon whereas others are separately regulated. The four assimilatory enzymes that appear to be coded on an operon are susceptible to catabolite repression. The remaining enzymes may also be encoded on a separate operon, but there is no evidence of this possibility. The metabolic "signal" involved in induction (assuming one is required) may not always be methanol itself, but perhaps a product of its oxidation.

A comparative study by McNerney and O'Connor (1980) has demonstrated a considerable diversity with respect to regulation at the genetic level in the facultative methylotrophs (see Table 28). The serine pathway enzymes in *Pseudomonas* 3A2 (another pink facultative methylotroph) and *Hyphomicrobium* X differed from those in *Pseudomonas* AM1 in not being susceptible to catabolite repression by succinate. After transfer of *Hyphomicrobium* from succinate to methanol, the onset of induction of all seven enzymes studied was almost simultaneous, whereas there were different lag periods before induction of some enzymes in *Pseudomonas* AM1 and 3A2. It was concluded from this that regulation may be coordinate in *Hyphomicrobium* but not in the other two bacteria. However, to demonstrate coordinate regulation it is necessary to show that the proportions of enzymes produced

TABLE 28
Regulation of synthesis of C_1 enzymes in facultative methylotrophs

Organism	Enzymes coordinately regulated	Enzymes susceptible to catabolite repression
Pseudomonas AM1	serine glyoxylate aminotransferase, glycerate kinase, hydroxypyruvate reductase, malyl-CoA lyase. (Not: methylamine dehydrogenase; serine transhydroxymethylase or PEP carboxylase)	Only those four enzymes that are coordinately regulated
Pseudomonas 3A2	Perhaps some; hydroxypyruvate reductase and PEP carboxylase are separate from the others	None
Hyphomicrobium X	All	None
Methylobacterium organophilum XX	All	None
Methylobacterium ethanolicum	Not tested	All

The enzymes studied were methanol dehydrogenase; serine, glyoxylate aminotransferase; serine transhydroxymethylase; hydroxypyruvate reductase; glycerate kinase; PEP carboxylase; malyl-CoA lyase (Dunstan *et al.*, 1972b; O'Connor and Hanson, 1977; McNerney and O'Connor, 1980; O'Connor, 1981). The coordinate regulation suggested for *M. organophilum* has been confirmed genetically (O'Connor and Hanson, 1978; O'Connor, 1981) (see text and Fig. 18, page 132).

are constant in a range of different growth conditions and this was not investigated; and furthermore, coordinate regulation of some of the enzymes might have been occurring (even if not all)—as has previously been concluded for *Pseudomonas* AM1 (Dunstan et al., 1972b). A more strictly valid conclusion from these very useful comparative data is that the enzymes whose onset of induction is simultaneous in *Hyphomicrobium* may be coordinately regulated, and that some but not all of the enzymes tested may be coordinately regulated in *Pseudomonas* AM1 and *Pseudomonas* 3A2 (Table 28).

Similar growth studies with two facultative methane-utilisers (*Methylobacterium organophilum* XX and *Methylobacterium ethanolicum*) by O'Connor (1981) have emphasised further the diversity in regulation at the genetic level in methylotrophs. In *M. organophilum* there was no catabolite repression by succinate whereas in *M. ethanolicum* all the C_1 enzymes studied were repressed during growth on methane plus ethanol. On transfer to methanol from succinate all the C_1 enzymes were induced together in *M. organophilum* perhaps suggesting a single operon, but this was not investigated in *M. ethanolicum*. A summary of regulation at the genetic level in facultative methylotrophs is given in Table 28.

That some enzymes are regulated by way of a C_1-specific operon in the facultative methylamine-utilising *Pseudomonas aminovorans* was suggested by the pattern of increasing specific activities measured during transfer from succinate medium to trimethylamine (Boulton and Large, 1977). Confirmation that such an operon does operate in *Ps. aminovorans* is indicated by the isolation of a number of pleiotropic (regulatory) mutants (Table 25) which have simultaneously lost the capacity to synthesise a number of oxidative enzymes and serine pathway enzymes including isocitrate lyase (Bamforth and O'Connor, 1979).

Regulation of isocitrate lyase synthesis has been investigated further (Bellion and Kim, 1978) in *Pseudomonas* MA which is similar to *Ps. aminovorans* in requiring this enzyme for growth on methylamine as sole source of carbon and energy. *Methylamine* can also be used as a sole nitrogen source during growth in the presence of "preferred" carbon substrates. Study of regulation of the isocitrate lyase is further complicated by the occurrence of two different enzymes, one produced during growth with methylamine and the other with acetate as carbon source (Bellion and Woodson, 1975). Bellion and Kim (1978) have shown that during growth on succinate, glycerol or glucose, the specific activity of isocitrate lyase is no more than 0.75% to 5% of that measured during growth on methylamine as sole source of carbon, nitrogen and energy. When any of these substrates was present during growth with methylamine as the sole nitrogen source, the specific activity of isocitrate lyase (type undetermined) increased to between 1.6%

and 37% of that on methylamine alone. The simplest interpretation of these results is that isocitrate lyase in *Pseudomonas* MA is induced by methylamine (or metabolic derivative), and also subject to catabolite repression by preferred carbon substrates. As is often the case, the extent of catabolite repression was inversely related to the growth rate with the alternative substrate. The best catabolite repressor was glucose, and the least effective was succinate during growth with methylamine as the sole nitrogen source. (N.B. Bellion and Kim (1978) have proposed a more complex interpretation.)

2. Investigations using genetic techniques

(*a*) *Preliminary experiments on gene transfer in methylotrophs.* Any sophisticated investigation of regulation at the genetic level requires techniques for chromosome mobilisation between one organism and another, but the development of any such techniques in methylotrophs has only recently-approached fruition. In the obligate methane-utilisers even the initial requirement of being able to induce mutation is difficult to satisfy; perhaps because they contain unusual DNA repair systems (Harwood *et al.*, 1972; Williams *et al.*, 1977; Williams and Shimmin, 1978). Transposon mutagenesis (Beringer *et al.*, 1978; Fennewald and Shapiro, 1979) might be an alternative method for obtaining mutants of obligate methanotrophs but results from studies of this method are not yet available.

A number of experiments on gene transfer in methylotrophs have used the promiscuous plasmids RP4 (from *Escherichia coli*) and R68.45 (from *Pseudomonas aeruginosa*). Using these plasmids, Warner *et al.* (1980) have successfully transferred antibiotic resistance from *E. coli* and *Ps. aeruginosa* to *Ps. extorquens* and *Methylosinus trichosporium* (but not *Methylococcus capsulatus*). Some indication of the presence of plasmid DNA in the methylotrophs was seen as satellite bands during gradient centrifugation and the antibiotic resistance could be transferred from the methylotrophs into antibiotic-sensitive strains of *E. coli* and *Ps. aeruginosa*. Jayaseelan and Guest (1979) have also shown that antibiotic resistance can be transferred from *Ps. aeruginosa* to three different pink facultative methylotrophs and into *Methylobacterium organophilum* XX using R68.45. Similar results have been obtained by Goodwin (1980) with *Pseudomonas* AM1 using the RP4 plasmid from *E. coli*. In this case the resistance markers were then transferred from the initial recipient into a second strain of *Pseudomonas* AM1. The function of the plasmids detected in wild-type *Pseudomonas* AM1 by Warner *et al.* (1977) is not yet known.

Mobilisation of methylotroph chromosomes by plasmids has not yet been demonstrated but work on gene transfer using hybrid DNA produced *in vitro* has been initiated. Gautier and Bonewald (1980) have shown that the plasmid

R1162 can be transferred from *E. coli* by RP4-induced mobilisation into *Pseudomonas* AM1 thus demonstrating the possibility of cloning experiments using this methylotroph. A hybrid plasmid was made containing R1162 DNA and DNA from *Pseudomonas* AM1 carrying the genetic information for methanol dehydrogenase synthesis. This plasmid was able to complement a *Pseudomonas* AM1 mutant (M15A), lacking methanol dehydrogenase thus enabling it to grow on methanol. It appears that active methanol dehydrogenase (plasmid-encoded) was not formed in *E. coli* and this is likely to be a hindrance to further development of the technique using this enzyme. This lack of activity is probably due to absence of the unusual prosthetic group of the dehydrogenase (PQQ) whose biosynthesis is unlikely to occur in *E. coli*.

A second example of using cloning techniques in the study of methylotrophs has been the successful transfer into *Methylophilus methylotrophus* of the modified RP4 plasmid from *E. coli* carrying the gene for glutamate dehydrogenase (Windass *et al.*, 1980). The modified RP4 plasmids were found to be unstable in *M. methylotrophus* and an alternative method was developed in which RP4 was used to mobilise the pTB70 plasmid carrying the glutamate dehydrogenase gene. The purpose of this work was to increase the cell yield of *M. methylotrophus* on methanol plus ammonia and this aspect of the work is discussed further in Chapter 12.

(b) Gene mapping in the facultative methanotroph, Methylobacterium organophilum XX. This work is described under a separate heading because, so far, it is the only information on the genetics of methylotrophs relating directly to the problem of regulation of gene expression. The gene transformation system first developed by O'Connor *et al.* (1977) has now been used to map a number of mutants of the facultative methanotroph. *M. organophilum* XX. The extent of linkage of the genes has been used together with measurements of enzyme levels in the mutants, to construct a model (Fig. 18) for regulation of some of the enzymes of methanol utilisation in this organism (based on O'Connor *et al.*, 1977; O'Connor and Hanson, 1977, 1978; O'Connor, 1981). The properties of the relevant mutants are described in Table 25. In this model, the genes coding for the enzymes of methanol assimilation, together with that for methanol dehydrogenase, are grouped on a single operon. The gene for cytochrome *c* (also required for methanol oxidation) is encoded separately, but the genetic evidence implies that it is regulated together with the main C_1 operon. The gene for formate dehydrogenase is assumed to be separate from this operon because this enzyme is not regulated coordinately with the other C_1 enzymes. By contrast with *Pseudomonas* AM1, the inducer molecule appears to be methanol itself, because a mutant lacking methanol dehydrogenase is still able to produce high levels of the other C_1 enzymes and of cytochrome *c*. In the presence of methanol, mutants of *M.*

organophilum lacking methanol dehydrogenase contain non-induced levels of formate dehydrogenase, and so it appears that a product of methanol utilisation is essential for induction of this enzyme. It has been suggested that the failure of *M. organophilum* to grow on formate is because this substrate cannot induce the serine pathway enzymes required for its assimilation.

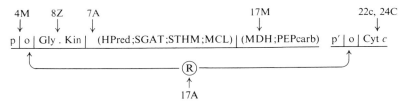

Fig. 18. Model for arrangement of genes coding for C_1 enzymes in *Methylobacterium organophilum* XX. This is based on the work of O'Connor *et al.* (1977); O'Connor and Hanson, (1977, 1978); O'Connor (1981). The properties of the mutants are described in Table 25 (page 108). Gly.Kin., glycerate kinase; HPred, hydroxypyruvate reductase; SGAT, serine-glyoxylate aminotransferase; STHM, serine transhydroxymethylase; MCL, malyl-CoA lyase; MDH, methanol dehydrogenase; PEP carb., PEP carboxylase.

B. Regulation at the level of enzyme activity

1. Introduction

The main factors controlling the fluxes of carbon compounds through the oxidative and assimilatory pathways are the availability of substrate, the removal of products, the availability of cofactors (particularly ATP, ADP, $NAD(P)^+$, $NAD(P)H$, coenzyme A and tetrahydrofolate) and the specific activation and inhibition of regulatory enzymes. The balance of flux through the following pathways must be considered:
(1) The formation of formaldehyde from the C_1 substrate;
(2) The further oxidation of formaldehyde to CO_2 yielding $NAD(P)H$ and reduced dehydrogenases;
(3) The oxidation of these reducing equivalents by an electron transport chain coupled to provide ATP;
(4) The assimilation of formaldehyde and CO_2 into "central precursors" for biosynthesis using $NAD(P)H$ and ATP.

The relative concentrations of ATP/ADP and $NAD(P)^+/NAD(P)H$ will be of considerable importance in regulating these fluxes. High rates of biosynthesis will be associated with high rates of utilisation of ATP and $NAD(P)H$, and thus high rates of production of ADP and $NAD(P)^+$. In turn, the rates of oxidation of substrates by way of NAD^+-linked dehydrogenases, and of electron transport coupled to phosphorylation of ADP to ATP, will be increased and hence able to supply the increased demand for ATP and NAD(P)H. The difficulty of understanding even this apparently straightforward relationship of fluxes and how it is controlled has been discussed in a comprehensive review of the regulation of respiration rate in growing bacteria by Harrison (1976). In many methylotrophs, consideration of this aspect of regulation is even more difficult than in typical heterotrophs because of their unusually high NAD(P)H requirement for carbon assimilation, and because of the prevalence of unusual dehydrogenases such as methanol dehydrogenase which is NAD^+-independent and is coupled to the electron transport chain at the level of cytochrome c. Because of this, the rate of oxidation of methanol to formaldehyde will depend on the concentration of oxidised cytochrome c, and not on the NAD^+ concentration. Furthermore, the first step in the oxidation of methane and methylated amines for eventual provision of NAD(P)H and ATP will often require NADH for hydroxylation reactions.

It is worth emphasising here the peculiar importance of formaldehyde, a potentially lethal metabolite reacting nonenzymically with many metabolites and polymers in living organisms. During growth on methane or methanol it is the product of methanol dehydrogenase, but it is also a substrate for this enzyme, which in some conditions can oxidise methanol to formate without the formation of detectable formaldehyde. Whether or not this means that the formaldehyde remains bound to the same active site between the two sequential oxidations is unknown; if it is, it may have to become hydrated before being oxidised to formate. If, on the other hand, it is released then it will compete with methanol for the active site of the dehydrogenase. Clearly some kind of regulation of this is likely to be necessary. After its formation, formaldehyde can be either oxidised or assimilated. Consideration of this important branch point in metabolism is unfortunately likely to remain difficult as long as the route for formaldehyde oxidation remains as unclear as it is at present.

After its non-enzymic formation from formaldehyde and tetrahydrofolate, methylene tetrahydrofolate may be used by the serine transhydroxymethylase or be oxidised by a dehydrogenase (Chapter 6). It has been suggested by Harder and Attwood (1978) that methylene tetrahydrofolate is a key intermediate in regulation, assimilation being favoured when all available tetrahydrofolate is bound to formaldehyde; only then will free formaldehyde become available for oxidation to CO_2 to provide the necessary NADH and ATP for the serine pathway, whose operation liberates the tetrahydrofolate for further condensation with formaldehyde.

Having stated the problem in general terms there is little merit in further discussion until more data are available. The most useful data would include the concentrations and ratios of $NAD(P)^+/NAD(P)H$, ATP/ADP/AMP, coenzyme A and tetrahydrofolate and derivatives during growth under a variety of continuous culture conditions. Such studies are necessary to supplement our fragmentary knowledge of specific regulatory enzymes, and to obtain a clear picture of the regulation of the serine pathway.

Besides regulation of pathways by availability of substrates, cofactors etc., regulation of metabolism also occurs by way of activation or inhibition of specific regulatory enzymes by effector molecules (allosteric effectors), not related structurally to the substrates or products of the enzymes. If an enzyme catalyses a reaction that is very close to equilibrium, a change in catalytic activity cannot regulate the flux through the reaction or the overall metabolic pathway. Hence regulatory enzymes usually catalyse "non-equilibrium" reactions; that is, the enzymes are insufficiently active to bring their substrates and products to equilibrium (or near to it), and their low catalytic activity will limit the flux through the pathway. Although not necessarily so, the non-equilibrium reactions that are important in regulation are often functionally irreversible and associated with a high negative free energy change. The activity of "non-equilibrium" enzymes (and their pathways) could be regulated by changes in substrate concentration, but a more limited and more usual definition of a regulatory enzyme is an enzyme catalysing a "non-equilibrium" reaction and whose activity is controlled by factors other than the substrate concentration. Such enzymes are characterised by their sensitivity to metabolites other than their substrate and reaction products, by their complex kinetics with respect to substrates, activators and inhibitors, and by their possession of a complex quaternary structure. (See Newsholme and Start (1973) for a comprehensive review of fundamental aspects of the regulation of metabolic pathways.)

There is remarkably little known about regulatory enzymes in the serine pathway, and what there is has been gleaned from a variety of different organisms. It is clearly hazardous to extrapolate conclusions from one organism to another; investigations of regulation in other groups of bacteria have shown that there is as great a diversity of regulatory systems as of actual biochemical pathways. It is also worth noting that the field of metabolic regulation is one in which complex and imaginative theories are relatively easier to produce than facts.

2. Regulatory enzymes of the serine pathway

Relatively few serine pathway enzymes have been suggested as being especially involved in regulation of assimilation. The first of these is the serine transhydroxymethylase from the facultative methanotroph *Methylobacterium organophilum* XX (O'Connor and Hanson, 1975). This tetrameric enzyme is activated by glyoxylate, and the significance of this observation is amplified by the observation that the equivalent isoenzyme formed during growth on succinate is a monomer and not activated by glyoxylate. Similar activation has been observed in *Pseudomonas* AM1, *Pseudomonas* 3A2 and *Hyphomicrobium* X (McNerney and O'Connor, 1980). Regulation of assimilation by glyoxylate is feasible because high concentrations will indicate a high potential for carbon assimilation and vice versa. Activation of serine transhydroxymethylase by glyoxylate is thus a suitable way of stimulating the operation of the serine pathway.

A second example of a regulatory enzyme in a serine pathway organism is the PEP carboxylase of *Pseudomonas* MA (Newaz and Hersh, 1975). This is a tetrameric enzyme showing complex allosteric kinetics, being activated by NADH and inhibited by ADP (see description of the enzyme, above). (The PEP carboxylases of methylotrophs usually differ from other PEP carboxylases in not being activated by acetyl-CoA.) When concentrations of NADH and ATP are high (and ADP is low) then sufficient energy will be available for biosynthesis by way of the serine pathway, and the PEP carboxylase will have a high activity. Conversely, if the concentration of NADH is low and ADP is high, then the activity of PEP carboxylase will be limited and the rate of operation of the serine pathway will be diminished, thus increasing the relative flow of formaldehyde into catabolism for the production of "energy". This control mechanism emphasises or amplifies the "general" regulatory effects of the ratios of $NADH/NAD^+$ and ATP/ADP discussed above.

This outline of the regulatory function of the PEP carboxylase during methylotrophic growth of *Pseudomonas* MA was first proposed by Newaz and Hersh (1975), who suggested that the NADH-activation of the enzyme indicated a novel route for formaldehyde oxidation in this organism. However, as described above, the regulatory characteristics of the PEP carboxylase make sense without their (perhaps unnecessary) complex proposals with respect to formaldehyde oxidation. These proposals arose because they were unable to detect a "separate enzyme system in *Pseudomonas* MA capable of producing energy by the direct oxidation of methylamine". "Therefore both energy production and carbon assimilation appear to proceed via common intermediates". They have suggested two possible energy-yielding pathways diverging from the main serine pathway, both involving pyruvate kinase and pyruvate instead of pyruvate carboxylase and oxaloacetate. In one scheme the pyruvate is oxidised by pyruvate dehydrogenase to acetyl-CoA and thence to glyoxylate by the enzymes of the glyoxylate cycle. In the alternative scheme the pyruvate is carboxylated to malate, and malyl-CoA derived from this is cleaved to glyoxylate plus acetyl-CoA which is then oxidised to CO_2 by the TCA cycle. Thus in both schemes, instead of carboxylation of PEP leading to biosynthesis, an overall decarboxylation occurs which merely leads to regeneration of glyoxylate with overall production of reducing equivalents and ATP. It is suggested that the activation by NADH of the PEP carboxylase diverts metabolism from one of these "oxidative bypasses" (when NADH levels are high) into the biosynthetic serine pathway.

Although interesting in concept, and maybe true, there is at present little evidence supporting either of these schemes, except that the enzymes are present (as they are in other methylotrophs). The evidence for the need for such a pathway has not been published and the regulation of PEP carboxylase makes sense regardless of the route for formaldehyde oxidation. A demonstration that such a pathway cannot be operating could be obtained by isolation of mutants lacking pyruvate dehydrogenase and/or 2-oxoglutarate dehydrogenase, as done for *Pseudomonas* AM1 which requires neither of these enzymes for growth on methanol (Taylor and Anthony, 1976a; Bolbot and Anthony, 1980b).

A third enzyme whose regulation should be mentioned is citrate synthase. In facultative methylotrophs this enzyme is inhibited by NADH, as found in other heterotrophs using the TCA cycle as a major source of NADH (Colby and Zatman, 1975b; Anthony and Taylor, 1975). This regulation of citrate synthase occurs even during methylotrophic growth when the TCA cycle functions in a solely biosynthetic capacity (Chapter 5). In bacteria having the icl$^+$-serine pathway (e.g. *Pseudomonas* MA) citrate synthase is essential during methylotrophic growth for the oxidation of acetyl-CoA to glyoxylate. In such organisms the enzyme might not be inhibited by NADH, as this would tend to counteract the positive effect of NADH on an earlier enzyme of the pathway.

5

The TCA cycle and growth of methylotrophic bacteria on multicarbon compounds

I. The operation of the TCA cycle in methylotrophs	137
II. The basis of obligate methylotrophy	142
III. The enzymes of the TCA cycle and their regulation	144
A. Citrate synthase	144
B. Isocitrate and malate dehydrogenases	145
C. Pyruvate and 2-oxoglutarate dehydrogenases	146
IV. The growth of *Pseudomonas* AM1 on ethanol, lactate and 3-hydroxybutyrate	147
V. The growth of *Pseudomonas* AM1 on propane 1,2-diol and 4-hydroxybutyrate	150
VI. The metabolism of trimethylsulphonium salts by *Pseudomonas* MS	151

This chapter discusses some aspects of the metabolism of methylotrophs that are germane to their methylotrophic metabolism but have no other place to go.

I. The operation of the TCA cycle in methylotrophs

The enzymes of the tricarboxylic acid (TCA) or Krebs cycle have two potential functions. The first is to catalyse, by a cyclic sequence of reactions, the oxidation of acetyl-CoA, providing NADH and then ATP for biosynthesis. The second function is to provide carbon precursors for biosynthesis. For this function 2-oxoglutarate dehydrogenase is unnecessary, but oxaloacetate must be replenished either by carboxylation of a C_3 compound or by the glyoxylate bypass or its equivalent.

During methylotrophic growth only the second (biosynthetic) function is essential but the complete cycle, including 2-oxoglutarate dehydrogenase, must operate if an organism is to grow on most multicarbon compounds.

Evidence for or against operation of the cycle in methylotrophs is based on enzyme measurements and on radioactive labelling experiments. Table 29 summarises work on the operation of the cycle in individual methylotrophs and the general conclusions derived from this work are listed below:

TABLE 29
The distribution of the TCA cycle in methylotrophs

Obligate methane-utilisers (Type I)
These have the RuMP pathway and an incomplete TCA cycle. Besides lacking 2-oxoglutarate dehydrogenase, the specific activities of succinate dehydrogenase, aconitase and malate dehydrogenase are usually low (malate dehydrogenase is very low in *M. capsulatus*).

Methylococcus sp.	Davey et al. (1972); Patel et al. (1969, 1975, 1977); Hazeu et al. (1980)
Methylomonas sp.	Davey et al. (1972); Trotsenko (1976)
Methylobacter sp.	Trotsenko (1976)

Obligate methane-utilisers (Type II)
These have the serine pathway and a complete TCA cycle.

Methylosinus sp.	Davey et al. (1972); Trotsenko (1976)
Methylocystis sp.	Davey et al. (1972); Trotsenko (1976)
Methanomonas methanooxidans	Wadzinski and Ribbons (1975b)

Facultative methane-utilisers (Type II)
These have the serine pathway and a complete TCA cycle.

Methylobacterium organophilum XX	Patt et al. (1974)
Methylobacterium R6	Patel et al. (1978b)
M. ethanolicum and *M. hypolimneticum*	Lynch et al. (1980)
(only 2-oxoglutarate dehydrogenase was measured)	

Obligate methylotrophs unable to use methane
These all have the RuMP pathway and an incomplete TCA cycle. Besides lacking 2-oxoglutarate dehydrogenase, the specific activities of aconitase, fumarate hydratase and the dehydrogenases for succinate and malate are also usually low.

Organism 4B6	Colby and Zatman (1972, 1975d)
Organism C2A1	Colby and Zatman (1975a, d)
Pseudomonas W6	Babel and Hoffman (1975); Hoffman and Babel (1980)
Pseudomonas W1	Dahl et al. (1972)
Methylophilus methylotrophus	Taylor (1977); Large and Haywood (1981)
Pseudomonas C	Ben-Bassat and Goldberg (1977)

More restricted facultative methylotrophs

These *Bacillus* sp. have the **RuMP** pathway and zero or negligible activities of 2-oxoglutarate dehydrogenase. The specific activities of other TCA cycle enzymes are similar to those in the facultative methylotrophs except for the low levels of isocitrate dehydrogenase in these *Bacillus* sp.

Bacillus sp. **S2A1**	Colby and Zatman (1975a, d)
Bacillus sp. **PM6**	Colby and Zatman (1975a, d)

Less restricted facultative methylotrophs

These are exactly as described for the obligate methylotrophs unable to use methane (above); they have the **RuMP** pathway and an incomplete TCA cycle.

Organism **W3A1**	Colby and Zatman (1975a, d)
Organism **W6A**	Colby and Zatman (1975a)

The Hyphomicrobia

These restricted facultative methylotrophs have the serine pathway and a complete TCA cycle. They resemble the majority of facultative methylotrophs (below) in having high levels of citrate synthase and malate dehydrogenase but they differ in having rather low levels of succinate dehydrogenase and isocitrate dehydrogenase; they have no pyruvate dehydrogenase (Attwood and Harder, 1974; Harder and Attwood, 1975, 1978).

Facultative methylotrophs unable to use methane

These bacteria all have a complete TCA cycle, the enzymes of which are induced to higher levels during growth on multicarbon compounds compared with those measured during methylotrophic growth. The majority have the serine pathway.

TABLE 29 (*continued*)

Bacteria with the serine pathway

Organism 5B1	Colby and Zatman (1972, 1975a, d)
Pseudomonas 3A2	Colby and Zatman (1972, 1975d)
Pseudomonas MA	Bellion and Hersh (1972)
"*Diplococcus*" PAR	Bellion and Kim (1979)
Pseudomonas AM1	Dunstan and Anthony (1973); Taylor and Anthony (1976a); Bolbot and Anthony (1980a)
Pseudomonas methylica	Babel and Müller-Kraft (1979)
Pseudomonas 1	Ben-Bassat and Goldberg (1977)

Bacteria with the RuMP pathway

Pseudomonas oleovorans	Babel and Muller-Kraft (1979); Loginova and Trotsenko (1979c)
Arthrobacter globiformis B-175	Loginova and Trotsenko (1976a)
Arthrobacter P1	Levering *et al.* (1981a)

Bacteria with the RuBP pathway (facultative autotrophs)

Achromobacter 1L	Loginova and Trotsenko (1979c)
Pseudomonas 8	Loginova and Trotsenko (1979c)
Mycobacterium 50	Loginova and Trotsenko (1979c)

A number of useful discussions of this topic have been published in reviews: Ribbons *et al.* (1970); Kelly (1971); Smith and Hoare (1977); Whittenbury and Kelly (1977); Zatman (1981).

(a) All facultative methylotrophs have a complete TCA cycle.
(b) Most obligate methylotrophs have an incomplete cycle, the exception being the Type II obligate methanotrophs; these use the serine pathway and have a complete TCA cycle.
(c) In all bacteria having the serine pathway the TCA cycle is complete.
(d) In all obligate methylotrophs having the RuMP pathway the TCA cycle is incomplete.
(e) The only tested facultative methylotroph having the RuMP pathway has a complete TCA cycle.
(f) The facultative methylotrophs that are also facultative autotrophs and have the RuBP pathway all have a complete TCA cycle.
(g) In bacteria lacking a complete cycle 2-oxoglutarate dehydrogenase is always absent.
(h) In the obligate methylotrophs and more restricted facultative methylotrophs which lack a complete cycle, the specific activities of the other TCA cycle enzymes are also very low (or zero).
(i) In the less restricted facultative methylotrophs (*Bacillus* sp.), 2-oxoglutarate dehydrogenase is absent and the specific activity of isocitrate dehydrogenase is usually low, but the levels of some other TCA cycle enzymes are high.
(j) Although the *Hyphomicrobia* have the potential for synthesis of all the TCA cycle enzymes, the specific activities of some of these are lower than in the typical facultative methylotrophs (e.g. the dehydrogenases for 2-oxoglutarate, succinate and isocitrate).
(k) The specific activities of the TCA cycle enzymes in methanotrophs having the serine pathway are usually slightly lower than in typical facultative methylotrophs measured during growth on C_1 compounds.
(l) In the facultative methylotrophs, the TCA cycle enzymes are present at higher specific activities during growth on multicarbon compounds than during methylotrophic growth.

In the presence of a complete TCA cycle, radioactive acetate should be oxidised completely to CO_2, it should be incorporated into all classes of compound in the cell and the C-1 and C-5 atoms of glutamate should become labelled during incubation with 1-^{14}C acetate. All these predictions have been confirmed in those bacteria whose enzymes have indicated the presence of an operational TCA cycle. These include the obligate methanotroph, *Methanomonas methanooxidans* (Wadzinski and Ribbons, 1975b), the less restricted facultative methylotroph *Bacillus* PM6 (Colby and Zatman, 1975d) and the typical facultative methylotrophs *Pseudomonas* AM1 (Dunstan *et al.*, 1972a; Dunstan and Anthony, 1973; Taylor and Anthony, 1976b) and organism 5B1, *Arthrobacter* 2B2 and *Pseudomonas* 3A2 (Colby and Zatman, 1975d).

By contrast, in an organism lacking 2-oxoglutarate dehydrogenase, radioactive acetate should not be oxidised to CO_2; it should be assimilated (only in the presence of an added energy source) predominantly into lipids and into amino acids derived directly from acetyl-CoA and 2-oxoglutarate (leucine, glutamate, arginine and proline); only the C-5 atom of glutamate should be labelled. All these predictions have been confirmed in bacteria lacking 2-oxoglutarate dehydrogenase and thus lacking an operational TCA cycle. These include the obligate methanotroph *Methylococcus capsulatus* (Patel *et al.*, 1975; Patel and Hoare, 1971), the obligate methanol-utilisers *Pseudomonas* W1 (Dahl *et al.*, 1972) and organism C2A1 (Colby and Zatman, 1975d), the obligate amine-utilising organism 4B6 (Colby and Zatman, 1975d) and the more restricted facultative methylotroph, organism W3A1 (Colby and Zatman, 1975d).

II. The basis of obligate methylotrophy

The obligate methylotrophs are nutritionally analogous to the obligate autotrophs and it is an obvious speculation that there might be a common basis for their lack of nutritional versatility. It appears, however, that there is no single basis for this amongst the methylotrophs and autotrophs (Smith and Hoare, 1977) and this should not be too surprising; it is perhaps to be expected that ability to achieve an objective (metabolic or otherwise) might depend on relatively few well-defined characteristics, whereas the diversity of reasons for failure is considerable. All the same, the search for a common basis for the specialist physiology of obligate autotrophs and methylotrophs is interesting in itself and particularly valuable reviews of this topic have been published by Kelly (1971), Smith and Hoare (1977) and Whittenbury and Kelly (1977). In essence the question is: why are many autotrophs and some methylotrophs unable to use multicarbon compounds as their sole source of carbon and energy? Possible reasons that have been proposed for this inability are:

(a) The presence of a lesion in a central metabolic pathway, the most likely being the absence of 2-oxoglutarate dehydrogenase.
(b) The lack of a system for coupling ATP synthesis to NADH oxidation.
(c) The lack of transcriptional control mechanisms.
(d) Toxicity of multicarbon compounds or products of their metabolism.
(e) The lack of suitable transport systems for multicarbon compounds.

The last two of these proposals are likely to be involved in limiting the nutritional ranges of all types of bacteria and certainly may be the reason for lack of growth of some obligate methylotrophs on some substrates. The electron transport chains of all methylotrophs studied are able to oxidise

NADH, and respiration with a number of substrates whose oxidation is coupled to NADH formation leads to proton translocation. Respiration-coupled ATP synthesis has not been demonstrated in extracts of obligate methylotrophs but this is probably for technical reasons and is unlikely to indicate a lack of such coupling during growth.

The only basis for obligate methylotrophy that might be generally applicable is the lack of 2-oxoglutarate dehydrogenase and a complete TCA cycle, but even this is not straightforward or all-embracing. As described above (Table 29) most obligate methylotrophs lack a complete TCA cycle. Furthermore, it has been shown that loss of 2-oxoglutarate dehydrogenase by mutation of a typical facultative methylotroph (*Pseudomonas* AM1) is sufficient to confer upon the mutant the property of obligate methylotrophy (Taylor and Anthony, 1976a). That there must be further reasons for obligate methylotrophy in the Type II methanotrophs (at least) is clear, because these bacteria do possess a complete TCA cycle. However, the specific activities of the TCA cycle enzymes are similar during growth on methane to those measured in typical facultative methylotrophs growing on C_1 compounds; they are very much lower than the activities induced during growth of these bacteria on multicarbon compounds. It thus remains possible that the failure to grow on multicarbon compounds may sometimes be because such compounds cannot induce the enzymes of the TCA cycle to sufficiently high levels for it to subserve both oxidative and biosynthetic functions. It should be noted that there may be more than one reason for obligate methylotrophy in the Type II methanotrophs. Besides the low levels of TCA cycle enzymes, they have low (or zero) levels of pyruvate dehydrogenase which is essential for growth on most compounds with 3 or more carbon atoms and it is not known whether they have the enzymes for conversion of C_4 compounds to C_3 compounds essential for growth on C_4 compounds such as succinate.

In this context the growth of restricted facultative methylotrophs on some multicarbon compounds is relevant. The more restricted organisms have no TCA cycle and yet grow very slowly on glucose whereas the less restricted organisms have low activities of the TCA cycle enzymes and are able to grow on a few multicarbon compounds besides their C_1 substrates. The *Hyphomicrobia* can also be considered to be restricted facultative methylotrophs; they can grow only on C_1 compounds or on compounds metabolised directly by way of acetyl-CoA; that is ethanol, acetate and 3-hydroxybutyrate. It has been demonstrated conclusively that these bacteria lack pyruvate dehydrogenase and it has been suggested that this is sufficient reason for their failure to grow on most C_3, C_4, C_5, and C_6 compounds (Harder *et al.*, 1975). The growth properties of a pyruvate dehydrogenase mutant of *Pseudomonas* AM1 have shown that loss of this enzyme is sufficient to confer upon a previously facultative methylotroph the properties of a restricted facultative methylotroph similar to the *Hyphomicrobia* (Bolbot and Anthony, 1980a).

III. The enzymes of the TCA cycle and their regulation

A. Citrate synthase

This enzyme catalyses the first condensation in the TCA cycle and is essential whether the cycle operates in the oxidative or biosynthetic mode. It is also an essential enzyme of the icl$^+$-serine pathway, being involved in the oxidation of acetyl-CoA to glyoxylate. Citrate synthases vary markedly with respect to their size and regulatory properties (Weitzmann and Danson, 1976). The smaller enzymes are inhibited by relatively low concentrations of ATP and are found in Gram-positive bacteria, yeasts, animals and higher plants. The larger enzymes found in Gram-negative bacteria and blue-green bacteria are regulated by NADH or 2-oxoglutarate or both. In bacteria having an incomplete TCA cycle the citrate synthase is inhibited by 2-oxoglutarate as an "end product" feedback inhibitor. By contrast, in bacteria in which a complete oxidative TCA cycle operates the citrate synthase is inhibited by the "end product" of the cycle, NADH. In the enteric bacteria, the cycle is usually complete during aerobic growth but is incomplete during aerobic growth with glucose, or during anaerobic growth; in these bacteria the citrate synthase is inhibited by both 2-oxoglutarate and NADH, the inhibition by NADH being unaffected by AMP. By contrast, the inhibition by NADH of the citrate synthases of most strictly aerobic Gram-negative bacteria is relieved by AMP.

The citrate synthases of a range of different methylotrophs investigated by Colby and Zatman (1975b) fit logically into this pattern. None were inhibited markedly by low concentrations of ATP (even that of the *Bacillus* sp.), although all were inhibited by higher amounts of both ATP and ADP. It might be expected that those methylotrophs having only the biosynthetic pathway would be sensitive to 2-oxoglutarate and indeed three of the obligate (or near obligate) methylotrophs (organisms 4B6, C2A1 and W3A1) were all inhibited by fairly high concentrations of 2-oxoglutarate (50% inhibition by about 10 mM). Colby and Zatman (1975b) have suggested, however, that this inhibition may be physiologically insignificant because of the high oxoglutarate concentration required and because the low concentrations of many of the other TCA cycle enzymes is likely to preclude oxoglutarate accumulation in these organisms. Neither of the types of obligate methanotrophs tested (*Methylomonas albus* and *Methylosinus trichosporium*) has oxoglutarate-sensitive citrate synthase and this is also true of the methylotrophs having complete TCA cycles (*Bacillus* PM6, organism 5B1 and *Pseudomonas* 3A2).

It appears that the only regulation of citrate synthase likely to be physiologically important in methylotrophs is the marked inhibition by NADH in those bacteria having high levels of the TCA cycle enzymes and in which a complete cycle operates as a major route for energy production; that is, in the typical facultative methylotrophs unable to grow on methane (the enzyme from facultative methanotrophs has not been investigated).

The facultative methylotroph *Pseudomonas oleovorans* is a facultative methanol-utiliser able to grow by way of the RuMP pathway. It appears to be able to produce two different citrate synthases; one during growth on acetate (inhibited by NADH) and the other produced at low specific activity during growth on methanol (insensitive to NADH) (Babel and Müller-Kraft, 1979).

The typical facultative methylotroph *Pseudomonas* AM1 has a typical large citrate synthase with a molecular weight between 250 000 and 300 000 (Anthony and Taylor, 1975; Cox and Quayle, 1976a) and it is allosterically inhibited by NADH, this inhibition being relieved by AMP (Anthony and Taylor, 1975; Taylor, 1976). It is not inhibited by 2-oxoglutarate, citrate, isocitrate, glutamate, succinate, malate, fumarate, NAD^+, $NADP^+$, NADPH, ATP or ADP (all at 5 mM). NADH is the only compound tested having an inhibitory effect although the K_i value for NADH (1mM) is much higher than recorded with citrate synthases of other bacteria (Weitzmann and Danson, 1976). Inhibition by NADH is competitive with respect to oxaloacetate, the K_m values for oxaloacetate being 10μM in the absence of NADH and 45 μM in its presence (1mM). Kinetics with respect to oxaloacetate are hyperbolic, the Hill coefficient for this substrate being almost 1·0. The kinetics with respect to acetyl-CoA are markedly sigmoidal in the presence of NADH, becoming hyperbolic in its absence. The Hill coefficient and K_m for acetyl-CoA were 1·8 and 220 μM respectively in the presence of NADH (1mM) and 0 8 and 124 μM in its absence. The 90% inhibition of citrate synthase by 3 mM NADH (measured with 100 μM acetyl-CoA) was greatly diminished (to 15%) by 0·3 mM ATP. The regulation of citrate synthase is similar in ethanol-grown and succinate-grown cells and was unchanged in the "artificial" obligate methylotroph (mutant 1CT41) derived from *Pseudomonas* AM1 (Anthony and Taylor, 1975; Taylor, 1976).

B. Isocitrate and malate dehydrogenases

In obligate methylotrophs having the RuMP pathway the activity of isocitrate dehydrogenase is usually coupled to reduction of NAD^+ although in *Pseudomonas* W6 there appear to be two isocitrate dehydrogenases, one using NAD^+ and the other $NADP^+$ (Hoffmann and Babel, 1980). *Pseudomonas oleovorans* is the only facultative methylotroph tested having the RuMP pathway, and high levels of an NAD^+-linked isocitrate dehydrogenase are produced during growth of this organism. In bacteria with the serine pathway, isocitrate dehydrogenase activity is coupled to synthesis of NADPH. This is presumably to provide reductant for biosynthesis which is achieved in bacteria having the RuMP pathway by $NADP^+$-linked glucose phosphate and phosphogluconate dehydrogenases. The $NADP^+$ specific isocitrate dehydrogenase of *Pseudomonas* AM1 was 90% inhibited in a concerted fashion by a combination of glyoxylate (1 mM) plus oxaloacetate (1 mM) (Taylor, 1976). Separately these compounds inhibited the enzyme by only 25% and less than 10% inhibition was measured with NADH, 2-oxoglutarate or malate (all at 1 mM); and none of these compounds increased the inhibition observed with glyoxylate (1 mM) or oxaloacetate (1 mM). These results are essentially similar to those observed with the $NADP^+$-specific enzymes from a number of different bacteria and from pig heart by Shiio and Ozaki (1968). The regulatory significance (if any) of these observations is unclear.

The malate dehydrogenases of nearly all methylotrophs are more active with $NADP^+$ than NAD^+. Specific activities of this dehydrogenase are usually much higher in typical facultative methylotrophs than in obligate methylotrophs.

C. Pyruvate and 2-oxoglutarate dehydrogenases

These complexes are large multimeric structures that catalyse the oxidative decarboxylation of oxoglutarate and pyruvate with the formation of succinyl-CoA and acetyl-CoA respectively (see Guest, 1978, for a review of the gene–enzyme relationships and subunit composition of these enzyme complexes). The dehydrogenases consist of at least 3 components. Usually there are E1 (dehydrogenase) and E2 (transacylase) components specific for each oxoacid, and a common E3 (lipoamide dehydrogenase) component.

As far as I know the only work characterising these enzymes in methylotrophs is that done in my laboratory (Taylor and Anthony, 1976a; Bolbot and Anthony, 1980b) and this work indicates that they are not markedly different from those in other bacteria. Two technical points of note are that we found the E2 (transacylase) components impossible to assay and that the E1 components appeared to sediment on centrifugation together with the overall complex but no seperate E1 activity could be detected in the resuspended pellet. A similar result was obtained with the pyruvate dehydrogenase complex of *E. coli* by Dietrich and Henning (1970). These workers found that addition of the supernatant to the resuspended pellet led to reactivation of the enzyme in the ferricyanide assay for the E1 component and they concluded that an unknown factor was involved as well as the usual 3 components of the complex. This could not be repeated with the *Pseudomonas* AM1 system and it appears that both the E1 and E2 components are unusual in being difficult to assay when present as part of the complex. Because of these problems, biochemical analysis of our 13 mutants lacking the dehydrogenases were not entirely conclusive (Bolbot and Anthony, 1980b). One mutant lacked the E1 component of the oxoglutarate dehydrogenase complex. None lacked the E3 component and it was concluded that the others had inactive E2 (transacylase) components. Some of these lacked overall pyruvate dehydrogenase activity and others, the oxoglutarate dehydrogenase activity, indicating that there are specific E2 components for each dehydrogenase. One mutant lacked both dehydrogenase activities and this might have been a double mutant. Reconstitution experiments such as those described for the *E. coli* enzyme by Guest and Creaghan (1973) were unsuccessful.

The growth properties of the mutants lacking 2-oxoglutarate dehydrogenase (growth only on C_1 compounds or oxalate) are similar to the obligate methylotrophs; and the growth properties of mutants lacking pyruvate dehydrogenase (growth only on C_1 and C_2 compounds and 3-hydroxybutyrate) are similar to those of the *Hyphomicrobia*.

IV. The growth of *Pseudomonas* AM1 on ethanol, lactate and 3-hydroxybutyrate

This topic is included here because it impinges upon the least understood part of the serine pathway for assimilation of C_1 compounds. During growth of typical heterotrophs on acetate, ethanol or 3-hydroxybutyrate these compounds are converted first to acetyl-CoA and then to malate by the glyoxylate bypass. Phosphoenolpyruvate (PEP) is formed from oxaloacetate by PEP carboxykinase and carbohydrates are produced from the PEP. During growth on C_4 compounds such as succinate the same PEP carboxykinase is involved in production of pyruvate and acetyl-CoA from the C_4 substrate. During growth on lactate or pyruvate, C_4 compounds are produced directly by carboxylation of pyruvate, or PEP is produced directly from pyruvate by way of PEP synthetase and then oxaloacetate produced by carboxylation of the PEP. These various metabolic relationships are described in Chapter 3 (Fig. 13).

Pseudomonas AM1 possesses PEP carboxykinase for conversion of C_4 to C_3 compounds during growth on C_4 compounds (Salem *et al.*, 1973a) but the route for biosynthesis from C_2 and C_3 compounds is unusual and not yet fully understood. The first remarkable aspect of the metabolism of *Pseudomonas* AM1 in this context is that it lacks isocitrate lyase and so must have an alternative route for assimilation of acetyl-CoA derived from ethanol or 3-hydroxybutyrate (for unknown reasons this organism grows poorly on acetate itself although closely related bacteria do grow on it) (Large and Quayle, 1963; Dunstan *et al.*, 1972b; Chapter 4).

148   The biochemistry of methylotrophs

The route for growth of *Pseudomonas* AM1 on pyruvate or lactate is also mysterious (Salem *et al.*, 1973a); there are no enzymes for direct phosphorylation of pyruvate to PEP, and pyruvate carboxylase is absent. The PEP carboxylase which functions in the serine pathway cannot be involved in the formation of C_4 compounds during growth on pyruvate in the absence of a route for PEP synthesis from the growth substrate. Experiments on the metabolic fate of radioactive C-1 and C-3 atoms of lactate showed conclusively that all assimilation of pyruvate occurs after its decarboxylation to acetyl-CoA; there is no direct assimilation route by which the carboxyl (C-1) group of lactate is assimilated into cell material. Thus *Pseudomonas* AM1 appears to use the same route for assimilation of pyruvate and lactate as it does for assimilation of ethanol and 3-hydroxybutyrate.

The isolation of a number of mutants unable to grow on C_1, C_2 or C_3 compounds, except in the presence of supplements of glycollate or glyoxylate, suggested that the same route for the oxidation of acetyl-CoA to glyoxylate is essential for growth on all these compounds (Dunstan *et al.*, 1972b; Dunstan and Anthony, 1973; Salem *et al.*, 1973a; Bolbot and Anthony, 1980a, b). Radioactive tracer experiments have shown that during growth on C_2 compounds a route exists for the rapid metabolism of acetate to malate and glycollate (probably derived from glyoxylate) (Dunstan *et al.*, 1972a; Dunstan and Anthony, 1973; Taylor and Anthony, 1976b). The results are all consistent with the operation of a malate synthase pathway analogous to the glyoxylate cycle (see Anthony, 1975a). This novel route consists essentially of the oxidation of acetyl-CoA to glyoxylate followed by condensation of the glyoxylate with a second molecule of acetyl-CoA in a reaction catalysed by malate synthase (see Fig. 19). In an attempt to confirm the role of malate synthase in *Pseudomonas* AM1, a mutant was obtained which was selected for its ability to grow on malate but failure to grow on a mixture of acetate plus glyoxylate. This mutant was found to lack malate synthase activity and, as expected, was unable to grow on ethanol, malonate or 3-hydroxybutyrate (Taylor and Anthony, 1976b).

This appeared to confirm the operation of the malate synthase pathway, but this apparent success was short-lived. It transpires that the malate synthase activity is not due to a single enzyme but is a result of the sequential activity of malyl-CoA lyase (acting in the opposite direction to that necessary during methylotrophic growth) plus a novel malyl-CoA hydrolase (Cox and Quayle, 1976a). Partial purification of this hydrolase has shown that it is distinct from citrate synthase which also hydrolyses malyl-CoA. The malyl-CoA hydrolase hydrolyses other CoA esters (of succinate, 3-hydroxybutyrate and acetate) at fairly similar rates but the K_m for malyl-CoA (7 μM) is 10- to 150-fold lower than for these other substrates. The mutant lacking malate synthase activity (ICT51) has a greatly increased K_m for malyl-CoA (about

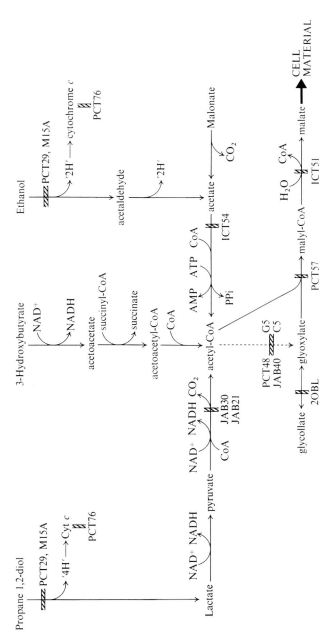

Fig. 19. Pathways for the assimilation of ethanol, malonate, pyruvate, lactate, propane 1,2-diol and 3-hydroxybutyrate in *Pseudomonas* AM1. The hatched bars indicate metabolic lesions in mutants: PCT29 and M15A lack methanol dehydrogenase; PCT 76 lacks cytochrome *c*; JAB21 and JAB30 lack pyruvate dehydrogenase; 20BL lacks hydroxypyruvate reductase; ICT54 lacks acetyl-CoA synthetase; PCT57 lacks malyl-CoA lyase; ICT51 lacks malyl-CoA hydrolase activity at normal substrate concentrations. The pathway for growth of *Pseudomonas* AM1 on C_1 compounds is given in Fig. 14. References to mutants are in Table 24.

200 μM) and hence a much lower apparent malate synthase activity. That this activity depends on coupling the hydrolase with malyl-CoA lyase was confirmed by showing that our mutant lacking malyl-CoA lyase (PCT57) also lacks malate synthase activity (Cox and Quayle, 1976a). (See Fig. 19 for metabolic lesions in these mutants.)

Although this explains the data described so far, two further problems must be considered. The first is that although some revertants of the "malate synthase" mutant (ICT51) regained normal hydrolase activity, some revertants regained the growth properties of the wild-type bacteria but retained the altered "inactive" malyl-CoA hydrolase (Cox and Quayle, 1976a). The second problem is that the mutants lacking malyl-CoA lyase (PCT57) unexpectedly grow well on ethanol and 3-hydroxybutyrate; and that mutants lacking malyl-CoA hydrolase (ICT51) grow perfectly well on pyruvate or lactate. Thus it appears that there is an alternative route for assimilation of pyruvate and other precursors of acetyl-CoA in mutant PCT57 and in some revertants of the "malate synthase" mutant which still lack this activity. It is not known whether this route is present (but not operating) in wild-type bacteria or if it is "made operative" in the mutants.

The proposed routes for biosynthesis of malate during growth of *Pseudomonas* AM1 on multicarbon compounds are summarised in Fig. 19.

V. The growth of *Pseudomonas* AM1 on propane 1,2-diol and 4-hydroxybutyrate

Neither of these compounds is oxidised by pure methanol dehydrogenase but both induce high levels of this enzyme and both are substrates in "methanol dehydrogenase" assays using crude extracts and, like methanol oxidation, their oxidation requires ammonia as activator (Cox and Quayle, 1976a; Bolbot and Anthony, 1980a). That methanol dehydrogenase is essential for their oxidation has been confirmed by the failure of mutants lacking this enzyme to grow on, or oxidise, these substrates. A further investigation of the metabolism of propanediol by *Pseudomonas* AM1 has shown it to be very unusual (Bolbot and Anthony, 1980a). Growth responses of a wide range of mutants are completely consistent with its oxidation and assimilation by way of pyruvate and acetyl-CoA (Fig. 19). The usual enzymes for the oxidation of propanediol are absent and it has been shown that it is oxidised by way of methanol dehydrogenase in the presence of a second protein (cofactor?) to L-lactate. No intermediate lactaldehyde was detected and it is possible that this oxidation of propane 1,2-diol to lactate may be analogous to the two-step oxidation of methanol to formate catalysed by some methanol dehydrogenases.

VI. The metabolism of trimethylsulphonium salts by *Pseudomonas* MS

Pseudomonas MS (Table 12) was first isolated by elective culture on trimethylsulphonium chloride, and the metabolism of this compound investigated by Wagner and his colleagues (Wagner, 1964; Wagner *et al.*, 1967; Kung and Wagner, 1970a; Hornig and Wagner, 1968; Wagner and Quayle, 1972). The first step in its metabolism involves transfer of a methyl group to tetrahydrofolate:

$(CH_3)_3S^+$ + tetrahydrofolate → $(CH_3)_2S$ + N^5-methyltetrahydrofolate + H^+

The dimethylsulphide is liberated into the growth medium and the "C_1-unit" oxidised to methylene tetrahydrofolate which is then oxidised or assimilated into cell material.

6

The bacterial oxidation of methane, methanol, formaldehyde and formate

I. Introduction	153
II. The oxidation of methane to methanol	153
A. Introduction	153
B. The methane monooxygenase from *Methylococcus capsulatus* (Bath)	155
C. The methane monooxygenase from *Methylosinus trichosporium* (OB3b)	158
D. Is there more than one type of methane monooxygenase?	159
E. Inhibitors of methane monooxygenase	161
1. Metal-chelating agents	161
2. The significance of acetylene inhibition in relation to nitrogenase assays	163
F. The substrate specificity of methane monooxygenases	163
G. The anaerobic oxidation of methane	167
III. Methanol dehydrogenase	167
A. Introduction	167
B. Substrate specificity of methanol dehydrogenase	171
C. Activators and inhibitors	173
D. The primary electron acceptor	174
E. Molecular weight, isoelectric point and amino acid composition	175
F. Serological relationships between methanol dehydrogenases	176
G. Localisation of the methanol dehydrogenase	176
H. The prosthetic group and mechanism of methanol dehydrogenase	177
1. Chemical structure of the prosthetic group (Pyrrolo-Quinoline Quinone, PQQ)	177
2. PQQ as the prosthetic group of other quinoproteins	182
3. Mechanism of methanol dehydrogenase	184
I. The interaction of methanol dehydrogenase with the cytochrome system	186
IV. Formaldehyde oxidation	187
A. Introduction	187
B. NAD^+-dependent formaldehyde dehydrogenase	187
C. Dye-linked aldehyde dehydrogenases	190
D. N^5, N^{10}-methylenetetrahydrofolate dehydrogenase	191
E. Methanol dehydrogenase	192
F. A cyclic route for the oxidation of formaldehyde	192
G. Oxidation of formaldehyde by way of the serine pathway	193
V. Formate oxidation	194

I. Introduction

The route for oxidation of methane and methanol to CO_2 is the obvious one:

$$CH_4 \xrightarrow[NADH + H^+]{O_2 \quad H_2O} CH_3OH \xrightarrow{} HCHO \xrightarrow{} HCOOH \xrightarrow{} CO_2$$
$$\qquad\qquad\qquad\qquad PQQH_2 \qquad\quad '2H' \qquad\quad NADH + H^+$$

The thermodynamic constants for this series of reactions are summarised in Table 30. The simplicity of the oxidation route is not reflected in the nature of the individual reactions, the enzymology of which is unusually interesting.

The first reaction in methane oxidation is a hydroxylation, catalysed by a monooxygenase which requires NADH as reductant. Methanol oxidation to formaldehyde is not coupled to NAD^+ reduction but is catalysed by a methanol dehydrogenase having a novel prosthetic group (PQQ). Formaldehyde oxidation is sometimes coupled to reduction of NAD^+, but it may be oxidised by the methanol dehydrogenase, or by other dye-linked dehydrogenases which may be flavoproteins. When the reductant for methane hydroxylation is NADH (probably always the case), the oxidation of formaldehyde must be coupled to NADH formation. Some bacteria lack a formate dehydrogenase and oxidise formaldehyde to CO_2 by a cyclic variant of their carbon assimilation pathway. The oxidation of formate is always coupled to reduction of NAD^+ and is sometimes the only oxidation step in this sequence to be so.

II. The oxidation of methane to methanol

A. Introduction

Work from many laboratories using a variety of methanotrophs (see Table 31) has led to the general conclusion that the first step in methane oxidation is catalysed by a mixed function monooxygenase system which hydroxylates methane to methanol using molecular oxygen and a reductant (AH_2), which is usually NADH:

$$CH_4 + O_2 + AH_2 \longrightarrow CH_3OH + A + H_2O$$

Before these studies had led to this general conclusion it was suggested that a free radical mechanism might be involved as an alternative to a

TABLE 30

Thermodynamic constants for reactions involved in the oxidation of C_1 compounds

Reaction	$\Delta G^{0'}$ (pH 7·0) (kJ mol^{-1})	Redox couple	$E^{0'}$ (pH 7·0) (Volts)
$CH_4 + 0.5\ O_2 \longrightarrow CH_3OH$	−109·7		
$CH_3OH + 0.5\ O_2 \longrightarrow HCHO + H_2O$	−188·2	$HCHO/CH_3OH$	−0·182
$HCHO + 0.5\ O_2 \longrightarrow HCOOH$	−240·0	$HCOO^- + H^+/HCHO$	−0·450
$HCOOH + 0.5\ O_2 \longrightarrow CO_2 + H_2O$	−244·7	$CO_2/HCOO^- + H^+$	−0·460
$CH_4 + 2.0\ O_2 \longrightarrow CO_2 + 2\ H_2O$	−782·6		
$CH_3OH + 1.5\ O_2 \longrightarrow CO_2 + 2\ H_2O$	−672·9		
$HCHO + 1.0\ O_2 \longrightarrow CO_2 + H_2O$	−484·7		
$NADH + H^+ + 0.5\ O_2 \longrightarrow NAD^+ + H_2O$	−236·8	$NAD^+/NADH + H^+$	−0·320

These values are taken from Ribbons et al. (1970).

6. Bacterial oxidation   155

conventional hydroxylase system (Hutchinson et al., 1976); this now appears less likely but, as emphasised by Higgins (1979a), an involvement of free radicals in a methane monooxygenase system is not yet ruled out. A second alternative unconventional mechanism, involving dimethyl ether as an intermediate, has been proposed (see Wilkinson, 1975), but recent elucidation of the methane monooxygenase system of methanotrophs, and analysis of such systems with respect to oxidation of dimethyl ether (Stirling and Dalton, 1980), confirms that this alternative is at best a remote possibility.

The conclusion that a mixed function monooxygenase is involved in methane oxidation was first implied by the work of Leadbetter and Foster (1959), and later confirmed by Higgins and Quayle (1970) who showed that whole cells of *Methylomonas methanica* and *Methanomonas methanooxidans* incorporate ^{18}O into methanol from $^{18}O_2$ but not from water containing ^{18}O. Further confirmation and understanding of the nature of the methane monooxygenase has depended on isolation of cell-free systems able to catalyse methane oxidation. Early work depended on measuring NADH oxidation and O_2 consumption occurring on addition of methane to particulate preparations derived from *Methylococcus capsulatus* (Texas strain) (Ribbons and Michalover, 1970). Because these preparations are also able to oxidise NADH, methanol, formaldehyde and formate (in the absence of methane) (Ribbons, 1975), interpretation and development of work with such preparations has presented considerable difficulties. These were eventually avoided by developing alternative methods of assay using analogues of methane (CO, bromomethane, ethylene) as substrate or by measuring the product, methanol.

Dalton's group at Warwick has provided the most definitive description of a methane monooxygenase from their work on the soluble, NADH-requiring system from *Methylococcus capsulatus* (Bath strain). They have also shown that a similar system operates in *Methylosinus trichosporium* OB3b. Some of the literature on this subject can appear rather confused on first acquaintance because Higgins and his group at Canterbury (Kent) have described a completely different particulate system from *Methylosinus trichosporium* which was said to use ascorbate or reduced cytochrome *c*, but not NADH, in solubilised, purified preparations. For convenience the system from *M. capsulatus* (Bath) will be described first and then that from *M. trichosporium*; then the two systems will be compared.

B. The methane monooxygenase from *Methylococcus capsulatus* (Bath)

This soluble methane monooxygenase complex catalyses *in vivo* the following reaction:

$$CH_4 + NADH + H^+ + O_2 \longrightarrow CH_3OH + NAD^+ + H_2O$$

TABLE 31
Studies on methane oxidation in bacteria

(a) *Reviews of methane oxidation*

Higgins (1979a) Microbial biochemistry of methane—a study in contrasts. Part 2—Methanotrophy.
Dalton (1980a) Transformations by methane monooxygenase.
Dalton (1980b) Oxidation of hydrocarbons by methane monooxygenases from a variety of microbes.
Dalton (1981a) Methane monooxygenases from a variety of microbes.
Higgins *et al*. (1981a) Hydrocarbon oxidation by *Methylosinus trichosporium*: metabolic implications of the lack of substrate specificity.

(b) *Methane oxidation in different methylotrophs*

Organism	References
Methylococcus capsulatus (Bath strain)	Colby and Dalton (1976, 1978, 1979); Colby *et al*. (1977); Dalton (1977a); Stirling and Dalton (1977, 1979a, 1980); Stirling *et al*. (1979); Dalton (1981a)
Methylococcus capsulatus (Texas)	Ribbons and Michalover (1970); Ribbons (1975); Ribbons and Wadzinski (1976); Stirling and Dalton (1977)
Methylomonas methanica (*Pseudomonas methanica*)	Leadbetter and Foster (1959); Ferenci (1974, 1976a, b); Ferenci *et al*. (1975); Colby *et al*. (1975); Stirling *et al*. (1979)
Methylomonas (other species)	Hubley *et al*. (1974, 1975)
Methanomonas methanooxidans	Higgins and Quayle (1970)
Methylosinus trichosporium	Hubley *et al*. (1974, 1975); Ferenci (1974); Patel *et al*. (1976); Tonge *et al*. (1975, 1977a); Thomson *et al*. (1976); Higgins *et al*. (1976a, 1979, 1981a, b); Hammond *et al*. (1979); Higgins (1979a, b); Stirling and Dalton (1979a); Stirling *et al*. (1979); Best and Higgins (1981); Scott *et al*. (1981)

After breakage of bacteria in a French pressure cell, the complex exists free in solution and has three components *A*, *B* and *C* which have been resolved and characterised (Colby and Dalton, 1978, 1979; Colby *et al.*, 1979; Dalton, 1981a).

Component A. Component *A* is a fairly stable protein of molecular weight about 220 000 and subunits of molecular weight 47 000 and 68 000. It contains two atoms of non-haem iron per molecule and some acid-labile sulphur; and an EPR signal is obtained after reduction with dithionite (but not NADH). Because this signal is enhanced in the presence of substrate (ethene), it is suggested that the reduced form of component *A* is responsible for binding substrate — probably by way of the non-haem iron species.

Component B. Component *B* is a small colourless protein of molecular weight about 15 000 whose function is uncertain. Recent work with purer preparations indicate (contrary to the first reports) that it is essential for activity (Dalton, 1980b). It requires phenylmethylsulphonylfluoride as stabilising agent (Dalton, 1981a).

Component C. Component *C*, purified to homogeneity, has a molecular weight of 44 000 and each molecule contains one molecule of FAD, two atoms of iron and two of acid-labile sulphide. Core extrusion and EPR studies have shown that the iron and sulphide are present as a single iron-sulphide centre of the [2 Fe − 2 S*(S-Cys)$_4$] type as found in spinach ferredoxin and putidaredoxin:

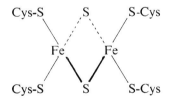

These properties suggest that the single protein has a function analogous to (for example) the combination of putidaredoxin plus NADH-putidaredoxin reductase (a flavoprotein) in the hydroxylation of camphor. Component *C* requires the presence of thioglycollate as stabilising agent (Dalton, 1981a).

Besides its apparent function in electron transport in methane hydroxylation, component *C* also has NADH-acceptor reductase activity. Thus its FAD is reducible with dithionite or NAD(P)H, and it can be oxidised by ferricyanide, 2,6-dichlorophenolindophenol, horse heart cytochrome *c* or by stoicheiometric amounts of component *A*. The optimum pH for this activity (assayed with indophenol) is 8·5–9·0 compared with 6·5–7·0 for methane monooxygenase activity. The K_m values for NADH and NADPH are 50 μM and 15·5 mM respectively, and the V_{max} values are 76 units (mg)$^{-1}$ for NADH and 7 units (mg)$^{-1}$ for NADPH, thus suggesting that NADH rather than NADPH is the natural electron donor. Unlike methane monooxygenase activity, the reductase activity is not inhibited by 8-hydroxyquinoline (1 mM) or by acetylene (0·5 mM), and some preparations of component C lose their oxygenase activity while retaining reductase activity. A possible scheme for the overall hydroxylation process is given in Fig. 20. The site of binding and activation of oxygen is not yet known but preliminary reports (see Dalton, 1981a) indicate that an oxenoid-type mechanism might operate. In such a mechanism O_2 is first converted to an electron-deficient, metal-bound monooxygen species which then reacts with the substrate. The non-haem iron species of components A and C are clearly prime candidates for the metal involved.

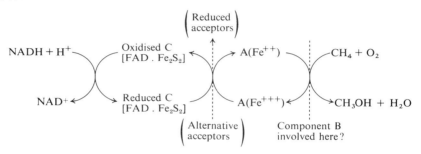

Fig. 20. Pathway of electron transfer between the components of the soluble methane monooxygenase complex during the oxidation of methane to methanol. (This is based on Dalton, 1981a.)

C. The methane monooxygenase from *Methylosinus trichosporium* (OB3b)

This is a difficult subject to summarise because of the markedly different results published from two laboratories. The earlier conclusions of Tonge *et al.* (1975, 1977a) appear to be inconsistent with the demonstration (Stirling and Dalton, 1979a; Stirling *et al.*, 1979) that a methane monooxygenase system can be isolated from *Methylosinus trichosporium* that is very similar to that in *Methylococcus capsulatus* described above. To facilitate discussion, that work of Tonge *et al.* (1975, 1977a) will first be described.

These workers purified a three component system from *Methylosinus trichosporium* able to catalyse the oxidation of methane to methanol, the first step in its isolation being its removal from cell membranes by phospholipase treatment. The purified components were: protein 1 (mol. wt, 47 000) containing one atom of copper per molecule; protein 2 (mol. wt, 9400); and soluble cytochrome *c* (mol. wt, 13 000) which contained variable amounts of copper and which was able to react with carbon monoxide (this cytochrome is discussed in Chapter 8). Equal amounts of each component were required for maximum activity (6 μmoles (min)$^{-1}$ (mg of protein)$^{-1}$) measured at the pH optimum for the system (pH7). All components were inactivated by freezing. The K_m for methane was 66 μM, and CO, ethane, propane and n-butane were also oxidised. Methane oxidation was highly sensitive to cyanide, chelating agents (especially those chelating copper), 2-mercaptoethanol and dithiothreitol.

In crude preparations NADH was able to act as electron donor, but after purification it was necessary to use ascorbate which is able to reduce the cytochrome c; this could be replaced by a mixture of methanol and partially purified methanol dehydrogenase. These results led to the suggestion that in intact bacteria the cytochrome c is not only an oxidase (see Chapter 8), but also the electron donor, passing back electrons from methanol oxidation to the initial methane hydroxylation reaction; it was also suggested that the cytochrome was able to bind methane during the hydroxylation reaction (Hammond et al., 1979).

If hydroxylation of methane to methanol requires NAD(P)H, then both oxidation steps in the subsequent oxidation of formaldehyde must generate NAD(P)H because methanol dehydrogenase is not NAD(P)$^+$-linked, and some NAD(P)H will be required for biosynthesis. The proposed involvement of cytochrome c as the source of reducing power for the initial hydroxylation of methane, as suggested for *Methylosinus trichosporium*, had the advantage of explaining how this bacterium could grow in the absence of NAD(P)$^+$-linked formaldehyde dehydrogenase; but recent observations (Stirling and Dalton, 1979a) have indicated the presence of an NAD(P)$^+$-linked formaldehyde dehydrogenase in this organism.

Further support for the idea of a cytochrome c-coupled methane monooxygenase in *Methylosinus trichosporium* came from a consideration of cell yields. If electrons are provided from methanol, by way of methanol dehydrogenase and cytochrome c, for the initial hydroxylation of methane, then cell yields should be higher than if NAD(P)H is the reductant (Chapter 9; Anthony, 1978b, 1980; Higgins, 1979a). However, it is probable (contrary to earlier statements) that yields of these bacteria on methane are not particularly high (Drozd and Linton, 1981); furthermore, the higher yields measured on methane compared with those on methanol are due to relatively poor yields on methanol rather than relatively high yields on methane (Anthony, 1980; Chapter 9).

D. Is there more than one type of methane monooxygenase?

The two most important characteristics of methane monooxygenases which differ from one source to another are the distribution of activity in cell-free extracts (soluble or membrane-bound), and the nature of the electron donor (NADH or reduced cytochrome c).

As far as the first point is concerned, it is possible that the soluble components A, B and C of the system from *Methylococcus capsulatus* (Bath) are associated with membranes in intact bacteria. Similar components to these may be tightly associated with membranes (and hence not readily solubilised) in those bacteria, such as *Methylococcus capsulatus* (Texas strain) and *Methylosinus trichosporium*, in which methane monooxygenase activity is particulate.

The nature of the electron donor is a more important characteristic because of its relevance to the bioenergetics of bacterial growth on methane. *Methylosinus trichosporium* is the only methanotroph in which cytochrome c (reduced by ascorbate or methanol plus methanol dehydrogenase) has been thought to be able to donate electrons in the hydroxylation of methane instead of the usual reductant (NADH) (Tonge *et al*., 1977a; Higgins, 1979a, 1980). However, this observation could not be repeated by Stirling and Dalton (1979a) who showed that the *Methylosinus trichosporium* contains a soluble NADH-requiring monooxygenase similar to (but not as stable as) that in *Methylococcus capsulatus* (Bath). Furthermore, after separation of the *M. trichosporium* extracts into two inactive fractions by ion-exchange chromatography, activity could be restored by addition of purified components from *Methylococcus capsulatus* (Bath).

The work of Higgins *et al.* (1981a) on M*ethylosinus trichosporium* indicates that some differences in their conclusions with respect to the cellular location of the monooxygenase and those of Dalton's group might be related to the conditions of growth of the bacteria. During growth under a variety of continuous culture conditions *M. trichosporium* contains soluble NADH-linked methane monooxygenase similar to that described in *M. capsulatus* (Higgins *et al.*, 1981a). The activity is NADH-linked, not coupled to cytochrome c, and it is relatively insensitive to inhibitors. By contrast, when grown in shake flasks under O_2-limiting conditions, between 30% and 66% of the total activity was found to be particulate. Although the inhibitor profile of this particulate activity was similar to that described previously (Higgins, 1979a), and different from that of the soluble monooxygenase, the electron donor was the same (NADH) in all growth conditions. It has been concluded by Higgins *et al.* (1981a) that "on balance the current evidence suggests that the soluble and particulate activities are different and that they are each associated in some way with morphological differences (see below) which are in turn dependent on growth conditions". The words "on balance" are critical in a summary of the available information on this subject which still requires much work before a satisfactorily complete picture is obtained.

6. Bacterial oxidation   161

It does appear that NADH is the usual reductant for the monooxygenase in all bacteria. The most critical questions remaining are: (a) Is it certain that cytochrome c is never involved directly as a reductant in the oxidation of methane to methanol? (b) What are the details of the interactions between the three components of the soluble monooxygenase from *M. capsulatus*? (c) Are these components (A, B and C) part of the membrane-bound system found in other methylotrophs? (d) Are they ever membrane-bound in *M. capsulatus* (Bath strain)? (e) What is the relationship (if any) between the internal membranes, methane oxidation and energy transduction in methanotrophs?

This last question arises because of the observations of Higgins *et al.* (1981a), Best and Higgins (1981) and Scott *et al.* (1981), which challenge previous conclusions that the internal membranes of methanotrophs are essential features of these organisms and are associated with methane oxidation. It was observed that membranes were not always present during growth of *M. trichosporium* on methane and that particulate methane monooxygenase activity only ever occurred in organisms possessing extensive intracellular membranes. These were never observed during growth in continuous culture under conditions of CH_4-, O_2- or nitrate-limitation. They were only formed during growth in shake flasks and their formation was encouraged by a low partial pressure of O_2. It was concluded that intracytoplasmic membranes are not obligatory for methane oxidation in *Methylosinus trichosporium* and that there is "no simple relationship between membrane content and any single physiological parameter although low growth rate and low O_2 tensions seem to favour membrane formation".

E. Inhibitors of methane monooxygenase

1. Metal-chelating agents

Earlier work on methane oxidation was limited to work with whole cells or unstable membrane preparations and one of the few useful observations possible with such systems was on the effect of inhibitors on methane oxidation (Hubley *et al.*, 1975; Ribbons, 1975; Colby *et al.*, 1975; Patel *et al.*, 1976). This work showed that methane oxidation is sensitive to a wide variety of inhibitors, especially metal-chelating agents indicating the importance of metal ions (most likely iron or copper) in the process and this observation has been amply confirmed in later work (Hou *et al.*, 1979b; Stirling and Dalton, 1979a). These inhibitors of methane oxidation (not necessarily in all bacteria) include acetylene (ethyne), 8-hydroxyquinoline, neocuproine, ferron, thiosemicarbazide, diethyldithiocarbamate, thiourea, thioacetamide, α,α'-dipyridyl, 1,10-phenanthroline and imidazole. Interpretation of such inhibitor studies are difficult because of the variety of potential sites of action in whole cell and particulate preparations; inhibitors may act on one of the components of the monooxygenase or on an enzyme involved in generating reductant for the hydroxylation, or on electron transport components — thus leading to inhibition of oxygen consumption. It might be expected that inhibitor studies with the purified methane monooxygenase from *Methylococcus capsulatus* (Bath) would help clarify this and they do seem to do so to some extent but some problems remain.

The soluble methane monooxygenase in cell-free extracts of *Methylococcus capsulatus* (Bath) is remarkably insensitive to a wide range of metal-chelators and other inhibitors, the only potent inhibitors being the metal-chelating agent 8-hydroxyquinoline and the acetylenic compounds ethyne (acetylene) and propyne which also inhibit methane oxidation by whole cells (Stirling and Dalton, 1977). In addition to these inhibitors five other compounds inhibit oxidation in whole cells; these are the metal-chelator diethyldithiocarbamate, *o*-aminophenol, ferron, cyanide and carbon monoxide. Most of these probably act by inhibiting the supply of NADH to the monooxygenase within the cells but the mode of action of CO is very complex (see Ferenci *et al.*, 1975; Stirling and Dalton, 1977). The insensitivity of the methane monooxygenase to metal-binding compounds other than 8-hydroxyquinoline suggests that if this is acting by chelating metal ions in the enzyme complex then these must be well shielded from attack by other metal-chelators.

The methane monooxygenase activity, measured with NADH as electron donor in the soluble fraction of extracts of *Methylosinus trichosporium*, has the same relative insensitivity to inhibitors as the soluble system from *Methylococcus capsulatus* (Bath) Stirling and Dalton (1979a). This is markedly different from results, mentioned above, observed with the purified system of Tonge *et al.* (1977a).

One problem remaining in this area is the relative insensitivity of methane oxidation in whole cells of *Methylococcus capsulatus* (Bath) to many inhibitors (mainly chelating agents) compared with methane oxidation by whole cells of other *Methylococcus capsulatus* strains, *Methylomonas methanica* and *Methylosinus trichosporium* (Tonge *et al.*, 1975, 1977a; Stirling and Dalton, 1977; Patel *et al.*, 1976; Hou *et al.*, 1979a, b). Many of the inhibitors appear to specifically affect the methane hydroxylation system in these bacteria and yet have no effect on the system in whole cells of *Methylococcus capsulatus* (Bath). It is significant that the soluble NADH-linked monooxygenase of *Methylosinus trichosporium* (produced in continuous culture) is sensitive to a narrower range of inhibitors than the particulate NADH-linked monooxygenase which is produced in flask-grown bacteria (Higgins *et al.*, 1981a).

2. The significance of acetylene inhibition in relation to nitrogenase assays

The powerful inhibition by acetylene (ethyne) of methane oxidation has been shown to complicate the use of the acetylene reduction assay for nitrogenase activity in methanotrophs. This assay depends on catalysis by nitrogenase of the reduction of added acetylene to ethylene. Because acetylene inhibits the methane monooxygenase, methane cannot be used as the source of reductant for this process. However, methanol, formaldehyde, formate, ethanol and hydrogen are all able to support the reduction of acetylene to ethylene indicating the presence of nitrogenase (Dalton and Whittenbury, 1976).

F. The substrate specificity of methane monooxygenases

The soluble methane monooxygenase of *Methylococcus capsulatus* (Bath) is very non-specific and many of its substrates show little or no structural resemblance to methane (see Table 32 taken from Colby *et al.*, 1977, in which there is extensive discussion of the significance of these results). The monooxygenase catalyses the hydroxylation of primary and secondary alkyl C–H bonds, the formation of epoxides from internal and terminal alkenes, the hydroxylation of aromatic compounds, the N-oxidation of pyridine, the oxidation of CO to CO_2 and the oxidation of methanol to formaldehyde. The list of products in Table 32 demonstrates that some substrates can be attacked at more than one position (e.g. the but-2-enes and toluene). In addition to the substrates listed in Table 32 the methane monooxygenase is also able to oxidise methyl formate to formaldehyde and formate, and ammonia to hydroxylamine (Dalton, 1977a; Stirling and Dalton, 1980).

The substrate specificity of the monooxygenase in crude extracts of *Methylosinus trichosporium* (measured with NADH as reductant) is very similar to that of *Methylococcus capsulatus* (Bath), whereas that of the system in *Methylomonas methanica* is more limited; aromatic, alicyclic and heterocyclic compounds are not oxidised (Stirling *et al.*, 1979).

The oxidation of all substrates by the monooxygenase requires added NADH in extracts, and in whole cells, likewise, NADH must be essential. Thus the only substrates oxidised by the monooxygenase in whole cells of *Methylococcus capsulatus* (of those listed in Table 32) are those whose further oxidation yields NADH; these are methane, methanol, ammonia, chloromethane, bromomethane, dimethylether, ethene and propene (Stirling and Dalton, 1979b). The last three substrates were oxidised more rapidly in the presence of formaldehyde whose oxidation yields NADH. This suggests that in the absence of an exogenous supply of reducing power the rate of initial hydroxylation of these substrates was limited by the poor generation of

TABLE 32

The substrate specificity of the soluble methane monooxygenase from *Methylococcus capsulatus* (Bath)

Substituted methane derivatives	% of rate with methane	C_1–C_8 n-alkanes	% of rate with methane	Products
Chloromethane	99	Ethane	81	ethanol
Bromomethane	78	Propane	82	propan 1-ol and propan 2-ol
Iodomethane	0	Butane	92	butan 1-ol and butan 2-ol
Dichloromethane	97	Pentane	87	pentan 1-ol and pentan 2-ol; not pentan 3-ol
Trichloromethane	41	Hexane	48	hexan 1-ol and hexan 2-ol; not hexan 3-ol
Tetrachloromethane	0	Heptane	87	heptan 1-ol and heptan 2-ol; not heptan 3-ol or heptan 4-ol
Cyanomethane	39			
Nitromethane	53	Octane	11	octan 1-ol and octan 2-ol; not octan 3-ol or octan 4-ol
Methanethiol	75			
Methanol	289			
Trimethylamine	0			
Carbon monoxide	72			

C_2–C_4 n-alkanes	% of rate with methane	Products	Alicyclic, aromatic and heterocyclic compounds	% of rate with methane	Products
Ethene	176	Epoxyethane	Cyclohexane	74	cyclohexanol
Propene	99	1,2-epoxypropane	Benzene	74	phenol
But-1-ene	58	1,2-epoxybutane	Toluene	63	benzyl alcohol and cresol
cis-But-2-ene	68	cis-2,3-epoxybutane and cis-buten 1-ol	Styrene	56	styrene epoxide
			Pyridine	35	pyridine N-oxide
trans-But-2-ene	168	trans-2,3-epoxybutane and trans-2-buten 1-ol	L-Phenylalanine	0	none
Ethers					
Dimethyl ether	295	not known			
Diethyl ether	54	ethanol and ethanal			

This table is taken from the results in Colby et al. (1977). The enzyme system was crude soluble extract; NADH was electron donor; O_2 was an absolute requirement and oxidation of all substrates was inhibited by the "specific" inhibitor ethyne (acetylene). The values are expressed as a percentage of the value with methane (85 nmoles of product formed (min)$^{-1}$ (mg protein)$^{-1}$). It should be noted that the relative rates are for the highest rates measured with various amounts of substrate; they are not V_{max} values, and K_m values were not determined. The products of oxidation of substituted methane derivatives were not identified; rates of hydroxylation of these substrates were determined from the rate of disappearance of the substrates. (See Table 60 in Chapter 12 for a list of substrates oxidised by whole organisms.)

NADH arising from their further oxidation. In addition to the eight substrates listed above, seven further substrates were oxidised by whole cells but only in the presence of formaldhyde as a source of reductant (NADH); these were carbon monoxide, diethyl ether, ethane, butane, 1-butene, *cis*-2-butene and *trans*-2-butene.

The oxidation of substrates that are unable to support growth has been termed co-oxidation, this being a special case of the phenomenon of co-metabolism. Stirling and Dalton (1979b) have discussed the ambiguities associated with these terms and have proposed that co-metabolism be redefined as " the transformation of a compound, which is unable to support cell replication, in the requisite presence of another transformable compound (co-substrate)". Thus, in the case of *Methylococcus capsulatus* (Bath) those seven compounds which are only oxidised in the presence of formaldehyde are co-metabolic substrates, the co-substrate being formaldehyde. It is suggested that oxidation of substrates (in the absence of co-substrate) that are unable to support growth (e.g. chloromethane, bromomethane, dimethyl ether, ethane and propene) is merely a reflection of the non-specific nature of the methane monooxygenase and that this oxidation should be termed "fortuitous oxidation". This phenomenon is quite common and is analogous, for example, to the oxidation by whole cells of *Pseudomonas* M27 of about 20 primary alcohols which are unable to support growth but which are good substrates for the methanol dehydrogenase (Anthony and Zatman, 1965).

With respect to the methane monooxygenase, Higgins *et al.* (1980b) have suggested that such "extraordinary lack of enzyme specificity would be extremely unusual if it were entirely fortuitous" and they have argued that "this phenomenon has developed and been retained because of its survival value to the species". It remains a matter of debate (see Stirling and Dalton, 1981; Higgins *et al.*, 1981b) whether or not these oxidations and co-oxidations are entirely fortuitous or whether some of them can be considered to be "supplementary metabolism" enabling these obligate methanotrophs to co-utilise other carbon and energy sources (Higgins *et al.*, 1981a).

Because of the wide substrate specificity of the methane monooxygenase, and because its substrates are relatively intractable to limited chemical oxidations, it offers a potentially valuable catalyst for effecting these oxidations. Because the hydroxylation requires an electron donor this must be added to extracts or reductant must be generated *in vivo*. This generation may occur by subsequent metabolism of the hydroxylation product or by oxidation of endogenous storage compounds or by addition of a co-substrate. A number of studies have explored the use of whole cells to effect limited chemical oxidations and these have confirmed that the specificity of the methane monooxygenase in a variety of methanotrophs is fairly similar, the differences in substrate specificities in these studies being probably related to the problem of *in vivo* generation of reductant. In some cases this was achieved by addition of formaldehyde over short periods (12 min) (Stirling and Dalton, 1977; Stirling *et al.*, 1979; Dalton, 1980a, b) or by addition of methane over long periods (15–24 h) (Higgins *et al.*, 1979, 1981b) while in some studies no source of reductant was added (Hou *et al.*, 1979a, b; Patel *et al.*, 1980a, b). The potential commercial exploitation of this aspect of methylotrophic metabolism is discussed in Chapter 12.

G. The anaerobic oxidation of methane

The hydroxylation of methane described above has an absolute requirement for molecular oxygen but there is considerable evidence that microbial oxidation of methane occurs under anaerobic conditions, the main bacteria involved in this process being the sulphate-reducing bacteria (Barnes and Goldberg, 1976; Reeburgh, 1976, 1980, 1981; Reeburgh and Heggie, 1977; Martens and Berner, 1977). The concentration of sulphate is a major factor determining the extent of this process which in some marine sediments consumes essentially all the upward methane flux. This oxidation of methane occurs in a thin subsurface anaerobic zone where the product $[CH_4][SO_4^{2-}]$ is at a maximum (Reeburgh, 1981). Although pure cultures capable of mediating anaerobic methane oxidation have not yet been isolated, Davis and Yarborough (1966) have shown that sulphate reducers are able to oxidise methane at low rates in a lactate medium. Furthermore, Panganiban *et al.*, (1979) have shown that enrichment cultures (N.B. not pure cultures) of organisms from Lake Mendota are capable of anaerobic methane oxidation; they required sulphate, acetate and methane for growth, assimilating (but not oxidising) the acetate while oxidising (but not assimilating) the methane to carbon dioxide. These organisms do not resemble *Desulfovibrio* in morphology and may be a new form of sulphate reducer.

Although physiologically and ecologically unimportant, it is of interest that methanogenic bacteria are also able to oxidise methane anaerobically; this only occurs during concomitant methane formation which is always at least 300 times the rate of oxidation (Zehnder and Brock, 1979).

III. Methanol dehydrogenase

A. Introduction

Methanol oxidation in bacteria is always catalysed by the NAD^+-independent alcohol dehydrogenase described originally in *Pseudomonas* M27 (Anthony and Zatman, 1964a, b, 1965, 1967a, b). Although not specific for

TABLE 33
Methanol dehydrogenase

Source of methanol dehydrogenase	References
Methanol-utilisers (facultative)	
Pseudomonas M27	Anthony and Zatman (1964a, b, 1965, 1967a, b); Patel *et al.* (1972, 1973)
Pseudomonas AM1	Johnson and Quayle (1964); Heptinstall and Quayle (1969); O'Keeffe and Anthony (1980a, b); Bolbot and Anthony (1980a)
Pseudomonas extorquens	Johnson and Quayle (1964)
Protaminobacter ruber	Johnson and Quayle (1964)
Pseudomonas PP	Ladner and Zatman (1969)
Pseudomonas RJ1	Mehta (1973b)
Pseudomonas TP-1	Sperl *et al.* (1974)
Pseudomonas J26	Michalik and Raczynska-Bojanowska (1976)
Pseudomonas 2941	Yamanaka and Matsumoto (1977a, b; 1979); Yamanaka (1981)
Pseudomonas S25	Yamanaka and Matsumoto (1977b, 1979); Yamanaka (1981)
Strain S50 (*Acinetobacter* sp.)	Yamanaka and Matsumoto (1977b, 1979); Yamanaka (1981)
Diplococcus PAR	Bellion and Wu (1978)
Paracoccus denitrificans	Bamforth and Quayle (1978a); Alefounder and Ferguson (1981)
Hyphomicrobium	Sperl *et al.* (1974); Harder and Attwood (1975); Duine *et al.* (1978); Duine and Frank (1980a, 1981a, a review); Duine *et al.* (1981)
Rhodopseudomonas acidophila	Sahm *et al.* (1976a); Bamforth and Quayle (1978b, 1979); de Beer *et al.* (1980)
Methanol-utilisers (obligate)	
Pseudomonas W1	Sperl *et al.* (1974)
Pseudomonas C	Goldberg (1976)
Methylomonas P11	Michalik and Raczynska-Bojanowska (1976); Drabikowska (1977)
Methylophilus methylotrophus	Ghosh and Quayle (1981); Cross and Anthony (1980b); Ghosh (1980)

Methanotrophs	
Methylococcus capsulatus	Patel and Hoare (1971); Patel *et al.* (1972, 1973); Wadzinski and Ribbons (1975a)
Methylomonas methanica	Johnson and Quayle (1964); Patel *et al.* (1978a)
Methylosinus sporium	Patel and Felix (1976)
Methylobacterium organophilum	Wolf and Hanson (1978)
Methylobacterium R6	Patel *et al.* (1978b)

The methanol dehydrogenases from these bacteria are all similar in most respects to that first described from *Pseudomonas* M27; they all oxidise a wide range of primary alcohols for which they usually have a very high affinity (low K_m values); they use phenazine methosulphate as primary hydrogen acceptor; they use ammonia or methylamine as activator; and they have a high pH optimum.

TABLE 34
Summary of properties of the methanol dehydrogenases listed in Table 33

Source of MDH	Molecular weight	Subunit mol. wt	Oxidation of secondary alcohols	Isoelectric point
Group A				
Pseudomonas M27	120 000		−	high
Pseudomonas AM1	120 000	60 000	−	8·8
Pseudomonas RJ1	120 000		−	
Pseudomonas 2941	128 000	62 000	−	7·38
Pseudomonas TP-1	120 000		−	
Pseudomonas S25	128 000	62 000	−	9·4
Pseudomonas W1	120 000		−	high
Hyphomicrobium	120 000		−	high
Methylophilus methylotrophus	115 000	62 000	−	high
Methylococcus capsulatus	120 000	62 000	−	high
Group B				
Methylobacterium organophilum	135 000	62 000	++	high
Pseudomonas C	128 000	60 000	++	—
Diplococcus PAR	112 000	56 000	++	—
Group C				
Paracoccus denitrificans	151 000	76 000	−	3·7
Strain S50	158 000	76 000	−	3·82
Group D				
Methylomonas methanica	60 000	60 000	−	high
Methylosinus sporium	60 000	60 000	−	high
Group E				
Rhodopseudomonas acidophila	116 000	63 000	+	9·35

The enzymes have been arranged in groups having similar properties. This division is rather arbitrary and sometimes based on incomplete or preliminary descriptions. Molecular weights of whole enzymes are based on gel filtration. A "high" isoelectric point is above 7·0 and is sometimes based only on observations during ion-exchange chromatography.

methanol its usual function is to catalyse methanol oxidation and so it will be referred to here as methanol dehydrogenase (MDH). It has sometimes been called primary alcohol dehydrogenase (PAD).

Typical methanol dehydrogenases (EC.1.1.99.8) oxidise a wide range of primary alcohols using phenazine methosulphate as electron acceptor and ammonia or methylamine as activator. The pH optima are pH 9 or higher and they are often stable at pH 4·0. They are usually dimers of identical subunits of 60 000 daltons. Possession of this enzyme is one feature that appears to be common to all methanol-oxidising bacteria; Table 33 gives references to the enzyme from a wide range of different methylotrophs and Table 34 summarises their properties.

The specific activity of MDH in crude extracts from different bacteria varies over a wide range (between 4 and 1300 nmol $(min)^{-1}$ $(mg)^{-1}$ but usually between 60 and 600 nmol $(min)^{-1}$ $(mg)^{-1}$). This reflects to some extent the variety of growth conditions and methods of cell breakage and enzyme assay, but it also suggests a genuine range of activities in methylotrophs. Certainly the specific activities of the purified enzymes vary considerably (0·3–18 μmol $(min)^{-1}$ $(mg\ protein)^{-1}$) (see Goldberg, 1976; Bamforth and Quayle, 1978a, b). A comparison of specific activities of crude extracts and pure enzymes shows that the methanol dehydrogenase usually constitutes between 5% and 15% of the soluble protein. This indicates its importance to the growth of methylotrophs as does the fact that it is induced to higher levels during growth on methane or methanol; this importance is confirmed by the isolation of mutants which lack the dehydrogenase and have lost the ability to grow on these substrates (Heptinstall and Quayle, 1970; Dunstan et al., 1972b; O'Connor and Hanson 1977).

B. Substrate specificity of methanol dehydrogenase

The dehydrogenase has a wide but well-defined specificity (Anthony and Zatman, 1965; Sperl et al., 1974). Results with the dehydrogenase from *Pseudomonas* M27 showed that only primary alcohols are oxidised and that their steric configuration is more important in determining whether or not they are oxidised than the presence or absence of atoms or groups producing electron displacement effects; and this has been confirmed with the *Hyphomicrobium* enzyme (Duine and Frank, 1980a). A second substituent on the C2 atom appears to prevent binding (Anthony and Zatman, 1965); the general formula for an oxidisable substrate is $R.CH_2OH$ where R may be H, OH (as in hydrated aldehydes), $R'.CH_2$- or $R'.R''C=CH$-. The rate of oxidation of most substrates is at least 30% of that with methanol (the best substrate) but the affinity for the enzyme often decreases with increasing size of the alcohol; the K_m for methanol is usually low (10–20 μM). Whole bacteria usually oxidise the same range of alcohols as are oxidised by the pure enzyme and their oxidation is inhibited by inhibitors of methanol oxidation in whole cells (EDTA, phenylhydrazines and high concentrations of phosphate).

As indicated above, 1,2-propanediol, having 2 substituents on its second carbon atom, is not oxidised by pure methanol dehydrogenase. In spite of this, recent studies on the metabolism of this alcohol by *Pseudomonas* AM1 have unexpectedly indicated that methanol dehydrogenase is involved in the oxidation of this substrate; mutants lacking methanol dehydrogenase or cytochrome *c* neither grow on propanediol nor oxidise it. Preliminary results indicate that a second protein modifies the pure dehydrogenase enabling it to bind alcohols with a substituent hydroxyl on the second carbon atom and to oxidise them in two consecutive steps to the carboxylic acid (Bolbot and Anthony, 1980a). A similar phenomenon may be involved in the oxidation of 4-hydroxybutyrate by *Pseudomonas* AM1 (see Cox and Quayle, 1976b).

Although most MDHs have a similar substrate specificity to that of *Pseudomonas* M27, there are some minor differences. For example, it was shown, using MDHs from three different methanol-utilisers (*Hyphomicrobium*, *Pseudomonas* TP-1 and *Pseudomonas* W1), that alcohols substituted with a methyl group on the second carbon atom have a relatively low affinity for these dehydrogenases but are still oxidised at high rates (Sperl *et al.*, 1974). Furthermore, preliminary observations with the MDH from the facultative methylotrophs *Methylobacterium organophilum* (Wolf and Hanson, 1978), from *Pseudomonas* C (Goldberg, 1976) and from an uncharacterised methanol-utiliser (*Diplococcus* PAR) (Bellion and Wu, 1978) suggest that some secondary alcohols may be substrates for these particular dehydrogenases.

The most unusual methanol dehydrogenase with respect to substrate specificity, and indeed with respect to other properties, is that from the photosynthetic methylotroph *Rhodopseudomonas acidophila* (Sahm *et al.*, 1976a; Bamforth and Quayle, 1978a, 1979); it oxidises primary and secondary alcohols at similar rates although the K_m values for the secondary alcohols are higher. This enzyme has the greatest affinity for ethanol (K_m, 6 μM) and the lowest affinity for methanol (K_m, 57 mM) when assayed in the oxygen electrode. Oxygen is an inhibitor of this dehydrogenase, competitive with respect to the alcohol substrate and so it has some effect on the measured K_m values for methanol; but the affinity for methanol is always remarkably low for an enzyme initiating the oxidative attack on the growth substrate.

A common characteristic of all MDHs tested (first shown by Ladner and Zatman, 1969 and Heptinstall and Quayle, 1969) is their ability to catalyse the oxidation of formaldehyde to formate; during the oxidation of methanol, formaldehyde is often undetectable, formate being the sole product of the reaction (e.g. Westerling *et al.*, 1979). The rate of formaldehyde oxidation is usually similar to that for methanol oxidation and the affinity of the enzyme for the two substrates is often similar. It has been suggested that the actual substrate during formaldehyde oxidation is the gem-diol hydrated aldehyde and that the extent to which other aldehydes are oxidised may be related to their degree of hydration (Sperl et al., 1974). This argument is based on the observation that although formaldehyde is often the only aldehyde oxidised, acetaldehyde, trifluoroacetaldehyde and trichloroacetaldehyde are sometimes substrates. The *Rhodopseudomonas* MDH is again unusual, catalysing the oxidation of formaldehyde, acetaldehyde and propionaldehyde. Although the rates are similar to those measured for ethanol (the best substrate), the affinity of the enzyme for aldehydes is relatively low (1 % of that for ethanol) (Sahm et al., 1976a; Bamforth and Quayle, 1978b). This enzyme has many similarities to the alcohol dehydrogenase from *Acinetobacter calcoaceticus* (Duine and Frank, 1981b).

Duine and Frank (1981a) have argued that the rather sharp break observed in the aldehyde substrate specificity spectrum, of those methanol and alcohol dehydrogenases able to oxidise aldehydes, cannot be accounted for by the extent of hydration because this follows a gradual change through the range of aldehydes. Instead, they suggest that the results can best be explained by assuming a dual substrate specificity, one for the alcohol and the other for the aldehyde substrate (Duine and Frank, 1981b). They emphasise that although a fully oxidised enzyme may be able to oxidise methanol directly to formic acid (four electrons being accepted by two prosthetic groups), this is unlikely to be the case in all organisms many of which have alternative, energetically more favourable formaldehyde dehydrogenases. They also point out that the enzymes having a molecular weight of 60 000 (and presumably only one prosthetic group per molecule) may not be able to oxidise methanol to formate but only to formaldehyde (Duine and Frank, 1981a).

C. Activators and inhibitors

When prepared aerobically, methanol dehydrogenase has an absolute requirement for ammonium salts. This requirement can be satisifed by methylamine but not by di- or tri-amines, nor by long-chain alkylamines. The relatively higher concentration of ammonium salt required at lower pH suggests that the free base is the active species (Anthony and Zatman, 1964b). Because the oxidised enzyme is reduced by substrate at pH 7·0 in the absence of activator, it appears that a high pH and ammonia activator are essential only for the reoxidation of the reduced enzyme (Duine and Frank, 1980a, 1981a).

The MDH from *Methylomonas* P11 is exceptional in being active only with ammonia (and not methylamine) and that from *Methylobacterium organophilum* has considerable ammonia-independent activity (20–40% of that in its presence) (Wolf and Hanson, 1978). The methanol dehydrogenase from *Rhodopseudomonas acidophila* is unusual in being activated by ammonia and a wide range of primary alkylamines; secondary or tertiary amines are not activators. Although the highest rate is obtained with ammonia the affinity for the amine activator increases with increasing chain length, the K_m for nonylamine being 26 μM and that for ammonia being 42 mM (this is similar to the alcohol dehydrogenase of *Acinetobacter calcoaceticus* described by Duine and Frank, 1981b). Cyanide and α,α' bipyridine inhibit the *Rhodopseudomonas* enzyme by competing with the amine activator (Bamforth and Quayle, 1978b). The MDH from *Hyphomicrobium* is inhibited by cyanide competitively with respect to methanol (K_i, 1 mM) and this may be true for other MDHs (Duine and Frank, 1980a, 1981a). A recent report by Duine and Frank (Proceedings of Sheffield Symposium, 1980) shows that the enzyme from *Hyphomicrobium* X is activated by esters of glycine or β-alanine but not by lysine-esters nor by aliphatic amines or amino acids. The K_m value for ethylglycine was 30 times lower than for ammonia. The dehydrogenase from *Rhodopseudomonas acidophila* was also able to use lysine esters in addition to those used by the *Hyphomicrobium* enzyme.

Although ammonia activation is such a well-defined and universal characteristic of methanol dehydrogenases, this may result from an alteration in enzyme function occurring during its aerobic preparation. When prepared anaerobically, ammonia is no longer required for activation of the *Hyphomicrobium* enzyme and the ammonia-independent enzyme is converted to the typical ammonia-dependent form on exposure to oxygen (Duine *et al.*, 1979a); this has been confirmed using crude extracts of *Pseudomonas* AM1 (O'Keeffe and Anthony, 1980a). This may not be the case with the enzyme from *Rhodopseudomonas acidophila* which still requires an amine activator when prepared anaerobically.

D. The primary electron acceptor

Electrons are eventually passed to cytochrome *c* (Chapter 8), but for assay of the extracted enzyme, it is necessary to use an artificial acceptor which is usually phenazine methosulphate (PMS) (alternative names: *N*-methylphenazinium methosulphate, *N*-methylphenazonium methosulphate):

6. Bacterial oxidation   175

In the enzyme assay the reoxidation of reduced PMS is coupled to reduction of 2,6-dichlorophenolindophenol (measured spectrophotometrically) or oxygen (measured in an O_2-electrode). The only conventional electron acceptor to replace the phenazine methosulphate is phenazine ethosulphate (PES) (alternative names: N-ethylphenazinium ethosulphate, N-ethylphenazonium ethosulphate). It has been suggested by Ghosh and Quayle (1979) that PES should always be used as electron acceptor in assaying dye-linked dehydrogenases rather than PMS because PES does not reduce the indophenol non-enzymically. They showed that both phenazine derivatives form free radicals in alkaline solution and it has been suggested that the free radical may be the true electron acceptor in the methanol dehydrogenase assay system (Duine *et al.*, 1978; Ghosh and Quayle, 1979). This suggestion is supported by the demonstration that alternatives to PMS and PES as electron acceptors are the free radicals produced by the one-electron oxidation of N,N,N',N'-tetramethyl-p-phenylenediamine (Wurster's blue) and of 2,2'-azino-di-(3-ethylbenzthiazoline-6-sulphonic acid) (Duine *et al.*, 1978).

E. Molecular weight, isoelectric point and amino acid composition

The majority of methanol dehydrogenases have molecular weights between 112 000 and 158 000 and can be dissociated by low pH or sodium dodecylsulphate to two identical subunits of 56 000–76 000 daltons. The MDHs from *Methylomonas methanica* and *Methylosinus sporium* are exceptions in being monomers of about 60 000 (Patel and Felix, 1976, Patel *et al.*, 1978a). Most MDHs have high isoelectric points (7–10·5) and the exceptions happen to be those having the largest molecular weights; the MDH from *Paracoccus denitrificans* has a molecular weight of 151 000 and an isoelectric point (pI) of 3·7 (Bamforth and Quayle, 1978a), and that from strain S50 has a molecular weight of 158 000 and a pI of 3·82 (Yamanaka and Matsumoto, 1977b, 1979). Most MDHs are acid-stable at pH 4·0, exceptions being those from *Methylosinus sporium* (Patel and Felix, 1976) and *Pseudomonas* W1 (Sperl *et al.*, 1974).

The amino acid compositions of the MDHs from the following bacteria have been published but, as might be expected, this information sheds little light on the activity of these enzymes: *Pseudomonas* M27 (Anthony and Zatman, 1967a); *Methylosinus sporium* (Patel and Felix, 1976); *Pseudomonas* C (Goldberg, 1976); *Methylomonas methanica* (Patel *et al.*, 1978a). As expected, the dehydrogenases with the lower lysine content tend to be those with lower isoelectric points (Yamanaka and Matsumoto, 1977b; Yamanaka, 1981).

F. Serological relationships between methanol dehydrogenases

Antisera prepared against pure MDH from *Methylococcus capsulatus* (a RuMP pathway, methane-utiliser) and from *Pseudomonas* M27 (a serine pathway, facultative methanol-utiliser) have been used to allocate a number of methane and methanol-utilising bacteria to various groups which were satisfyingly similar to those arrived at by conventional methods (Patel *et al.*, 1973, 1978a; Patel and Felix, 1976). The Type I (RuMP pathway) methane-utilisers were similar to one another but different from Type II (serine pathway) methane-utilisers which showed some similarity to the facultative methanol-utilisers (serine pathway). The MDH from the obligate methanol-utiliser (*Pseudomonas* W1) (RuMP pathway) was serologically distinct from all other MDHs. A similar study using antisera produced against pure MDH from the facultative methane-utiliser *Methylobacterium organophilum* (serine pathway) showed, as expected, that this enzyme is more similar to the MDHs from the facultative methanol-utilisers and from serine pathway methanotrophs than to the MDHs from *Methylococcus capsulatus* (an RuMP pathway methanotroph) or from the facultative autotroph *Rhodopseudomonas acidophila* (Wolf and Hanson, 1978).

G. Localisation of the methanol dehydrogenase

Although the MDH from most bacteria is found in the soluble fraction after cell breakage, it is probable from its function that it binds to the membrane in which the components of the electron transport chain are situated. As the MDH often constitutes about 10% of the protein in the cell extracts, it is perhaps unlikely that it all arises from the membrane during cell breakage. It is possible that the soluble cytochrome c might be reduced by the soluble methanol dehydrogenase and this cytochrome c then be oxidised by the electron transport chain. Certainly all of the cytochrome c, both soluble and membrane-bound, is able to be reduced by methanol in whole cells (Anthony, 1975b; Cross and Anthony, 1980a, b).

Breakage of cells in a French Press or by sonication usually provides sufficient enzyme for purification and the membrane fraction is often discarded and not assayed for the presence of MDH. When it is measured, some is usually found on the membrane; the amount remaining is affected by the treatment of the membrane fraction. MDH has been found on membranes, or perhaps enclosed in membrane vesicles in a wide variety of bacteria including *Methylococcus capsulatus* (Wadzinski and Ribbons, 1975a), *Hyphomicrobium* X (Duine *et al.*, 1978), *Pseudomonas* AM1 (Netrusov and Anthony, 1979) and *Methylophilus methylotrophus* (Cross and Anthony, 1980b).

A particularly important study of membrane-bound dehydrogenase is that of Wadzinski and Ribbons (1975a; Ribbons and Wadzinski, 1976) who have shown that the MDH released by detergent from membranes (about 60% of the total MDH) of *Methylococcus capsulatus* has identical properties after purification to those of the soluble form. Probably all "methanol oxidase" activities reported for preparations of methylotrophs are due to membrane-bound MDH. A comparison of the binding of MDH to membranes of Type I and Type II methanotrophs showed that about 60% remained on the membranes prepared by French Press extraction of Type I methanotrophs (*Methylococcus*, *Methylobacter* and *Methylomonas*), whereas all is found in the soluble fraction of the Type II methanotrophs (*Methylosinus* and *Methylocystis*) (Patel and Felix, 1976). Once again, this result probably reflects the methods used for extraction, because other workers have found that some of the MDH of *Methylosinus trichosporium* remains bound to membranes (Tonge *et al.*, 1975, 1977a).

Indirect observations have indicated that the MDH might react with cytochrome *c* on the outer side of the cytoplasmic membrane (O'Keeffe and Anthony, 1978; Dawson and Jones, 1981a; Chapter 8), and the recent observation that most of the methanol dehydrogenase is found in the periplasmic space seems to support this (Alefounder and Ferguson, 1981).

H. The prosthetic group and mechanism of methanol dehydrogenase

1. Chemical structure of the prosthetic group (Pyrrolo-Quinoline Quinone, PQQ)

The pure methanol dehydrogenase has a characteristic absorption spectrum (Fig. 21a) which was first described in 1967 (Anthony and Zatman, 1967b) and is a feature of all methanol dehydrogenases described since then. The absorption due to the protein has a peak about 280 nm and a shoulder at 290 nm; that due to the prosthetic group has a peak at about 345 nm and a shoulder at about 400 nm. Although the enzyme has only a typical protein fluorescence, on boiling or treatment with acid or alkali a green fluorescent compound is released with concomitant loss of enzyme activity. This prosthetic group is reddish-brown in colour, highly polar, acidic and it has a low molecular weight. Maximum fluorescence occurs at low pH, excitation maxima are at 255 nm and 365 nm and the fluorescence maximum is at 470 nm. (See Fig. 21b for its absorption spectrum.) Because these fluorescence characteristics were typical of no known compounds except for pteridines it was originally concluded that the novel prosthetic group of methanol dehydrogenase might be a pteridine (Anthony and Zatman,

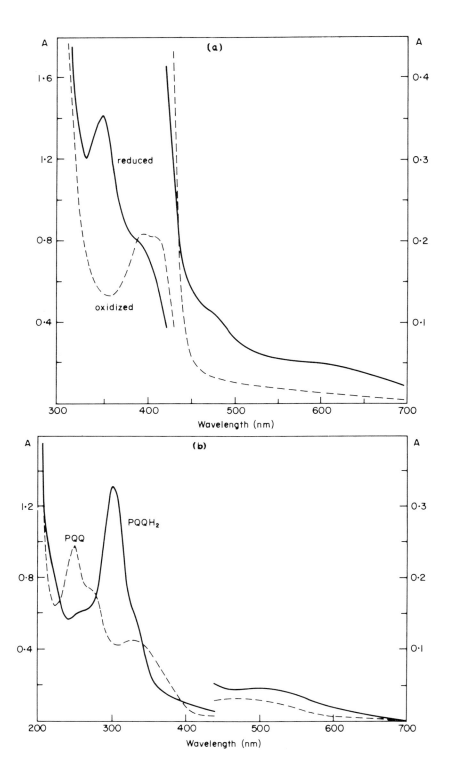

1967b), but the structure of this molecule has resisted all attempts at elucidation for more than a decade since its first description. During this time, the similarity of the fluorescence characteristics of the prosthetic group to those of pteridines was confirmed (Urushibara et al., 1971; Sperl et al., 1973) and the prosthetic groups from a range of methylotrophs shown to be similar or identical (Duine and Frank, 1981a).

The chemical characterisation of the prosthetic groups has now been achieved by two independent groups of workers (see Table 35 and Duine and Frank, 1981a for a review of this achievement). Duine, Frank and their colleagues, using a wide range of chemical and physical techniques, have shown that it is a multicyclic ring compound with an inner ring orthoquinone, two nitrogen atoms and one or more carboxyl groups; NMR measurements indicate the presence of two uncoupled aromatic protons and ESR measurements of the prosthetic group and ENDOR measurements of the whole enzyme indicate the presence of three protons one of which (on the nitrogen atom) is exchangeable (Table 35). These proposals are consistent with the structure first proposed by Salisbury et al. (1979) (Fig. 22) based on X-ray diffraction analysis of a crystalline acetonyl derivative of the presumed prosthetic group extracted from whole cells of *Pseudomonas* TP-1. Their structure shows that the prosthetic group is a novel and complex orthoquinone derivative of fused quinoline and pyrrole rings. They suggested the trivial name methoxatin but a more informative name is pyrrolo-quinoline quinone (PQQ).

The possibility of achieving reconstitution of active enzyme has facilitated the testing of analogues of the PQQ prosthetic group for activity, and such studies have shown that the pyrrolo ring and the 9-carboxylic acid are not essential for activity as they can be replaced by a pyridinol ring and a 9-hydroxy group respectively (in phenanthroline-dione analogues of PQQ). In similar experiments, the orthoquinone in PQQ was shown to be essential for activity by showing that the enzyme is inactive when reconstituted with the 4,5-dihydro, 4,5-diol derivative of PQQ (Duine et al., 1980).

Fig. 21. (a) Absorption spectra of different forms of methanol dehydrogenase. This figure is taken (with permission) from Duine et al. (1981). The enzyme concentration is 5 mg (ml)$^{-1}$. The reduced form (———) corresponds to MDHred in Fig. 23; it contains PQQH$_2$ and has a similar appearance to the spectra of methanol dehydrogenases as they are usually isolated. The oxidised form (------) corresponds to MDHox$_2$ in Fig. 23. It is produced by addition of electron acceptor to MDHred in the presence of activator (NH$_4$Cl) and cyanide. (b) The absorption spectrum of the quinol and quinone forms of the prosthetic group of methanol dehydrogenase. The quinol form (PQQH$_2$) was obtained by reducing the quinone form (PQQ) with hydrogen in the presence of platinum oxide. The spectra were measured in anaerobic conditions in 50 mM-potassium phosphate, pH 7·0. These spectra were kindly provided by Drs Duine and Frank who used MDH from *Hyphomicrobium* X.

TABLE 35
The PQQ prosthetic group of methanol dehydrogenase

X-ray crystallography and structure proposal	Salisbury et al. (1979)
UV/Vis absorption spectra of MDH	Anthony and Zatman (1967b); Duine et al. (1979a, 1980, 1981)
Fluorescence and absorption spectra of PQQ	Anthony and Zatman (1967b); Duine et al. (1978, 1980); Duine and Frank (1980b)
HPLC characterisation and quantitative analysis	Duine et al. (1980, 1981); Duine and Frank (1980b, 1981b)
Reconstitution of a PQQ enzyme	Duine et al. (1980)
ESR and ENDOR spectrometry	Duine et al. (1978, 1981); Westerling et al. (1979); de Beer et al. (1979, 1980)
NMR and mass spectrometry	Duine et al. (1980, 1981)
Formation of adducts with water, methanol, aldehydes and acetone	Salisbury et al. (1979); Duine and Frank (1980b); Duine et al. (1981)
The mechanism of quinoprotein dehydrogenases	Duine et al. (1980, 1981); Forrest et al. (1980); Duine and Frank (1981a)

This novel prosthetic group was first described by Anthony and Zatman (1967b); its structure (Fig. 22) was determined by Salisbury et al. (1979) and Duine and Frank and their colleagues (references above). The relationship between the structure of PQQ and its function has been reviewed by Duine and Frank (1981a).

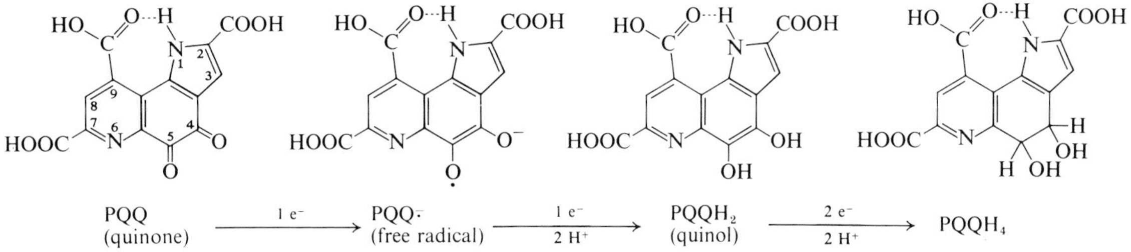

Fig. 22. The prosthetic group of methanol dehydrogenase and derivatives of it. The full name of PQQ is 2,7,9-tricarboxy-1H-pyrrolo[2,3-f]quinoline-4,5-dione. The trivial name methoxatin was initially proposed (Salisbury et al., 1979) but the abbreviation PQQ (Pyrrolo-Quinoline Quinone) emphasises the functional importance of the orthoquinone part of the structure. PQQ, PQQ$^-$, PQQH$_2$ and PQQH$_4$ are all involved in the reaction cycle but PQQH$_4$ has no known biological function (see review of Duine and Frank, 1981a). Adducts with water, methanol, acetaldehyde and acetone are formed by addition at C$_5$. The midpoint redox potential of the PQQ/PQQH$_2$ couple is +90 mV at pH 7·0 and +419 mV at pH 2·0 indicating that PQQ is a 2e$^-$/2H$^+$ redox carrier (Duine et al., 1981).

PQQ can be determined quantitatively by its absorption at 249 nm ($\mathscr{E} =$ 18 400 M^{-1} cm^{-1}). Use of this method led to the suggestion that there is one PQQ molecule in each molecule of methanol dehydrogenase. Subsequent knowledge of the properties of $PQQH_2$, however, necessitated a re-evaluation of this result. Although $PQQH_2$ is sufficiently stable to permit analysis by HPLC at low pH, it is less polar than PQQ and it has a low absorbance at 254 nm; under the conditions used for analysis of PQQ it was not observed. By choosing other chromatographic conditions and measuring at 313 nm, it has now been established that two prosthetic groups can be extracted from each molecule of the dimeric methanol dehydrogenase; one being the oxidised form (PQQ) and the other the quinol form ($PQQH_2$) (Duine et al., 1981).

2. PQQ as the prosthetic group of other quinoproteins

Although thought for many years to be both novel and unique, Duine, Frank and their colleagues have now shown that the novel PQQ is the prosthetic group of a range of different dehydrogenases which they have called quinoproteins (Table 36). They have achieved this by chemical studies of the extracted and purified prosthetic groups and by reconstituting pure identified prosthetic groups with apoprotein to yield active enzyme (Duine et al., 1979b). This approach was impossible with the methanol dehydrogenase because release of the prosthetic group always denatures the enzyme (Anthony and Zatman, 1967b), but it has now become feasible by using the glucose dehydrogenase from *Acinetobacter calcoaceticus;* the apoprotein from this dehydrogenase is able to form active enzyme when mixed with its own purified prosthetic group or with prosthetic group from methanol dehydrogenase. Similar reconstitution studies, supported by chemical data, have shown that PQQ is also present in extracts of *Acetobacter pasteurianum s.s loviensis* (possibly arising from the NAD^+-independent alcohol dehydrogenase) and of *Gluconobacter oxydans* (possibly arising from the glucose dehydrogenase and the polyol dehydrogenases). PQQ is also the prosthetic group of the primary amine dehydrogenase of *Pseudomonas* AM1 (de Beer et al., 1980) but in this case it is covalently bonded to the polypeptide chain, part of which remains attached to the prosthetic group during its isolation.

The quinoprotein alcohol dehydrogenase of *Acinetobacter* has now been purified and characterised (Duine and Frank, 1981b). Although methanol is a very poor substrate it resembles methanol dehydrogenases in other respects; it has a pH optimum of 9·5; it has a requirement for an amine activator; it oxidises primary alcohols; it shows the characteristic absorption spectrum; it has a PQQ prosthetic group; and it has a molecular weight of about 120 000. The activator and, to some extent, the substrate specificity can be compared

TABLE 36
Dehydrogenases having a PQQ prosthetic group (quinoproteins)

Enzyme	Organism	References
Methanol dehydrogenase	Methylotrophs	Duine and Frank (1980b, 1981a)
Primary amine dehydrogenase (the PQQ is covalently bound)	*Pseudomonas* AM1	de Beer *et al.* (1980)
	Thiobacillus A2	Duine and Frank (personal communication)
Alcohol dehydrogenase	*Acinetobacter calcoaceticus*	Duine and Frank (1981b)
	Acetobacter pasteurianum	Duine *et al.* (1979b)
	Gluconobacter suboxydans	Ameyama *et al.* (1981)
Polyol dehydrogenases (?)	*Gluconobacter oxydans*	Duine *et al.* (1979b)
Alcohol (long-chain) dehydrogenase	Alkane-grown *Ps. aeruginosa*	Duine *et al.* (1979b)
Glucose dehydrogenase	*Acinetobacter calcoaceticus*	Duine *et al.* (1979b, 1980)
	Gluconobacter oxydans	Duine *et al.* (1979b)
	Gluconobacter suboxydans	Ameyama *et al.* (1981)
	Pseudomonas aeruginosa	Ameyama *et al.* (1981)
	Pseudomonas fluorescens	Ameyama *et al.* (1981)
Aldehyde dehydrogenase	*Gluconobacter suboxydans*	Ameyama *et al.* (1981)
Lactate dehydrogenase (?)	*Propionibacterium pentosaceum*	Duine and Frank (personal communication)

with those of *Rhodopseudomonas acidophila* methanol dehydrogenase but the physico-chemical properties of the two enzymes are quite different.

3. Mechanism of methanol dehydrogenase

After the first description of the methanol dehydrogenase and its prosthetic group, further elucidation of its mode of action has been hampered by the fact that the pure enzyme is always in the reduced state and that oxidation with phenazine methosulphate inactivates the enzyme; thus spectral changes occurring on addition of substrate cannot be observed. This problem has now been overcome by Duine and Frank (1980a,b) who have oxidised the pure dehydrogenase by addition of phenazine methosulphate and activator (ammonia) in the presence of KCN which, being an inhibitor, competitive with respect to methanol, binds to the active site and protects it against inactivation. On oxidation the characteristic 345 nm peak of the reduced form is shifted to about 400 nm and additions of stoicheiometric amounts of substrate change the spectrum back to that of the initial reduced form (Fig. 21a). Similar changes occur on reduction by mercaptoethanol or catalytic hydrogenation of the fluorescent purified prosthetic group; the fluorescence is lost and the higher absorption peak decreases in wavelength (Fig. 21b). The reduced prosthetic group can be re-oxidised by molecular oxygen back to the oxidised fluorescent form. The absorption maxima in the whole enzyme are 20–50 nm higher than in the isolated prosthetic group.

An important characteristic of the reduced enzyme is the possession of a quinone free radical which is lost on oxidation of the enzyme. An identical free radical is obtained with the half-reduced form of the prosthetic group.

The three forms of PQQ involved in catalysis are shown in Fig. 22. In analysing the catalytic cycle it would be ideal to know the form of the prosthetic group and related absorption spectra of the enzyme during various phases of catalysis. At present this ideal cannot be achieved, and a single straightforward picture is not possible. This is related to the fact that the enzyme contains two prosthetic groups each able to be in one of the three forms shown in Fig. 22; and the fact that the two prosthetic groups can probably interact with one another on the enzyme. Figure. 23 is a summary of the tentative *in vitro* reaction cycle proposed by Duine and Frank (1981a).

In summary, although PQQ, PQQ$^{\cdot-}$ and PQQH$_2$ can all be extracted from methanol dehydrogenase, it is not absolutely certain that they occur as such in the enzyme, because comproportionation and disproportionation reactions may take place. Having said this, ENDOR measurements do show that the free radical PQQ$^{\cdot-}$ is present as such in the enzyme and all the other evidence (Table 35) implicates *o*-quinone reduction to the quinol during the catalytic cycle. The reactivity of the methanol dehydrogenase, with one-electron acceptors during its assay and its oxidative titration, suggests that electron transfer from the reduced dehydrogenase proceeds by way of one-electron steps, and this is consistent with kinetic studies. It is not known, however, if the reduction of enzyme by its substrate occurs by one-electron steps or by hydride or hydrogen transfer. Furthermore, it is not known if the two prosthetic groups react together with one substrate molecule; or separately, each with one substrate molecule. As there are indications that the prosthetic groups can interact with one another, it seems plausible to consider that they might act together during catalysis (see Duine *et al.*, 1981). This would be consistent with the observed 4-electron oxidation of methanol to formate and with the possibility that the enzyme shuttles between either the redox forms 2 PQQ/2PQQ· or between 2PQQ·/2PQQH$_2$ (2-electron oxidation steps) (Duine and Frank, 1981a).

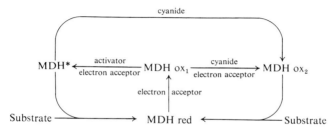

Fig. 23. A tentative reaction cycle for *in vitro* activity of methanol dehydrogenase. This is based on a scheme presented in the review of Duine and Frank (1981a). The isolated enzyme is a mixture of MDH ox$_1$ and MDH red. MDH ox$_2$ is produced by oxidation of this isolated enzyme by electron transfer to an electron acceptor in the presence of KCN and activator (NH$_4$Cl). MDH red is produced from MDH ox$_2$ by titration with substrate (see Fig. 21a for spectra of these two species). MDH ox$_1$ is produced from MDH red by titration with electron acceptor, and MDH* (unstable) is produced from MDH red by a similar titration but in the presence of NH$_4$Cl. (MDH ox$_2$ is a cyanide adduct of the labile form of oxidised dehydrogenase—MDH*). HPLC analysis shows that an extract of MDH red contains only PQQH$_2$: that of MDH ox$_1$ contains equimolar concentrations of PQQ and PQQH$_2$. This indicates that MDH ox$_1$ contains either one PQQ plus one PQQH$_2$ and/or two PQQ$^-$ (which may disproportionate to one PQQ and one PQQH$_2$ during extraction).

Important questions remaining include the nature of the binding of methanol to the enzyme and the role of the activator molecules (ammonia or amines). It is not known if the ease of formation of adducts (including alcohols and aldehydes) at the C$_5$ position has any significance in the catalytic activity of the dehydrogenase. Forrest *et al.* (1980) have proposed a mechanism in which activation by ammonia or amines involves its covalent bonding at C$_4$. The alcohol is then bonded to the same C$_4$ atom and then released as the aldehyde, the 2H being passed to the prosthetic group which is then oxidised by the electron acceptor (Fig. 24). This scheme does not appear to be consistent with the observation that the ammonia or amine activator is readily removed by dialysis nor with the conclusion (Duine and Frank, 1980a; Duine et al., 1981) that a mechanism involving a PQQ-alcohol adduct, which is subsequently oxidised by an electron acceptor, is not possible. Rather similar intermediates to those in Fig. 24 are considered in Chapter 7 (Fig. 28) for the involvement of PQQ in catalysis by methylamine dehydrogenase.

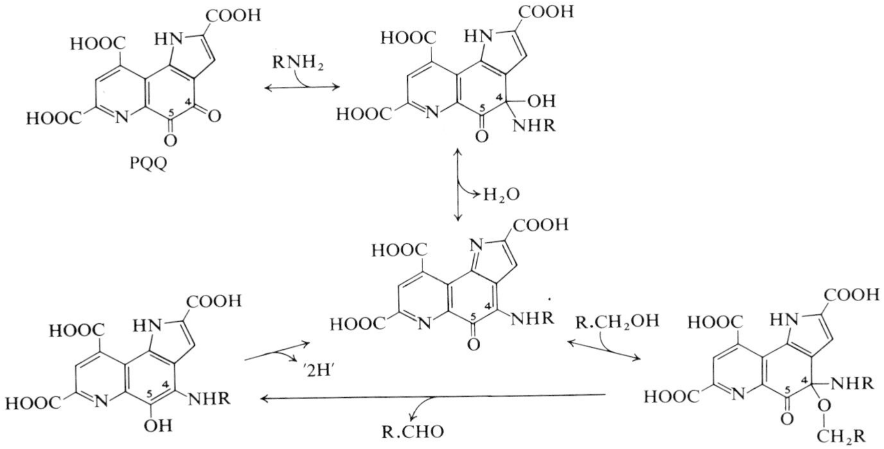

Fig. 24. A mechanism proposed for the involvement of PQQ in catalysis by methanol dehydrogenase (based on Forrest et al., 1980).

I. The interaction of methanol dehydrogenase with the cytochrome system

There is considerable evidence that electrons from MDH pass to the electron transport system chain at the level of cytochrome c, thus bypassing cytochrome b (Chapter 8). Because the pure MDH reduces pure cytochrome c even in the absence of added methanol, it has been impossible to show whether or not MDH catalyses the reduction of cytochrome c by methanol (Anthony, 1975b). The observed reduction in the absence of methanol does not necessarily involve electron transfer from MDH to the cytochrome c, because this cytochrome is able to undergo an intramolecular "autoreduction" reaction and the MDH appears to stimulate this intramolecular autoreduction (O'Keeffe and Anthony, 1980a). Although the observed reduction of cytochrome c by MDH may thus be explained in terms of this autoreduction phenomenon, the catalysis by pure MDH of reduction of pure cytochrome c by methanol has still not been demonstrated. This may be because oxygen alters the enzyme during its preparation (Duine et al., 1979a). When prepared anaerobically the MDH activity is independent of ammonia and it is able to catalyse the reduction by methanol of the cytochrome c present in crude extracts (Duine et al., 1979a; O'Keeffe and Anthony, 1980a). Whether anaerobically-purified MDH will catalyse the reduction of pure cytochrome c by methanol as well as catalyse autoreduction of the cytochrome c has yet to be determined. A speculative mechanism for involvement of the autoreduction mechanism in the transfer of electrons from methanol to cytochrome oxidase catalysed by methanol dehydrogenase and cytochrome c is given in Chapter 8 (Fig. 31).

IV. Formaldehyde oxidation

A. Introduction

Formaldehyde is produced by oxidation of methanol or methylamines, and then it is either assimilated into cell material or oxidised completely to CO_2. This oxidation may be by way of dehydrogenation to formate and thence to CO_2 by formate dehydrogenase or it may involve a cyclic route. Table 37 lists the various types of dehydrogenase that may be involved in formaldehyde oxidation in methylotrophs. Few have been fully characterised and the list is rather tentative because full descriptions are not available. A comprehensive list of specific activities of various formaldehyde dehydrogenases has been published by Zatman (1981).

The most important consideration in formaldehyde oxidation is whether or not NAD(P)H is produced, because in some methylotrophs this reductant is required in relatively large amounts for biosynthesis (in the serine pathway bacteria) or for NAD(P)H-dependent hydroxylation of methane or substituted methylamines. Furthermore, relatively less ATP is likely to be produced during formaldehyde oxidation if this process is coupled to the electron transport chain by way of flavoprotein dehydrogenase or methanol dehydrogenase than if NAD(P)H is produced.

B. NAD^+-dependent formaldehyde dehydrogenase

The four NAD^+-dependent types given in Table 37 are distinguished on the basis of their cofactor requirements and whether or not they are specific for formaldehyde. The reduced glutathione (GSH)-dependent, formaldehyde-specific enzyme is often stated to be responsible for formaldehyde oxidation but the requirement for GSH and the substrate specificity are rarely tested. In some bacteria this enzyme is induced to relatively higher levels during methylotrophic growth indicating that it is likely to be important in formaldehyde oxidation in these particular bacteria (see Table 37).

TABLE 37
The oxidation of formaldehyde

Type of dehydrogenase	NAD(P)$^+$-dependent	Dye-dependent	GSH-dependent	Specific for formaldehyde	Comments
1	+	−	+	+	*Rhodopseudomonas* enz. uses methylglyoxal
2	+	−	−	+	
3	+	−	−	−	Needs peptide cofactor
4	+	−	−	−	Methylenetetrahydrofolate deH$_2$
5	−	+	−	+	Only in *Pseudomonas* RJ1
6	−	+	−	+	
7	−	+	−	−	
8	−	−	−	−	Methanol dehydrogenase
(9)	(+)	−	−	+	Cyclic route

Organism	Type of dehydrogenase (other than 5, 8 or 9)	References
Methylococcus capsulatus (Bath)	4	Stirling and Dalton (1978)*
Methylomonas methanica	(1)	Johnson and Quayle (1964)*; Ferenci *et al.* (1975)
Methylobacterium organophilum	7	O'Connor and Hanson (1977)
Methylobacterium R6	6, 7 or 8^t	Patel *et al.* (1978b)
Methylosinus trichosporium	7	Stirling and Dalton (1979a); Patel *et al.* (1980b)*
Pseudomonas C	2 or 3	Ben-Bassat and Goldberg (1977)
Methylomonas methylovora	1 (low)	Patel *et al.* (1979b)*
Organism 4B6	6 or 7	Colby and Zatman (1973)
Methylophilus methylotrophus	7	Large and Haywood (1981)
Pseudomonas AM1	3	Johnson and Quayle (1964)*; Large and Quayle (1963); Marison and Attwood (1980)
	1	
	7^c	

Organism			References
Pseudomonas RJ1	1	6	Mehta (1975c)*; Marison and Attwood (1980)
Pseudomonas 1 and 135	1	6 or 7	Rock et al. (1976); Ben-Bassat and Goldberg (1977)
Pseudomonas extorquens		7	Johnson and Quayle (1964)
Pseudomonas extorquens 16	2 or 3^c	6 or 7^c	Marison and Attwood (1980)
Pseudomonas methylica		6 or 7^c	Loginova and Trotsenko (1977b); Marison and Attwood (1980)
Protaminobacter ruber	1	7	Johnson and Quayle (1964)
Pseudomonas aminovorans	1^i	6 or 7^i	Boulton and Large (1977); Bamforth and O'Connor (1979)
Pseudomonas MS	2 or 3^i		Kung and Wagner (1970a, b)
Organism 5H2	1	6 or 7	Hampton and Zatman (1973)
Hyphomicrobium X		7^c	Marison and Attwood (1980)*
Hyphomicrobium vulgare 3	2 or 3	6 or 7^c	Marison and Attwood (1980)*
Paracoccus denitrificans	1^i	7	Cox and Quayle (1975); Bamforth and Quayle (1978a); Marison and Attwood (1980)
Rhodopseudomonas acidophila	1^i		Sahm et al. (1976a)

Most possible systems are listed above; those for which there are more complete descriptions are marked with an asterisk. Many descriptions do not state that GSH-dependence or substrate specificity was actually tested. Any organism with methanol dehydrogenase (type 8), methylenetetrahydrofolate dehydrogenase (type 5) or the Entner-Duodoroff enzymes of the RuMP pathway plus 6-phosphogluconate dehydrogenase (type 9) have the potential for oxidising formaldehyde by way of these enzymes. *i* shown to be inducible; *c* shown to be constitutive.

190 The biochemistry of methylotrophs

The only NAD^+-dependent formaldehyde dehydrogenase that has been purified and characterised from a methylotroph is that from *Methylococcus capsulatus* (Bath strain) (Stirling and Dalton, 1978). It is independent of GSH or any other thiol for activity but requires a small, heat-sensitive, trypsin-sensitive, component (probably peptide or protein) for activity, as well as the major protein component which has a molecular weight of 115 000 and is composed of two equal subunits of 57 000 daltons. This enzyme is not absolutely specific for formaldehyde; it also oxidises glyoxal and glycolaldehyde but it does not oxidise those aldehydes oxidised by other non-specific aldehyde dehydrogenases (acetaldehyde, propionaldehyde etc.).

C. Dye-linked aldehyde dehydrogenases

These enzymes are usually non-specific aldehyde dehydrogenases and they are unlikely to be essential or solely responsible for the oxidation of formaldehyde because they are rarely induced to higher levels during methylotrophic growth. Furthermore, the specific activities are often too low to account for the very high rates of formaldehyde oxidation observed with most methylotrophs. However, this may reflect the necessity of assaying the enzyme with an artificial electron acceptor and the possibility that some of the enzyme may be membrane-bound. All these dehydrogenases use phenazine methosulphate as electron acceptor and some can use other dyes such as 2,6-dichlorophenolindophenol.

The first dye-linked aldehyde dehydrogenase purified (partially) from a methylotroph was that from *Pseudomonas* AM1 (Johnson and Quayle, 1964) and since then the enzymes from *Methylomonas methylovora* (an obligate methanol-utiliser) and from *Hyphomicrobium* X have been partially purified and characterised. They all oxidise a wide range of aldehydes, and formaldehyde is not usually the best substrate, either in terms of binding, or maximum velocity. Although they are broadly similar in some respects, lack of information makes detailed comparisons impossible. An exceptional enzyme is the specific formaldehyde dehydrogenase from *Pseudomonas* RJ1 (Mehta, 1975c); the K_m for formaldehyde is very high (27 mM) and the enzyme is only produced during growth on methylamine and oxalate, being completely absent during growth on methanol.

By analogy with aldehyde dehydrogenases from other sources, the dye-linked dehydrogenases might be expected to be flavoproteins but there is no direct evidence for this. After reduction of the enzyme from *Methylomonas methylovora* with dithionite or propionaldehyde, the characteristic spectrum of reduced cytochrome *c* is seen. This has led to the proposal that this enzyme is a haemoprotein (Patel *et al.*, 1979b), but the evidence is also compatible with contamination of the enzyme with cytochrome *c*, particularly as the enzyme was purified only 13-fold and as the molar ratio of haem to protein monomer appears to give a value of only 0·1. (N.B. My calculation gives a lower value than that of the authors.) The dye-linked dehydrogenase from *Methylosinus trichosporium* has also been purified and it too appears to contain haem (Patel *et al.*, 1980b). Once again this could be due to contamination by cytochrome *c*. The haem/monomer ratio is very low (0·08 — my calculation based on an extinction coefficient at 550 nm of 20 mM^{-1} cm^{-1}).

The partially-purified enzyme from *Hyphomicrobium* X contains no bound haem but uses cytochrome *c* preferentially as the primary electron acceptor (Marison and Attwood, 1980). If the *Methylomonas* enzyme is similar to this, then any contaminating cytochrome *c* would be reduced by added aldehyde and the enzyme would give the appearance of being a haemoprotein.

Clearly the nature of the prosthetic group of this enzyme is of some importance in considering the bioenergetics of methylotrophs, because this will determine the point at which electrons enter the cytochrome chain and hence the ATP yield during formaldehyde oxidation. In this context, it should be noted that a mutant of *Pseudomonas* AM1 completely lacking cytochrome *c* is still able to oxidise formaldehyde suggesting that in this organism the formaldehyde dehydrogenase is able to donate electrons to cytochrome *b* or to an oxidase directly, without the intermediacy of cytochrome *c* (Anthony, 1975b).

D. N^5, N^{10}-methylenetetrahydrafolate dehydrogenase

The substrate for this enzyme is formed by the non-enzymic reaction of formaldehyde with tetrahydrofolate. The dehydrogenase reaction produces NAD(P)H and N^5,N^{10}-methenyltetrahydrofolate which is required for purine biosynthesis and which is an intermediate in the reduction of formate to the level of formaldehyde during growth on formate (Large and Quayle, 1963; Attwood and Harder, 1978). Probably the majority of bacteria possess this enzyme but whether or not it is ever of significance in the oxidation of formaldehyde during growth on C_1 compounds is uncertain. It should be noted that in order to act as a formaldehyde dehydrogenase, high levels of an enzyme capable of releasing the bound formate from its product (formyl-tetrahydrofolate) would also be essential. *Pseudomonas* AM1 has high (induced) levels of the dehydrogenase during growth on methanol but the specific activity of the formyl tetrahydrofolate synthetase appears to be too low for this route to be important in formaldehyde oxidation (Large and Quayle, 1963).

E. Methanol dehydrogenase

This enzyme has been discussed extensively above. Besides oxidising methanol to formaldehyde it is also able to catalyse the oxidation of formaldehyde to formate, and formaldehyde is not always detectable during the two-step oxidation of methanol to formate. Whether or not the methanol dehydrogenase ever functions as a formaldehyde dehydrogenase is not known, but it is certainly able to do so in the absence of any other system. In *Pseudomonas* AM1 it cannot be important in formaldehyde oxidation because mutants lacking the methanol dehydrogenase oxidise formaldehyde at the same rates as wild-type bacteria and they are also able to grow on methylamine during which formaldehyde oxidation is essential (Heptinstall and Quayle, 1970; Dunstan *et al.*, 1972b).

F. A cyclic route for the oxidation of formaldehyde

This route (Fig. 25), proposed by Strøm *et al.* (1974) and by Colby and Zatman (1975c), uses some of the enzymes of the KDPG aldolase/transaldolase version of the RuMP pathway for formaldehyde assimilation (see Chapter 3). The only extra enzyme required to produce a cycle effecting the complete oxidation of formaldehyde to CO_2 plus 2 NAD(P)H is 6-phosphogluconate dehydrogenase. This enzyme has been found in a number of methylotrophs with the RuMP pathway (Quayle and Ferenci, 1978; Ben-Bassat and Goldberg, 1980; Beardsmore *et al.*, 1982), and isotopic evidence for this dissimilatory cycle during growth on methanol has been found in *Pseudomonas* C (Ben-Bassat and Goldberg, 1977; Ben-Bassat *et al.*, 1980) and *Pseudomonas oleovorans* (Sokolov and Trotsenko, 1977, 1978a). The operation of this cycle obviates the necessity for formaldehyde and formate dehydrogenases. which are indeed absent or very low in some

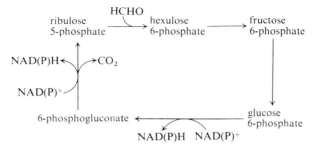

Fig. 25. The cyclic route for oxidation of formaldehyde. This is probably the main route for oxidation of formaldehyde in *Pseudomonas* C, *Methylophilus methylotrophus* and *Pseudomonas oleovorans*. This cyclic pathway is discussed further in Chapter 3.

bacteria having the RuMP pathway. Besides providing a route for oxidation of formaldehyde, the cycle provides a system for control between assimilation of formaldehyde and its complete oxidation to CO_2, with the branch point at the level of 6-phosphogluconate dehydrogenase. As might be expected for an enzyme with such a regulatory role, it is inhibited by its "end products" NADH and ATP (Beardsmore *et al.*, 1982; Ben-Bassat and Goldberg, 1980). The energetics of the KDPG aldolase/transaldolase version of the RuMP pathway are marginally less favourable than alternative versions, and it has been suggested that the opportunity for having this control system might be the rationale for the "choice" of this particular version of the pathway (Quayle, 1980). Zatman has emphasised that not all methylotrophs assimilating formaldehyde by the RuMP pathway are also able to oxidise it by the dissimilatory cycle; the suggested pattern is that those RuMP pathway bacteria unable to grow on methane use this dissimilatory cycle for formaldehyde oxidation, and usually lack formate dehydrogenase, whilst the obligate methane utilisers which assimilate formaldehyde by the RuMP pathway do not use the dissimilatory cycle for oxidation and these bacteria must oxidise formaldehyde by way of NAD-linked formaldehyde and formate dehydrogenases (Zatman, 1981).

G. Oxidation of formaldehyde by way of the serine pathway

It has been proposed that the oxidation of formaldehyde to CO_2 in *Pseudomonas* MA may be achieved by one of two unconventional routes involving most of the enzymes of the serine pathway (except PEP carboxylase), plus either pyruvate kinase and dehydrogenase, or plus pyruvate kinase, malic enzyme and the decarboxylating enzymes of the TCA cycle (Newaz and Hersh, 1975). However, no investigation of possible conventional routes for formaldehyde oxidation has been published and the only evidence for this suggestion is the interesting and important observation that the PEP carboxylase of this organism is activated by NADH. However, activation of this biosynthetic enzyme by a necessary "precursor" for biosynthesis (NADH) is perfectly logical whatever the mechanism of formaldehyde oxidation, and this unconventional route remains, in my view, an interesting speculation. It should be noted that, in such an integrated system for oxidation and assimilation, the specific activity of nearly all the enzymes would have to be high enough for both requirements. This is also true of the cyclic RuMP pathway described above but the cases are different because relatively less of the substrate must be oxidised for provision of ATP and NAD(P)H (about 25%) in the energetically favourable RuMP pathway than in the less favourable serine pathway (50%) (Chapter 9).

V. Formate oxidation

Two types of formate dehydrogenase have been described in bacteria; one is a soluble, NAD^+-linked enzyme and the other is membrane-bound and donates electrons to the cytochrome chain at the level of cytochrome b. Except for *Pseudomonas oxalaticus* and *Achromobacter parvulus*, which appear to have both types (Dijkhuizen *et al.*, 1978, 1979; Rodionov and Zakharova, 1980), formate is oxidised in methylotrophs by a soluble dehydrogenase which is specific for formate and NAD^+ and is often induced to higher levels during methylotrophic growth (Kaneda and Roxburgh, 1959a; Johnson and Quayle, 1964; Kung and Wagner, 1970a; Mehta, 1973a, b; Harder and Attwood, 1975; Rodionov *et al.*, 1977a, b; Attwood and Harder, 1978; Dijkhuizen *et al.*, 1978; Müller *et al.*, 1978; Stirling and Dalton, 1978, 1979a; Egorov *et al.*, 1979). The distribution and specific activities of formate dehydrogenases from a variety of methylotrophs have been presented and discussed in a review by Zatman (1981). In many methylotrophs, the formate dehydrogenase appears to be the only enzyme providing NADH for biosynthesis during growth on C_1 compounds. It is not essential in those bacteria able to oxidise formaldehyde by the cyclic route described above, but bacteria lacking this enzyme are unlikely to be able to grow on formate.

Because the formate dehydrogenases are often unstable and probably because they appear to be typical NAD^+-linked dehydrogenases, they have aroused little interest amongst enzymologists until recently when Müller *et al.* (1978) showed the formate dehydrogenase from *Pseudomonas oxalaticus* to be exceptionally interesting. This soluble enzyme is complex, having a large molecular weight (315 000), and it is sensitive to O_2. It has at least 2 different types of subunit, 2 FMN prosthetic groups, about 20 non-haem iron atoms and acid-labile sulphides per molecule, by way of which it is able to transfer electrons from formate not only to NAD^+ but also to O_2, ferricyanide or redox dyes.

Whether or not this unusual dehydrogenase is unique to *Pseudomonas oxalaticus* remains to be seen, but slightly unexpected characteristics have also been observed in formate dehydrogenases from other methylotrophs. For example, that from *Pseudomonas* AM1 (Johnson and Quayle, 1964) is sensitive to cyanide, ferrous and cuprous ions; that from an unidentified methylotroph (strain 1) is a dimer, unstable except in the presence of mercaptoethanol or EDTA (Egorov *et al.*, 1979); while the formate dehydrogenase from *Hyphomicrobium* X is also able to transfer electrons from formate to ferricyanide (Marison, 1980). It has been suggested that this formate dehydrogenase may play a role in regulating the carbon flux between assimilation and oxidation in *Hyphomicrobium* X because it is inhibited by both NADH and ATP (Marison, 1980).

7
The oxidation of methylated amines

I. Introduction 195
II. The oxidation of tetramethylammonium salts. 197
III. The oxidation of trimethylamine to dimethylamine and formaldehyde . . 197
   A. Trimethylamine dehydrogenase 197
   B. The indirect route for trimethylamine oxidation 200
      1. Trimethylamine monooxygenase 200
      2. Trimethylamine N-oxide demethylase 200
IV. The oxidation of dimethylamine to methylamine and formaldehyde . . 201
   A. Dimethylamine monooxygenase 201
   B. Dimethylamine dehydrogenase 203
V. The oxidation of methylamine to formaldehyde 204
   A. Methylamine dehydrogenase 204
      1. Enzyme assay, substrates and inhibitors of methylamine dehydrogenase . 205
      2. The catalytic mechanism of methylamine dehydrogenase . . 205
      3. The subunit structure of methylamine dehydrogenase. . . 206
   B. Methylamine oxidase 208
   C. The oxidation of methylamine by way of methylated amino acids . . 210
      1. The synthesis of N-methylglutamate 210
      2. The oxidation of N-methylglutamate 211
VI. The distribution of metabolic routes for the oxidation of methylated amines . 217
   A. The oxidation of trimethylamine 217
   B. The oxidation of dimethylamine 217
   C. The oxidation of methylamine 217

I. Introduction

Many methylotrophic bacteria are able to grow on methylated amines as their sole source of carbon and energy, this property being at least as widespread as the ability to grow on methane or methanol (Meiberg, 1979; Large, 1981; Chapter 1). The methylated amines are all oxidised to formaldehyde which can be further oxidised to provide metabolic energy, or assimilated into cell material by the same routes as operate during growth on methane, methanol or formate. Each methyl group of the methylated amines is oxidised to one molecule of formaldehyde, some of these oxidation reactions producing reducing equivalents and others requiring NAD(P)H as reductant. Some oxidations, therefore, provide metabolic energy, whereas other oxidations utilise it. A summary of the oxidative pathways is presented in Fig. 26 and the individual reactions are described below.

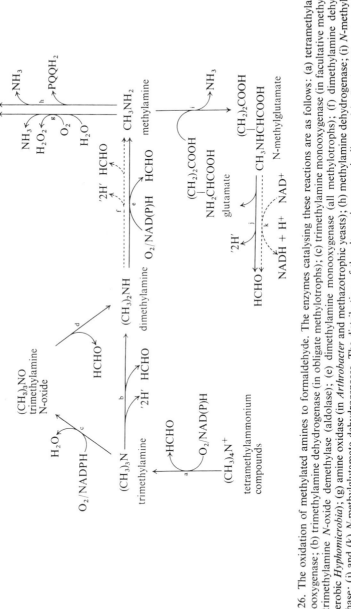

Fig. 26. The oxidation of methylated amines to formaldehyde. The enzymes catalysing these reactions are as follows: (a) tetramethylammonium monooxygenase; (b) trimethylamine dehydrogenase (in obligate methylotrophs); (c) trimethylamine monooxygenase (in facultative methylotrophs); (d) trimethylamine N-oxide demethylase (aldolase); (e) dimethylamine monooxygenase (all methylotrophs); (f) dimethylamine dehydrogenase (anaerobic *Hyphomicrobia*); (g) amine oxidase (in *Arthrobacter* and methazotrophic yeasts); (h) methylamine dehydrogenase; (i) N-methylglutamate synthase; (j) and (k) N-methylglutamate dehydrogenases. The distribution of the alternative routes is discussed in Section VI and summarised in Tables 39 and 40.

II. The oxidation of tetramethylammonium salts to trimethylamine

Although a number of methylotrophs are able to grow on tetramethylammonium salts, the metabolism of only one, organism 5H2, has been investigated in any detail (Hampton and Zatman, 1973). The enzyme complement of bacteria grown on this substrate indicates that it is oxidised to trimethylamine by a monooxygenase and that the further oxidation of this substrate involves trimethylamine monooxygenase, trimethylamine N-oxide demethylase and dimethylamine monooxygenase (see Fig. 26). Organism 5H2 does not contain methylamine dehydrogenase. The tetramethylammonium monooxygenase is unstable, but the six-fold purified enzyme has been characterised to some extent with respect to its sensitivity to inhibitors. The results of this investigation suggest that it may be an iron-containing, thiol-dependent monooxygenase, that it is different from trimethylamine monooxygenase, and that it is not a cytochrome P-450 type of monooxygenase.

III. The oxidation of trimethylamine to dimethylamine and formaldehyde

As shown in Fig. 26, trimethylamine may be oxidised directly to dimethylamine by trimethylamine dehydrogenase, or indirectly by way of trimethylamine N-oxide which may also act as a growth substrate for those bacteria having this indirect route.

A. Trimethylamine dehydrogenase

This is a novel type of dehydrogenase, occurring in obligate methylotrophs and the more restricted facultative methylotrophs. It was first described in organism 4B6 and purified to homogeneity by Colby and Zatman (1973, 1974), who showed that it catalyses the anaerobic, oxidative demethylation of trimethylamine so producing dimethylamine and formaldehyde:

$$(CH_3)_3N + PMS + H_2O \longrightarrow (CH_3)_2NH + HCHO + PMSH_2$$

The pH optimum is 8·5; the molecular weight is about 160 000; the only electron acceptors used (in order of preference — out of 15 tested) are phenazine methosulphate (PMS), brilliant cresyl blue and methylene blue; and trimethylamine is the best substrate.

A comprehensive investigation of its substrate specificity showed that only secondary and tertiary amines having N-methyl or N-ethyl groups are oxidised. Primary amines, quaternary ammonium salts and diamines are not substrates. The best substrates are tertiary amines and diethylamine which has, however, a low affinity for the enzyme (K_m 1·7 mM) compared with that for tertiary amines (usually 2–10 μM).

The activity of the dehydrogenase is unaffected by chelating agents and carbonyl reagents, but is inhibited by some thiol-binding reagents, by heavy metal ions and by some substrate analogues. A particularly important observation, indicating that trimethylamine dehydrogenase must have an unusual mechanism, is its inhibition by monoamine oxidase inhibitors of the substituted hydrazine and non-hydrazine types. Inhibition by substituted hydrazines is irreversible, prevented by trimethylamine and sometimes requires the electron acceptor PMS (Colby and Zatman, 1974).

The unusual nature of the trimethylamine dehydrogenase, as shown by Colby and Zatman (1974), has been extensively confirmed by Steenkamp and his colleagues using the enzyme from a closely related methylotroph, organism W3A1. In all respects tested this dehydrogenase resembles that of organism 4B6 (Steenkamp and Mallinson, 1976). It has a molecular weight of 147 000 and consists of two non-identical subunits of 70 000 to 80 000 daltons. The mechanism most consistent with the kinetic data is the Ping Pong mechanism;

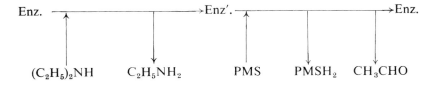

Diethylamine was used in these experiments because of the low accuracy obtainable with trimethylamine which has a very high affinity for the enzyme (Steenkamp and Mallinson, 1976). The absorption spectrum of the enzyme in the visible region has a maximum at 445 nm; addition of substrate leads to a decrease in absorption at this wavelength and a concomitant increase at 360 nm (Steenkamp and Mallinson, 1976).

Early experiments on the pure enzyme from organism W3A1 indicated that it might have two prosthetic groups, one containing nonhaem iron and acid-labile sulphide, and the other an unknown covalently-bound chromophore whose preliminary characterisation indicated that it might be a phosphorylated pteridine (lumazine) derivative (Steenkamp and Mallinson, 1976; Steenkamp and Singer, 1976). The prosthetic groups from the trimethylamine dehydrogenase of *Hyphomicrobium* X are the same as for this enzyme from organism W3A1 (Steenkamp, 1979).

7. The oxidation of methylated amines

One of the prosthetic groups has now been identified as a $Fe_4S_4^*$ core unit, probably present as a ferredoxin-type, cysteine-ligated cluster $[Fe_4S_4^*(S-Cys)_4]$ (Hill et al., 1977). There is one of these iron-sulphur centres in each molecule of enzyme. It has been pointed out by Hill et al. (1977) that although more than 20 enzymes are known to contain iron-sulphur centres in addition to an organic prosthetic group (usually flavin), succinate dehydrogenase is the only member of the group to contain an $Fe_4S_4^*$ core unit.

The second prosthetic group has now been isolated, purified and characterised by means of absorption, fluorescence and proton NMR spectroscopy; by chemical analysis of the isolated prosthetic group; and by chemical synthesis of its photolysis product (Steenkamp et al., 1978b, c). The prosthetic group is a flavin, but instead of being substituted at the 8 position, as in all other covalently-bound flavins, the prosthetic group is substituted at the 6-position. The prosthetic group is 6-S-cysteinyl-FMN, the cysteine residue being part of the polypeptide chain in the complete dehydrogenase (Fig. 27).

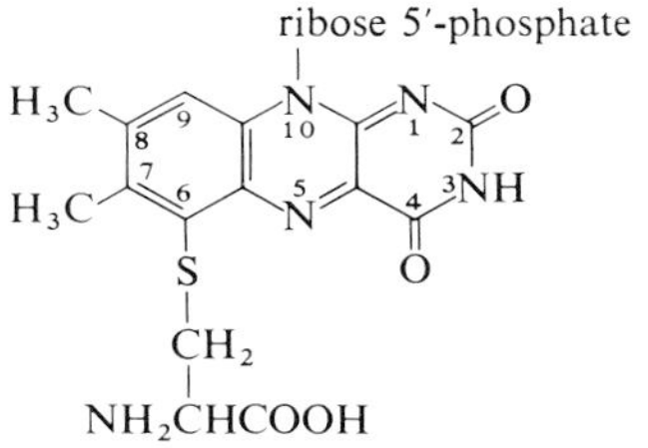

Fig. 27. The prosthetic group of trimethylamine dehydrogenase. This structure is from Steenkamp et al. (1978b, c). The prosthetic group is 6-S-cysteinyl-FMN, the cysteine residue being part of the polypeptide chain in the complete dehydrogenase. Flavins are usually bonded to enzyme proteins by an 8-methylene bridge (or non-covalently bonded).

As a first step in understanding the reaction mechanism of the dehydrogenase, Steenkamp et al. (1978a) have examined the catalytic intermediates by a combination of rapid-freeze ESR and stopped-flow spectrophotometry. These techniques have demonstrated two unusual properties of the enzyme; a very intense spin–spin interaction between its flavin semiquinone and its iron-sulphur centre on reduction with substrate, and the fact that the rate of formation of this species from the reduced flavoprotein seems to be the rate-limiting step in catalysis. It was shown that the initial reduction of the flavin to the quinol form, and the re-oxidation of the substrate-reduced enzyme by PMS are too rapid to be rate-limiting.

The trimethylamine dehydrogenase from Hyphomicrobium vulgare NQ-521 has been used as the basis of an enzymic method for microestimation of trimethylamine (Large and McDougall, 1975) (Chapter 12).

The natural electron acceptor from trimethylamine dehydrogenase appears to be an electron transferring flavoprotein having a molecular weight of

77 000 and composed of 2 dissimilar subunits of 38 000 and 42 000. This has been purified and partially characterised (Steenkamp and Gallup, 1978). It contains one mole of FAD per mole of protein and it cycles between the oxidised and anionic semiquinone forms during catalysis, rather than between the hydroquinone and semiquinone forms. By analogy with other electron transferring flavoproteins, this is likely to interact with the cytochrome chain prior to cytochrome b.

B. The indirect route for trimethylamine oxidation

This route involves a trimethylamine monooxygenase whose product is trimethylamine N-oxide, followed by a demethylase which converts this compound to dimethylamine plus formaldehyde.

1. Trimethylamine monooxygenase

This enzyme catalyses the following reaction:

$$(CH_3)_3N + NADPH + H^+ + O_2 \longrightarrow (CH_3)_3NO + H_2O + NADP^+$$

The inducible enzyme from *Pseudomonas aminovorans* has been purified (about 6-fold) and characterised by Boulton *et al.* (1974). NADPH (K_m 14 μM) is preferred as electron donor to NADH (K_m about 2 mM). The K_m for O_2 is too low to be determined. The enzyme oxidises a wide range of secondary and tertiary amines, but the best substrate is trimethylamine. All active substrates are seen to bear at least two unsubstituted alkyl groups, and all active tertiary amines bear at least two methyl or ethyl groups. The products of dimethylamine oxidation by trimethylamine monooxygenase are N-methylhydroxylamine plus formaldehyde, rather than the methylamine plus formaldehyde produced by the more specific dimethylamine monooxygenase present in the same organism; it appears that the oxidation of dimethylamine by trimethylamine monooxygenase has no physiological function (Boulton and Large, 1975). Although not purified extensively nor characterised in detail, it can be concluded that the monooxygenases for trimethylamine and dimethylamine are completely different types of enzyme; the trimethylamine monooxygenase is almost insensitive to carbon monoxide whereas the dimethylamine monooxygenase is one of the most CO-sensitive enzymes ever described.

2. Trimethylamine N-oxide demethylase

This enzyme, which is an aldolase, was first discovered in *Bacillus* PM6 by Myers and Zatman (1971) and shown to catalyse the non-oxidative, non-hydrolytic cleavage of trimethylamine N-oxide to dimethylamine plus formaldehyde:

$$(CH_3)_3NO \longrightarrow (CH_3)_2NH + HCHO$$

It is essential for growth on trimethylamine N-oxide and on trimethylamine in those bacteria lacking trimethylamine dehydrogenase. The enzyme from *Bacillus* PM6 has been purified (175-fold) to homogeneity. It has a pH optimum of 7·5 and consists of a single polypeptide chain of between 37 000 and 50 000. The K_m for its only substrate trimethylamine N-oxide is 2·85 mM. The demethylase shows no absorption in the visible region of the spectrum and is strongly stimulated by ferrous ions, glutathione and L-ascorbate. The less stable demethylase from *Pseudomonas aminovorans* is rather different from this; it has a lower pH optimum (6·0) and is not stimulated by ferrous ions or glutathione (Large, 1971; Boulton and Large, 1979b), has been purified 25-fold and shown to have a molecular weight of between 240 000 and 280 000 and a low isoelectric point (pH 4·7); cyanide is a powerful inhibitor (C.A. Boulton and P.J. Large, unpublished observations).

IV. The oxidation of dimethylamine to methylamine and formaldehyde

The same routes operate for oxidation of dimethylamine when it is a growth substrate as operate when it is an intermediate in the oxidation of higher methylated amines. It is usually oxidised by dimethylamine monooxygenase but in the *Hyphomicrobia* when grown under low (or zero) O_2 tension a dimethylamine dehydrogenase is induced.

A. Dimethylamine monooxygenase

This enzyme catalyses the oxidation of dimethylamine to methylamine plus formaldehyde in a reaction requiring NAD(P)H and O_2:

$$(CH_3)_2NH + NAD(P)H + H^+ + O_2 \longrightarrow CH_3NH_2 + HCHO + NAD(P)^+ + H_2O$$

It was first described in *Pseudomonas aminovorans* (Eady *et al.*, 1971) and has subsequently been shown to be the only enzyme oxidising dimethylamine during aerobic growth of methylotrophs on methylated amines.

The enzyme from *Ps. aminovorans* (purified about 5-fold) has a pH optimum of about 7·0. The V_{max} is the same with NADH (K_m, 6·5 μM) and NADPH (K_m 13·2 μM) as electron donors. Dimethylamine is the best substrate K_m, 60 μM) and some other secondary amines are substrates. The enzyme shows no activity with primary, tertiary or quaternary amines. It is inhibited by mercurials, thiol compounds, cyanide and carbon monoxide. A preliminary analysis of the enzyme showed the presence of flavin, acid-labile iron and non-acid-extractable iron and it showed an ESR signal at about $g = 1·946$ on reduction with NADH. Subsequent work showed that the preparation used for this analysis was only about 10% pure and so these results should be treated with caution.

A kinetic investigation by Brook and Large (1976) using a 100-fold purified enzyme has indicated the following reaction sequence (a Hexa Uni Ping Pong (triple transfer) reaction):

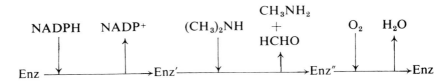

Using the more purified enzyme, these authors (Brook and Large, 1975) have characterised the most important unusual feature of this enzyme — its extreme sensitivity to carbon monoxide. The partition constant (the $CO:O_2$ ratio causing 50% inhibition) was about 10^{-3} and the light sensitivity (L – the reciprocal of the intensity of light required to double the dark partition constant) was calculated to be $2·5 \times 10^8$ cm^2 (mole quantum)$^{-1}$ at 417 nm. The low partition constant is lower than any previously reported (being 5-fold lower than that for reaction of CO with haemoglobin), and it emphasises the extreme sensitivity of the enzyme to CO. The value for light sensitivity (L) again shows greater similarity to haemoglobin than to microsomal monooxygenases.

The absorption spectrum of the NADH-reduced (oxygenated) enzyme has peaks at 422 nm, 540 nm and 575 nm, which is similar to other reduced haemoproteins that have the capacity for binding O_2 (cytochrome P-450, cytochrome o and haemoglobin). The spectrum of the dithionite-reduced enzyme (peaks at 422 nm, 526 nm and 557 nm) is similar to typical b-type cytochromes but differs from cytochrome P-450. Reaction with CO displaces the O_2 and leads to formation of the carboxy-enzyme having peaks at 426 nm, 534 nm and 565 nm, which is similar to the carboxy-derivative of other O_2-reactive haemoproteins (some cytochromes o, some cytochromes P-450 and haemoglobin). That the absorption spectra described here are due to the prosthetic group of the monooxygenase, has been confirmed by showing that the photochemical action spectrum for the light reversal of CO inhibition has a clearly defined peak at about 420 nm.

It thus appears that dimethylamine monooxygenase is a b-type, CO-binding cytochrome that probably acts in a manner analogous to cytochrome P-450 or cytochrome P-420 in other hydroxylation systems.

B. Dimethylamine dehydrogenase

Hyphomicrobium X is able to grow anaerobically on trimethylamine or dimethylamine in the presence of nitrate as terminal electron acceptor. Under these conditions the O_2-requiring dimethylamine monooxygenase, usually involved in dimethylamine oxidation, cannot function and a dehydrogenase must therefore replace it. The trimethylamine dehydrogenase in *Hyphomicrobium* X is able to oxidise dimethylamine but only very slowly. These considerations have led to the discovery of a dimethylamine dehydrogenase in this organism which is specific for secondary amines (Meiberg, 1979; Meiberg and Harder, 1978, 1979) and which catalyses the following reaction;

$$(CH_3)_2NH + PMS + H_2O \longrightarrow CH_3NH_2 + HCHO + PMSH_2$$

This dehydrogenase replaces the monooxygenase during growth on trimethylamine or dimethylamine at low O_2 tensions and in the absence of O_2 with nitrate as terminal electron acceptor (Meiberg *et al.*, 1980).

The dehydrogenase from *Hyphomicrobium* X has been purified (16-fold) and characterised by Meiberg and Harder (1979). This enzyme was 90% pure and had a specific activity of 364 nmoles $(min)^{-1}$ $(mg)^{-1}$. It has a molecular weight of 176 000 and consists of two, probably identical, subunits of 91 000. The pH optimum is 8·1 and suitable artificial electron acceptors besides phenazine methosulphate (PMS) are phenazine ethosulphate, methylene blue and Wurster's blue. Dimethylamine is the best substrate (K_m 16 μM); and, of a wide range of potential substrates tested, only secondary amines are oxidised. Trimethylamine is not oxidised but has a high affinity for the dehydrogenase, acting as a potent competitive inhibitor (K_i 7·1 μM); this leads to accumulation of dimethylamine in the growth medium during anaerobic growth of *Hyphomicrobium* X on trimethylamine.

Dimethylamine dehydrogenase has an absorption spectrum very similar to that of trimethylamine dehydrogenase (above). A peak in the oxidised form at 441 nm is diminished on reduction with substrate which leads to appearance of a new peak at 356 nm. The dehydrogenase has two prosthetic groups which are the same as those in trimethylamine dehydrogenase; a single tetrameric iron-sulphur centre and an unusual covalently-bound flavin, 6-S-cysteinyl-FMN (Steenkamp, 1979) (Fig. 27).

Large et al. (1979) have shown that neither pure dimethylamine dehydrogenase nor partially pure trimethylamine dehydrogenase is able to reduce a partially-purified, soluble cytochrome c from *Hyphomicrobium* X. Because the prosthetic groups are the same as those of trimethylamine dehydrogenase, it is probable that the dimethylamine dehydrogenase reacts with an electron-transferring flavoprotein which in turn probably reacts with the cytochrome chain at, or prior to, cytochrome b.

V. The oxidation of methylamine to formaldehyde

As can be seen from Fig. 26 methylamine may be oxidised directly by methylamine dehydrogenase or oxidase; or indirectly by way of *N*-methylglutamate and a dehydrogenase oxidising it to formaldehyde plus glutamate. This is the most important indirect route but others may occur which involve other *N*-methylated amino acids (Fig. 29). The distribution of these various routes is summarised in Tables 39 and 40 and the individual reactions discussed below.

A. Methylamine dehydrogenase

This enzyme (also called primary amine dehydrogenase) catalyses the oxidation of methylamine to formaldehyde and ammonia in the presence of an artificial electron acceptor (usually phenazine methosulphate, PMS). The prosthetic group, which donates electrons to the PMS, is almost certainly the same as that of methanol dehydrogenase (PQQ) so the overall reaction is:

$$CH_3NH_2 + PQQ + H_2O \longrightarrow HCHO + NH_3 + PQQH_2$$

This inducible methylamine dehydrogenase was first described in the pink facultative methylotroph *Pseudomonas* AM1 and thoroughly investigated by Eady and Large (1968, 1971), and later by Matsumoto and his colleagues (Shirai et al., 1978; Matsumoto et al., 1980). An almost identical enzyme has been described in detail in the obligate methylotroph *Pseudomonas* sp. J. (Matsumoto, 1978); and the methylamine dehydrogenases from other obligate methylotrophs (*Pseudomonas* RJ3 and *Methylomonas methylovora*) appear to be similar to these enzymes (Mehta, 1976, 1977). This dehydrogenase is completely different from the primary amine dehydrogenase, which has a haem c prosthetic group, found in *Pseudomonas putida* but not found in methylotrophs (Durham and Perry, 1978a, b, c).

1. Enzyme assay, substrates and inhibitors of methylamine dehydrogenase

The methylamine dehydrogenase constitutes 3–5% of the soluble protein of the methylamine-grown cells and the completely pure enzyme has a specific activity between 2 and 6 μmoles methylamine oxidised per minute per mg protein measured at the pH optimum of 7·5.

The substrate specificity depends on the source of the enzyme. The dehydrogenases from obligate methylotrophs oxidise a very limited range of primary amines, with methylamine being the best substrate. By contrast the enzyme from *Pseudomonas* AM1 oxidises a wide range of primary aliphatic amines and diamines at rates as high, and sometimes higher than, that observed with methylamine; these include, for example, *n*-pentylamine, 1,3-diaminopropane, spermidine, ethanolamine and histamine. Although some substrates are oxidised at a greater rate than methylamine, the affinity of the enzyme for substrate is probably greatest for methylamine (K_m 5·2 μM). Secondary and tertiary amines are not oxidised (Eady and Large, 1968).

The best primary electron acceptor is phenazine methosulphate (PMS) (K_m 18–28 μM). Alternative electron acceptors give much lower rates and they have much lower affinities for the enzyme; these include; mammalian cytochrome *c*, ferricyanide, 2,6-dichlorophenolindophenol, brilliant cresyl blue, Wurster's blue and resazurin.

The methylamine dehydrogenase is relatively insensitive to inhibitors that react with metal ions or with thiol groups, confirming the conclusion that metals and thiol groups are not important in enzyme catalysis. The most powerful inhibitors are those reacting with carbonyl groups (semicarbazide, hydroxylamine, isoniazid). This is consistent with the earlier suggestion that the prosthetic group may be a pyridoxal derivative, but it is also consistent with the more recent suggestion of the *o*-quinone prosthetic group (PQQ).

2. The catalytic mechanism of methylamine dehydrogenase

The methylamine dehydrogenase is greenish-yellow, becoming less coloured on addition of substrate. The absorption spectrum of the oxidised enzyme has a peak at about 280 nm, and a peak due to the prosthetic group at 430 nm with a shoulder at 460 nm. When reduced by addition of methylamine or borohydride the 430 nm peak diminishes and a new peak at 325 nm appears. The spectrum of the oxidised enzyme returns after addition of the electron acceptor PMS. The spectrum is also affected by reaction with carbonyl reagents. The reduced form of the enzyme-bound prosthetic group is fluorescent with an excitation maximum at about 330 nm and an emission maximum at 380 nm.

Investigations of the kinetics of methylamine dehydrogenase with respect to amines and PMS concentrations, measured in the direction of methylamine oxidation, have shown that catalysis occurs by way of a Ping Pong mechanism. In such a mechanism one substrate combines with enzyme to produce a modified enzyme and one product is released. The second substrate then combines with the enzyme to give the second product and the enzyme in its original form. This evidence, together with the spectral evidence quoted above, has led to the following postulated reaction sequence (Eady and Large, 1971):

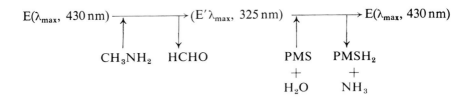

The prosthetic group is covalently bound to the enzyme and so cannot be readily released for comparison with the prosthetic group of methanol dehydrogenase (PQQ) (Fig. 22, Chapter 6) but the ESR and ENDOR spectra are sufficiently similar to conclude that the prosthetic group of methylamine dehydrogenase is also PQQ. It has been suggested that some differences in the ENDOR spectra are because either the structure of the prosthetic group or its interaction with the protein is different in methanol and methylamine dehydrogenases (de Beer *et al.*, 1980). If PQQ is the prosthetic group of methylamine dehydrogenase then a number of questions arise about its mechanism; perhaps the most important is whether or not the amine substrate binds to the prosthetic group (as in my speculation in Fig. 28), or whether the *o*-quinone of PQQ merely acts as an electron acceptor.

The natural electron acceptor from the methylamine dehydrogenase is not known but, like the PQQ-containing methanol dehydrogenase, it may pass its electrons (directly or indirectly) to one of the cytochromes *c* present in these bacteria. The preliminary results described by Matsumoto (1978) suggest that if cytochrome *c* is the acceptor then electron transfer is probably indirect.

3. The subunit structure of methylamine dehydrogenase

This section is a summary of the work of Matsumoto and his colleagues on the subunit structure of the methylamine dehydrogenases of *Pseudomonas* AM1 and *Pseudomonas* sp. J (Matsumoto *et al.*, 1978, 1980; Matsumoto and Tobari, 1978b).

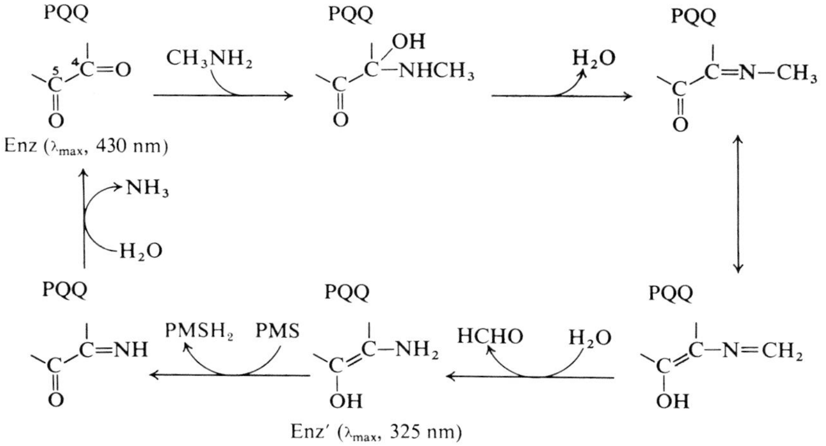

7. The oxidation of methylated amines   207

Fig. 28. A speculative mechanism for methylamine dehydrogenase. In this mechanism the prosthetic group PQQ is involved in binding the methylamine substrate. Reaction with methylamine might occur at positions 4 (as here) or 5. Alternative mechanisms in which PQQ alternates between the o-quinone forms, without binding substrate in a Schiff's base, can also be envisaged. The full structure of the PQQ prosthetic group is given in Fig. 22. The kinetic mechanism is given on page 206.

Both enzymes have a molecular weight of 105 000; each consists of two light subunits (13 000 daltons) and two heavy subunits (40 000 daltons) arranged in a $\alpha_2\beta_2$ configuration. Both types of subunit are required for activity but they can be separated and reconstituted to study their individual properties. The prosthetic group is bound to the light subunits and the amino acid composition of these subunits is very similar in the two enzymes. By contrast, the amino acid composition of the heavy subunit form of the enzyme from *Pseudomonas* sp. J differs from that of the *Pseudomonas* AM1 enzyme and this leads to other differences in the properties of the two enzymes. For example, there is a higher lysine content in the "J" enzyme and this leads to a high isoelectric point for the enzyme (pI, 9·0) compared with that of the "AM1" enzyme (pI, 5·2). The circular dichroism spectra show that there is some β structure (35%) in the heavy subunit of the AM1 enzyme but no ordered structure in the J enzyme. This relatively more ordered structure in the AM1 enzyme is probably the cause of its greater stability to extremes of heat and pH and to its relative ease of crystallisation. This stability makes it particularly suitable for immobilisation studies and Boulton and Large (1979a) have shown that when immobilised on agarose its kinetic properties are almost unchanged. The functions of the subunits have been further investigated by producing an active hybrid enzyme, consisting of the light subunit from the J enzyme plus the heavy subunit from the AM1 enzyme. No hybrid could be prepared from the light AM1 subunit plus the heavy J subunit. The conclusion drawn from these studies is that the substrate specificity and catalytic activity of methylamine dehydrogenase are determined by the light subunit, containing the prosthetic group; and that affinity for substrate, electrophoretic mobility and thermal stability are endowed by the heavy subunit (Matsumoto *et al.*, 1980).

The methylamine dehydrogenase from *Pseudomonas* AM1 has been used as the basis for an enzymic method for the microestimation of primary amines (Large *et al.*, 1969) (Chapter 12).

B. Methylamine oxidase (amine oxidase)

This primary amine oxidase uses molecular oxygen and produces formaldehyde, ammonia and hydrogen peroxide:

$$CH_3NH_2 + O_2 + H_2O \longrightarrow HCHO + NH_3 + H_2O_2$$

The energy produced in this reaction is not harnessed in a metabolically usable form, and high concentrations of catalase are required to remove the H_2O_2. Cytochemical studies have shown that in *Arthrobacter* P1 the amine oxidase and catalase are located on invaginations of the cytoplasmic membrane (Levering *et al.*, 1981). The importance of catalase was shown by the inhibitory effect of a catalase inhibitor (aminotriazole) during growth on glucose with methylamine, but not ammonia, as nitrogen source. The amine oxidase from *Arthrobacter* P1 is probably a copper-containing enzyme (van Vliet Smits *et al.*, 1981).

An amine oxidase is also involved in the oxidation of methylamine to formaldehyde in yeasts. Although most yeasts can use one or more of the methylated amines as nitrogen source they are unable to use them as sole source of carbon and energy (van Dijken and Bos, 1981). It has been suggested by Zatman (1981) that these yeasts, and the many bacteria that also use methylated amines as their sole source of nitrogen (but not of carbon) (Bicknell and Owens, 1980), could be called methazotrophs. Presumably all such methazotrophs must be able to oxidise (or perhaps excrete) the potentially lethal formaldehyde produced during oxidation of methylated amines to ammonia; but whether or not energy is available to the organisms from this process is not known.

In methazotrophic yeasts, the oxidase and catalase are contained within peroxisomes which are induced together with dehydrogenases for formaldehyde and formate during growth on glucose with methylamine as nitrogen source (Zwart *et al.*, 1980; Veenhuis *et al.*, 1981). The purified amine oxidase from *Candida boidinii* oxidises primary alkylamines from C_1 to C_{10} (with decreasing efficiency) but not secondary or tertiary amines. It is very sensitive to some carbonyl reagents but not to typical "monoamine oxidase inhibitors" (Large et al., 1980).

It is not obvious why methylotrophic yeasts should be able to grow on methanol by assimilating the formaldehyde produced by oxidation of methanol by a peroxisomal methanol oxidase but be incapable of growing on methylamine by assimilating the formaldehyde produced by oxidation of methylamine by an analogous peroxisomal methylamine oxidase.

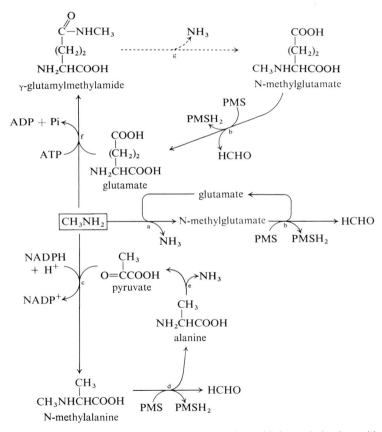

Fig. 29. The possible involvement of N-methylated amino acids in methylamine oxidation. (a) N-methylglutamate synthase (see text for references); (b) N-methylglutamate dehydrogenase (see text for references); (c) N-methylalanine synthase (Kung and Wagner, 1970b); (d) N-methylalanine dehydrogenase (Lin and Wagner, 1975); (e) deamination of alanine; (f) γ-glutamylmethylamide synthetase (Kung and Wagner, 1969; Trotsenko et al., 1974; Loginova et al., 1976; Levitch 1977a, b; Meiberg and Harder, 1978); (g) postulated sequence (no evidence) (Trotsenko et al., 1974; Loginova et al., 1976; Meiberg and Harder, 1978).

210   The biochemistry of methylotrophs

C. The oxidation of methylamine by way of methylated amino acids

The following N-methylated amino acids have been implicated in the metabolism of methylated amines by methylotrophs; N-methylglutamate, N-methylalanine and γ-glutamylmethylamide. Although there have been suggestions that these compounds may be intermediates in the assimilation of formaldehyde, it is more likely that, when physiologically important, they are intermediates in the oxidation of methylamine as indicated in Fig. 29. The system most extensively investigated and whose operation has been demonstrated most widely is that involving N-methylglutamate synthase and dehydrogenase, and this is described below. An almost analogous system involving N-methylalanine synthesis and oxidation has also been described in *Pseudomonas* MS (Kung and Wagner, 1970b; Lin and Wagner, 1975). However, this system may not be involved in growth on methylated amines because more methylalanine dehydrogenase is present during growth on glucose or succinate than on amines (Lin and Wagner, 1975; Boulton *et al.*, 1980); and mutants of *Pseudomonas aminovorans*, unable to grow on amines because of the loss of a whole series of enzymes involved in C_1 oxidation and assimilation, still retained this enzyme (Bamforth and O'Connor, 1979).

1. The synthesis of N-methylglutamate

The description of N-methylglutamate synthase from *Pseudomonas* MA was the first demonstration of the importance of N-methylglutamate in any biological system (Shaw *et al.*, 1966; Shaw and Stadtman, 1970). The inducible synthase catalyses the reversible synthesis of N-methyl-L-glutamate from glutamate plus methylamine, with the production of ammonia. The reaction proceeds by a direct displacement mechanism and not by a transmethylation:

$$CH_3NH_2 + \underset{\underset{\underset{COOH}{|}}{\underset{(CH_2)_2}{|}}}{NH_2CHCOOH} \longrightarrow \underset{\underset{\underset{COOH}{|}}{\underset{(CH_2)_2}{|}}}{CH_3NHCHCOOH} + NH_3$$

The enzyme has been purified to homogeneity, and characterised by Pollock and Hersh (1971, 1973). In the presence of glutamate its molecular weight is 350 000, but in its absence the enzyme dissociates into 12 subunits of 30 000 to 35 000 daltons. The synthase is specific for glutamate, but methylamine (the best substrate) can be replaced by other amines. The

7. The oxidation of methylated amines

enzyme, whose pH optimum is about 9, is inactivated by carbonyl- and thiol-binding reagents. Resolution and reconstitution studies have shown that the synthase is a flavoprotein whose FMN prosthetic group undergoes an oxidation-reduction cycle during catalysis. Kinetic and mechanistic studies have confirmed that this enzyme has a particularly unusual mechanism. The first step in the reaction sequence (Fig. 30) is an oxidative attack on the glutamate by the FMN prosthetic group at the α-carbon atom, which is thus activated towards nucleophilic attack by methylamine. After displacement of ammonia, reduction of the α-carbon by the reduced FMN completes the catalytic cycle.

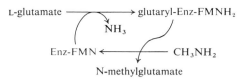

Fig. 30. Proposed reaction sequence for N-methylglutamate synthase (from Pollock and Hersh, 1973).

Studies with whole cells and extracts of some bacteria (e.g. *Pseudomonas* MS and *Hyphomicrobium*) have indicated that an alternative route for synthesis of *N*-methylglutamate from methylamine may occur (Fig. 29). This route involves the formation of γ-glutamylmethylamide from glutamate plus methylamine, catalysed by an inducible γ-glutamylmethylamide synthetase which requires ATP for activity (Kung and Wagner, 1969; Loginova *et al.*, 1976; Loginova and Trotsenko, 1976b; Meiberg and Harder, 1978).

Another novel reaction catalysed by a methylotroph (*Pseudomonas* MA) is the synthesis of 5-hydroxy-*N*-methylpyroglutamate from methylamine plus 2-oxoglutamate, but the biological significance of this reaction is unknown (Hersh *et al.*, 1969).

2. The oxidation of N-methylglutamate

(a) N-*methylglutamate dehydrogenase* (NAD^+-*independent*). This enzyme, first described by Hersh *et al.*, (1971, 1972), catalyses the oxidative demethylation of *N*-methylglutamate, producing glutamate and formaldehyde:

N-methylglutamate + PMS + H_2O ⟶ glutamate + HCHO + $PMSH_2$

The enzyme is usually assayed with the artificial electron acceptors phenazine methosulphate (PMS) or 2,6-dichlorophenolindophenol. The properties of the soluble enzyme from the pink facultative methylotroph *Pseudomonas* AT2, and the solubilised enzymes from two non-pigmented "pseudo-

TABLE 38
The properties of N-methylglutamate dehydrogenase

	Pseudomonas MA	Pseudomonas aminovorans	Pseudomonas AT2
References	Hersh et al. (1971, 1972)	Bamforth and Large (1977a, b)	Boulton et al. (1980)
Solubility	particulate	particulate	soluble
Purification	5·6-fold (after solubilisation)	22-fold (15–20% pure) (after solubilisation)	6·6-fold (nearly homogeneous)
Specific activity (when purified)	50·8 nmoles $(\min)^{-1}$ $(mg)^{-1}$	55·4 nmoles $(\min)^{-1}$ $(mg)^{-1}$	113 nmoles $(\min)^{-1}$ $(mg)^{-1}$
Molecular weight (subunit)		550 000 (tetramer) (130 000)	407 000 (tetramer) (108 000)
Electron acceptors	phenazine methosulphate, 2,6-dichlorophenolindophenol, ferricyanide	phenazine methosulphate, 2,6-dichlorophenolindophenol, cytochrome c	phenazine methosulphate, 2,6-dichlorophenolindophenol, Wurster's blue and the radical cation of ABTS
Inhibitors	thiol-binding agents	thiol-binding agents	
Kinetic mechanism	Ping Pong	Ping Pong	Ping Pong
Substrates	N-methyl derivatives of: glutamate, aspartate, alanine, valine, isoleucine, phenylalanine, serine, glycine	N-methyl derivatives of: glutamate, aspartate, alanine and glycine	N-methyl derivatives of: glutamate, aspartate, alanine and glycine
Prosthetic group	not determined	Flavin	Flavin (non-covalently bound FAD)

monads" (*Pseudomonas* MA and *Pseudomonas aminovorans*) are summarised in Table 38. Although the differences in preparation — necessary because of the differences in solubility of the enzymes — makes some comparisons difficult, the dehydrogenases are similar in most of their characteristics including: pH optima (6·5–7·5), large molecular weight, tetrameric structure, substrate specificity and sensitivity to inhibitors. *N*-methylglutamate is always the best substrate.

In terms of their physiological function, the most important consideration is the nature of their prosthetic groups which appear to be flavin derivatives. The spectra of the more purified enzymes demonstrate the presence of substrate-reducible flavin groups and some cytochrome or haem. In the case of the solubilised enzyme from *Pseudomonas aminovorans*, the low concentration of a *b*-type cytochrome is not reducible by substrate and is probably a contaminant. By contrast, the spectrum of the purified, soluble dehydrogenase from *Pseudomonas* AT2 indicates the presence of a substrate-reducible *c*-type cytochrome (Boulton *et al*., 1980). It was concluded from this observation that *N*-methylglutamate dehydrogenase may be similar to methanol dehydrogenase in reacting with the cytochrome chain at the level of cytochrome *c*. However, assuming the usual extinction coefficients for flavoprotein cytochromes, it can be calculated that there is only about 10% as much cytochrome *c* as substrate-reducible flavin in the enzyme preparation, perhaps indicating that reduction of the cytochrome *c* by the enzyme may be fortuitous. Clearly, further investigation is required before firm conclusions are possible. This is particularly so in the light of the conclusion, based on inhibitor studies of the particulate *N*-methylglutamate dehydrogenase from *Pseudomonas* MA, that this dehydrogenase, like most flavoproteins, interacts with the cytochrome chain at, or prior to, the level of cytochrome *b* (Hersh *et al*., 1971, 1972).

The specific activities of *N*-methylglutamate dehydrogenases in crude extracts of methylotrophs are usually too low to account for the growth rates measured on methylated amines. However, these low activities are probably the result of measuring membrane proteins with artificial electron acceptors; and the fact that the synthesis of the dehydrogenase is induced during growth on methylated amines is indicative of its physiological importance (see Boulton and Large, 1977; Bamforth and Large, 1977a, b; Bamforth and O'Connor, 1979).

(*b*) *N-methylglutamate dehydrogenase* (NAD^+-*linked*). This unstable, soluble enzyme, first described by Netrusov (1975), is present in a number of methylotrophs including some *Pseudomonas* sp. (e.g. *Ps. methylica* sp. 2) and some *Hyphomicrobia* (Loginova and Trotsenko, 1974; Loginova *et al*., 1976, 1977, 1978).

TABLE 39
Distribution of enzymes involved in methylamine oxidation in facultative methylotrophs

Organism	Methylamine dehydrogenase	N-methylglutamate dehydrogenase	References
Pink facultative methylotrophs (serine pathway, icl−; Table 11)			
Pseudomonas AM1	+	nd	Eady and Large, (1968 1971); Shirai et al. (1978)
Pseudomonas sp. M27, 1 and 135	+	nd	Rock et al. (1976)
Pseudomonas extorquens	+	nd	Marison and Attwood (1980)
Pseudomonas 3A2	−	(+)	Colby and Zatman (1973)
Pseudomonas methylica sp. 2	−	+(N)	Loginova and Trotsenko (1974); Netrusov (1975); Marison and Attwood (1980)
Pseudomonas AT2	−	+	Boulton et al. (1980)
Non-pigmented 'pseudomonads' (serine pathway, icl+; Table 12)			
Pseudomonas aminovorans	−	+	Bamforth and Large (1977a, b); Bamforth and O'Connor (1979)
Pseudomonas MS	−	+	Kung and Wagner (1970b); Hersh et al. (1971)
Pseudomonas MA	nd	+	Hersh et al. (1971, 1972)
The Hyphomicrobia (serine pathway, icl−)			
Hyphomicrobium X	−	+	Meiberg and Harder (1978); Marison and Attwood (1980)
Hyphomicrobium ZV	−	+(N)	Loginova et al. (1976)
Hyphomicrobium 3	−	+(N)	Trotsenko et al. (1974); Loginova et al. (1977); Meiberg and Harder (1978); Marison and Attwood (1980)

Organism		Reference
Gram-negative (or variable) non-motile rods (RuMP pathway; Table 13)		
*Arthrobacter globiformis B-175	−	Loginova and Trotsenko (1976b)
*Arthrobacter P1	nd	Levering et al. (1981)
Organism 5H2	−	Hampton and Zatman (1973)
Organism 5B1	−	Colby and Zatman (1973)
Gram-positive bacteria (RuMP pathway; Table 14)		
Bacillus sp. PM6 and S2A1	nd	Colby and Zatman (1975c)
Brevibacterium 24	+	Loginova and Trotsenko (1977b)
Mycobacterium 10	nd	Loginova and Trotsenko (1977b)
The facultative autotrophs (RuBP pathway; Table 15)		
Bacterium 7d	nd	Loginova and Trotsenko (1977b)
Paracoccus denitrificans	nd	Marison and Attwood (1980)
Thiobacillus A2	nd	van Dijken et al. (1981b)
Bacteria which may have two routes (Table 12)		
Pseudomonas oleovorans	+	Loginova and Trotsenko (1977b)
Pseudomonas 20	+	Loginova and Trotsenko (1977b)

In many cases only one of the possible enzymes was measured or reported; it has sometimes been implicitly assumed that the presence of one precludes the operation of another (e.g. Large, 1981). Unless there is a published statement that the presence of an enzyme has been investigated it has been recorded here as not determined (nd); in many cases this is probably a negative result which was not published by the investigators. (+) indicates that there is presumptive evidence for the presence of a given enzyme.

* The Arthrobacter species oxidise methylamine by a primary amine oxidase (Levering et al., 1981a); there are no reports of the presence or absence of this enzyme in the other methylotrophs listed here.

(N) NAD(P)⁺-linked dehydrogenase.

TABLE 40

The oxidation of methylated amines by obligate methylotrophs

Organism	Growth substrates			Enzymes for oxidation of methylated amines		
	Methylamine	Other methylated amines	Methanol	Methylamine	Dimethylamine	Trimethylamine
Pseudomonas sp. W1, RJ3, J	+	−	+	MD	−	−
Methylomonas methylovora	+	−	+	MD	−	−
Methylophilus methylotrophus	+	+	+	MD	MO	TMD
Organism C2A1	+	+	+	MD	MO	TMD
Organisms W3A1 and W6A (more restricted facultative methylotrophs)	+	+	+	MD	MO	TMD
Organism 4B6	+	+	−	NMGD	MO	TMD

These organisms are biochemically very similar to one another; they are described, with references to their oxidising enzymes, in Tables 9 and 10. All assimilate formaldehyde by the RuMP pathway and all have a methylamine dehydrogenase except organism 4B6 which probably has the N-methylglutamate dehydrogenase. All those growing on trimethylamine and dimethylamine oxidise the trimethylamine by a dehydrogenase and dimethylamine by a monooxygenase. MD, methylamine dehydrogenase; MO, dimethylamine monooxygenase; TMD, trimethylamine dehydrogenase; NMGD, N-methylglutamate dehydrogenase.

VI. The distribution of metabolic routes for the oxidation of methylated amines

The main routes for oxidation of methylated amines are summarised in Figs 26 and 29.

A. Oxidation of trimethylamine

Trimethylamine is oxidised by way of the NAD^+-independent dehydrogenase in the obligate (and more-restricted facultative) methylotrophs (Table 9) (Colby and Zatman, 1973, 1975c; Large and Haywood, 1981) and in the *Hyphomicrobia* (Meiberg, 1979). All other methylotrophs (including the less-restricted *Bacillus* sp.) use the monooxygenase (Jarman and Large, 1972; Colby and Zatman, 1973, 1975c; Hampton and Zatman, 1973; Loginova and Trotsenko, 1976b).

B. Oxidation of dimethylamine

Dimethylamine is oxidised by the monooxygenase in all aerobic methylotrophs (Eady *et al.*, 1971; Loginova *et al.*, 1976; Colby and Zatman, 1973, 1975c; Large and Haywood, 1981). It is present in *Hyphomicrobia* when growing under aerobic conditions, but the alternative dimethylamine dehydrogenase is induced in *Hyphomicrobia* during anaerobic growth (Meiberg *et al.*, 1980).

C. Oxidation of methylamine

There are at least three possible systems for oxidation of methylamine, and some generalisations can be made about the distribution of these systems amongst the methylotrophic bacteria (Tables 39 and 40). The picture with respect to enzymes involved in methylamine oxidation is confused by the fact that few studies have considered the possibility of more than one system operating in a single organism. Although this is probably a valid conclusion, the possible presence of two systems in *Pseudomonas* sp. 20 and *Ps. oleovorans* (not described in detail) is a warning that this may not always be the case. The most important conclusions to be derived from the survey in Tables 39 and 40 are as follows:

(1) Most obligate methylotrophs (and the more restricted facultative methylotrophs) oxidise methylamine by the methylamine dehydrogenase and are similar in all other aspects of methylated amine metabolism (Table 40). This further supports the suggestion in Chapter 1 that all these bacteria should be included in the same genus.

(2) The only bacteria known to oxidise methylamine by the amine oxidase are the *Arthrobacter* sp.; the yeasts also have this enzyme.
(3) In the facultative autotrophs, methylamine is oxidised by the methylamine dehydrogenase.
(4) The three non-pigmented "pseudomonads" all oxidise methylamine by way of the NAD^+-independent, N-methylglutamate dehydrogenase.
(5) The *Hyphomicrobia* all oxidise methylamine by way of N-methylglutamate dehydrogenase, some having the NAD^+-dependent and some the NAD^+-independent enzyme.
(6) Some pink facultative methylotrophs have the methylamine dehydrogenase and some have N-methylglutamate dehydrogenase (NAD^+-dependent or independent).
(7) Although most bacteria probably have only one system for methylamine oxidation, *Pseudomonas* 20 and *Ps. oleovorans* appear to have both dehydrogenases.
(8) There is no correlation between the possession of particular pathways of carbon assimilation and the pathway for oxidation of methylated amines.

8
Electron transport and energy transduction in methylotrophic bacteria

I. Introduction 219
II. The cytochromes c of methylotrophs 224
   A. Introduction 224
   B. Reaction of cytochromes c with CO 226
   C. The autoreduction of cytochrome c in methylotrophs 227
III. The role of cytochrome c in the oxidation of methanol 229
   A. Introduction 229
   B. The reaction between methanol dehydrogenase and cytochrome c . 230
IV. Electron transport and proton-translocating systems in methylotrophs . 232
   A. Introduction 232
   B. Electron transport and proton translocation in *Pseudomonas* AM1 . . 233
   C. Electron transport and proton translocation in *Methylophilus methylotrophus* 236
   D. Electron transport and proton translocation in *Paracoccus denitrificans* . 239
   E. Electron transport and proton translocation in *Methylosinus trichosporium* and other methanotrophs 240
V. The coupling of methanol oxidation to ATP synthesis 242

I. Introduction

This chapter is concerned with how the energy available from the oxidation reactions described in Chapters 6 and 7 is harnessed as ATP by way of proton-translocating electron transport chains. The next chapter will deal with wider aspects of bioenergetics and growth yields in methylotrophs.

The electron transport chains effect the oxidation of NADH and the reduced prosthetic groups of dehydrogenases; they consist of flavoproteins, iron-sulphur proteins, quinones (Coenzyme Q), cytochromes and cytochrome oxidases. The flavoproteins and iron-sulphur proteins that are involved in the oxidation of C_1 compounds have been described in Chapters 6 and 7. The type of Coenzyme Q operating in electron transport depends on the type of methylotroph. All those studied are Gram-negative and contain ubiquinone; the obligate methylotrophs have Coenzyme Q_8 (eight isoprenoid units); the *Hyphomicrobia* have Coenzyme Q_9; the pink facultative methylotrophs and

220 The biochemistry of methylotrophs

Microcyclus species have Coenzyme Q_{10} (Drabikowska, 1977, 1981; Natori *et al.*, 1978; Urakami and Komagata, 1979.

In aerobic bacteria using O_2 as terminal electron acceptor, cytochrome chains do not vary greatly in their composition. Such variations that do occur concern the presence or absence of cytochrome c; the nature of the terminal oxidase, which is usually cytochrome a_3 or the *b*-type cytochrome oxidase, cytochrome o; and the branch point to alternative oxidases which occurs at the level of cytochrome b or cytochrome c (Jones, 1977; Haddock and Jones, 1977). That methylotrophs may differ from this normal pattern and that some variation is to be expected within this diverse group of bacteria is indicated by the following special features of the metabolism of methane, methanol and methylated amines:

(1) The methanol dehydrogenase is a novel type of dehydrogenase, having an unusual prosthetic group (Chapter 6, Section III.H). All methanol is oxidised by way of this enzyme including that which is eventually assimilated into cell material. In some bacteria, formaldehyde may also be oxidised by this dehydrogenase. Methylamine dehydrogenase has a similar prosthetic group. This raises the following questions: How does methanol dehydrogenase interact with the electron transport chain? Is a separate electron transport chain involved? Is methanol (and methylamine) oxidation coupled to proton translocation and to ATP synthesis and, if so, how much ATP is available from this oxidation step?

(2) Between 50% and 90% of the oxygen used by electron transport chains is for oxidation of the dehydrogenases for methanol or methylated amines (Chapter 9, Section V).

(3) Some methylotrophs have a particularly high requirement for NAD(P)H; these bacteria will tend to be NAD(P)H-limited and very little electron transport will occur from NADH as electron donor compared with that in typical heterotrophic bacteria (Chapter 9, Section V).

(4) The hydroxylation of methane and methylated amines might require "reversed electron transport" to provide NADH or it might involve "recycling" of electrons from methanol to the oxygenase by way of cytochrome c.

Investigations of electron transport systems in methylotrophs have followed the usual pattern of such investigations and have included the following aspects:

(1) Identification and characterisation of the dehydrogenases for growth substrates and intermediates in their oxidation pathways.

(2) Identification of components of their electron transport systems; this has been mainly limited to cytochromes because these are readily measured without prior purification. This includes the identification of CO-binding cytochromes as a preliminary guide to potential oxidases.

8. Electron transport and energy transduction

(3) Correlation of variations in the nature and amounts of cytochromes with variations in growth substrates and growth conditions.
(4) Isolation and characterisation of mutants lacking particular cytochromes in order to identify better their functions.
(5) Characterisation of electron transport components (especially cytochromes) by determination of their midpoint redox potentials, in order to propose the likely sequence operating between substrate and O_2 in the electron transport chain.
(6) Analysis of electron flow by measurement of substrate-reducible components and by determining the sensitivity to inhibitors of respiration.
(7) Purification and characterisation of electron transport components such as dehydrogenases and cytochromes in order to study their interactions *in vitro*.
(8) Measurement of respiration-coupled proton translocation, membrane potentials and ATP synthesis.
(9) Determination of the configuration of the dehydrogenases and other electron transport components with respect to one another and to the cytoplasmic membrane.

Although no methylotroph has been fully investigated with respect to all the points listed above, some generalisations can be made about their electron transport chains (see Table 41 for references).

All methylotrophs grown on methane and methanol have cytochromes b and c and usually cytochrome a/a_3; but the cytochrome c is often present in exceptionally high concentrations which mask the cytochrome b in spectra of whole bacteria when recorded at room temperature.

In methylotrophs it is usually found that all three types of cytochrome (a, b and c) are able to react with CO to some extent, as shown by measurements of spectra of whole bacteria. Because it is assumed that the reaction of haem iron with CO is analogous to its reaction with O_2, the reaction of CO with cytochromes is used as a preliminary indication of the presence of cytochrome a_3 and cytochrome o (a b-type oxidase). This is, in itself, insufficient evidence for the operation of a particular type of oxidase and some demonstrations of "cytochrome o" in the literature are probably erroneous; it is especially difficult to be certain that a CO-binding b-type cytochrome is cytochrome o if any other oxidase is also present. The cytochromes c of all methylotrophs examined are able to react with CO; this is an unusual feature for a cytochrome c and is discussed further below.

Although the cytochrome complements of different methylotrophs grown on methane or methanol are fairly similar, the proportion of cytochrome types measured in a single species depends on the growth substrate and growth conditions. For example, the concentration of soluble cytochrome c

TABLE 41
Cytochromes, electron transport systems and proton-translocation in methylotrophs

(a) Cytochromes and electron transport systems

Organism	References
Methanotrophs	
Methylomonas albus	Davey and Mitton (1973)
Methylomonas agile	Monosov and Netrusov (1975)
Methylomonas methanica	Tonge et al. (1974); Ferenci et al. (1975); Ferenci (1976a); Babel and Steudel (1977); Patel et al. (1979a)
Methylococcus capsulatus	Ribbons et al. (1970)
Methylosinus trichosporium	Davey and Mitton (1973); Weaver and Dugan (1975); Monosov and Netrusov (1975); Tonge et al. (1974, 1975, 1977a); Higgins et al. (1976a); Higgins (1979a, 1980); Hammond et al. (1979)
Methylosinus sp. GB2	Babel and Steudel (1977)
Methylobacterium organophilum	O'Connor and Hanson (1978); Wolf and Hanson (1978)
Obligate methanol-utilisers	
Methylomonas P11	Drabikowska (1977)
Methylophilus methylotrophus	Cross and Anthony (1980a, b); Dawson and Jones (1981c)
Pseudomonas W6 (MB53)	Babel and Steudel (1977)
Facultative methanol or methylamine-utilisers	
Pseudomonas extorquens	Tonge et al. (1974, 1977b); Higgins et al. (1976a, b)
Pseudomonas AM1	Tonge et al. (1974); Anthony (1975b); Widdowson and Anthony (1975); O'Keeffe and Anthony (1978, 1980a, b); Netrusov and Anthony (1979); Keevil and Anthony (1979a, b); Ivanovsky et al. (1980)

Pseudomonas methylica sp. 2	Netrusov *et al.* (1977)
Pseudomonas MA	Hersh *et al.* (1971)
Pseudomonas MS	Widdowson and Anthony (1975)
Hyphomicrobium	Hirsch *et al.* (1963); Tonge *et al.* (1974); Widdowson and Anthony (1975); Babel and Steudel (1977); Large *et al.* (1979); Duine *et al.* (1979a)
Paracoccus denitrificans	Bamforth and Quayle (1978a); van Verseveld and Stouthamer (1978a); Willison and John (1979); Boogerd *et al.* (1980); Porte and Vignais (1980); Willison and Haddock (1981); Willison *et al.* (1981a, b)

(b) Proton translocation and ATP synthesis in methylotrophs
Papers indirectly related to this topic are included in square brackets

Methylosinus trichosporium	Tonge *et al.* (1977b)
Pseudomonas extorquens	Hammond and Higgins (1978); Higgins (1980); (Hammond *et al.* 1981)
Pseudomonas AM1	O'Keeffe and Anthony (1978); Keevil and Anthony (1979b); Netrusov and Anthony (1979); Netrusov (1981); (Hammond *et al.* 1981)
Methylophilus methylotrophus	Cross and Anthony (1978); Dawson and Jones (1981a, b, c)
Pseudomonas EN (NCIB 11040)	Drozd and Wren (1980)
Paracoccus denitrificans	van Verseveld and Stouthamer (1978a, b); van Verseveld *et al.* (1978, 1981); Alefounder and Ferguson (1981); [Kell *et al.*, 1978; McCarthy *et al.*, 1981; Willison and Haddock, 1981]

For reviews see Higgins (1980) and Anthony (1981).

is higher in methylotrophically-grown *Pseudomonas extorquens* and *Pseudomonas* AM1 than when they are heterotrophically grown; furthermore, the concentrations of all cytochromes (*a*, *b* and *c*) on membranes is markedly lower during methylotrophic growth (Tonge *et al.*, 1974; Widdowson and Anthony, 1975; Higgins *et al.*, 1976a; Keevil and Anthony, 1979b). This is probably related to the greater capacity for NADH oxidation required during heterotrophic growth and to the particular importance of cytochrome *c* during methanol oxidation.

In the facultative methylotroph *Paracoccus denitrificans* there is some indication that the proportions of cytochromes *a*, *b* and *c* may vary with conditions of growth; in particular, it is suggested that alternative oxidases (*o* and a_3) may differ in relative importance during growth on methanol compared with growth on succinate (van Verseveld and Stouthamer, 1978a; Section IV.D).

There is an unexpected effect of growth conditions on the oxidases of the obligate methanol-utiliser *Methylophilus methylotrophus* during growth in continuous culture; cytochrome a_3 is only present in methanol-limited conditions, and it is replaced in methanol-excess conditions with an inducible cytochrome *o* (Cross and Anthony, 1980a; Section IV.C).

Pseudomonas MS is exceptional in containing very low (or zero) concentrations of cytochrome *c* (Widdowson and Anthony, 1975). The only C_1 substrates able to support growth are methylated amines whose oxidation (in this organism) does not involve methylamine dehydrogenase. Its lack of cytochrome *c* does not, therefore, argue against the conclusion that this cytochrome is essential for electron transport by way of methanol and methylamine dehydrogenases (see below).

The only components of the electron transport chains of methylotrophs that have been isolated and characterised in detail apart from the dehydrogenases are the soluble cytochromes *c*. As has been mentioned above these are rather unusual and appear to have a particularly important role during methylotrophic growth. They are therefore discussed in detail in the following section.

II. The cytochromes *c* of methylotrophs

A. Introduction

All methylotrophs able to grow on methane or methanol contain at least one cytochrome *c* having a midpoint redox potential of about 300 mV or greater; usually some of this is membrane-bound and some is soluble. Table 42 summarises the properties of cytochromes *c* purified from methylotrophs; these are from the obligate methanotroph, *Methylosinus trichosporium*,

TABLE 42
Properties of the cytochromes c of methylotrophs

Organism	Isoelectric point	Molecular weight	Midpoint redox potential, E_{m7}
Methylosinus trichosporium	(high)	13 000	+ 310 mV
Pseudomonas extorquens	(high)	13 000	+ 295 mV
Pseudomonas AM1			
cytochrome c_H (72% of total)	8.8	11 000	+ 294 mV
cytochrome c_L (28% of total)	4.2	20 900	+ 256 mV
Methylophilus methylotrophus			
cytochrome c_H (50% of total)	8.9	8 500	+ 373 mV
cytochrome c_{LM} (8% of total)	4.6	16 800	+ 336 mV
cytochrome c_L (42% of total)	4.0–4.4	21 000	+ 310 mV

The material in this table is taken from the following references: *Methylosinus trichosporium* (Tonge *et al.*, 1975, 1977a; Higgins *et al.*, 1976a); *Pseudomonas extorquens* (Tonge *et al.*, 1974; Higgins *et al.*, 1976a, b); *Pseudomonas* AM1 (O'Keeffe and Anthony, 1980a, b); *Methylophilus methylotrophus* (Cross and Anthony, 1980a). The subscripts c_H and c_L refer to high and low isoelectric points respectively. The "high" isoelectric points (pI unspecified), suggested for some of the cytochromes, is based on their failure to bind to DEAE-cellulose at pH 7·0. All these cytochromes are monomers with one haem per molecule of protein. During purification of the cytochrome c_L from *Methylophilus methylotrophus* a 4000 dalton fragment was lost thus giving a cytochrome of lower molecular weight (17 000) but with otherwise identical properties. The absorption spectra were typical of cytochromes c, with α-absorption maxima between 549 and 552 nm. All the cytochromes are autoxidisable to some extent. The purified cytochrome c from *Methylosinus trichosporium* appears to have a small amount of copper associated with it. The cytochromes c from *Methylophilus methylotrophus* differed from all the others in showing split α-absorption bands at low temperature (77 K).

from the obligate methanol-utiliser *Methylophilus methylotrophus* and from the pink facultative methylotrophs, *Pseudomonas* AM1 and *Pseudomonas extorquens*. The soluble cytochrome *c* from *Methylomonas methanica* (mol. wt 18 000) has also been described (Patel *et al.*, 1979a) but its spectrum ($A_{280}/A_{550} = 10$) indicates that it is impure.

The first point to note is that in two cases studied there are at least two soluble cytochromes *c* differing markedly in their isoelectric points, molecular weights and redox potentials. The two cytochromes *c* from *Pseudomonas* AM1 have completely different amino acid compositions (O'Keeffe and Anthony, 1981). The absence of reports of more than one cytochrome *c* in *M. trichosporium* and *Ps. extorquens* might merely reflect a similar oversight that led to an erroneous description of a single cytochrome *c* in *Pseudomonas* AM1 (Anthony 1975a). Certainly there are at least 2 soluble cytochromes *c* in all the methylotrophs that we have tested, including *Hyphomicrobium* and *Paracoccus denitrificans* (Beardmore-Gray and Anthony, 1981). Whether or not the two or more cytochromes *c* of the methylotrophs have entirely separate functions has not yet been unequivocally determined, but it is conceivable that one (cytochrome c_L) might interact with the methanol dehydrogenase while the other(s) might mediate between cytochrome *b* and the cytochrome oxidase(s) (see p. 232). Also, whether or not the soluble cytochromes *c* are the same as those observed tightly bound to membranes is not known but the similarity of the midpoint redox potentials of the membrane-bound cytochromes *c* to the soluble cytochromes in *Methylophilus methylotrophus* suggests that they may be the same. This bacterium is unusual in releasing up to about 40% of its cytochromes *c* into the culture medium during growth (Cross and Anthony, 1980b), and in the high midpoint potentials of its cytochromes which are more like those of photosynthetic bacteria.

The cytochrome *c*-deficient mutant of *Pseudomonas* AM1 (Anthony, 1975a) lacks both soluble, and membrane-bound cytochromes *c*.

That the mutation is not in a gene responsible for haem biosynthesis is indicated by the normal levels of cytochromes *a* and *b* in the mutant (Anthony, 1975a; Widdowson and Anthony, 1975). It is thus possible that the mutation is in a gene affecting the synthesis of two different cytochromes *c*. It is of interest that all cytochromes *c*, both soluble and membrane-bound, were also lost by a single mutation in *Paracoccus denitrificans* (Willison and John, 1979) and in *Rhodopseudomonas capsulatus* (Michels and Haddock, 1980).

B. Reaction of cytochromes *c* with CO

A remarkable characteristic of the cytochromes *c* of methylotrophs is their reaction with CO which has led to them being referred to as cytochromes c_{CO}. This terminology has not been used here because I wish to avoid attaching too great a functional significance to this characteristic.

The slow reaction with CO is unlikely to be due to damage to the cytochrome because neither the rate, nor extent of reaction, changes during purification, and it is observable to the same extent in whole bacteria. The estimated CO-binding of less than 100% is not a reflection of a mixed population of cytochromes c, some binding CO and some not; it is probably because the cytochrome reacts slowly with CO to form a complex (absorption maximum 412 nm) with a high dissociation constant (Widdowson and Anthony, 1975; O'Keeffe and Anthony, 1980a).

The CO-binding of the cytochrome c of *Methylosinus trichosporium* has led to the speculation that it might have an oxygenase or oxidase function and this has been extended to propose an oxidase function for the cytochrome c of the facultative methanol-utiliser, *Pseudomonas extorquens* (Tonge *et al.*, 1975, 1977a, b; Higgins, 1979a, 1980). There is little evidence, however, supporting the suggestion that it functions as an oxidase, and in *Pseudomonas* AM1 and *Methylophilus methylotrophus* all the evidence is against such a function (Widdowson and Anthony, 1975; O'Keeffe and Anthony, 1980a, b; Cross and Anthony, 1980a, b). The reaction with CO of these cytochromes might merely reflect the structure around the haem pocket that allows a more readily dissociable iron-methionine bond. That the haem environment might be slightly unusual is perhaps indicated by the unusual response of the midpoint potential of the cytochromes c of *Pseudomonas* AM1 to changing pH (O'Keeffe and Anthony, 1980a). Both cytochromes c have two ionizing groups affecting the redox potentials, the pK values being 3·5 and 5·5 in the oxidised forms, and 4·5 and 6·5 in the reduced forms. If these dissociations arise from the haem, then the higher of the pK values is likely to be due to the rear (inner) haem propionate in the hydrophobic environment of the haem cleft, and the lower pK due to the front (outer) propionate in its more hydrophilic environment. These pK values are sufficiently different from the pH within the bacteria to preclude a proton-translocating function for the cytochrome c. (Similar results with the cytochrome c_{551} of *Pseudomonas aeruginosa* have recently been reported by Moore *et al.*, 1980).

C. The autoreduction of cytochrome c in methylotrophs

A further unusual feature of the cytochromes c of methylotrophs is their capacity for rapid autoreduction (O'Keeffe and Anthony, 1980b). Autoreduction is the reduction of the haem iron of ferricytochrome c occurring in the absence of added reducing reagent. This phenomenon occurs in horse heart cytochrome c (Brady and Flatmark, 1971), but we have shown that it occurs at a very much higher rate in both cytochromes *c* of *Pseudomonas* AM1, and at an even greater rate in the cytochromes *c* of *Methylophilus methylotrophus*. In the methylotroph cytochromes *c* the autoreduction process is a first-order intra-molecular reaction which occurs at high pH values, the pK for this process being greater than pH 10. A mechanism involving electron transfer between a dissociable group (XH) and the haem iron of ferricytochrome *c* that is consistent with all the available evidence is presented in Fig. 31. The weakly acidic group (XH) dissociates at a high pH to give a negatively charged species able to donate an electron to the haem.

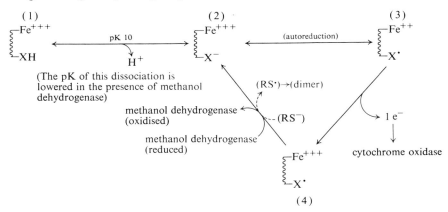

Fig. 31. A speculative mechanism for the autoreduction of cytochrome *c* and its involvement in reaction with methanol dehydrogenase (from O'Keeffe and Anthony, 1980b). The electron donor in autoreduction must be a weakly acidic group (XH) dissociating at high pH to give a negatively charged species able to donate an electron to the haem. Species (1) is the undissociated cytochrome *c*; this becomes dissociated to species (2) at high pH values (or in the presence of methanol dehydrogenase); species (3) is the radical complex of ferrous iron, isoelectronic with the ferric species (2); species (4) is the ferric form of this radical; it is reduced to species (2) by mercaptoethanol (RS^-) or methanol dehydrogenase. If the autoreduction process is involved in the physiological reaction with methanol dehydrogenase, and with cytochrome oxidase, then the ferrous radical species (3) would be oxidised by cytochrome oxidase to the ferric radical (4) which would then be reduced by methanol dehydrogenase to the ferric ion species (2).

An essential feature of this scheme is that the free radical produced by this process is stabilised by sharing an electron with the haem iron; this proposal has the advantage that electron transfer to the iron, and stabilisation of the resulting radical, do not have to be explained separately. The electron-donating group must be within the usual atomic distance to the iron, and it is conceivable that it might indeed replace the usual methionine as the sixth ligand to the iron. That the sixth ligand is methionine (at pH 7·0), as is usually the case in cytochrome *c*, is indicated by the 695 nm absorption band shown to be present in the methylotroph ferricytochromes *c*. This absorption band is lost (as expected) on reduction with dithionite at pH 7·0. Autoreduction of the *Pseudomonas* AM1 cytochrome c_L at high pH also leads to loss of the 695 nm absorption band, concomitant with appearance of the usual α-band absorbance at 550 nm (O'Keeffe, Beardmore-Gray and Anthony, unpublished data). Our preliminary results, using magnetic circular dichroism (MCD) spectrometry, have indicated that this cytochrome *c* is similar to that from horse heart with respect to haem-ligation (O'Keeffe, Beardmore-Gray, Thompson and Anthony, unpublished results). These studies also indicate that autoreduction at high pH of ferricytochrome *c* does not lead to a marked change in haem-ligation. The iron in this cytochrome *c* appears to be low spin in both oxidised and reduced states at pH 7·0; when autoreduced at pH 10; and when reacted with methanol dehydrogenase at pH 7·0. The MCD results show clearly that if the sixth ligand does change during autoreduction at high pH, or in the presence of the dehydrogenase, then the methionine must be replaced by another strong-field ligand.

The possible involvement of the intramolecular redox process described here in the reaction of cytochrome *c* with methanol dehydrogenase is discussed in the following section.

III. The role of cytochrome *c* in the oxidation of methanol

A. Introduction

One of the most important conclusions from work on electron transport in methylotrophs is that electrons from methanol are donated by methanol dehydrogenase to the electron transport chain at the level of cytochrome *c*. The evidence for this is extensive but often indirect and must be gleaned from studies with a wide range of methylotrophs. This is summarised below:

(a) Mutants lacking cytochrome *c* no longer oxidise or grow on methanol or methylamine, but oxidise and grow on other substrates; such mutants have been isolated from *Pseudomonas* AM1 (Anthony, 1975a; Widdowson and Anthony, 1975), *Methylobacterium organophilum* (O'Connor and Hanson, 1977) and *Paracoccus denitrificans* (Willison and John, 1979). All oxidisable substrates, but not methanol or methylamine, are able to reduce the cytochromes *b* and a/a_3 in the cytochrome *c*-deficient mutant of *Pseudomonas* AM1; whereas all substrates, including methanol and methylamine, are able to reduce both cytochromes *c* and a/a_3 in the wild-type bacteria.

(b) Membrane vesicles of *Pseudomonas* AM1 oxidise NADH, succinate and methanol and these substrates all reduce cytochromes *c* and a/a_3

(the cytochrome b is masked by the cytochrome c and so it cannot be measured). Respiration and reduction of cytochromes c and a/a_3 by NADH and succinate is inhibited by antimycin A, but oxidation of methanol, and the reduction by methanol of cytochromes c and a/a_3 is unaffected by this inhibitor. Antimycin A, and usually n-heptyl-quinoline N-oxide (HQNO), inhibit the oxidation of cytochrome b and thus of substrates donating electrons to sites prior to cytochrome b in the electron transport chain; so these results suggest that methanol donates electrons to the chain after the level of cytochrome b (Netrusov and Anthony, 1979). Similar studies using whole cells or extracts have shown that the oxidation of NADH and succinate, but not of methanol, is inhibited by Antimycin A and HQNO in *Pseudomonas extorquens* (Higgins *et al.*, 1976a, b; Tonge *et al.*, 1977b), *Paracoccus denitrificans* (van Verseveld and Stouthamer, 1978a), *Methylosinus trichosporium* (Higgins *et al.*, 1976a) and *Pseudomonas* sp.2 (Netrusov *et al.*, 1977).

(c) During the respiration-coupled ATP synthesis catalysed by membrane vesicles of *Pseudomonas* AM1, the P/O ratio with methanol is the same as measured with ascorbate/N,N,N'N'-tetramethylphenylenediamine but it is markedly more when succinate or NADH is the substrate (Netrusov and Anthony, 1979; Netrusov, 1981).

(d) Pure methanol dehydrogenase and cytochrome c are able to react. The published evidence is not unequivocal because methanol is not usually required for the reaction (O'Keeffe and Anthony, 1980b), but methanol-dependent cytochrome c reduction has been demonstrated (see the following Section). When tested it has always been possible to demonstrate the reduction by methanol of cytochrome c in whole cells of methylotrophs (Anthony, 1975a; Widdowson and Anthony, 1975; Cross and Anthony, 1980b). The soluble fraction of extracts of *Methylosinus trichosporium* (Tonge *et al.*, 1977a) and of *Methylophilus methylotrophus* (Cross and Anthony, 1980b), containing methanol dehydrogenase and soluble, slowly autoxidisable cytochrome c (the only cytochrome present) are able to oxidise methanol.

The problems associated with demonstrating direct electron transfer between methanol dehydrogenase and cytochrome c are discussed in the following section.

B. The reaction between methanol dehydrogenase and cytochrome c

If electrons from methanol are donated to the electron transport chain at the level of cytochrome c, it becomes necessary to consider whether the methanol dehydrogenase reacts directly with the cytochrome c, or whether intermediate electron transport components are required. To demonstrate a direct reaction it is essential to purify the two components, and it appears that the purification procedure affects some of the properties of the dehydrogenase (see Chapter 6, Section III).

The high isoelectric points and high solubilities of methanol dehydrogenase (MDH) and cytochrome c_H lead to co-purification of these proteins during ammonium sulphate fractionation and anion-exchange chromatography. The final step in purification of dehydrogenase is often the removal of contaminating cytochrome c by gel filtration. During purification, the cytochrome c is always in the reduced form when associated with MDH. Furthermore, if oxidised cytochrome c is added to pure MDH then the cytochrome becomes reduced — even in the absence of added methanol. Although at first sight this implies that electrons do pass from MDH to the cytochrome c, this is not necessarily the case because the cytochromes c of many methylotrophs are now known to be autoreducible (O'Keeffe and Anthony, 1980b; this Chapter, section II.C). It is suggested that an electron, made available by dissociation of a proton from a group on the ferricytochrome c, reduces the iron in the haem in an intramolecular redox reaction. Methanol dehydrogenase appears to lower the pK for the initial dissociation, thus allowing the intramolecular autoreduction of the cytochrome c to occur at a lower pH than it otherwise would (pH 7·0). Consistent with this proposal is the demonstration that the 695 nm absorption band of the ferricytochrome c disappears on reaction with methanol dehydrogenase at pH 7·0; and that the MCD spectrum of the ferricytochrome c is the same whether it is autoreduced at pH 10 or reacted with methanol dehydrogenase at pH 7·0.

Whether or not the autoreduction process is involved in the physiological function of cytochrome c is not yet proved, but a speculative mechanism for this is presented in Fig. 31 (from O'Keeffe and Anthony 1980b). That free radicals may be involved in this process is perhaps supported by the observation that the oxidation of methanol dehydrogenase is probably by way of free radicals and single-electron transfer reactions (Chapter 6, Section. H3).

It appears from the work of Duine and Frank and their co-workers that the failure to demonstrate a direct methanol-dependent, dehydrogenase-mediated reduction of cytochrome c using pure proteins might be because the dehydrogenase is damaged in some way during its aerobic preparation. The evidence for this is that reduction by methanol of cytochrome c can be demonstrated using crude, anaerobically-prepared extracts containing methanol dehydrogenase and cytochrome c from *Hyphomicrobium* X (Duine *et al.*, 1979a; Duine and Frank, 1981a) or *Pseudomonas* AM1 (O'Keeffe and Anthony, 1980b).

Although the involvement of other factors in the passage of electrons from methanol by way of dehydrogenase to cytochrome c has not been unequivocally ruled out, we have no evidence for the existence of such factors. At present we consider that the balance of evidence is in favour of methanol dehydrogenase being the only protein necessary for electron transport from methanol to cytochrome c. Our preliminary observations suggest that methanol dehydrogenase reacts far more rapidly with cytochrome c_L than with cytochrome c_H (O'Keeffe and Anthony, 1981; Beardmore-Gray and Anthony, 1981). By contrast methylamine dehydrogenase reacts preferentially with cytochrome c_H (Lawton and Anthony, 1981).

IV. Electron transport and proton-translocating systems in methylotrophs

A. Introduction

As mentioned previously, all methylotrophs grown on methane or methanol have cytochromes of the a, b and c types and usually a cytochrome a_3. Likewise, it appears to be a general conclusion that methanol dehydrogenase reacts (directly or indirectly) with cytochrome c, thus bypassing cytochrome b in the electron transport chain. This leaves the following questions to consider. Are the cytochromes arranged in the conventional mitochondrial sequence as found in many heterotrophic bacteria? Are the electron transport chains arranged to translocate protons as conventionally described? Is there a typical ATP synthetase coupled to a proton motive force? Is electron transport by way of the unusual methanol dehydrogenase/cytochrome system arranged to create a proton motive force and if so how?

Before going further it should be borne in mind that all the wisdom of the ages (about four decades) has failed to produce a clear answer to similar questions with respect to the most intensively-studied system, the mitochondrial electron transport chain. The sequence of some components is uncertain; the stoicheiometry and mechanism of proton translocation is uncertain; the mechanism of ATP synthesis is uncertain; how electrons move to, from or through the haem of cytochromes is uncertain. What is certainly clear is that it would be unwise to be too dogmatic about the significance of some of the results summarised below.

What I have done below is to select some examples of electron transport pathways in methylotrophs in order to highlight any points that might make them especially interesting. In the light of my comments above there is probably little point in debating the detailed significance of the actual values of H^+/O ratios, P/O ratios etc. in methylotrophs. Instead, I have discussed proton translocation studies in relation to the work on electron transport in the same organism, where these two approaches can complement each other. I have assumed for convenience (as have most workers) that $2H^+$ may be translocated outwards per electron pair transferred through NADH dehydrogenase; and that $4H^+$ are ejected during the further transfer of these electrons to O_2 if cytochrome c is involved, or only $2H^+$ if cytochrome c is not involved. I have also assumed for convenience that a $H^+/2e^-$ ratio of 2 is equivalent to a P/O ratio of 1.

The significance of, and problems associated with, transmembrane proton gradients in the harnessing of energy from the special system for oxidising methanol to formaldehyde is discussed in a separate last section below.

B. Electron transport and proton translocation in *Pseudomonas* AM1

Pseudomonas AM1 is a pink facultative methylotroph able to grow on methanol but not on methane. It assimilates methanol by the serine pathway, and formate dehydrogenase is the only NAD^+-linked enzyme involved in oxidation of methanol to CO_2. Growth thus tends to be NADH-limited and less than 5% of electron transport to O_2 during growth on methanol is from NADH; 50% is from methanol dehydrogenase and the remainder from formaldehyde dehydrogenase and flavoproteins (see Chapter 9). Its electron transport system (and that of its cytochrome c-deficient mutant) has been investigated using a range of approaches including cytochrome characterisation (Anthony, 1975a; Widdowson and Anthony, 1975; Keevil and Anthony, 1979b; O'Keeffe and Anthony, 1980a, b); studies of proton translocation (O'Keeffe and Anthony, 1978; Anthony, 1978; Keevil and Anthony, 1979a); measurements of ATP synthesis in vesicle preparations (Netrusov and Anthony, 1979; Netrusov, 1981); determination of growth yields (Goldberg et al., 1976; Keevil and Anthony, 1979a). Some of the results with respect to the site of interaction of methanol dehydrogenase and the cytochrome chain have already been discussed above.

Pseudomonas AM1 contains at least two soluble cytochromes c, some membrane-bound cytochrome c, two cytochromes b and cytochrome a/a_3, which is probably the oxidase. In the presence of sufficient KCN to inhibit cytochrome a_3, some KCN-insensitive respiration can occur by way of one of the cytochromes b. Presumably this is the cytochrome b which binds CO, but there is no evidence that it has a physiological function as an oxidase in the absence of KCN.

Because all oxidisable substrates are able to reduce all of the cytochrome c and cytochrome a/a_3, it was initially concluded that the "conventional" cytochrome sequence ($b,c, a/a_3$) occurs in this organism (Fig. 32a). However, in the mutant lacking cytochrome c, all substrates except alcohols and methylamine are oxidised and they are all capable of reducing cytochrome

b and a/a_3. From this it appeared that cytochrome c is not normally involved in the oxidation of NADH or of cytochrome b (Fig. 32b). If cytochrome c is not involved in the oxidation of NADH, then a maximum H^+/O ratio of only 4, equivalent to 2 proton-translocating segments, would be expected during measurements of respiration-coupled proton translocation (Jones, 1977 and Jones *et al.*, 1977b). This was indeed the result found for substrates oxidised by way of NADH, and the value was the same in the mutant lacking cytochrome c; this appeared, therefore, to confirm the pathway shown in Fig. 32b (O'Keeffe and Anthony, 1978). Further confirmation was obtained by measuring ATP synthesis in membrane vesicles of batch-grown *Pseudomonas* AM1 and its cytochrome c-deficient mutant (Netrusov and Anthony, 1979; Netrusov, 1981). The actual P/O ratios in such experiments are difficult to interpret because of the presence of some uncoupled electron transport (due to wrongly-oriented vesicles), but the general conclusions of these experiments are plain; the oxidation of methanol to formaldehyde is coupled to the synthesis of ATP, the P/O ratio being similar to that observed during oxidation of ascorbate plus TMPD, and lower than with NADH and succinate. A second important conclusion that appears to support the scheme in Fig. 32 is the observation that the P/O ratios are identical in vesicles prepared from wild-type bacteria and from the mutant lacking cytochrome c (Netrusov, 1981). This all supports the scheme indicated in Fig. 32b in which cytochrome c is not involved in proton translocation in the cytochrome b/O_2 part of the electron transport chain. That *all* substrates can reduce *all* of the cytochrome c in whole cells and membrane preparations, suggests either that cytochrome c (when present) is always part of the electron transport chain, but that it is not always involved in proton translocation and ATP synthesis from NADH; or that the measured level of cytochrome c reduction merely reflects the level of reduction of the cytochrome a/a_3 (presumably in reversible equilibrium with it). This simple (but unconventional) conclusion cannot hold for all conditions, however. This is because higher H^+/O ratios, equivalent to three proton-translocating segments, are measurable in wild-type bacteria grown in carbon-limited conditions. This contrasts with the lower H^+/O ratios (two proton-translocating segments) found in the wild-type bacteria grown in carbon-excess conditions, and in the cytochrome c-deficient mutant in all growth conditions. These results suggest that the cytochrome c may be involved in electron transport and proton translocation from NADH in carbon-limited conditions, but not during growth with an excess of carbon substrate (Keevil and Anthony, 1979a). That cytochrome c may be involved in ATP synthesis coupled to NADH oxidation during carbon-limited growth on succinate is indicated by the higher yields measured on this substrate in wild-type bacteria compared with the yields of the cytochrome c-deficient mutant.

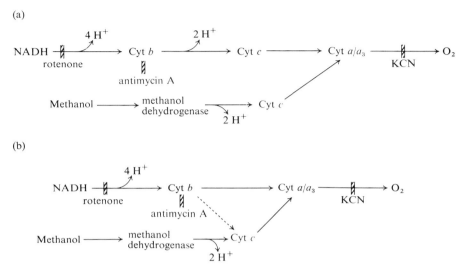

Fig. 32. Electron transport and proton translocation in *Pseudomonas* AM1 (based on the work of the author and his colleagues). Cytochrome *c* may be able to mediate between cytochrome *b* and the oxidase in all conditions but it does not appear to be involved in proton translocation in carbon-excess conditions (see text). It is not known if the cytochrome *c* mediating between cytochrome *b* and a/a_3 is the same as that involved in methanol oxidation. (a) Electron transport in wild-type bacteria (carbon-limited conditions). (b) Electron transport in wild-type bacteria (carbon-excess conditions; O_2- or nitrogen-limited) and in a cytochrome *c* deficient mutant.

It should be noted that although incorporation of cytochrome *c* into the electron transport chain between cytochrome *b* and a/a_3 gives these higher yields on succinate, an increase in the P/O ratio for NADH oxidation would give little benefit to the organism during growth on methanol, because relatively little NADH is oxidised during growth on this substrate (Chapter 9).

In summary, all of the experiments described here are associated with considerable difficulties of execution and interpretation. The simplest conclusion is that in *Pseudomonas* AM1, cytochrome *c* is essential for the oxidation of methanol, and for coupled ATP synthesis, and that it may also be involved in electron transport, proton translocation and ATP synthesis during oxidation of NADH, but perhaps only under carbon-limiting conditions.

Pseudomonas extorquens, a pink facultative methanol-utiliser, is similar in most respects to *Pseudomonas* AM1 except that it might have an NAD^+-linked formaldehyde dehydrogenase (Johnson and Quayle, 1964). Preliminary reports on electron transport in this organism support the conclusion that electrons from methanol dehydrogenase enter the electron transport chain at the level of cytochrome c, and that during growth on methanol this substrate and NADH are oxidised by way of the same terminal oxidase (Higgins et al., 1976a; Tonge et al., 1977b). It has been proposed that this oxidase may be the CO-binding cytochrome c as suggested for *Methylosinus trichosporium* (see below), but from the preliminary evidence available this appears unlikely (see Fig. 35).

C. Electron transport and proton translocation in *Methylophilus methylotrophus*

Methylophilus methylotrophus is the obligate methanol-utiliser which, because it grows at high growth rates and yields, has been selected by ICI as the organism of choice for production of single cell protein. It assimilates methanol by the RuMP-pathway and produces two molecules of NAD(P)H during oxidation of methanol to CO_2; less than 35% of electron transport to oxygen is likely to be from NAD(P)H and more than 65% from methanol dehydrogenase (Chapter 9). The work described below on electron transport in this organism was published by Cross and Anthony (1980a, b).

M. methylotrophus has three soluble cytochromes c (Table 42), remarkable for their high midpoint redox potentials (310 mV–375 mV). Of the cytochrome c measured in these bacteria, 30% is membrane-bound and redox potential measurements indicate that the two or three cytochromes c on the membranes are the same as found in solution. There is about twice as much cytochrome c on membranes as cytochrome b. A curious characteristic of this organism (and of some similar bacteria described in Chapter 1) is that about 40% of the total cytochromes c produced are released into the growth medium.

Membranes of *M. methylotrophus* always have at least two cytochromes b (midpoint potentials, $+60$ mV and $+110$ mV). In carbon-excess conditions (O_2-limited or nitrogen-limited) a third, high potential cytochrome b (midpoint potential, $+260$ mV), which binds CO, is induced ten-fold, concomitant with a similar increase in the rate of oxidation of ascorbate plus TMPD. There is no cytochrome a/a_3 in these growth conditions and it is reasonable to conclude that this cytochrome b is a genuine alternative oxidase and hence it is called cytochrome o. It is sensitive to n-heptyl-hydroxyquinoline N-oxide (HQNO) and it probably reacts directly with TMPD (midpoint potential, $+225$ mV), its turnover number during oxidation of this substrate being typical of cytochrome oxidases ($220\ s^{-1}$).

Analysis of the electron transport system of *M. methylotrophus* has been facilitated by the fact that whole cells are permeable to NADH and NADPH; and that the concentration of alternative oxidases can be determined by choice of growth conditions, and that the cytochrome o is ten times more sensitive to azide than the cytochrome a_3. This analysis, together with the results summarised above, has led to proposal of the electron transport chains in Fig. 33.

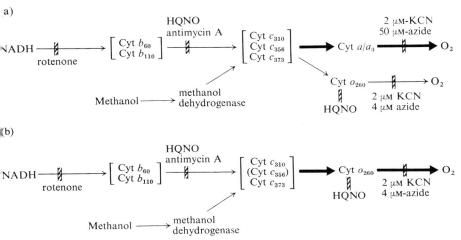

Fig. 33. Electron transport in *Methylophilus methylotrophus* (based on Cross and Anthony, 1980a, b). The hatched bars indicate the site of action of inhibitors (conc. required for 50% inhibition). The subscripts of the cytochromes refer to the midpoint redox potentials at pH 7·0. The thickness of the lines indicates the relative importance of the alternative oxidases. The cytochrome o is induced ten-fold during growth under methanol-excess conditions (when cytochrome a/a_3 is absent). Whether or not the three cytochromes c have separate roles is not yet known. (a) Methanol-limited conditions. (b) Methanol-excess conditions (O_2- or nitrogen limited).

The inhibitor-insensitive respiration, continuing after cytochromes o and a_3 are completely inhibited, is thought to be by way of the autoxidisable cytochrome c, but this respiration is probably physiologically insignificant and may only occur in the presence of inhibitors.

All studies with NADH and NADPH in this bacterium suggest that they are oxidised by way of the same, or very similar, enzymes and there is no evidence for an "energy-linked" transhydrogenase.

Both oxidases (o and a_3) appear to be involved in the oxidation of methanol, NADH and NADPH and all the evidence indicates that cytochrome c is the point of entry of electrons from methanol dehydrogenase into the electron transport chain. These two facts, supported by evidence from the inhibitor studies, suggest that the branch point for electron flow to the alternative oxidases is at the level of cytochrome c.

Why *M. methylotrophus* has alternative oxidases, whose synthesis is regulated by the carbon supply, is not clear. It is more usual for regulation of oxidase synthesis to be determined by oxygen concentration, but even here the rationale for possession of alternative oxidases is not particularly obvious

(Harrison, 1976; Jones, 1977; Haddock and Jones, 1977). The affinity for both oxidases is high, although the K_m values could not be determined because they were below the limit of sensitivity of the conventional oxygen electrode. The majority of cytochromes o have relatively low midpoint redox potentials (less than $+130$ mV) and they are often synthesised in response to decreasing O_2 concentration. The unusually high midpoint potential of the cytochrome o of *M. methylotrophus* may be related to its function in accepting electrons from cytochromes c also having exceptionally high midpoint potentials, but why it is regulated in an unusual manner (with high levels under O_2- or nitrogen-limitation) is not clear. If the terminal part of the electron transport chain with cytochrome o as oxidase is not coupled to proton translocation and ATP synthesis, then in carbon-excess conditions (O_2- or nitrogen-limitation), when cytochrome a_3 is absent, the cell yield would be lower — as has indeed been demonstrated (Brooks and Meers, 1973) — but the advantage to the organism is unclear unless it allows a higher growth rate in the presence of plentiful carbon substrate.

By analogy with other proton-translocating electron transport chains, the involvement of cytochrome c as proposed in Fig. 33 would indicate that up to three proton-translocating segments might be operating during the oxidation of NADH, and one during the oxidation of methanol. This has now been confirmed in a thorough investigation of respiration-coupled proton translocation in *M. methylotrophus* by Dawson and Jones (1981a, b). In this study the bacteria were grown under methanol-limitation, and so presumably had the electron transport chain shown in Fig. 33a, in which cytochrome a/a_3 is the predominant oxidase. The stoicheiometries of proton translocation were determined by using both the O_2-pulse and the initial rate methods. The latter was also used to measure K^+/O ratios, in order to determine the charge/O ratios. It was concluded that $6H^+/O$ are translocated during NADH oxidation, and that $2H^+/O$ are translocated during the oxidation of methanol to formaldehyde. There was no evidence for underestimation of the H^+/O ratios due to H^+/anion symport, except by the movement of formic acid during formate oxidation (as shown in *Pseudomonas* AM1 by O'Keeffe and Anthony, 1978). By comparing their results with the known growth efficiencies of this organism, a H^+/ATP ratio of close to 2 moles of H^+/mole of ATP was calculated. It was thus proposed that the respiratory chains of *M. methylotrophus* are arranged such that there are three sites of energy conservation for NADH oxidation, each translocating $2H^+$ and each linked to the synthesis of one molecule of ATP; and that only the third site of energy conservation is involved in methanol oxidation. This conclusion is further supported by the work reported in Dawson and Jones (1981c).

D. Electron transport and proton translocation in *Paracoccus denitrificans*

The facultative autotroph, *Paracoccus denitrificans*, grows on methanol as sole source of carbon and energy, oxidising the methanol to CO_2 and assimilating this by the ribulose bisphosphate pathway (Chapter 2). Methanol is oxidised by methanol dehydrogenase, and formaldehyde and formate by NAD^+-linked dehydrogenases. It can be estimated that about 40% of electron transport to O_2 will be from NADH and 60% from methanol dehydrogenase during growth on methanol (Chapter 9). This organism is also capable of typical heterotrophic growth on succinate etc., and of typical autotrophic growth on a mixture of H_2 and CO_2. It is perhaps the wide variety of growth conditions that may be used that has led to the rather confused picture of electron transport in this bacterium (see Table 41 for references).

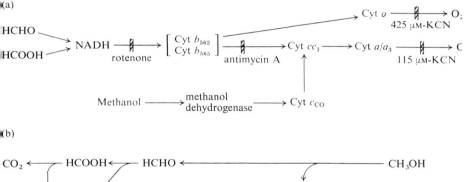

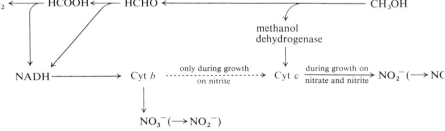

Fig. 34. The respiratory chain of *Paracoccus denitrificans*. (a) From van Verseveld and Stouthamer (1978a, b); (b) from Bamforth and Quayle (1978a). The hatched bars indicate the site of action of inhibitors (with concentrations required for 50% inhibition). It should be noted that no evidence has been published to support the involvement of more than one cytochrome c in methanol oxidation. The cytochrome o is probably not involved in methanol oxidation. (a) Aerobic electron transport. (b) Anaerobic electron transport.

A generally-accepted conclusion is that this organism contains cytochrome c, at least two cytochromes b, cytochrome a/a_3 and a CO-binding cytochrome o. *P. denitrificans* differs from *M. methylotrophus* in the role of cytochrome *o*. This oxidase is involved in methanol oxidation in *M. methylotrophus* (in some conditions), but cytochrome a_3 appears to be the only oxidase involved in the oxidation of methanol and NADH during growth of *P. denitrificans* on methanol (van Verseveld and Stouthamer, 1978a; Bamforth and Quayle, 1978a). The scheme presented in Fig. 34a (based on van Verseveld and Stouthamer, 1978a) shows the electron transport chain branching from cytochrome *b* to the alternate oxidases. This differs again from the pathway in *M. methylotrophus*, in which the branch point to the alternative oxidases is at cytochrome *c* rather than cytochrome *b*.

The demonstration that a cytochrome *c*-deficient mutant of *P. denitrificans* (Willison and John, 1979) is unable to grow on methanol, but is able to grow on other substrates, indicates a similarity between electron transport in *P. denitrificans* and *Pseudomonas* AM1, particularly in the ability to bypass cytochrome *c*.

Measurements of cell yields and proton translocation in *P. denitrificans* suggest that this organism differs from other methylotrophs in translocating between three and four protons during methanol oxidation (van Verseveld and Stouthamer, 1978a). This high number is probably because the cytochrome oxidase also pumps protons (van Verseveld *et al.*, 1981). The results of these studies confirm that the oxidation of methanol to formaldehyde is coupled to synthesis of one molecule of ATP per molecule of methanol oxidised.

During anaerobic growth of *P. denitrificans*, methanol is oxidised by the same dehydrogenase as operates during aerobic growth, and it has been proposed that this dehydrogenase reduces the electron transport chain at the level of cytochrome *c*, which is oxidised by nitrite as the terminal electron acceptor (Fig. 34b). Because nitrate usually accepts electrons from cytochrome *b*, it is proposed that during growth on nitrate the reductant for nitrate reductase is NADH, produced during subsequent oxidation of formaldehyde and formate; the nitrite produced from this process is used as the terminal electron acceptor for methanol dehydrogenase (Bamforth and Quayle, 1978a).

E. Electron transport and proton translocation in *Methylosinus trichosporium* and other methanotrophs

M. trichosporium is an obligate methane-utiliser, assimilating its carbon substrate by the serine pathway and having a Type II internal membrane system. As with all methane-utilisers, less than 40% of the O_2 consumed during growth is by way of oxidases, the rest being used in the initial hydroxylation of methane. A second general point is that in all methane utilisers which use NADH as the reductant for the methane monooxygenase, between 80% and 90% of electron transport is coupled to oxidases by way of methanol

8. Electron transport and energy transduction

dehydrogenase, NADH dehydrogenase being relatively unimportant (Chapter 9). All studies of electron transport during methane oxidation are very difficult to interpret because there are at least two O_2-consuming sites, the oxygenase and the oxidase, and these may be sensitive to the same or different inhibitors. A further complication is that electron donors to the oxygenase are required, and their production may depend on the further metabolism of methanol arising from the methane hydroxylation. Thus midchain inhibitors such as amytal or antimycin A may inhibit a conventional electron transport chain between NADH and O_2, or they may inhibit the production of reductant for the oxygenase.

The relatively few studies of electron transport in methanotrophs have emphasised some points of similarity with the methanol-oxidisers. In particular, they contain cytochromes b and c and have cytochrome a_3 as a potential oxidase; the methanol dehydrogenase is typical and some, at least, is bound to membranes in the intact bacteria; the methanol dehydrogenase almost certainly donates electrons to the cytochrome chain at the level of cytochrome c (see Table 41 for references).

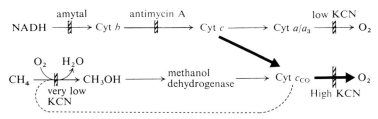

Fig. 35. Tentative scheme for electron transport in *Methylosinus trichosporium* and *Pseudomonas extorquens*. This is based on Higgins (1980). A notable feature of this scheme is that the cytochrome c is proposed as electron donor to the methane monooxygenase of *M. trichosporium* and as the predominant oxidase during growth on methanol by both bacteria. As emphasised in the text there is little evidence for an oxidase function for cytochrome c_{CO} in methylotrophs. (N.B. *Ps. extorquens* does not oxidise methane.)

The electron transport scheme proposed (tentatively) for *Methylosinus trichosporium* is shown in Fig. 35. This is based on inhibitor studies, and on the properties of the soluble cytochrome c, whose CO-binding characteristics have led to the suggestion that it is the main oxidase during methanol oxidation (Higgins *et al.*, 1976a; Tonge *et al.*, 1977a). In considering an oxidase function for the cytochrome c it should be recalled that, provided it is reducible by substrate (methanol, NADH or ascorbate), and also slightly autoxidisable, then an apparent (non-physiological) activity is bound to be observed when the activity of the normal oxidase such as cytochrome a_3 is inhibited by low concentrations of cyanide. In the absence of any further evidence it is probably not justified to ascribe an oxidase function to the cytochrome c of *Methylosinus trichosporium*.

Proton translocation measurements with *M. trichosporium* have shown that the highest H^+/O ratio obtainable is 2 with methane, methanol, formaldehyde or formate as respiratory substrates (Tonge *et al.*, 1977b). Whether or not this reflects a maximum P/O ratio of only one for each oxidation step is not known, but the failure to measure any respiration-coupled proton translocation at all in *Methylococcus capsulatus* (which has the alternative internal membrane arrangement) might suggest the existence of special technical and interpretive difficulties in measuring proton translocation in bacteria having complex internal membrane systems (O'Keeffe, Dalton and Anthony, unpublished results).

V. The coupling of methanol oxidation to ATP synthesis

The results summarised above have led to the conclusion that methanol is oxidised to formaldehyde by way of methanol dehydrogenase, cytochrome *c* and a cytochrome oxidase (o or a_3). This oxidation is coupled to a measurable acidification of the external suspending medium (usually $2H^+$ per mole of methanol oxidised). This in turn is coupled to synthesis of one molecule of ATP. This summary raises 2 questions:
(1) How are the components of the "methanol oxidase" system arranged on or in the membrane to effect the appropriate proton translocation?
(2) Why do bacteria have this system? Why is the "end product" of methanol oxidation only one ATP?

(1) The two main options for arrangement of the "methanol oxidase" system are shown in Fig. 36. The first option (Fig. 36a) is a classical proton-translocating loop. The key point in this arrangement is that methanol in the cytoplasm reacts with the methanol dehydrogenase which must be integrated with the membrane and actually move the protons across it. An unlikely variant of this scheme, possible only if methanol dehydrogenase and cytochrome *c* do not react directly, would be to have an intermediary hydrogen carrier between methanol dehydrogenase on the inside and cytochrome *c* on the outer side of the membrane.

The second option (Fig. 36b) is one which we first proposed for the methanol dehydrogenase-cytochrome *c* interaction in *Pseudomonas* AM1 (O'Keeffe and Anthony, 1978). This is similar to the first option in having the cytochrome *c* on the outer side of the membrane, but in this case the protons do not actually move across it. They are released from the $PQQH_2$ prosthetic group of the dehydrogenase when it reacts with the cytochrome *c* on the outer side of the membrane in the same way that ascorbate plus TMPD releases its proton(s) when used as a reductant for cytochrome *c*.

8. Electron transport and energy transduction

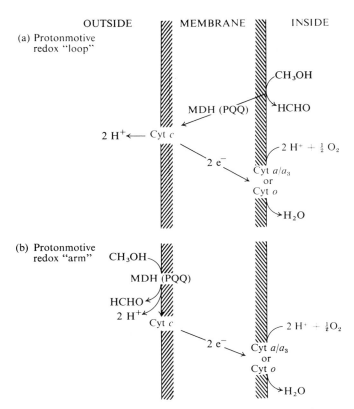

Fig. 36. Possible arrangements of methanol dehydrogenase and cytochrome c in the bacterial membrane. In both schemes 2 H^+ are effectively moved from the inside to the outside for every molecule of methanol oxidised to formaldehyde. In (a) the MDH is transmembranous and in (b) it is on the outer surface.

The evidence available at present all supports the second of these options (Fig. 36b) (O'Keeffe and Anthony, 1978; Dawson and Jones, 1981a, b, c; van Verseveld et al., 1981; Alefounder and Ferguson, 1981). The methanol dehydrogenase and its prosthetic group are very hydrophilic and it is unlikely, therefore, to span a lipoprotein membrane. That the hydrophilic cytochrome c is on the outer surface is indicated by its ease of release into the culture medium during growth of some methylotrophs, and by the acidification of the suspending medium during the oxidation of ascorbate plus TMPD. Strong evidence supporting this conclusion (in *P. denitrificans*, at least) is the demonstration that both the methanol dehydrogenase and the cytochrome c are located on the outer side of the cytoplasmic membrane (very loosely bound), or in the periplasmic space of *Paracoccus denitrificans* (Alefounder and Ferguson, 1981).

If the methanol dehydrogenase is on the outer side of the bacterial membrane then the fate of its product, formaldehyde, must be considered. Although some bacteria might be able to oxidise formaldehyde at the outer surface, many oxidise it by soluble NAD^+-linked dehydrogenases (presumably internal) and, furthermore, the assimilatory enzymes are likely to be on the inside. It will thus be of interest to determine whether or not a transport system operates for formaldehyde, and if it does, then to consider whether it might perhaps also be involved in the regulation of oxidation and assimilation of this critical "branch point" intermediate.

(2) The second question is — Why make only one ATP during the oxidation of methanol to formaldehyde? The redox potential for the $HCHO/CH_3OH$ couple is -0.182 V, which is sufficiently low to support the synthesis of two ATP per molecule of methanol oxidised by O_2.

For some methylotrophs the answer is obvious. In those methylotrophs that are NADH-limited rather than ATP-limited (this includes most methanotrophs), increasing the P/O ratio for methanol oxidation would have a negligible effect on cell yields (Chapter 9). In methanol-utilisers having the RuMP pathway, however, such an increase in P/O ratio would increase the yield (although by less than 15%; see Chapter 9). The yield would not increase as much as might be expected, because these bacteria are to some extent carbon-limited and they would become very carbon-limited if the yield for P/O ratio for methanol oxidation were to be increased from 1 to 2.

This discussion raises the question of what good is efficiency to an organism. High efficiency (with respect to utilisation of carbon substrate) is probably not what is selected in nature, but high growth rates are. It is therefore possible that the system that appears to have evolved in all methylotrophic bacteria is one that permits the most rapid oxidation of methanol at very low methanol concentrations. Although free energy changes of reactions are completely independent of rates of reactions, classical thermodynamics do not apply to open systems such as growing bacteria. It may well be that the rate of methanol oxidation, and its coupled ATP synthesis, can be greater if the ATP yield is only one per mole (plus a large free energy change) than if the yield were 2ATP per mole of methanol (plus a smaller free energy change) (See Rottenberg, 1979; Stucki, 1980).

9
Growth yields and bioenergetics of methylotrophs

I. The concept of Y_{ATP}	245
II. Assumptions and methods for developing theoretical assimilation equations	249
III. The limitation of cell yields by ATP, NADH and carbon supply	251
A. ATP-limited bacteria	257
B. Carbon-limited bacteria	258
C. NADH-limited bacteria	258
IV. The prediction of growth yields in methylotrophs	260
A. Growth of non-photosynthetic organisms on methanol	260
B. Growth of *Rhodopseudomonas acidophila* on methanol	262
1. Aerobic growth in the dark	263
2. Anaerobic growth in the light	263
C. Growth on methane	264
V. The effect on electron transport systems of the oxidative and assimilatory pathways of methylotrophs	266

The purpose of this chapter is to explore the effects on cell yields of a wide range of metabolic variables in order to emphasise those aspects of the biochemistry and physiology of methylotrophs that differ markedly from more typical heterotrophs. An imporant conclusion is that for highly reduced substrates, such as long-chain hydrocarbons, methane, methanol and methylated amines, growth yields are often limited predominantly by the supply of NAD(P)H or carbon, instead of the ATP supply as is usually the case with more conventional substrates. This derives mainly from the nature of the unusual oxidising enzymes involved, and not from thermodynamic considerations.

I. The concept of Y_{ATP}

It is usually true that growth yields of microorganisms reflect the efficiency with which they conserve energy (as ATP) by oxidative- and substrate-level phosphorylation; and subsequently use this ATP for growth. The yield of cell material per mole of ATP available for biosynthesis (the Y_{ATP}) is a parameter that reflects this generalisation. The Y_{ATP} is not a true yield but a composite parameter, produced by dividing the measured yield of cells per mole of substrate used (Y_s) by the assumed yield of ATP produced per mole of substrate metabolised (N). In the case of those fermentative bacteria in which all ATP is synthesised by substrate-level phosphorylation, the value of N may be known, but this is not so in most aerobic microorganisms in which ATP is produced by oxidative phosphorylation.

The Y_{ATP} occurs as a "constant" in many equations pertaining to measurements and predictions of growth yields. An important example is the relationship between the yield on O_2 and the P/O ratio (i.e. $Y_O = Y_{ATP} \times$ P/O ratio). This will be true for normal organisms whose ATP supply determines the growth yield. This equation may, therefore, be used for comparing P/O ratios operating in bacteria growing on closely related substrates by similar assimilation pathways. However, although the ATP requirement for biosynthesis of cell polymers from monomeric precursors is likely to be fairly constant, the amount required for conversion of the carbon substrate to these monomers will vary. The actual value of Y_{ATP} will, therefore, depend on the nature of the growth substrate.

A theoretical Y_{ATP} can be calculated from known metabolic pathways for assimilation of substrate carbon into the monomeric precursors of biosynthesis, and from knowledge of the ATP requirement for biosynthesis from monomers. Such theoretical Y_{ATP}^{max} values are usually about twice that estimated from growth yields and known routes for ATP synthesis. A minor reason for this discrepancy may be the additional requirement for ATP for "maintenance" of the organism, but uncoupling of energy generation and growth is the major cause of this discrepancy. These considerations have led to development by Harder and van Dijken (1976) of the following relationship between the Y_{ATP}, theoretical Y_{ATP}, maintenance energy (m_e), specific growth rate (μ) and a "coupling constant", k (assumed to be about 0·5):

$$\frac{1}{Y_{ATP}} = \frac{m_e}{\mu} + \frac{1}{k \cdot Y_{ATP}^{max}}$$

Such relationships have been used by microbial physiologists as a basis for relating assimilation pathways, electron transport systems, proton translocation, solute transport systems, P/O ratios and measured cell yields in bacteria. In particular, growth yields have been used for estimating P/O ratios operating in growing bacteria, which are otherwise almost impossible to determine (see Table 43 for reviews of this topic.)

These relationships have been used in this way in the study of methylotrophs, and they have also been used for predicting which methylotrophs are most likely to give the highest yields on C_1 substrates. This topic is complex because of the diversity of metabolism in methylotrophs, and because of the unusual nature of some of their assimilation pathways and systems for

TABLE 43
The relationship between microbial physiology and cell yields

General reviews

1. Forrest and Walker (1971). The generation and utilisation of energy during growth.
2. Stouthamer (1976). Yield studies in microorganisms.
3. Stouthamer (1977a). Energetic aspects of the growth of microorganisms.
4. Stouthamer (1977b). Theoretical calculations of the influence of the inorganic nitrogen source on parameters for aerobic growth in microorganisms.
5. Stouthamer (1978). Energy-yielding pathways.
6. Stouthamer (1979). The search for correlation between theoretical and experimental growth yields.
7. Litchfield (1977). Comparative technical and economic aspects of single cell protein processes.
8. Linton and Stephenson (1978). A preliminary study of growth yields in relation to the carbon and energy content of various organic growth substrates.
9. Ho and Payne (1979). Assimilation efficiency and energy contents of prototrophic bacteria.
10. Mateles (1979). The physiology of single cell protein production.
11. Babel (1979). Bewertung von Substraten für das mikrobielle Wachstumm auf der Grundlage ihres Kohlenstoff/Energie-Verhaltnisses
12. Roels (1980). Simple models for the energetics of growth on substrates with different degrees of reduction.
13. Tempest and Neijssel (1980). Growth yield values in relation to respiration.

Reviews and discussions with special emphasis on methylotrophs

1. Harrison et al. (1972). Yield and productivity in single cell protein production from methane and methanol.
2. van Dijken and Harder (1975). Growth yields of microorganisms on methanol and methane. A theoretical study.
3. Harder and van Dijken (1975). A theoretical study of growth yields of yeasts on methanol.
4. Harder and van Dijken (1976). Theoretical considerations on the relation between energy production and growth of methane-utilising bacteria.
5. Barnes et al. (1976). Process considerations and techniques specific to protein production from natural gas.
6. Goldberg et al. (1976). Bacterial yields on methanol, methylamine, formaldehyde and formate.
7. Rokem et al. (1978a). Maintenance requirements for bacteria growing on C_1 compounds.
8. Anthony (1978b). The prediction of growth yields in methylotrophs.
9. Anthony (1980). Methanol as substrate; theoretical aspects.
10. Drozd and Wren (1980). Growth energetics in the production of bacterial single cell protein from methanol.
11. Drozd and Linton (1981). Single cell protein from methane and methanol in continuous culture.
12. Harder et al. (1981). Utilisation of energy in methylotrophs.

oxidising C_1 compounds. Many of these compounds are more highly reduced than cell material and their high heats of combustion (more than about 11 Kcal (g carbon)$^{-1}$) are not reflected in correspondingly high yields (Linton and Stephenson, 1978). These authors suggest that investigations of problems of coupling between ATP production and ATP utilisation in cells growing on energy-rich substrates will be confused by the excess of potential energy available to fix the substrate carbon, and that investigations of microbial energetics should be undertaken using poorer substrates in which it is likely that the process is energy-limited. This supports other suggestions that relationships in which Y_{ATP} is used as the basis of calculation must be used with care, because they assume by definition that growth yields are limited predominantly by the ATP supply (Anthony, 1978b, 1980).

A number of theoretical discussions of the relationship between energy production and growth yields in methylotrophs have been published since the germinal papers on this subject by Harrison *et al.* (1972) and van Dijken and Harder (1975) (Table 43). These have all attempted to provide methods of predicting the effects on cell yields of carbon pathways, oxidising systems and P/O ratios. Because some assumptions must be made (consciously or not) for all methods, it is probably a good thing that a range of different approaches has been used. Although differing in their chosen assumptions, they have tended to lead to similar conclusions.

The different methods also have different uses; for example, that of Barnes *et al.* (1976) is straightforward to use and only makes a few well-defined assumptions, but it is not suitable for testing a wide range of variables. By contrast, the methods of Stouthamer are exhaustingly comprehensive, whereas the earlier method of van Dijken and Harder (1975) is particularly adaptable. This adaptability is important when considering the wide range of variables that have accumulated with our increasing knowledge of the diversity of methylotrophic biochemistry, and it is this method that I have adopted previously and in the discussion below.

Before embarking on this discussion it should be emphasised that useful growth yield studies must include measurements of yield with respect to O_2 and carbon consumed, and CO_2 produced, and that the cell composition and full carbon balance should be determined (see Goldberg *et al.*, 1976; Rokem *et al.*, 1978a; Linton and Vokes, 1978; Drozd and Wren, 1980; Drozd and Linton, 1981). It must, therefore, be acknowledged that any method of predicting the effects of physiological variables on cell yields is very much easier than actually measuring them.

II. Assumptions and methods for developing theoretical assimilation equations

The majority of assumptions and methods used for the following calculations are derived from the work of van Dijken and Harder (1975):

(1) The formula used for cell material is $C_4H_8O_2N$ (47% carbon) which is close to the composition measured for a number of methylotrophs (Maclennan *et al.*, 1971; Goldberg *et al.*, 1976). Only major differences from this could alter markedly conclusions with respect to factors affecting cell yields.

(2) It is assumed that the reductant for biosynthesis is NAD(P)H.

(3) It is assumed that all substrate is converted into cell material or CO_2. This is usually the case when bacteria are grown in continuous culture with the carbon source as rate-limiting nutrient. During measurements of yields this assumption should, of course, be tested.

(4) The nitrogen source is ammonia. The use of nitrate or molecular N_2 would require more NAD(P)H for assimilation of these more oxidised nitrogen sources. The use of molecular N_2 would also require the expenditure of a considerable amount of ATP for the initial complex reduction process. A detailed analysis of the effect of the nature of the nitrogen source on cell yields of methylotrophs and other microorganisms has been published by Stouthamer (1977b).

(5) All cell material is assimilated by way of 3-phosphoglycerate (PGA). During preliminary metabolism of some multicarbon substrates to PGA, a decarboxylation reaction is essential. All substrates metabolised by way of acetyl-CoA (most alkanes and long chain fatty acids) are first converted by the glyoxylate cycle (or equivalent enzymes) into malate, which must be decarboxylated to give PGA. This decarboxylation is also necessary for assimilation of many C_4-carboxylic acids and it is why some assimilation equations (Table 44) include production of CO_2 as well as production of cell material. Because lipid and some amino acids (up to 20% of those required) are synthesised without decarboxylation, there is a small potential error in the amount of CO_2 assumed to be produced.

(6) It is assumed that no energy is required for active transport of alkanes, methanol, formate and methylated amines. The energy requirement for active transport is probably the area of greatest uncertainty in these sorts of calculations. If more ATP is specifically required for transport than that allowed for, then the relative extent of growth limitation by ATP supply will tend to increase. Many substrates are transported by systems depending on the proton motive force across the membrane, and so the energy required for their transport will depend on the relative concentrations of substrate inside and outside the cell. Unless otherwise stated, it has been assumed that

one ATP per mole of substrate transported is required. This assumption is the greatest source of potential error when the assimilation of small molecules is being considered.

(7) It is assumed that bacteria are growing near their maximum growth rate, so that the effect of maintenance requirements on cell yields is minimised.

(8) It is assumed that the assimilation of each PGA molecule into cell material requires 7·25 ATP. This value is based on the "measured" ATP requirement for synthesis of cell material from PGA in anaerobes in which the amount of ATP produced by substrate level phosphorylation is known. The same value may be calculated by multiplying a theoretical Y_{ATP}^{max} (Harder and van Dijken, 1976; Stouthamer, 1979) by 0·5 to allow for inefficient coupling of ATP generation and utilisation. In deriving the overall assimilation equations, the ATP required for synthesis of PGA from C_1 compounds is added to the estimated ATP requirement for assimilation of PGA. In effect, the ATP requirements "chosen" in my method are equivalent to assuming various Y_{ATP} values for growth on C_1 substrates of 5, 8·3 to 10·6, and 7·5 for assimilation by the RuBP, RuMP and serine pathways respectively. That the ATP requirement for biosynthesis of one gram of cell material can be expressed in the form of a Y_{ATP} value does not mean that this value can necessarily be used as a constant or that yields will be ATP-limited.

It has been pointed out by Harder et al. (1981) that my method for arriving at a value for the ATP requirement for biosynthesis from C_1 substrates treats the two parts of the calculations differently, and that the ATP requirement for synthesis from C_1 substrates should perhaps be more than that used here. Certainly, if the total ATP requirement for synthesis from C_1 compounds has a coupling factor of 0·5 applied to it, then the requirement for biosynthesis by the RuMP and serine pathways would increase by values between zero and 30%. When these higher ATP requirements are used instead of those chosen here, the predicted yields of some methylotrophs decrease but the overall conclusions with respect to NAD(P)H- and carbon-limitation of growth yields remain essentially the same.

The assumptions listed above lead to the following equation for assimilation of phosphoglycerate (PGA) into cell material (in these equations H^+, H_2O, NAD^+, ADP and Pi are omitted for clarity):

$$4 \text{ PGA} + 29 \text{ ATP} + 5\cdot5 \text{ NADH} + 3 \text{ NH}_3 \longrightarrow 306\text{g cells } (C_4H_8O_2N)_3$$

By taking into consideration the biosynthesis of PGA from growth substrate, an overall assimilation equation may be derived.

For example, if the formula for glucose transport and metabolism to PGA is:

$$\text{glucose} + \text{ATP} \longrightarrow 2 \text{ PGA} + 2 \text{ NADH}$$

9. Growth yields and bioenergetics

then the assimilation equation for glucose will be:

2 glucose + 31 ATP + 1·5 NADH + 3NH$_3$ ⟶ 306g cell material

The NADH is produced by oxidising glucose to CO_2 (FPH$_2$ is reduced succinate dehydrogenase):

0·15 glucose ⟶ 0·9 CO_2 + 1·5 NADH + 0·6 ATP + 0·3 FPH$_2$

By adding the last two equations, the overall assimilation equation (simplified) for predicting the effect of varying P/O ratios on growth yields on glucose is obtained:

2·15 glucose + 30·6 ATP ⟶ 306g cell + 0·9 CO_2 + 0·3 FPH$_2$

All the discussions below are based on similar equations. Although a single final equation for assimilating each multicarbon compound may be derived, this is not so with the C_1 compounds because of the variety of assimilation routes and oxidising enzymes.

The overall assimilation equations for a wide range of substrates are summarised in Table 44. The more detailed equations for predicting cell yields on C_1 compounds, alkanes and other multicarbon compounds have been presented previously (Anthony, 1978b, 1980). The equations permit calculation of the effect of varying the P/O ratios on the yield with respect to utilisation of carbon substrate and oxygen (Y- and Y-), and with respect to production of CO_2. The carbon conversion efficiency (CCE) (% of substrate carbon converted to cell material), and the proportion of substrate oxidised to provide NADH and ATP can also be calculated for various assumed P/O ratios; the CCE is the parameter most affected by changing P/O ratios in ATP-limited bacteria. The same equations can also be used for calculating the proportion of electron transport occurring from NADH and from reduced methanol dehydrogenase (PQQH$_2$), and also the proportion involved in hydroxylation reactions.

It can be seen from Table 44 that the ATP requirement does not vary markedly from one substrate to another. The main differences are in the NADH requirement for carbon assimilation, and the amount and nature of reductant produced during oxidation of the more reduced substrates to the level required for assimilation.

III. The limitation of cell yields by ATP, NADH and carbon supply

To avoid confusion it must first be mentioned that in the terminology of continuous culture, a growth-limiting nutrient in the supply medium is that nutrient whose concentration determines the growth *rate*. By contrast,

TABLE 44
Summary of assimilation equations

Growth substrates (heats of combustion, Kcal g^{-1} substrate carbon)	Moles required during assimilation		Moles produced during assimilation				
	ATP	NAD(P)H	NAD(P)H	FPH$_2$	PQQH$_2$	YH$_2$	MH$_2$
Conventional substrates (6·6–11·0)	29–41	0–5·5	0–6·5	0–4	0	0	0
Alkanes and derivatives (13–14·2)	30–41	0	9·5–18·5	4–12	0	0	0
Methane (17·56)							
Serine pathway	41	21·5	0	4	12	4	0
RuMP pathway (KDPG)	37	13·5	0	0	12	0	0
RuMP pathway (FBP)	29	13·5	0	0	12	0	0
Methanol (14·8)							
Serine pathway	41	9·5	0	4	12	4	0
RuMP pathway (KDPG)	37	1·5	0	0	12	0	0
RuMP pathway (FBP)	29	1·5	0	0	12	0	0
RuBP pathway	61	13·5	0	0	12	0	0
DHA pathway (yeasts)	37	1·5	0	0	0	0	0
Methylamine (see note below)							
Dimethylamine (oxidation involving one hydroxylation)							
Serine pathway	41	15·5	0	4	0	4	6
RuMP pathway (KDPG)	37	7·5	0	0	0	0	6

Trimethylamine (oxidation involving two hydroxylations)							
Serine pathway	41	17·5	0	4	0	4	4
RuMP pathway (KDPG)	37	9·5	0	0	0	0	4
Trimethylamine (oxidation involving one hydroxylation)							
Serine pathway	41	13·5	0	4	0	4	8
RuMP pathway (KDPG)	37	5·5	0	0	0	0	8
Formate (5·2)							
Serine pathway	49	17·5	0	4	0	0	0
RuBP pathway	61	13·5	0	0	0	0	0

The equations are for synthesis of 306 g cell material ($C_4H_8O_2N$)$_3$. The reductant for hydroxylation reactions is assumed to be NAD(P)H; this may be replaced by YH_2 or MH_2 if the oxidation of formaldehyde or methylated amines is NAD$^+$-linked. *Abbreviations*: PQQH$_2$, YH$_2$ and MH$_2$ represent the reduced dehydrogenases for methanol, formaldehyde and methylated amines (or *N*-methylglutamate) respectively. KDPG indicates the KDPG aldolase/transaldolase variant of the RuMP pathway; FBP indicates the alternative FBP aldolase/TA variant (see Table 17). The figures in parentheses are the heats of combustion (Kcal (g substrate carbon)$^{-1}$); this is a measure of the level of reduction of each substrate. The conventional substrates considered include malate, succinate, lactate, gluconate, glucose, mannitol and glycerol. Four moles of CO_2 are produced during assimilation of alkanes and 0–4 moles during assimilation of conventional substrates (see text).

Note. The assimilation equations for methylamine are usually the same as for methanol except that the reductant produced may be either PQQH$_2$, NADH or reduced *N*-methylglutamate dehydrogenase (see Chapter 7). In bacteria having an amine oxidase the oxidation of methylamine to formaldehyde produces no reductant.

254　The biochemistry of methylotrophs

in the present context, growth-limitation refers to limitation of growth *yield* by the internal ATP, NADH or carbon supply. An ATP- or NADH-limited organism is one in which a hypothetical increase in the supply of ATP or NADH would increase the growth yield. Carbon-limitation is rare and arises when ATP and NADH are potentially in excess, and the only way to increase cell yields would be by providing a more oxidised carbon source in addition to the more reduced substrate.

In order to facilitate discussion it is useful to have a means of quantifying the extent of yield-limitation by NADH etc. For this purpose the metabolism of each substrate is divided into three parts (Fig. 37):

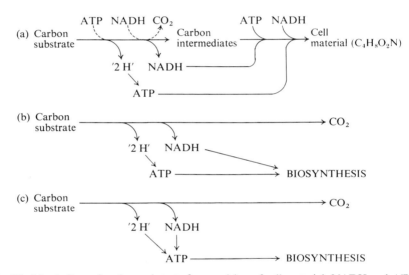

Fig. 37. Metabolism of carbon substrate for provision of cell material, NADH and ATP. (a) Corresponds to the assimilation equations in Table 44; (b) is the oxidation of substrate to CO_2 to provide NADH for biosynthesis, this route may also produce reduced flavoproteins, reduced PQQ etc.; (c) is the oxidation of substrate to provide ATP.

(a) corresponds to the equations summarised in Table 44; this assimilatory route may also produce CO_2 as well as cell material.

(b) is the oxidation of substrate to provide NADH (if necessary) for biosynthesis; if a high proportion of substrate is oxidised for this purpose then NADH-limitation is likely.

(c) is the oxidation of substrate to provide ATP, and the proportion oxidised for this purpose will depend on the P/O ratios.

Thus the proportion of substrate metabolised by (a), (b) and (c) is a rough indication of the extent of limitation by carbon, NADH and ATP respectively. These proportions (presented in Table 45) do not permit a *direct* conclusion with respect to growth limitation in a given organism. What matters is the proportion of a particular substrate metabolised for provision of cell carbon, NADH and ATP, compared with the proportions estimated for conventional growth substrates on which growth is ATP-limited. If for a given substrate, the proportion oxidised for ATP synthesis is markedly less than this norm, then growth is (to some extent) NADH- or carbon-limited. Growth on a given substrate can be said to be completely NADH- or carbon-limited if the range of values given in Table 45 is very narrow; that is, if large changes in P/O ratios lead to only very slight changes in the proportions of substrate metabolised for assimilation and energy production.

A. ATP-limited bacteria

Most bacteria growing on typical multicarbon compounds are limited predominantly by the ATP supply. The level of reduction of these substrates (having heats of combustion of less than 11 Kcal (g substrate carbon)$^{-1}$) are less than (or similar to) that of cell material. Predicted and measured growth yields are closely related to the level of reduction of the substrate (see Linton and Stephenson, 1978; Anthony, 1980). Growth on the more oxidised substrates is more ATP-limited, whereas growth on more reduced substrate tends to be more carbon-limited. During growth on these conventional substrates the direct relationship between the yield on O_2 and the P/O is valid (that is, $Y_O = Y_{ATP} \times$ P/O ratio).

ATP-limited methylotrophs will usually have some or all of the following characteristics: NADH-requiring hydroxylations will not be present; the oxidation of formaldehyde to CO_2 will yield 2 NADH; carbon will be assimilated by the RuMP or RuBP pathway. Bacteria with low P/O ratios will be more ATP-limited than those with higher P/O ratios. Some RuMP pathway bacteria (such as *Methylophilus methylotrophus*) can be considered to be both ATP and carbon-limited (see below).

Growth on formate is both ATP- and NADH-limited ("energy limited"), as is the growth of autotrophs. Because NADH is the only reductant produced during formate oxidation, and ATP is only produced by oxidation of NADH, the predicted growth yield (Y_O) is directly proportional to the P/O ratio.

The carbon assimilation pathway in yeasts (the DHA pathway) is energetically very similar to the RuMP pathway, but the yeasts are not carbon-limited because methanol is not oxidised by methanol dehydrogenase, but by an oxidase. Although ATP-limited, the predicted yield on O_2 (Y_O) is not proportional to the P/O ratio because of the O_2 requirement of the oxidase which is not coupled to ATP synthesis. The same will be true for bacteria such as *Arthrobacter* P1 in which methylamine is oxidised by an amine oxidase.

TABLE 45
The limitation of cell yields by ATP, NADH and carbon supply

	Percentage substrate metabolised for:			
	Assimilation	Energy production		
Substrate, carbon pathway and oxidative pathway	Cell carbon ($+CO_2$)	$NAD(P)H$	ATP
Conventional substrates	32–76 (+0–20)	0–12	22–54
Alkanes and derivatives	54–75 (+18–25)	0	0–28
Formate			
Serine pathway	16–28	24–41	31–60
RuBP pathway	14–26	16–29	45–75
Methanol (and usually methylamine)			
Serine pathway: $YH_2 \neq$ NADH: (e.g. *Pseudomonas* AM1)	53–56	42–44	5
$YH_2 =$ NADH	55–71	11–14	15–34
RuMP pathway; (KDPG var.) (e.g. *M. methylotrophus*)	58–74	3–5	21–39
(FBP variant)	67–80	4–5	15–29
RuBP pathway; $YH_2 \neq$ NADH;	42–47	47–53	0–11
$YH_2 =$ NADH; (e.g. *Pa. denitrificans*)	41–59	5–7	24–54
DHA pathway (e.g. yeasts)	38–64	3–4	32–59
Methylamine (oxidation by amine oxidase) ($YH_2 =$ NADH)			
RuMP pathway (FBP variant); (e.g. *Arthrobacter* P1)	44–68	4	28–52
Methane (hydroxylation requires NADH)			
Serine pathway (Type II methanotrophs)	36–39	53–57	4–11
RuMP pathway (KDPG var.); (Type I methanotrophs)	38–42	43–47	11–19

Substrate / Pathway			
Methane (hydroxylation requires cytochrome *c*)			
Serine pathway; $YH_2 \neq NADH$	34–54	27–41	7–39
$YH_2 = NADH$	36–59	8–14	27–56
RuMP pathway (KDPG var.)	38–63	2–4	33–59
Dimethylamine (hydroxylation involved)			
Serine pathway; ($YH_2 \neq NADH$); (e.g. some pink pseudomonads and *Hyphomicrobia*)	28	72	0
Serine pathway; ($YH_2 \neq NADH$); (e.g. some pink pseudomonads and non-pigmented pseudomonads)	36–53	23–33	14–41
RuMP pathway (KDPG var.) $YH_2 \neq NADH$;	37–44	44–56	0–17
$YH_2 = NADH$; (a) methylamine oxidised by dehydrogenase	39–58	16–24	17–45
(b) methylamine oxidised by oxidase	29–48	12–20	32–59
Trimethylamine; serine pathway			
1 hydroxylation involved; $YH_2 \neq NADH$	37	63	0
$YH_2 = NADH$	42–62	20–30	8–38
2 hydroxylations involved; $YH_2 \neq NADH$;	19	81	0
$YH_2 = NADH$	30–46	26–39	15–44
Trimethylamine; RuMP pathway (KDPG var.)			
1 hydroxylation involved; $YH_2 \neq NADH$	45–55	31–38	7–24
$YH_2 = NADH$	45–60	12–16	24–43
2 hydroxylations involved; $YH_2 \neq NADH$ (obligate methylotrophs)	30	70	0
$YH_2 = NADH$	32–49	19–30	21–49

A rough measure of the extent of limitation by carbon supply is given by the percentage of substrate metabolised to cell material (the carbon conversion efficiency, CCE) (plus any CO_2 produced during assimilation). The percentage substrate oxidised to CO_2 to provide NADH or ATP is a measure of the extent of limitation by NADH or ATP respectively. The range of values is for a range of P/O ratios between 1 and 3 for NADH oxidation, between 1 and 2 for oxidation of flavoproteins and only 1 for oxidation of methanol and methylamine. When the P/O ratio is low the carbon conversion efficiency (CCE) is low; the percentage substrate oxidised to provide ATP will therefore be high and so cells will be more ATP-limited. By contrast, high P/O ratios lead to high CCEs and to a greater chance of limitation by NADH or carbon supply. Abbreviations etc. are as for Table 44. Predicted values for bacteria oxidising formaldehyde by a cyclic route generating 2 moles of NAD(P)H are similar to those in which formaldehyde dehydrogenase is NAD^+-linked ($YH_2 = NADH$). The estimated proportion for individual alkanes and conventional substrates have been presented elsewhere (Anthony, 1980).

B. Carbon-limited bacteria

These are rare. The most extreme examples are not methylotrophs but the alkane-utilising bacteria, many of which are predominantly, or completely, carbon-limited. Because alkanes and derivatives such as long-chain alcohols and fatty acids are highly reduced, sufficient NADH and ATP is produced for biosynthesis during the initial oxidation of substrate to acetyl-CoA; thus very little substrate needs to be oxidised "merely" to provide ATP. At first sight this might suggest that all of the substrate should be converted into cell carbon material (100% carbon conversion efficiency). However, this is not so because the acetyl-CoA is assimilated by way of the glyoxylate bypass to malate, and thence by decarboxylation to PGA. Thus a high proportion (say 15–20%) of substrate carbon must be "wasted" as CO_2. This is the reason why yields of alkane-utilising yeasts are not markedly different from those of bacteria even though some steps in alkane oxidation by yeasts are coupled to flavoprotein oxidases in the peroxisomes and are not coupled to ATP synthesis.

The only methylotrophs likely to be at all carbon-limited are some of the methanol- or methylamine-utilisers assimilating carbon by the RuMP pathway (e.g. *Methylophilus methylotrophus*, *Pseudomonas* C, organism C2A1 etc.). In these bacteria, the magnitude of the P/O ratio has a relatively small effect on predicted yields (see Anthony, 1978b, 1980; Harder *et al.*, 1981). The reason for this partial carbon-limitation in methylotrophs is similar to that operating in alkane utilisers; the NADH requirement for assimilation is low and oxidation of substrate to formaldehyde provides most of the ATP needed. The amount of substrate oxidised "merely" for ATP synthesis is therefore relatively low.

C. NADH-limited bacteria

As can be seen from Table 45 many methylotrophs are, to some extent, NADH-limited. Three aspects of their biochemistry contribute to this. The first is that oxygenases catalysing the initial attack on the methyl groups of methane and methylated amines require NADH as reductant. The second is the high NADH-requirement for assimilation of formaldehyde by the serine pathway. The third is that dehydrogenases for catalysing the oxidation of methanol and methylamine (and sometimes dimethylamine and trimethylamine) are not coupled to production of NADH but only to production, by way of electron transport, of ATP. Oxidation of these substrates to the level of formaldehyde, prior to biosynthesis, thus produces a high yield of ATP. Similarly, more ATP is produced during he oxidation of substrate to CO_2 for provision of NADH for biosynthesis. Thus oxidation of more substrate "merely" for ATP production is often unnecessary.

9. Growth yields and bioenergetics

In some bacteria, oxidation of formaldehyde to CO_2 yields only one molecule of NADH, and this will aggravate the problem in those bacteria having a high NADH-requirement.

The only methanol-utilisers likely to be NADH-limited are bacteria having the serine pathway and no NAD^+-linked formaldehyde dehydrogenase (e.g. *Pseudomonas* AM1). The same applies to those bacteria growing on methylated amines and having "dye-linked" dehydrogenases for oxidising these substrates to formaldehyde; in this case the assimilation formulae are almost identical to those for methanol assimilation. In these NADH-limited methylotrophs, increasing the P/O ratio for methanol or amine oxidation from one to two would have a negligible effect on cell yields.

Assuming that the reductant for methane hydroxylation is NADH, then all methanotrophs are predominantly NADH-limited, regardless of their carbon assimilation pathway. If methanol (by way of cytochrome *c*) is the reductant then growth will be ATP-limited.

Growth yields of most methylotrophs on methylated amines, whose oxidation is catalysed by oxygenases, are limited by NADH to a large extent. As always, the extent is greater in serine pathway bacteria and in those bacteria in which oxidation of formaldehyde to CO_2 yields only one molecule of NADH.

I have concluded here that many methylotrophs are NADH-limited and that a few are likely to be carbon-limited. This conclusion is based on one particular method and set of assumptions. It must be emphasised that although different assumptions and methods might lead to slight differences in predicted growth yields, all methods lead to the same general conclusions with respect to those factors limiting cell yields. The apparent disagreement of Harder *et al.* (1981) with this conclusion should therefore be mentioned. These authors conclude that their calculations "indicate that NADH limitation of growth by organisms on methanol is not as common as predicted by Anthony". However, when I proposed that for the majority of methylotrophs growth yield is determined by NADH supply as well as ATP supply (Anthony, 1978b), I was considering all methylotrophs in my generalisations — including the many methylotrophs growing on methane and methylated amines, in which NADH-requiring hydroxylations necessarily lead to greater NADH-limitation. Harder *et al.* (1981) considered only growth on methanol and with respect to this substrate we come to identical conclusions;

(a) High P/O ratios increase the tendency to carbon- or NADH-limitation.
(b) The only methanol-utilisers tending towards carbon-limitation are those growing by the RuMP pathway (organisms such as *Methylophilus methylotrophus*).
(c) The only methanol-utilisers likely to be NADH-limited are those serine pathway bacteria in which the oxidation of formaldehyde to CO_2 yields only one molecule of NADH.

260   The biochemistry of methylotrophs

A final point with respect to NADH-limitation is the possibility of "reversed electron transport" systems which might use ATP or the proton motive force to drive reduction of NAD^+ by methanol or methylated amines. Any bacteria having such a system will be energy limited (effectively ATP-limited) rather than NADH-limited. There is no evidence for the operation of such systems in any methylotroph except for the facultatively photosynthetic bacteria (see below).

IV. The prediction of growth yields in methylotrophs

It is not the purpose of this section to predict actual growth yields for particular methylotrophs, but to investigate the different factors likely to affect yields in different types of methylotroph. As emphasised previously, our knowledge about the coupling of ATP generation and utilisation in bacteria is insufficiently complete to predict growth yields with a high degree of accuracy. This is balanced by the technical difficulties of accurately determining growth yields. In spite of this, predicted values and measured yields are remarkably consistent. They are so consistent that any marked divergences of measured from predicted yields is likely to be sufficiently significant to be worth investigating. Table 43 gives references to work on prediction and measurement of growth yields in methylotrophs; and the most important theoretical conclusions have been discussed already in this chapter. Specific predictions are illustrated in Table 46. These are based on the method discussed in this chapter and published previously (Anthony, 1978b, 1980). The values differ (usually slightly) if alternative methods are used, but the general conclusions are similar (for comparison with alternative methods see van Dijken and Harder, 1975; Harder and van Dijken, 1975, 1976; Harder et al., 1981). The actual values predicted will, of course, be lower for ATP-limited bacteria if growth and energy generation are more uncoupled than has been assumed. By contrast, in those bacteria whose growth yields are not ATP-limited, changes in efficiency of energy coupling will lead to relatively little change in predicted growth yields.

A. Growth of non-photosynthetic organisms on methanol

The specific conclusions arising from consideration of Table 46 are as follows:

(1) Measured yields are within the ranges predicted. These are: 15·7–19·5 g $(mole)^{-1}$ for RuMP pathway bacteria, and 9·8–14·6 g $(mole)^{-1}$ for serine pathway bacteria (Goldberg et al., 1976; Rokem et al., 1978a); 13·4 g $(mole)^{-1}$

TABLE 46
Predicted yields on methanol and methane

Carbon pathway	Y_s values	Y_o values	Carbon conversion efficiency (%)
Methanol as substrate			
Serine pathway			
(a) HCHO oxidation yields 1 NADH	13·6–14·2 (14·2)	7·3– 7·9 (7·9)	53–56 (56)
(b) HCHO oxidation yields 2 NADH	13·9–20·7 (18·0)	7·4–15·5 (11·6)	55–81 (71)
RuMP pathway (KDPG aldolase/TA variant)	14·7–21·6 (18·9)	8·3–17·8 (13·2)	58–85 (74)
RuMP pathway (FBP aldolase/SBPase variant)	17·0–23·4 (20·4)	10·6–21·7 (15·4)	67–92 (80)
RuBP pathway (HCHO oxidation yields 2 NADH)	10·4–17·2 (15·1)	5·0–11·7 (9·0)	41–67 (59)
DHA pathway (yeasts)	9·8–16·2 (16·2)	4·7– 9·8 (9·8)	38–64 (64)
Methane as substrate (hydroxylation coupled to NADH)			
Serine pathway	9·2–10·4 (10·0)	2·8– 3·3 (3·2)	36–41 (39)
RuMP pathway (KDPG aldolase/TA variant)	9·7–12·0 (10·7)	3·0– 4·0 (3·5)	38–47 (42)
Methane as substrate (hydroxylation coupled to cytochrome c)			
Serine pathway: $YH_2 \neq$ NADH	8·7–13·2 (13·2)	2·7– 4·6 (4·6)	34–54 (54)
$YH_2 =$ NADH	9·2–15·1 (15·1)	3·0– 5·9 (5·9)	36–59 (59)
RuMP pathway (KDPG var.)	9·8–16·2 (16·2)	3·1– 6·1 (6·1)	38–63 (63)

Predicted yields are presented as Y_s values (g dry wt (mole substrate used)$^{-1}$) or Y_o values (g(g atom oxygen)$^{-1}$). The range of predicted values is for P/O ratios between 1 and 3 for NADH, and between 1 and 2 for other substrates. The values in parentheses are for the highest probable P/O ratios (3 for NADH, 2 for flavoproteins and only 1 for methanol dehydrogenase). Low P/O ratios lead to low predicted yields and carbon conversion efficiencies.

for RuBP pathway bacteria (van Verseveld and Stouthamer, 1978b), and about 16 g (mole)$^{-1}$ for yeasts having the DHA pathway (Harder and van Dijken, 1975). The only bacteria in which measured yields on methanol are markedly different from those predicted are those initially isolated on methane but also capable of growth on methanol. In these bacteria, measured yields on methanol (about 12·5 g (mole)$^{-1}$ for methanotrophs having the RuBP pathway) are much lower than predicted. Possible explanations for this are discussed below.

(2) It has been predicted that yields of bacteria growing by the RuMP pathway should be considerably higher than those growing by way of the alternative serine pathway; this was implied by Quayle and his colleagues when first describing these pathways and it has been extensively elaborated in later theoretical discussions (van Dijken and Harder, 1975; Anthony, 1978b). This prediction has been amply confirmed by the work of Goldberg, Mateles and their colleagues (Goldberg *et al.*, 1976; Rokem *et al.*, 1978a).

(3) The relatively low yields with respect to methanol and oxygen for yeast, compared with most bacteria, is because the first step in methanol oxidation in yeast involves a flavoprotein oxidase which is not coupled to ATP synthesis.

(4) Alterations in the ATP or NADH requirement for assimilation, resulting from changes in the nitrogen source or the route for its assimilation, will lead to changes in the predicted growth yield. The large ATP requirement for nitrogen fixation will have relatively little effect on yields of NADH- or carbon-limited bacteria compared with that likely in ATP-limited bacteria (see Stouthamer, 1977b).

A change in the pathway for ammonia assimilation, which causes a decreased ATP requirement for growth (Windass *et al.*, 1980), leads to a change in yields of *Methylophilus methylotrophus*; but because this organism is to some extent carbon-limited the predicted increase (2–5%) is only about half that predicted for predominantly ATP-limited bacteria. It is impressive that such small changes can be accurately determined, but clearly during large scale production of SCP such small increases in yield are economically significant (Senior and Windass, 1980).

B. Growth of *Rhodopseudomonas acidophila* on methanol

Rh. acidophila, growing on methanol, assimilates cell carbon at the level of CO_2 by way of the RuBP pathway and obtains energy by oxidation of methanol or from light. Reductant for biosynthesis is produced by oxidation of methanol.

1. Aerobic growth in the dark

The low yield measured during growth in the dark (7·4 g (mole methanol)$^{-1}$) (Seifert and Pfennig, 1979) suggests that the first step in methanol oxidation, catalysed by methanol dehydrogenase, may not be coupled to ATP synthesis.

2. Anaerobic growth in the light

Measurements of growth yield and cell composition during anaerobic growth of *Rh. acidophila* in the light shows the following carbon balance (Quayle and Pfennig, 1975):

$$7 CH_3OH + 3 CO_2 + 2 NH_3 \longrightarrow (C_5H_8O_2N)_2 + 9 H_2O$$

All the cell carbon is assimilated after oxidation of the methanol to CO_2, whose assimilation then requires NADH as reductant (Fig. 38). All NADH for biosynthesis must be obtained by oxidising methanol to CO_2 by way of methanol dehydrogenase, and NAD^+-linked dehydrogenases for formaldehyde and formate. In the equation above, seven molecules of methanol are oxidised, thus yielding 14 molecules of NADH and seven of reduced methanol dehydrogenase ($PQQH_2$). This raises two problems. The first is that the $PQQH_2$ must be reoxidised (in the absence of O_2) and the second is that seven further molecules of NADH are required to assimilate the CO_2 into cell material. This extra NADH must be produced by reducing NAD^+ with reduced methanol dehydrogenase (Fig. 38). The $HCHO/CH_3OH$ redox couple is more positive than the $NAD^+/NADH$ couple and the electron acceptor from methanol dehydrogenase is thought to be cytochrome *c*. This process is thus an example of "reversed" electron transport and it is presumably driven by ATP or a proton motive force produced by a light-driven electron transport system.

```
                         7 H₂O           3 CO₂    2 NH₃
                                                  │ 16 H₂O
7 CH₃OH  ───→  7 HCHO  ───────→  7 CO₂  ─────────→  (C₅H₈O₂N)₂
         │                                ↗
         ↓                  14 NADH ───→ 21 NADH
         7 PQQH₂  ········→ 7 NADH  ─┘
              "reversed electron
                 transport"
```

Fig. 38. Assimilation of methanol and CO_2 into *Rhodopseudomonas acidophila* during anaerobic growth in the light. The assimilation equation is from Quayle and Pfennig (1975). ATP is provided exclusively by photophosphorylation. $PQQH_2$ is the reduced form of the prosthetic group of methanol dehydrogenase.

C. Growth on methane

This topic is discussed at some length because methanotrophs are the only organisms whose measured growth yields sometimes differ markedly from those predicted, perhaps indicating some unexpected feature in the metabolism of methane. A further reason is that comparisons of growth yields on methanol and methane have led to proposals about the mechanism of methane hydroxylation which are probably wrong. Because methane is the most highly reduced carbon compound (heat of combustion, 17·56 Kcal (g carbon)$^{-1}$), it might be expected to be an ideal substrate for producing high yields of single cell protein (SCP). However, it is so stable that, in common with other alkanes, an initial activation by molecular oxygen is necessary for its metabolism. The hydroxylation mechanism by which this is achieved requires an electron donor for reduction of the second atom of oxygen to water; this reductant must be produced by the further oxidation of methanol. Thus half of the reducing potential of methane is "lost" during its oxidation to formaldehyde. If no energy in the form of ATP is available from this oxidation then the "useful" level of reduction of methane is the same as that of formaldehyde (heat of combustion, 11·37 Kcal (g carbon)$^{-1}$).

The possible mechanisms of methane hydroxylation differ mainly in the nature of the reductant for the reaction (Chapter 6). In *Methylococcus* this is NADH, and it has recently been proposed that this is the only reductant ever used for methane hydroxylation in bacteria. In an alternative system proposed by Higgins and his colleagues (Higgins, 1979a), the electrons for hydroxylation come from methanol dehydrogenase by way of cytochrome *c*. Unless otherwise stated the following discussion assumes that NADH is the reductant. The predicted effects on cell yields of the nature of this reductant for methane hydroxylation have been fully discussed elsewhere (Anthony, 1978b).

Three specific conclusions arise from consideration of Table 46 and two of these (points 2 and 3 below) raise important questions of interpretation:

(1) Predicted yields on methane are similar for bacteria growing by way of the serine and RuMP assimilation pathways, because yields of these bacteria are similar with respect to limitation by NADH. This contrasts with bacteria growing on methanol where the difference in extent of ATP- and NADH-limitation during growth by the two pathways leads to considerable differences in predicted yields.

(2) The highest published yields on methane (about 17 g (mole methane)$^{-1}$) (Higgins, 1979a) are much higher than those predicted for bacteria with the more likely NADH-coupled system (about 12 g (mole)$^{-1}$). These high yields are more consistent with the values predicted for bacteria having a cytochrome

9. Growth yields and bioenergetics 265

c-coupled system (16·2–17·4 g (mole)$^{-1}$) (Anthony, 1978b). That the difference between predicted and measured yields is not because of the particular methods of prediction is indicated by the fact that measured yields for all other substrates considered here (and in Anthony, 1980) are within the ranges predicted, and that Harder and van Dijken (1976), using a different method have predicted almost identical yields. If more recent measured values of about 12·5 g (mole)$^{-1}$ (cited in Linton and Vokes, 1978) are more reliable than the previously published higher values (this is probable), then much of the following discussion of reversed electron transport is spurious. It must be emphasised that the low predicted maximum yields of about 12 g (mole)$^{-1}$ cannot be altered by changing the assumptions made with respect to the ATP requirement or ATP production in these bacteria. For example, coupling of hydroxylation of methane to ATP synthesis (Higgins, 1979a) would not increase the predicted yields; even if it is assumed that the ATP requirement for biosynthesis is zero, that the P/O ratio is 3 and that hydroxylation is coupled to ATP synthesis, then the maximum predicted yields would still be only about 12 g (mole methane)$^{-1}$. If the high measured yields of more than 15 g (mole)$^{-1}$ are verified in bacteria having the NADH-requiring methane monooxygenase then the only explanation must be that methanol is able to reduce NAD$^+$ by a process of "reversed electron transport" by way of methanol dehydrogenase. This is thermodynamically feasible because the midpoint redox potential (at pH 7·0) for the formaldehyde/methanol couple is $-0·182$ V while that of the NAD$^+$/NADH couple is $-0·32$ V. However, all the evidence available indicates that methanol dehydrogenase donates electrons to the electron transport chain at the level of cytochrome c and this has a very high midpoint potential (about $+0·3$ V at pH 7·0). Although this does not preclude the involvement of cytochrome c in the transfer of electrons from methanol dehydrogenase to NAD$^+$, this electron transport chain would be rather remarkable and there is no evidence for the operation of such reversed electron transport systems in methanotrophs.

(3) It is predicted in Table 46 that yields should always be lower on methane than methanol, but in practice this is not so. In methane-utilising bacteria the measured yields on methane are similar or higher than those of the same organism on methanol. This has led to the suggestion (Linton and Vokes, 1978) that growth of these bacteria may be carbon-limited. Reference to Table 45, however, indicates that growth of bacteria having the RuMP pathway (and giving yields of about 12 g per mole of methane or methanol) must be predominantly ATP- or NADH-limited.

The alternative, erroneous, conclusion that usable energy may be produced during the oxidation of methane to methanol arises from the false assumption that yields are inexplicably *high* on methane rather than inexplicably *low* on methanol. In fact, the measured yields of methanotrophs growing on methanol (about 12·5 g (mole methanol)$^{-1}$) are much lower than any predicted values for bacteria growing on methanol by the RuMP pathway (Table 46). Even more important, the yields of these methanotrophs growing on methanol are very much lower than the measured yields of typical methanol-utilisers assimilating methanol by the RuMP pathway (15·7–19·5 g (mole)$^{-1}$) (Goldberg et al., 1976; Rokem et al., 1978a). Even with a P/O ratio of only one for each oxidation step, a yield of about 15 g (mole methanol)$^{-1}$ would be expected (Table 46).

There are two possible explanations for these low growth yields of methanotrophs growing on their "unnatural" substrate methanol. The first is related to the fact that very little electron transport by way of NADH dehydrogenase is necessary during growth on methane, compared with that required during growth on methanol (see the following section). If the "extra" NADH dehydrogenase required during growth on methanol is not well-coupled to ATP synthesis, then low yields on methanol will result. If, for example, the P/O ratio for NADH oxidation is only 0·5 then the predicted yield on methanol would be 12·3 g (mole)$^{-1}$ (the same as the measured value).

A second (perhaps more likely) explanation is that during growth on methanol the methane monooxygenase is still present and active. This enzyme oxidises methanol at a high rate (Colby et al., 1977; Stirling et al., 1979) and if some methanol oxidation occurs by this route then lower growth yields must result.

V. The effect on electron transport systems of the oxidative and assimilatory pathways of methylotrophs

During aerobic growth on most conventional substrates 75–80% of electron transport to oxygen involves NADH dehydrogenase, the remainder being from the flavoprotein succinate dehydrogenase. By contrast, during methylotrophic growth relatively few dehydrogenases are NAD$^+$-linked, in spite of the fact that NADH is often in exceptionally high demand for hydroxylation reactions and biosynthesis.

The equations summarised in Table 44 can be used to estimate the proportions of oxygen consumed by methane monooxygenases, and by way of dehydrogenases for NADH, succinate, methanol and formaldehyde. The proportions will be affected by the P/O ratio but the main determining factors will be the assimilation pathway, the mechanism of methane hydroxylation, the nature of the methanol dehydrogenase, and whether or not the formaldehyde dehydrogenase generates NADH. Similar arguments to those used here will apply to the nature of electron transport in bacteria growing on methylated amines.

TABLE 47

Electron flow from each dehydrogenase expressed as a proportion of total electron transport

Growth substrate	Percentage of electron flow from each dehydrogenase			
	Methanol	Formaldehyde	Flavoprotein	NADH
Methane				
Serine pathway bacteria (e.g. *Methylosinus trichosporium*)	78–88	(NAD$^+$-linked)	9–12	0–13
RuMP pathway bacteria (e.g. *Methylococcus capsulatus*)	84–92	(NAD$^+$-linked)	0	8–16
Methanol				
Serine pathway bacteria				
(a) HCHO oxidation yields 1 NADH (e.g. *Pseudomonas* AM1)	53–56	35	10	0–3
(b) HCHO oxidation yields 2 NADH	53–64	(NAD$^+$-linked)	9–15	21–40
RuMP pathway bacteria (e.g. *Methylophilus methylotrophus*)	57–70	(NAD$^+$-linked)	0	30–43
RuBP pathway bacteria (e.g. *Paracoccus denitrificans*)	52–62	(NAD$^+$-linked)	0	38–48
Conventional substrates (including formate)	0	0	0–26	74–100

The range of values predicted is for a range of assumed P/O ratios; the lowest is assumed to be one for each oxidation step and the highest P/O ratios are assumed to be 1, 2 or 3 for methanol dehydrogenase, flavoprotein and NADH respectively. Higher P/O ratios lead to a lower proportion of electron transport from NADH dehydrogenase and a higher proportion from other dehydrogenases. During growth on methane only 25–40% of the total oxygen consumed is by way of electron transport and oxidases, the remainder being used in the initial hydroxylation reaction (assumed here to be NADH-linked). In serine pathway bacteria the oxidation of acetyl-CoA to glyoxylate is assumed to involve a flavoprotein.

Three main conclusions arise from consideration of Table 47:
(1) During growth on methane, between 60% and 75% of the oxygen consumed is used in the initial hydroxylation reaction.
(2) Between 50% and 92% of the oxygen used in electron transport is coupled to the dehydrogenases for methanol or methylated amines.
(3) NADH dehydrogenase is relatively less important during methylotrophic growth (except on formate) than it is during heterotrophic growth. NADH dehydrogenase is less important at higher P/O ratios, and in some cases the proportion of electron transport involving this dehydrogenase is negligible.

10
Metabolism in the methylotrophic yeasts

I. The methylotrophic yeasts 269
II. Oxidative metabolism in yeasts 276
   A. The oxidation of methanol to formaldehyde 276
      1. The alcohol oxidase of methylotrophic yeasts 276
      2. The role of catalase in methanol oxidation 279
      3. Peroxisomes in yeasts 281
         (a) Peroxisomes in alkane-utilising yeasts 281
         (b) Peroxisomes in methanol-utilising yeasts 283
   B. The formaldehyde dehydrogenase of yeasts 283
   C. Hydrolysis of S-formylglutathione, and the oxidation of formate . 284
III. The dihydroxyacetone (DHA) cycle of formaldehyde assimilation in yeasts . 285
   A. Description of the DHA cycle 285
   B. Oxidation of formaldehyde by a dissimilatory DHA cycle . . 287
   C. Evidence for the DHA cycle of formaldehyde assimilation . . 288
   D. The reactions of the DHA cycle 290
IV. Regulation of methanol metabolism in yeasts 292
   A. Regulation of synthesis of methanol-metabolising enzymes . . 292
   B. Regulation of enzyme activity 294

The C_1 substrate used predominantly by methylotrophic yeasts is methanol. These yeasts differ fundamentally from methylotrophic bacteria in being eucaryotic organisms, in their enzymes for oxidising methanol and in their carbon assimilation pathway. For this reason it has seemed best to confine them to this one chapter, although some aspects of their bioenergetics are discussed in Chapter 9. Valuable reviews of the metabolism of methanol by yeasts have been published by Sahm (1977), by Tani *et al.* (1978a) and by van Dijken *et al.* (1981a).

I. The methylotrophic yeasts

Although three species of mycelial fungi able to use methanol have been described (Table 48), the majority of eucaryotic methylotrophs are yeasts; these are able to grow at a high rate and to a high density during laboratory

TABLE 48
Eucaryotic methylotrophs

Mycelial fungi (unable to grow on methane)	
Gliocladium deliquescens (grows on methanol, formaldehyde and formate)	Sakaguchi et al. (1975)*
Paecilomyces varioti (grows on methanol, formaldehyde and formate)	Sakaguchi et al. (1975)*
Trichoderma lignorum (grows on methanol, methylamine, formate)	Tye and Willets (1973)*
Yeasts unable to grow on methane	
Hansenula polymorpha DL-1 (ATCC 26012)*	Levine and Cooney (1973); Kato et al. (1976)
H. polymorpha (CBS 4732)	Hazeu et al. (1972); van Dijken and Harder (1974)
H. capsulata (CBS 1993)	Hazeu et al. (1972)
H. glycozyma (CBS 5766)	Hazeu et al. (1972)
H. henricii (CBS 5765)	Hazeu et al. (1972)
H. minuta (CBS 1708)	Hazeu et al. (1972)
H. nonfermentans (CBS 5764)	Hazeu et al. (1972)
H. philodendra (CBS)	Hazeu et al. (1972)
H. wickerhamii (CBS 4307)	Hazeu et al. (1972)
H. ofuaensis	Asai et al. (1976)
Candida boidinii (ATCC 32195)*	Sahm and Wagner (1972)
C. boidinii (CBS 2428, 2429)	Hazeu et al. (1972)
C. boidinii KM-2*	Kato et al. (1974c)
C. boidinii (10 strains)*	van Dijken and Harder (1974)
C. boidinii NRRL Y-2332	Mehta (1975a, b); Yokote et al. (1974)

C. boidinii S-1, S-2	Shimizu *et al.* (1977b, c)
C. boidinii 25-A	Yamada *et al.* (1979)
C. alcamigas	Uragami *et al.* (1973) (cited in Ogata *et al.*, 1975)
C. methanolica*	Oki *et al.* (1972)
C. parapsilosis	Okomura *et al.* (1970) (cited in Sahm, 1977)
C. utilis (ATCC 26387)	Patel *et al.* (1979d)
Candida sp. N-16, N-17*	Fujuii and Tonomura (1972, 1975a); Tonomura *et al.* (1972)
Kloeckera sp. 2201* (this may be a strain of *C. boidinii*)	Ogata *et al.* (1969)
Kloeckera sp. A2*	Patel *et al.* (1979d)
Pichia pinus (26 strains)*	van Dijken and Harder (1974)
P. pinus (CBS 5098)	van Dijken and Harder (1974)
P. pinus (CBS 744)	Hazeu *et al.* (1972)
P. pinus NRRL YB-4025	Mehta (1975a, b)
P. haplophila (CBS 2028)	Okomura *et al.* (1969)
P. pastoris (CBS 704)	Hazeu *et al.* (1972)
P. pastoris*	Kato *et al.* (1974c)
P. pastoris (IFP 206)	Couderc and Baratti (1980a)
P. trehalophila (CBS 5361)	Hazeu *et al.* (1972)
P. lidnerii*	Henninger and Windisch (1975)
P. methanolica (5 strains)*	Kato *et al.* (1974c)
P. methanothermo (thermotolerant)*	Minami *et al.* (1978)*
Pichia sp. NRRL-Y-11328*	Patel *et al.* (1979d)
Saccharomyces H-1*	Tonomura *et al.* (1972) (cited in Ogata *et al.*, 1975)
Torulopsis pinus (CBS 970)	Hazeu *et al.* (1972)

TABLE 48 (continued)

T. nitatophila (CBS 2027)	Hazeu et al. (1972)
T. nemodendra (CBS 6280)	Hazeu et al. (1972)
T. molishiana (all strains in the CBS collection grow on methanol except the type strain)	Hazeu et al. (1972)
T. methanolovescens*	Oki et al. (1972)*
T. glabrata*	Asthana et al. (1971)*
T. enoki and *T. methanophiles**	Uragami et al. (1973) (cited in Ogata et al., 1975)
T. methanosorbosa; *T. methanodomercquii**	Yokote et al. (1974)
T. nagoyaensis*	Asai et al. (1976)
Torulopsis sp. A1*	Patel et al. (1979d)
Rhodotorula sp.	Asano et al. (1973) (cited in Ogata et al. 1975)
Yeasts able to grow on methane	
Rhodotorula glutinis (strain cy)	Wolf and Hanson (1979); Wolf et al. (1980); Wolf (1981)
Sporobolomyces roseus (strain y)	Wolf and Hanson (1979); Wolf et al. (1980); Wolf (1981)

These are all facultative methylotrophs able to grow on methanol. The asterisk indicates that the strain was isolated on methanol; other strains were from culture collections. No species of *Kloeckera* from the Central Bureau Voor Schimmel Cultures (CBS culture collection) (Delft, the Netherlands) was able to grow on methanol (Hazeu et al., 1972); *Kloeckera* sp. 2201 may apparently be a strain of *Candida boidinii*. None of the 16 type strains of *Torulopsis glabrata* grew on methanol (Hazeu et al., 1972). All strains of *Candida boidinii* (18 strains), *Pichia pinus* (14) and *Hansenula capsulata* in the CBS collection grew on methanol. The classification of methylotrophic yeasts has been discussed by Komagata (1981).

10. Metabolism in the methylotrophic yeasts 273

culture on methanol. The study of the utilisation of methanol by yeasts began with the investigations of *Kloeckera* sp. 2201, isolated on methanol by Ogata *et al.* (1969). Since then many yeasts have been isolated on methanol (Table 48), and a few yeasts in culture collections shown to grow upon this substrate (only 30 out of about 700 strains tested) (Oki *et al.*, 1972; Hazeu *et al.*, 1972). Only six or seven of the 39 genera of yeasts are able to grow on methanol, the majority of methylotrophic yeasts being species of *Hansenula*, *Candida*, *Pichia* and *Torulopsis*.

All the methylotrophic yeasts divide by budding; in most cases this occurs at various sites on the cell surface and not just at the poles of the cells. *Hansenula*, *Pichia* and *Saccharomyces* are Ascomycetous yeasts producing sexually-derived ascospores, whereas *Candida*, *Kloeckera*, *Torulopsis* and *Rhodotorula* are asporogenous yeasts, not producing sexually-derived spores. The majority of yeasts require vitamins, most commonly biotin and/or thiamine.

A few yeasts grow on methane; they are unable to grow on methanol or formate but some grow on methylamine and higher alkanes. They were originally isolated on methane but grow very slowly on this substrate with a generation time of more than two days (Wolf and Hanson, 1979; Wolf *et al.*, 1980; Wolf, 1981).

Although it has been suggested that, in the presence of 0·1% yeast extract, some methanol-utilising yeasts can also grow on methylamine or methylformate (Patel *et al.*, 1979d), it is possible that the methylamine was being used solely as nitrogen source. This is suggested tentatively because a study of 461 strains of yeast by van Dijken and Bos (1981) has shown that none is able to use methylamine, dimethylamine, trimethylamine, tetramethylammonium chloride or a range of other primary amines as sole source of carbon and energy. By contrast, more than half of the strains are able to use methylamine, dimethylamine or trimethylamine as nitrogen source when glucose is provided as carbon source. In *Hansenula polymorpha* and *Candida boidinii* the utilisation of methylamine as nitrogen source is paralleled by development of formaldehyde and formate dehydrogenases, and by the synthesis of amine oxidase-containing peroxisomes; this may well be true of all yeasts able to utilise amines as nitrogen source (Zwart *et al.*, 1980).

Van Dijken and Harder (1974) have shown that for enrichment and isolation of yeasts on methanol, an acidic growth medium is required (pH 4·5 — the optimum pH for growth being 5·0–6·0); vitamins, especially biotin and thiamine, were found to be necessary, and the presence of antibiotics successfully diminished competition by bacteria during the isolation procedure. The optimum concentration of methanol was 0·1–0·5% and a temperature of 28°C was more effective than 37°C, at which temperature no strains were isolated. Ability to grow at 37°C appears to be uncommon in methylotrophic yeasts; of 55 methylotrophic strains isolated from 600 samples by Ogata et al. (1975), only seven were able to grow at 37°C. *Hansenula polymorpha* D-L1 is unusual in being a thermotolerant methylotrophic yeast; it grows well at 45°C and its temperature optimum is about 40°C. This strain was isolated by Levine and Cooney (1973) after enrichment in continuous culture on methanol at 37°C. The best sources for isolation of yeasts are those rich in organic material (van Dijken and Harder, 1974; Ogata et al., 1975; Kato et al., 1974c; Oki et al., 1972).

By contrast with methylotrophic bacteria, there is relatively little diversity among the methylotrophic yeasts with respect to the biochemistry of methanol oxidation and assimilation.

II. Oxidative metabolism in yeasts

The oxidative metabolism of yeasts differs from that in bacteria in many ways including the nature of the oxidising enzymes, their compartmentation in peroxisomes and the involvement of mitochondria in electron transport and energy transduction. The overall process of oxidation of methanol to CO_2 is shown in Fig. 39a. Methanol is oxidised to formaldehyde in the peroxisomes, and its subsequent oxidation and assimilation occurs in the cytosol. None of the energy from oxidation of methanol to formaldehyde is harnessed as NAD(P)H or ATP. Assuming that formaldehyde and formate are not oxidised to any extent by the peroxidative activity of catalase, then the overall reaction is:

$$CH_3OH + 0.5\, O_2 + 2\, NAD^+ \longrightarrow CO_2 + 2\, NADH + 2\, H^+$$

For oxidative phosphorylation to occur the NADH must enter the mitochondria and be oxidised by the electron transport chain. If the entry of NADH is by an indirect energy-consuming process, as occurs during alkane oxidation (Fukui and Tanaka, 1979a, b; Kawamoto et al., 1979, 1980), then the maximum number of molecules of ATP likely to be produced per NADH oxidised is two, because the NADH dehydrogenase is bypassed. A P/O ratio of only 2 is also likely if the NADH is oxidised by an external NADH dehydrogenase (van Dijken et al., 1981a).

A. The oxidation of methanol to formaldehyde

1. The alcohol oxidase of methylotrophic yeasts

Yeasts do not contain the typical bacterial methanol dehydrogenase. Instead, methanol is initially oxidised to formaldehyde in a reaction catalysed

TABLE 49
The alcohol oxidase of yeast

Prosthetic group	FAD (8 moles (mole enzyme)$^{-1}$; not covalently bound)
Molecular weight	500 000–700 000 (in *Candida* 25-A it is 210 000)
Sub-unit mol. wt	65 000–83 000
Number of sub-units (mole)$^{-1}$	8
Optimum pH	7·5–9·5
Optimum temperature	30–38°C (45–50°C in *H. polymorpha*—thermotolerant)
Inhibitors	Thiol reagents and some metal-chelating agents
Absorption maxima (nm) of oxidised enzyme (approx.)	280, 375, 460 (shoulder at 395)
Substrates	Oxygen; primary aliphatic alcohols up to about C_5; formaldehyde
Apparent K_m values	
methanol	0·2–2·0 mM (20 mM for oxidase from *Candida* 25-A)
oxygen (at high methanol conc.)	0·4–1·0 mM (*H. polymorpha* and *P. pastoris*)

The alcohol oxidase from the following yeasts have been described in some detail (usually after crystallisation): *Kloeckera* sp. 2201 (Tani et al., 1972a, b; Kato et al., 1976); *Candida* N-16 and N-17 (Fujuii and Tonomura, 1972); *Candida boidinii* (Sahm and Wagner, 1973a); *Candida* 25-A (Yamada et al., 1979); *Hansenula polymorpha* (Kato et al., 1976; van Dijken, 1976); *Pichia pastoris* (Couderc and Baratti, 1980a). Similar oxidases have been described in *Basidiomycetes* (Janssen and Ruelius 1968; Bringer et al., 1979). The methane-utilising yeasts do not contain this oxidase (Wolf et al., 1980).

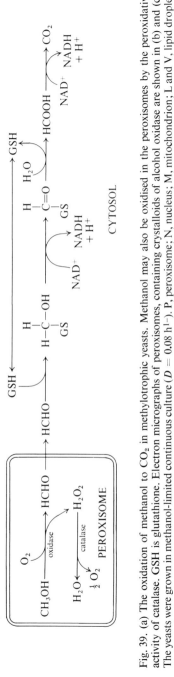

Fig. 39. (a) The oxidation of methanol to CO_2 in methylotrophic yeasts. Methanol may also be oxidised in the peroxisomes by the peroxidative activity of catalase. GSH is glutathione. Electron micrographs of peroxisomes, containing crystalloids of alcohol oxidase are shown in (b) and (c). The yeasts were grown in methanol-limited continuous culture ($D = 0.08$ h^{-1}). P, peroxisome; N, nucleus; M, mitochondrion; L and V, lipid droplet. (b) KMnO$_4$ fixation; 21,500 X. (c) stained with 1% uranyl acetate; 150,000 X. I am indebted to Dr Veenhuis for these electron micrographs of *Hansenula polymorpha*. (See Table 50 for further references which include further pictures of peroxisomes and their crystalline inclusions).

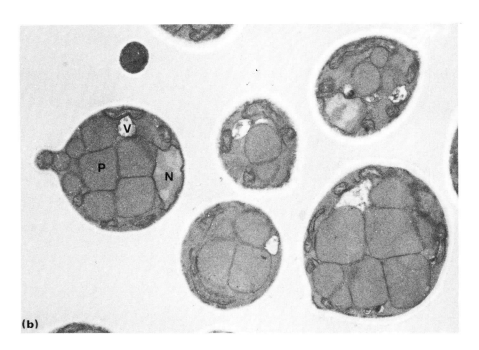

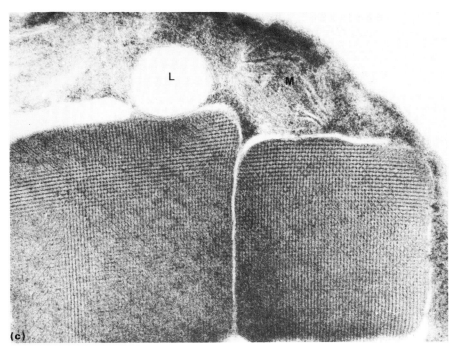

278   The biochemistry of methylotrophs

by a non-specific alcohol oxidase similar to that found in *Basidiomyces* (Janssen and Ruelius, 1968). Oxygen is the electron acceptor and the product is hydrogen peroxide:

$$CH_3OH + O_2 \longrightarrow HCHO + H_2O_2$$

This enzyme was first described in methanol-utilising yeasts by Tani *et al.* (1972a, b,), and since then a number of similar enzymes have been described (Table 49). Its importance during growth on methanol is indicated by the much higher levels on this substrate compared with the levels on ethanol, glycerol or glucose (Sahm and Wagner, 1973b; Kato *et al.*, 1974b). The specific activity of the oxidase was between 0·07 and 5 μmoles methanol oxidised $(min)^{-1}$ (mg protein)$^{-1}$ in crude extracts of yeasts grown on methanol, and less than 1% of this value during growth on other substrates (see Tani *et al.*, 1978a; Veenhuis *et al.*, 1978a; Egli *et al.*, 1980). That the oxidase is only essential for growth on methanol has been confirmed by the demonstration that a mutant of *Candida boidinii*, lacking the oxidase, is unable to grow on methanol, but grows well on ethanol and glucose (Sahm and Wagner, 1973a).

The alcohol oxidase usually consists of eight subunits having a total molecular weight of about 600 000, the subunits being arranged as an octad aggregate, composed of two tetragons, face to face (Kato *et al.*, 1976; van Dijken *et al.*, 1976a). The prosthetic group is non-covalently-bound FAD; there are eight FAD molecules per molecule of enzyme, and so probably one per subunit. The absorption spectrum is typical of flavoproteins having peaks in the visible range at about 375 nm and 460 nm, the latter being diminished by reduction with methanol, and returning on passage of oxygen through the solution. The enzyme is usually yellow, but when grown under methanol-limitation in a chemostat, the purified enzyme from *Hansenula polymorpha* was red instead of yellow, having absorption peaks at 395 nm and 495 nm; this form of enzyme was almost identical in other respects to the yellow form purified from batch-grown yeasts (van Dijken, 1976). The amino acid composition of the oxidases from *Kloeckera* sp., *Hansenula polymorpha* and *Pichia pastoris* are very similar and the *N*-terminal amino acid is alanine in these three enzymes (Kato *et al.*, 1976; Couderc and Baratti, 1980a).

The enzyme oxidises only short-chain aliphatic primary alcohols (C_1–C_5), including unsaturated and substituted alcohols. The V_{max} decreases, and the K_m increases with increasing chain length. The K_m value for methanol (0·2–2·0 mM) of alcohol oxidase is 10–100 times that measured for the methanol dehydrogenases of bacteria. These enzymes are similar in being able to oxidise formaldehyde to formate, the substrate possibly being the predominant hydrated form, $HC(OH)_2$. The K_m of the oxidase for formaldehyde is usually 3–10 times higher than that for methanol, and the V_{max}

10. Metabolism in the methylotrophic yeasts

about 20% of that for methanol (van Dijken, 1976; Kato et al., 1976). The K_m for oxygen of the pure enzyme from *H. polymorpha* was between 0·24 mM and 0·4 mM O_2 and was dependent on the methanol concentration (between 1 mM and 100 mM) (van Dijken et al., 1976a). As the oxygen concentration in air-saturated water is about 0·2 mM, this characteristic might be important in limiting the growth rate of this (and maybe all) yeasts in methanol-limited continuous culture (van Dijken, 1976; Middlehoven et al., 1976).

The nature of the alcohol oxidase has an important effect on the bioenergetics of yeasts. None of the energy available from this first step in the oxidation of methanol is harnessed as a usable reductant (NADH) or as ATP, and this will affect growth yields with respect to both methanol and oxygen (Harder and van Dijken, 1975; Anthony, 1980; Chapter 9).

The demonstration that extracts of *Pichia pinus*, *Hansenula polymorpha*, *Kloeckera* sp. 2201 and *Candida boidinii* catalysed the reduction of NAD^+ by methanol led to the proposal that these yeasts contain, besides the usual alcohol oxidase, low activities of a primary alcohol dehydrogenase (specific activity 20-100 nmoles methanol oxidised $(min^{-1}$ $(mg$ $protein)^{-1})$ (Mehta, 1975a, b; Dudina et al., 1977). The requirement for reduced glutathione (GSH) shown by this enzyme is similar to that for the oxidation of formaldehyde by yeasts and this, together with the observation that a mutant lacking the normal alcohol oxidase also lacks the NAD^+- and GSH-dependent alcohol dehydrogenase, suggests that this enzyme activity results from the sequential action of the usual alcohol oxidase plus the NAD^+-linked formaldehyde dehydrogenase (Sahm, 1977). The oxidation of primary alcohols other than methanol is catalysed by a constitutive NAD^+-dependent alcohol dehydrogenase (Sahm and Wagner, 1973a; Kato et al., 1974b). It should be noted that if yeasts contain an NAD^+-linked methanol dehydrogenase, then the first step in the oxidation of methanol would provide energy for biosynthesis, and cell yields could be higher than if the oxidase is the only enzyme responsible for methanol oxidation.

2. The role of catalase in methanol oxidation

Catalase has two potential roles in methanol oxidation in yeasts. The most obvious is the removal of toxic H_2O_2 produced during the oxidation of methanol to formaldehyde, catalysed by the alcohol oxidase (Fujuii and Tonomura, 1972; Sahm and Wagner, 1973a; Kato et al., 1974b). The products of the catalase reaction in this case are water and oxygen:

$$2H_2O_2 \longrightarrow 2 H_2O + O_2$$

During growth of *C. boidinii* on methanol the specific activity of catalase was 20 times higher than on glucose and 80 times higher than on ethanol; full activity was reached within 5 h of transfer from a glucose to a methanol medium (Roggenkamp et al., 1974; Sahm, 1977). Similar demonstrations of the importance of catalase during growth on methanol of a range of different yeasts have been summarised by Tani et al. (1978a). The specific activity of catalase during growth on methanol is usually very high (e.g. 1·45 mmoles $(min)^{-1}$ (mg protein)$^{-1}$ — from van Dijken et al., 1975a) and is between 10^3 and 10^4 times greater than that of the preceding enzyme in methanol metabolism, alcohol oxidase. The catalase purified from methylotrophic yeasts does not appear to be different from that purified from other sources (van Dijken et al., 1975a; Fujuii and Tonomura, 1975b).

A second proposed role for catalase is as a peroxidase, oxidising methanol, formaldehyde or formate with H_2O_2 as oxidant, as first proposed by Keilin and Hartree (1945). It has been suggested that the catalase in methylotrophic yeasts might act in a similar way, but it is extremely difficult to prove this one way or the other. The stoicheiometry of the overall reaction for oxidation of methanol to formaldehyde is the same whether the catalase is acting catalatically or peroxidatively:

Catalatic: $\quad 2\ CH_3OH + 2\ O_2 \longrightarrow 2\ HCHO + 2\ H_2O_2$
$\qquad\qquad\ \ 2\ H_2O_2 \longrightarrow 2\ H_2O + O_2$

TOTAL: $2\ CH_3OH + O_2 \longrightarrow 2\ HCHO + 2\ H_2O$

Peroxidative: $\ CH_3OH + O_2 \longrightarrow HCHO + H_2O_2$
$\qquad\qquad\quad\ CH_3OH + H_2O_2 \longrightarrow HCHO + 2\ H_2O$

TOTAL: $2\ CH_3OH + O_2 \longrightarrow 2\ HCHO + 2\ H_2O$

The demonstration that the stoicheiometry (oxygen consumed per mole of formaldehyde produced) changes, when catalase is inhibited (Roggenkamp et al., 1974), does not distinguish between the modes of operation of catalase, because the resulting change in stoicheiometry would be the same in both cases. Van Dijken et al. (1975a) have shown that methanol, formaldehyde and formate may be oxidised to some extent by the peroxidative action of purified catalase from *H. polymorpha*, but they point out that the extent of this depends on the concentration of carbon substrate, H_2O_2 and catalase, and on the affinity of catalase for H_2O_2 and carbon substrates. They have shown that when the ratio of H_2O_2 production to the concentration of catalase is low, the catalase functions peroxidatively; whereas when this ratio is high the catalatic function (producing O_2 from H_2O_2) predominates. They concluded that it is impossible to acquire sufficient data to determine the importance of the peroxidative potential of catalase in oxidising methanol in yeasts.

Because the stoicheiometry of the two potential modes of operation of catalase in methanol oxidation is the same, then energy yields are the same, and growth yields are unaffected by the extent of peroxidative oxidation of methanol to formaldehyde. However, if formaldehyde or formate were to be oxidised peroxidatively instead of by NAD^+-linked dehydrogenases, then the energy yield and hence growth yields would be lower; but the possibility of this appears unlikely (van Dijken et al., 1975a).

3. Peroxisomes in yeasts

Peroxisomes or microbodies are organelles containing flavin-linked oxidases whose reactions with oxygen produce hydrogen peroxide, and catalase which catalyses its rapid removal (Baudhuin et al., 1965). Presumably the compartmentation of reactions producing H_2O_2 within the cell protects sensitive enzymes from its toxic effects.

(a) *Peroxisomes in alkane-utilising yeasts.* In 1972, Fukui, Osumi and their colleagues first demonstrated the importance of peroxisomes containing catalase in the dissimilation of alkanes by yeasts (see Fukui and Tanaka, 1979a, b for excellent reviews of peroxisomes in alkane- and methanol-utilising yeasts). During growth on alkanes, the peroxisomes contain the enzymes for β-oxidation of long-chain acyl-CoA, and the enzymes involved in transfer out of the peroxisomes of the NADH and acetyl-CoA produced during the β-oxidation process. They also contain isocitrate lyase and malate synthase, essential for the assimilation of the acetyl-CoA. The first oxidation step differs from that occurring in mitochondria; in peroxisomes the reduced flavoprotein is reoxidised by molecular oxygen producing H_2O_2 which is subsequently removed by catalase. The reducing equivalents from the NADH produced in the second step in β-oxidation are transported from the peroxisomes into the mitochondria as glycerol 3-phosphate; this is then oxidised by a flavoprotein in the electron transport chain, thus giving an effective P/O ratio of only 2 for NADH oxidation. The dihydroxyacetone phosphate then passes back from the mitochondria into the peroxisome. This way of oxidising long-chain acyl-CoA is not as wasteful as might at first appear, because cell yields of microorganisms growing on long chain alkanes, alcohols and acids are not predominantly energy-limited but carbon-limited. This is because about 20% of the carbon in substrates assimilated exclusively by way of acetyl-CoA is lost by decarboxylation during assimilation of the oxaloacetate produced by the glyoxylate cycle (Anthony, 1980; Chapter 9).

TABLE 50
Structure and function of peroxisomes in methanol-utilising yeasts

Localisation of alcohol oxidase and catalase (cytochemical techniques)

Candida boidinii	Fukui *et al.* (1975a, b, c)
Hansenula polymorpha	van Dijken *et al.* (1975c); Veenhuis *et al.* (1976, 1978a); Fukui *et al.* (1975a, b, c)
Kloeckera sp.	Fukui *et al.* (1975b); Osumi and Sato (1978)
Pichia and *Torulopsis*	Fukui *et al.* (1975a, b, c); Hazeu *et al.* (1975)

Localisation of alcohol oxidase and catalase by subcellular fractionation

Candida boidinii	Sahm *et al.* (1975); Roggenkamp *et al.* (1975)
Kloeckera sp.	Fukui *et al.* (1975b); Tanaka *et al.* (1976)

Regulation of peroxisome synthesis

Candida boidinii	Fukui *et al.* (1975a); van Dijken *et al.* (1975b); Bormann and Sahm (1978)
Hansenula polymorpha	van Dijken *et al.* (1975b); Fukui *et al.* (1975a); Veenhuis *et al.* (1978); Eggeling and Sahm (1978); Egli *et al.* (1980)
Kloeckera sp.	Fukui *et al.* (1975a); Yasuhara *et al.* (1976); Tanaka *et al.* (1976); Egli *et al.* (1980)
Pichia sp.	van Dijken *et al.* (1975b); Fukui *et al.* (1975a)

Regulation of peroxisome degradation

Candida boidinii	Bormann and Sahm (1978); Sahm (1977)
Hansenula polymorpha	Veenhuis *et al.* (1978b); Egli *et al.* (1980)
Kloeckera sp.	Egli *et al.* (1980)

Demonstration of crystalline inclusions

Candida boidinii	Sahm *et al.* (1975); Fukui *et al.* (1975a)
Hansenula polymorpha	van Dijken *et al.* (1975c); Fukui *et al.* (1975a); Veenhuis *et al.* (1976, 1978a, b)
Kloeckera sp.	Fukui *et al.* (1975a, b); Tanaka *et al.* (1976)
Pichia sp.	Fukui *et al.* (1975a)
Torulopsis sp.	Fukui *et al.* (1975a); Hazeu *et al.* (1975)

Good reviews of this subject have been published by Sahm (1977), Tani *et al.* (1978a) and Fukui and Tanaka (1979a, b).

10. Metabolism in the methylotrophic yeasts

(b) *Peroxisomes in methanol-utilising yeasts.* Peroxisomes in methanol-utilising yeasts were first described in 1975 by three independent groups of workers (van Dijken *et al.*, 1975b, c; Fukui *et al.*, 1975a, b; Sahm *et al.*, 1975). All types of methylotrophic yeasts appear to have similar peroxisomes (see Table 50), but during growth on glucose they are either absent or few in number and very small. There are usually beween 5 and 12 peroxisomes per cell after growth on methanol, and they vary in size between 0·4 μm and 1·0 μm in diameter (Figs 39b and c). During growth and bud formation, small peroxisomes are produced by separation from larger mature peroxisomes in the mother cell (Veenhuis *et al.*, 1978a). In batch culture the number of peroxisomes per cell increases during early growth, but later their increasing size becomes more important, until they may fill up to 80% of the cell volume in the stationary phase. In continuous culture, the peroxisomes tend to be always large and, especially at low growth rates, tend to fill the cells. Whenever this occurs they appear to be flat-sided cubes (up to 1·5 μm across) (Figs 39b and c).

The alcohol oxidase and catalase responsible for the first step in methanol oxidation are localised within the peroxisomes as shown cytochemically, and by characterisation of the isolated peroxisomes (Table 50). After transfer from a glucose medium to methanol the oxidase and catalase levels in the cells increase with the increasing size and number of peroxisomes.

The peroxisomes of methylotrophic yeasts are bound by a unit membrane and they form crystalline inclusions or crystalloids; these increase in size as the peroxisomes increase in size until they completely fill the peroxisomes in the stationary phase and in continuous culture (Veenhuis *et al.*, 1978a) (Figs 39 b and c). The crystalloids consist of crystalline alcohol oxidase (Veenhuis *et al.*, 1976a), the most direct evidence for this being their absence from a mutant lacking the alcohol oxidase; catalase was present to the same extent as in wild-type cells but crystalloids were absent (Sahm *et al.*, 1975; Eggeling *et al.*, 1977).

The peroxisomes of methane-utilising yeasts are more like those from alkane-utilising yeasts than those from methanol-utilisers; they are relatively small and contain catalase but not alcohol oxidase (Wolf *et al.*, 1980; Wolf, 1981).

B. The formaldehyde dehydrogenase of yeasts

Early studies of methanol dissimilation in a number of methylotrophic yeasts indicated that formaldehyde is oxidised by an inducible NAD^+-dependent dehydrogenase (Fujuii and Tonomura, 1972; Kato *et al.*, 1972, 1974b; Sahm and Wagner, 1973b), which has since been purified and characterised. The enzymes from the following yeasts are similar to one another and to the formaldehyde dehydrogenase from human liver (described by Uotila and Koivusalo, 1974): *Kloeckera* sp. (Kato *et al.*, 1972), *Candida boidinii* (Schütte *et al.*, 1976) and *Hansenula polymorpha* (van Dijken *et al.* 1976b).

The formaldehyde dehydrogenase is absolutely specific for NAD^+ and reduced glutathione, and it is inhibited by thiol reagents. Of the aldehydes tested, formaldehyde and methylglyoxal were the only suitable substrates. The K_m values for NAD^+, glutathione, formaldehyde and methylglyoxal were 0·02–0·15 mM, 0·13–0·18 mM, 0·2–0·3 mM and 1–3 mM respectively. The enzyme from *Candida boidinii* has been shown to be a dimer consisting of two equal subunits of molecular weight about 40 000.

The actual substrate for the yeast enzyme is the thiohemiacetal formed spontaneously from reduced glutathione and formaldehyde (*S*-hydroxymethylglutathione) (Fig. 39a). The product is not free formate but *S*-formylglutathione, which can act as the substrate for the reverse reaction using NADH (van Dijken *et al.*, 1976b; Schütte *et al.*, 1976); formate itself does not act as substrate for the reverse reaction.

Kinetic data obtained with the formaldehyde dehydrogenase from *Candida boidinii* are consistent with an ordered Bi–Bi mechanism in which NAD^+ becomes bound to the enzyme before *S*-hydroxymethylglutathione, and *S*-formylglutathione is released before the NADH (Kato *et al.*, 1979c). The same study suggested that the inhibition of formaldehyde dehydrogenase by NADH and ATP may be important in regulating formaldehyde oxidation.

Formaldehyde is also oxidised in yeast by a dissimilatory cycle, yielding CO_2 and two molecules of NADPH for biosynthesis (page 287 and Fig. 41).

C. Hydrolysis of S-formylglutathione, and the oxidation of formate

The product of the formaldehyde dehydrogenase reaction is *S*-formylglutathione and this must be first hydrolysed to formate and glutathione before oxidation to carbon dioxide (Fig. 39a).

In *Candida boidinii*, this hydrolysis is catalysed by an *S*-formylglutathione hydrolase which is induced to very high levels during growth on methanol (specific activity, 50 μmoles $(min)^{-1}$ $(mg)^{-1}$) (Schütte *et al.*, 1976; Neben *et al.*, 1980). The purified enzyme has a low K_m for *S*-formylglutathione, and hydrolyses other *S*-acylglutathiones at less than 1·3% of the rate measured with *S*-formylglutathione. The hydrolase is composed of two non-identical subunits of molecular weights 25 000 and 35 000.

In *Hansenula polymorpha* the hydrolysis of *S*-formylglutathione appears to be catalysed by the formate dehydrogenase, prior to oxidation of the enzyme-bound formate produced by its hydrolytic activity (van Dijken *et al.*, 1976b). The K_m for oxidation by formate dehydrogenase of *S*-formylglutathione (1·1 mM) is much lower than the K_m for formate oxidation (40 mM). The presence of NAD⁺ is required for the hydrolysis, indicating that this is not due to contamination of the dehydrogenase by a highly active hydrolase.

The inducible formate dehydrogenases from *Kloeckera* (Kato et al., 1974a), *Hansenula* (van Dijken et al., 1976b) and *Candida* (Fujuii and Tonomura 1972; Sahm and Wagner, 1973b; Schütte et al., 1976) have been purified and shown to have similar properties.

The dehydrogenase is specific for formate and NAD⁺, the K_m values for these substrates being 10–40 mM and 0·07–0·1 mM respectively. The pH optima are between 7 and 8, and the temperature optima between 50°C and 60°C. Cyanide and *p*-chloromercuribenzoate are inhibitors. The dehydrogenase from *Candida boidinii* is a dimer of identical subunits of molecular weight 36 000. Kinetic data obtained with this enzyme are consistent with an ordered Bi–Bi mechanism in which NAD⁺ is bound to the enzyme before formate, and carbon dioxide released before NADH.

III. The dihydroxyacetone (DHA) cycle of formaldehyde assimilation in yeasts

This pathway, also called the xylulose monophosphate pathway, occurs only in yeasts and was the last of the main C_1 assimilation pathways to be elucidated in aerobic microorganisms. The first section below is a brief description of the pathway; this is followed by a chronological survey of the evidence for the pathway, descriptions of its special key enzymes and a discussion of its regulation.

A. Description of the DHA cycle

The only enzyme required exclusively for formaldehyde assimilation in yeast is the initial formaldehyde-fixing enzyme, dihydroxyacetone (DHA) synthase, all other enzymes being also required for growth on carbohydrates, ethanol or glycerol. As seen in Fig. 40 the DHA cycle of formaldehyde assimilation achieves the net synthesis of one molecule of triose phosphate from three molecules of formaldehyde, the summary equation being:

3 HCHO + 3 ATP → dihydroxyacetone phosphate + 3 ADP + 2 Pi

For comparison with other pathways, the summary equation for production of 3-phosphoglycerate from formaldehyde is:

3 HCHO + 2 ATP + NAD⁺ ⟶ 3-phosphoglycerate
$$+ \ 2\ \text{ADP} + \text{NADH} + \text{H}^+ + \text{Pi}$$

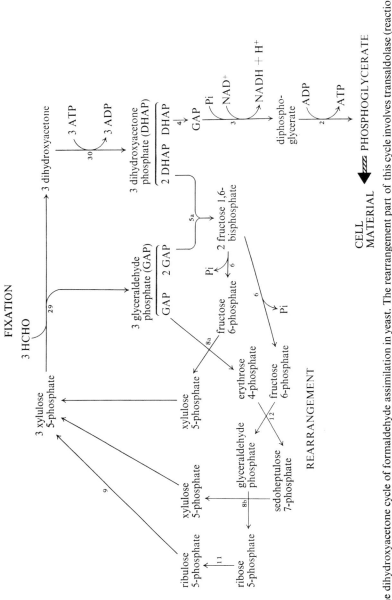

Fig. 40. The dihydroxyacetone cycle of formaldehyde assimilation in yeast. The rearrangement part of this cycle involves transaldolase (reaction 12), but an alternative rearrangement of the triose phosphates might involve sedoheptulose bisphosphate aldolase and phosphatase instead of transaldolase (as in Fig. 7, page 64). The enzyme numbers refer to those listed in the text (this chapter and Chapter 3): (2) phosphoglycerate kinase; (3) glyceraldehyde phosphate dehydrogenase; (4) triose phosphate isomerase; (5a) fructose bisphosphate aldolase; (6) fructose bisphosphatase; (8) transketolase; (9) pentose phosphate epimerase; (11) pentose phosphate isomerase; (12) transaldolase; (29) dihydroxyacetone synthase; (30) triokinase.

10. Metabolism in the methylotrophic yeasts

The first enzyme of the DHA cycle transfers a glycolaldehyde moiety from xylulose 5-phosphate to formaldehyde, giving glyceraldehyde 3-phosphate and dihydroxyacetone. This is then phosphorylated by a specific triokinase to dihydroxyacetone phosphate (DHAP), which is condensed with glyceraldehyde phosphate in an aldolase reaction, giving fructose 1,6-bisphosphate. Fructose bisphosphatase then catalyses the hydrolysis of this to fructose 6-phosphate. Two molecules of fructose 6-phosphate formed in this way are then rearranged with a glyceraldehyde phosphate to regenerate three molecules of xylulose 5-phosphate.

For production of PEP, pyruvate and C_4 carboxylic acids from DHAP, the enzymes for oxidation of triose phosphate are also required; these operate in the direction of glycolysis at the same time as the fructose bisphosphate aldolase and phosphatase act in the 'gluconeogenic' direction.

B. Oxidation of formaldehyde by a dissimilatory DHA cycle

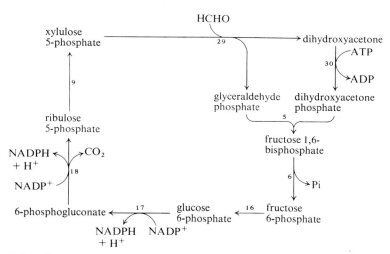

Fig. 41. The dissimilatory DHA cycle of formaldehyde oxidation. This was first investigated in *Candida boidinii* by Kato *et al.* (1979b). The summary equation is:

$$HCHO + 2\,NADP^+ + ATP \rightarrow CO_2 + 2\,NADPH + 2\,H^+ + ADP + Pi$$

The enzyme numbers refer to those in the text (this chapter and Chapter 3): (5) fructose bisphosphate aldolase; (6) fructose bisphosphatase; (9) pentose phosphate epimerase; (16) glucose phosphate isomerase; (17) glucose 6-phosphate dehydrogenase; (18) 6-phosphogluconate dehydrogenase; (29) dihydroxyacetone synthase; (30) triokinase.

Although yeasts contain oxidases and dehydrogenaes able to oxidise methanol directly to CO_2, a cyclic route is also possible, analogous to that operating in some bacteria with the RuMP assimilation pathway (Kato *et al.*,

1979b). Such an oxidative cycle in yeasts would involve, in addition to the enzymes of the assimilation cycle, hexose phosphate isomerase, and dehydrogenases for glucose 6-phosphate and 6-phosphogluconate. These enzymes are all present in *Candida boidinii* (Sahm, 1977), and probably also in other methylotrophic yeasts. In order to study this problem further, the dehydrogenases of this potential cycle have been purified and characterised (Kato *et al.*, 1979b) (Table 21 and page 89). During exponential growth on methanol the specific activities (0·1–0·3 μmoles (min)$^{-1}$ (mg)$^{-1}$) of the dehydrogenases are 25–50% of those measured during growth on glucose. The enzymes differ from those of bacteria in being completely specific for NADP$^+$ and in being inhibited by NADPH (not NADH). This occurs at the concentration of NADPH found in whole cells, which is similar to the K_i values for inhibition of the dehydrogenases by NADPH. It has therefore been suggested that the dissimilatory cycle depicted in Fig. 41 is only concerned in production of NADPH for biosynthesis and that it is unimportant in oxidising formaldehyde for energy production in yeast (Kato *et al.*, 1979b).

C. Evidence for the DHA cycle of formaldehyde assimilation

The first investigations of methanol assimilation in yeasts by Fujuii and Tonomura (1973) showed that ^{14}C methanol is rapidly incorporated into hexose phosphates, thus suggesting the operation of a pathway similar to the RuMP pathway of methylotrophic bacteria, and earlier reports of enzyme assays using crude extracts of yeasts appeared to confirm this (Fujuii and Tonomura, 1974; Diel *et al.*, 1974; Trotsenko *et al.*, 1973). However, the assimilation of formaldehyde into hexoses, catalysed by crude extracts, differed from that in bacterial extracts in requiring ATP, and it became evident (Kato *et al.*, 1977a; van Dijken *et al.*, 1978) that the bacterial hexulose phosphate synthase is not the enzyme responsible for the initial fixation of formaldehyde in yeasts (see Table 51 for references to this earlier exploratory work). In the first description of the pathway as it now stands (van Dijken *et al.*, 1978) (Fig. 40), an alternative enzyme was proposed for the initial assimilation of formaldehyde. This was a transketolase able to transfer a glycolaldehyde moiety from xylulose 5-phosphate to formaldehyde, yielding glyceraldehyde 3-phosphate and dihydroxyacetone. A somewhat similar enzyme had been described by Dickens and Williamson (1958), and was first considered as a potential route for formaldehyde assimilation during growth of *Arthrobacter* 2B2 on trimethylamine (Cox and Zatman, 1974). Although shown to be unimportant in bacterial metabolism, it is now known to be the key formaldehyde-fixing enzyme in yeast. The inducible transketolase, subsequently known as dihydroxyacetone (DHA) synthase, was first described in *Kloeckera* sp. 2201 by Kato *et al.* (1979a), and similar

TABLE 51
The assimilation of methanol in yeasts; the DHA pathway

Formaldehyde-fixing enzymes in crude extracts (early studies before discovery of DHA synthase)	
Candida sp. N-16	Fujui and Tonomura (1974)
Candida boidinii	Diel *et al.* (1974); Sahm and Wagner (1974); van Dijken *et al.* (1978;) Sahm (1977); Babel and Loffhagen (1979)
Hansenula polymorpha	van Dijken *et al.* (1978); Kato *et al.* (1977a)
Kloeckera sp. 2201	Kato *et al.* (1977a)
Torulopsis ingeniosa	Babel and Loffhagen (1979)
DHA synthase	
Kloeckera sp. 2201	Kato *et al.* (1979a)
Hansenula polymorpha	O'Connor and Quayle (1980)
Candida boidinii	O'Connor and Quayle (1980); Waites and Quayle (1980, 1981); Bystrykh *et al.* (1981); Lindley *et al.* (1981)
Triokinase	
Hansenula polymorpha	van Dijken *et al.* (1978); O'Connor and Quayle (1979)
Candida boidinii	van Dijken *et al.* (1978); O'Connor and Quayle (1979); Lindley *et al.* (1981)
Candida methylica	Hoffman and Babel (1980)
Rearrangement enzymes	
Hansenula polymorpha	van Dijken *et al.* (1978); O'Connor and Quayle (1979)
Candida boidinii	van Dijken *et al.* (1978); O'Connor and Quayle (1979); Sahm (1977); Babel and Loffhagen (1979); Lindley *et al.* (1981)
Torulopsis ingeniosa	Babel and Loffhagen (1979)
Incorporation of ¹⁴C methanol by whole cells	
Candida N-16	Fujuii and Tonomura (1973)
Hansenula polymorpha	van Dijken *et al.* (1978); Lindley *et al.* (1980); Waites *et al.* (1981)
Mutant isolation and characterisation	
Hansenula polymorpha	O'Connor and Quayle (1979)
Candida boidinii	O'Connor and Quayle (1979)

enzymes have now been purified and thoroughly characterised from *Candida boidinii* and *Hansenula polymorpha* (Waites and Quayle, 1980; O'Connor and Quayle, 1980). The subsequent metabolism of the DHA requires its phosphorylation, and the induction of a triokinase specific for DHA was a key observation in the first description of the pathway (van Dijken *et al.*, 1978). The presence of triokinase explains the ATP-requirement for the synthesis of hexose phosphates from radioactive formaldehyde described in earlier reports. This is because the next step in the pathway involves fructose bisphosphate (FBP) aldolase, which catalyses condensation of the radioactive DHAP with glyceraldehyde 3-phosphate to give FBP. This is subsequently hydrolysed to fructose 6-phosphate by an FBPase, which is induced about ten-fold during growth on methanol (van Dijken *et al.*, 1978; Babel and Loffhagen, 1979).

The importance of triokinase during growth on methanol was conclusively demonstrated by the isolation of mutants of *C. boidinii* and *H. polymorpha* having low (or zero) levels of this enzyme, and which were thus unable to grow on methanol or dihydroxyacetone. In the same investigation it was also shown that decreased levels of fructose bisphosphatase in a mutant of *H. polymorpha* led to lower growth rates and yields than were measured in the wild-type yeasts (O'Connor and Quayle, 1980).

Earlier studies with whole cells had shown that radioactive methanol is rapidly incorporated into hexose phosphates, as would be expected if the DHA cycle is operating, but labelled dihydroxyacetone was not detected (Fujuii and Tonomura, 1973, 1974; van Dijken *et al.*, 1978). However, by introducing a novel, elegant sampling technique for such experiments, Lindley *et al.* (1980) have now been able to show that within two seconds of incubation of methanol-limited cells of *H. polymorpha* with ^{14}C methanol, 55% of the radioactivity is found in DHA. A final confirmation of the operation of the cycle was the pattern of labelling found in the DHA and glucose phosphate present after a two second incubation with ^{14}C methanol, in similar experiments to these. Exactly as predicted by the proposed pathway (Fig. 40), less than 10% of the radioactivity was in the C-2 of DHA and less than 1% in the C-2 and C-5 of glucose (Waites *et al.*, 1981).

D. The reactions of the DHA cycle

As mentioned above, all but one of the reactions of the DHA cycle are also involved in growth of yeast on multicarbon compounds. The rearrangement enzymes are similar to those described for the bacterial RuMP pathway (Chapter 3) and are not included again here.

Dihydroxyacetone synthase (29)

DHA synthase is a transketolase transferring a glycolaldehyde moiety from xylulose 5-phosphate to formaldehyde:

$$\text{HCHO} + \begin{array}{c} \text{CH}_2\text{OH} \\ | \\ \text{C=O} \\ | \\ \text{HO-C-H} \\ | \\ \text{H-C-OH} \\ | \\ \text{CH}_2\text{-OP} \end{array} \longrightarrow \begin{array}{c} \text{CHO} \\ | \\ \text{H-C-OH} \\ | \\ \text{CH}_2\text{O-P} \end{array} + \begin{array}{c} \text{CH}_2\text{OH} \\ | \\ \text{C=O} \\ | \\ \text{CH}_2\text{OH} \end{array}$$

xylulose 5-phosphate   glyceraldehyde 3-phosphate   dihydroxyacetone

This enzyme is absent during growth on multicarbon compounds, but is induced during growth on methanol to sufficiently high levels to account for the growth rate on this substrate (Kato *et al.*, 1979a; Waites and Quayle, 1980, 1981; O'Connor and Quayle, 1980). It has been partially purified from *Kloeckera* sp. 2201 (Kato *et al.*, 1979a) and from *C. boidinii* and *H. polymorpha* (O'Connor and Quayle, 1980, Bystryck *et al.*, 1981). It has a pH optimum of about 7·5 and it is unstable when purified. Although originally reported to be specific for formaldehyde and xylulose 5-phosphate it has since been shown to have some classical transketolase activity with formaldehyde being replaced by ribose 5-phosphate as the glycolaldehyde acceptor (Waites and Quayle, 1981; Lindley *et al.*, 1981). The DHA synthase is also able to use hydroxypyruvate as the glycolaldehyde donor instead of xylulose 5-phosphate (Waites and Quayle, 1981). The fact that DHA synthase is only synthesised during growth on methanol, whereas the classical transketolase is synthesised to high levels on all substrates tested, indicates that the activities are due to two different enzymes; and this has been confirmed by separating them by ion-exchange chromatography (O'Connor and Quayle, 1980; Waites and Quayle, 1981). The purified enzyme from *Kloeckera* has apparent K_m values for formaldehyde of 1·43 mM (with 2 mM XuMP), and for xylulose 5-phosphate of 1·25 mM (with 3 mM HCHO); activity was doubled by 1 mM $MgSO_4$ (Kato *et al.*, 1979a).

Triokinase (dihydroxyacetone kinase) (30)

This enzyme catalyses the phosphorylation of DHA to DHAP:

$$\begin{array}{c} CH_2OH \\ | \\ C=O \\ | \\ CH_2OH \end{array} + ATP \longrightarrow \begin{array}{c} CH_2O\text{-}P \\ | \\ C=O \\ | \\ CH_2OH \end{array} + ADP$$

It is very much more active with DHA than with other trioses (glycerol and glyceraldehyde), and is induced to high specific activity during growth on methanol or glycerol (van Dijken *et al.*, 1978; Hoffman and Babel, 1980). Studies with permeabilised cells of *C. methylica* have failed to find any activators or inhibitors of the enzyme, except for ADP (above 0·5 mM) and ATP (above 5·0 mM) (Hoffman and Babel, 1980).

IV. Regulation of methanol metabolism in yeasts

Although general aspects of regulation in yeasts will be similar to those discussed for bacteria (page 132), some special aspects must be considered. These include the regulation of synthesis and degradation of peroxisomes, compartmentation of metabolites between peroxisomes, cytosol and mitochondria, and the fact that before formaldehyde can be oxidised it must diffuse from the peroxisome into the cytosol and react with glutathione to give the substrate for formaldehyde dehydrogenase, S-hydroxymethylglutathione (Fig. 39).

A. Regulation of synthesis of methanol-metabolising enzymes

Little is known about the regulation of synthesis of the key assimilatory enzymes, DHA synthase and triokinase, except that they are induced to much higher specific activities during growth on methanol (see Table 51 for references). Similarly, during growth on methanol, the rate of synthesis of all of the following enzymes involved in the oxidation of methanol to CO_2 is increased; alcohol oxidase, catalase, formaldehyde dehydrogenase and formate dehydrogenase (Kato *et al.*, 1972, 1974a, b; Roggenkamp *et al.*, 1974; van Dijken *et al.*, 1976a; Yasahura *et al.*, 1976; Eggeling *et al.*, 1977; Eggeling and Sahm, 1978; Sahm, 1977; Shimizu *et al.*, 1977b; Tani *et al.*, 1978a; Egli *et al.*, 1980). That methanol is not necessarily the inducing molecule is shown by the work on peroxisomes described below, and by studies of the oxidising enzymes of *Candida boidinii* and *Kloeckera* sp.

2201, which are induced by formaldehyde and formate (Eggeling and Sahm, 1977; Shimizu et al., 1977b; Sahm, 1977). However, this does not show that methanol must be oxidised to these products for increased synthesis of the oxidising enzymes to occur, because methanol is still able to induce their synthesis even in a mutant completely lacking the alcohol oxidase (Eggeling et al., 1977).

Growth conditions affect not only the synthesis of the oxidising enzymes, but also the synthesis of the peroxisomes containing them. The regulatory mechanism may differ slightly from one yeast to another, but it appears that in no case is their synthesis dependent merely on the presence or absence of methanol (induction). Thus, the peroxisomes, alcohol oxidase, catalase, formaldehyde dehydrogenase and formate dehydrogenase are all absent from *Hansenula polymorpha* growing exponentially on glucose, but their synthesis occurs after depletion of the glucose during the stationary phase, even in the absence of methanol (Eggeling and Sahm, 1978). Furthermore, these enzymes and the peroxisomes are all synthesised during exponential growth on glycerol, xylose, ribose and sorbitol. The relative levels of alcohol oxidase during growth on these compounds (compared with that during growth on methanol) was 62%, 56%, 24% and 10%, respectively, and the catalase, formaldehyde and, usually, formate dehydrogenase, were synthesised to similar relative extents. There was no well-defined relationship between the growth rates on a particular substrate and the levels of enzyme synthesised (Eggeling and Sahm, 1978). These results, and some similar results using *Candida boidinii* (Sahm, 1977), suggest that synthesis of methanol-dissimilating enzymes in yeasts is controlled by catabolite repression and derepression rather than by an induction mechanism, and that the level of metabolite responsible (possibly, by analogy with some bacterial systems, cyclic AMP) varies with growth substrate. This conclusion is supported by the work of Harder's group on the regulation of synthesis of peroxisomes and catabolic enzymes in *Hansenula polymorpha* and *Kloeckera* sp. grown in glucose-limited continuous culture (Egli et al., 1980). At low dilution rates (i.e. low growth rates) at which the external glucose concentration is low, the synthesis of peroxisomes and oxidative enzymes is derepressed. That this is not merely an expression of the low growth rate itself was shown by the strong repression of alcohol oxidase and catalase synthesis, which occurred when high concentrations of glucose were added to nitrogen-limited cultures growing at low rates.

Although the main mechanism of regulation of synthesis of peroxisomes, and of methanol-assimilating enzymes, in yeasts appears to involve catabolite repression and derepression, the details of this may vary from one yeast to another; in particular, the response of the yeast to mixtures of methanol and other substrates may vary. It should be noted that this variation may merely reflect differences in growth conditions, such as batch or continuous culture, growth rates etc. (see Sahm and Wagner, 1973a; Egli et al., 1980).

When yeasts (*Candida, Hansenula, Kloeckera*) are transferred from methano to a substrate such as glucose, where the peroxisomes and the enzymes contained within them are no longer required, then they are actively destroyed (Bormann and Sahm, 1978; Veenhuis et al., 1978b; Egli et al., 1980). It has been suggested that this active degradation of the peroxisomes, alcohol oxidase and catalase is due to proteolysis, and this process has been called catabolic inactivation (Bormann and Sahm, 1978; Veenhuis et al., 1978b; Egli et al., 1980). By contrast with these peroxisomal enzymes, the cytoplasmic formaldehyde dehydrogenase is not actively destroyed on transfer of yeast from methanol to a glucose growth medium.

Because the inducible FAD-containing alcohol oxidase of yeasts forms such a high proportion of the total soluble protein of methylotrophic yeasts (van Dijken et al., 1976a), a higher rate of synthesis of FAD is required during growth on methanol. In a number of different yeasts this is achieved by increasing the levels of enzymes responsible for FAD biosynthesis. During growth on methanol the specific activity of riboflavin synthase and riboflavin kinase are induced 2- to 3-fold, and the last enzyme of the biosynthetic sequence, FMN adenyltransferase, is induced 8- to 10-fold (Shimizu et al., 1977b; Eggeling et al., 1977; Sahm, 1977). Using mutants completely lacking alcohol oxidase, it has been shown that the induction of FMN adenyltransferase is correlated with the synthesis of the alcohol oxidase protein, but results with other mutants indicate that there is no *stringent* connection between the increase in specific activity of the two enzymes (Eggeling et al., 1977).

B. Regulation of enzyme activity

When considering regulation at the level of enzyme activity, the main problem is that of controlling the oxidation of methanol to formaldehyde, and of balancing the flow of formaldehyde into the oxidative pathway (by direct and cyclic routes), and into the assimilation pathway. If more methanol is oxidised than is required, then accumulation of formaldehyde in the peroxisomes might lead to its oxidation by catalase or alcohol oxidase; although this would afford protection to the yeast from the potentially lethal formaldehyde, it is probable that this energetically inefficient oxidation is avoided by some as yet unknown regulatory mechanism. The assimilation pathway requires free formaldehyde, whereas the oxidation of formaldehyde is by way of hydroxymethylglutathione; it is thus possible that the levels of free and bound glutathione may play some regulatory role, as suggested for tetrahydrofolate in regulation of the serine pathway (page 133). In

Candida boidinii the oxidation of formaldehyde to CO_2 by way of formate is regulated by the "end-products" NADH and ATP. Both formaldehyde and formate dehydrogenases are inhibited by concentrations of NADH and ATP similar to the cellular concentrations in this yeast (Kato *et al.*, 1979b, c). Operation of the cyclic route for formaldehyde oxidation provides NADPH for biosynthesis and is regulated by inhibition of the two $NADP^+$-specific dehydrogenases by their "end-product" NADPH. Nothing is known about the regulation of the assimilatory pathway but it is most likely to be by way of the DHA synthase and triokinase.

11
Methanogens and methanogenesis

I. Introduction	296
II. The methanogens	300
III. Structure, cell wall structure and lipid composition of methanogens	301
IV. The RNAs and DNA of methanogens	304
V. Biosynthesis in methanogenic bacteria	304
VI. Energy coupling in methanogens	306
A. Introduction	306
B. Novel coenzymes from methanogens	308
1. Coenzyme F_{420}, a deazaflavin	308
2. Methanopterin and its derivatives	311
3. Factor F_{430}; a nickel-containing tetrapyrrole	312
4. Coenzyme M	312
5. Compounds F_A and F_C	314
6. Component B of methyl Coenzyme M reductase	314
C. ATP synthesis in methanogens	315
1. ATP synthetase	315
2. The establishment of a proton motive force (pmf) in methanogens	316
D. The reduction of CO_2 to CH_4	319
1. The role of methyl transfer reactions and vitamin B_{12} derivatives (corrinoids) in methanogenesis	319
2. The methyl-Coenzyme M reductase complex and the reduction of "XCH_2OH" to CH_4	320
3. The RPG effect and the reduction of CO_2 to "$X-CH_2OH$"	322
E. Methanogenesis from acetate, methanol, methylated amines and formate	323
1. Methanogenesis from acetate	324
2. Methanogenesis from methanol	325
3. Methanogenesis from methylated amines	326
4. Methanogenesis from formate	327

I. Introduction

A few methanogenic bacteria grow on formate, methanol or methylated amines, and so this chapter is not completely out of place in this book on methylotrophs. This will not deal exclusively with the methylotrophic methanogens but will aim to cover the whole subject of methanogenesis. As I am not an authority on methanogenic bacteria this chapter can do little more than summarise the authoritative and stimulating reviews listed in Table 52a. References to individual aspects of methanogens and methanogenesis are listed in Table 52b.

TABLE 52
(a) Reviews of methanogens and methanogenesis

Balch et al. (1979). Methanogens: re-evaluation of a unique biological group.
Barker (1956). Bacterial fermentations; biological formation of methane.
Ellefson and Wolfe (1981). Biochemistry of methylreductase and evolution of methanogens.
Hungate (1966). The rumen and its microbes.
Hungate (1976). The rumen fermentation.
Kell et al. (1981). Energy coupling in methanogens.
Keltjens and Vogels (1981). Novel coenzymes of methanogens.
Mah et al. (1977). Biogenesis of methane.
Mah et al. (1981). Methanogenesis from $H_2 + CO_2$, methanol and acetate by *Methanosarcina*.
Stadman (1967). Methane fermentation.
Shlegel et al. (1976). Microbial production and utilisation of gases (H_2, CH_4, CO).
Thauer et al. (1977). Energy conservation in chemotrophic anaerobic bacteria.
Thauer and Fuchs (1981). Biosynthesis in methanogenic bacteria.
Walther et al. (1981). Growth of methanogens on methylamines.
Vogels (1979). The global cycle of methane.
Wolfe (1971). Microbial formation of methane.
Wolfe (1979a). Methanogenesis.
Wolfe (1979b). The methanogens: a surprising microbial group.
Wolfe (1980). Respiration in methanogenic bacteria.
Zeikus (1977). The biology of methanogenic bacteria.

(b) The methanogens

Ultrastructure
Langenberg et al. (1968); Zhilina (1971); Zeikus and Wolfe (1973); Ferry et al. (1974); Zeikus and Bowen (1975a, b); Zeikus (1977); Kandler and Konig (1978); Doddema et al. (1979a); Kandler (1979); Romesser et al. (1979); Sauer et al. (1980b).

Cell wall composition
Jones et al. (1977a); Kandler and Hippe (1977); Kandler and Konig (1978); Konig and Kandler (1979a, b); Hammes et al. (1979); Labischinski et al. (1980); Schönheit and Thauer (1980).

Lipid composition
Makula and Singer (1978); Tornabene and Langworthy (1978); Tornabene et al. (1978, 1979).

Biosynthesis
(a) In *Methanobacterium thermoautotrophicum*: Taylor et al. (1976); Zeikus et al. (1977); Daniels and Zeikus (1978); Fuchs and Stupperich (1978, 1980); Fuchs et al. (1978a, b, 1979b); Oberlies et al. (1980); Thauer and Fuchs (1981).
(b) In *Methanosarcina*: Daniels and Zeikus (1978); Weimer and Zeikus (1978a, b, 1979); Smith and Mah (1978); Mah et al. (1978); Thauer and Fuchs (1981).

Growth yields
Stadtman (1967); Robertson and Wolfe (1970); Taylor et al. (1976); Taylor and Pirt (1977); Weimer and Zeikus (1978a); Smith and Mah (1978); Schönheit et al. (1980); Walther et al. (1981).

298 The biochemistry of methylotrophs

The importance of methanogenesis is indicated by the thousands of millions of tonnes of biologically-produced methane released annually into the atmosphere: 45% of this is from swamps, 20% from ruminants, 25% from paddy fields and most of the remainder from river and lake mud (Schlegel et al., 1976; Vogels, 1979; Hanson, 1980). This is only part of the methane produced, because the methanotrophs probably oxidise the greater part of all methane produced before it reaches the atmosphere.

Methanogenic bacteria have the most stringent anaerobic requirements among anaerobes; not only must O_2 be excluded, but methanogenesis only occurs where the redox potential is lower than -330 mV. In most habitats O_2 is removed by the activity of aerobic bacteria, and a reducing potential is formed by the production of sulphide and H_2 by anaerobic bacteria. Although methanogenesis does occur at low pH (e.g. in peat bogs), in pure culture methanogens are most active between pH 6·7 and pH 8·0.

The methanogens are predominantly chemolithotrophic in their metabolism, assimilating cell carbon at the level of CO_2, and obtaining the energy for this process by the anaerobic reduction of CO_2 to methane, using H_2 as electron donor. Many strains, however, can grow on formate as sole carbon and energy source, and *Methanosarcina barkeri* can also use CO (poorly), methanol, methylated amines or acetate (Mah et al., 1978, 1981; Weimer and Zeikus, 1978a, b, 1979; Hutten et al., 1980; Walther et al., 1981). A few other methanogens also use acetate and some of these "acetate organisms" are exceptional in being unable to use $H_2 + CO_2$ as growth substrate (Zinder and Mah, 1979; Zehnder et al., 1980). The only substrates of major importance for methanogenesis in nature are $H_2 + CO_2$, formate and acetate; these substrates are produced by fermentative bacteria in the two main types of natural habitat summarised in Fig. 42. These are the habitats which account for most of the methane produced, and in which methanogenesis is the final stage of anaerobic fermentation. Type A habitats include all those that are rich in organic nutrients, excluding the alimentary tracts (rumen, caecum and intestine) of ruminant animals (Type B habitats). A third type of habitat includes volcanic environments, geothermal springs, and lakes supplied by such springs. Here the predominant source of methane is $H_2 + CO_2$, and thermophilic methanogens are especially important (Zeikus, 1977; Zeikus et al., 1980).

An important phenomenon observed in both habitats described in Fig. 42 is that of "interspecies hydrogen transfer" (Ianotti et al., 1973; Wolin, 1976; Mah et al., 1977; Vogels et al., 1980). The evolution of H_2 from NADH and protons is thermodynamically unfavourable under standard conditions, and the removal of H_2 by the methanogens is essential for the growth of such proton-reducers (Thauer et al., 1977). It is this close relationship ("thermodynamic symbiosis!") that lay behind the earlier conclusion that

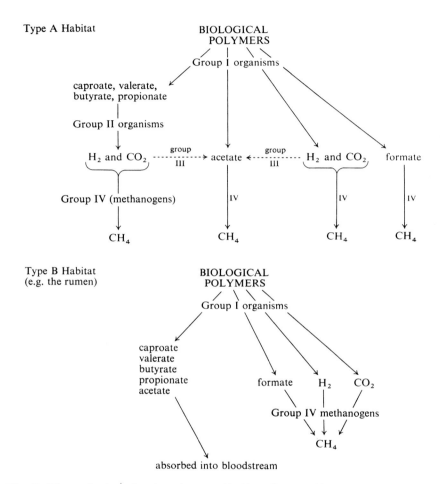

Fig. 42. The production of methane in natural habitats (from Wolfe, 1979a). The reactions summarised here produce methane plus CO_2 as the final products. *Type A habitats* include aquatic sediments, paddy fields, swamps, bogs, tundra, decaying heartwood and sludge digesters. In these habitats acetate plays a major role in the flow of carbon from polymers to methane and CO_2. *Type B habitats* are the alimentary tracts (the rumen, caecum and intestine) of animals. Acetate is not a methanogenic substrate in these habitats. Group I organisms include those anaerobes that excrete polymer-hydrolysing extracellular enzymes, and those that ferment the hydrolysis products to fatty acids, CO_2 and H_2. Group II organisms are the "obligate proton-reducers" that oxidise fatty acids or alcohols to acetate, with the concomitant reduction of protons to H_2 (Thauer et al., 1977). Group III organisms (*Acetobacterium* and *Clostridium aceticum*) oxidise H_2 anaerobically with concomitant reduction of CO_2 to acetate (Balch et al., 1977; Braun et al., 1981). Group IV organisms are the methanogens.

"*Methanobacillus omelianskii*" obtains energy from oxidation of ethanol to acetate, coupled to reduction of externally-supplied CO_2 to methane. In fact, this "organism" consists of two mutually interdependent bacteria; *Methanobacterium* strain M.o.H., a H_2-utilising methanogen, and the "S organism", which oxidises ethanol to acetate plus H_2 (Bryant et al., 1967; Wolfe, 1971).

II. The methanogens

We are concerned here solely with bacteria; there are no eukaryotic methanogens. The names of methanogenic bacteria have the prefix "*Methano-*", contrasting with the typical methylotrophs which may have the prefix "*Methylo*". Although some methanogens use methanol or methylated amines as a precursor for methane synthesis, and can thus be considered methylotrophs, no typical methylotroph (Chapter 1) produces methane; the methylotrophs and the methanotrophs are (taxonomically) totally dissimilar.

The methanogenic bacteria and their place in phylogenetic schemes have been comprehensively considered in a recent review; "Methanogens; re-evaluation of a unique biological group" (Balch et al., 1979). This provides histories of important methanogens, proposals for a complete taxonomy of methanogens, and determinative keys to species based on simple phenotypic characters. The following section is a summary of this radical review.

If considered in a general organisational sense, the methanogenic bacteria are clearly procaryotes, thus representing a variety of sizes and shapes (and responses to the Gram strain) typical of bacteria. They lack nuclear membranes and organelles, and their ribosomes appear more like bacterial ribosomes than their eucaryotic counterparts. In relation to their 16S rRNA homologies, however, they are no nearer to ordinary bacteria than they are to eucaryotes, and they represent a coherent phylogenetic grouping quite distinct from other typical bacteria. Just how distinct they may be is indicated by the fact that enteric bacteria, *Bacillus* spp. and cyanobacteria are more closely related to one another than to any of the methanogens. Furthermore, it is apparent that the methanogens as a group are very diverse; the deepest branches are as different from one another as are the enteric bacteria and *Bacillus* spp. These conclusions, based on 16S rRNA sequence homologies, are supported by many other characteristics, some of which are described below. The taxonomic significance of all these characteristics is indicated by the proposed separation of the methanogens, together with the halobacteria and two thermoacidophiles, into a new "primary kingdom" — the "archaebacteria" (Woese et al., 1978).

The archaebacteria do not have murein cell walls, the lipid components of their membranes are unusual ether-linked polyisoprenoid (branched-chain) lipids, and they possess characteristic transfer RNAs and ribosomal RNA. They are as ecologically and biochemically diverse as the eubacteria.

III. Structure, cell wall structure and lipid composition of methanogens

Almost all types of shape and structure are found amongst the methanogens: some are Gram-positive, some Gram-negative; some are motile, others not; some form single cells, others form chains, spiral filaments or packets of cells; some are spherical, irregular or lancet-shaped cocci, and others form rods which may be short or long, straight, curved or crooked (Tables 52a and 52b).

Methanogens are represented by a diverse range of both Gram-positive and Gram-negative cell wall types (Table 52b). Gram-positive strains have a thick, rigid sacculus, whereas Gram-negative strains have no rigid sacculus. Table 53 summarises the major features associated with cell wall structure and composition. Only the Gram-positive *Methanobacterium* and *Methanobrevibacter* contain peptidoglycan cell walls. Typical Gram-positive bacteria have peptidoglycan which contains muramic acid and D-amino acids. (This is called murein.) By contrast, the peptidoglycan in methanogens (pseudomurein) contains only L-amino acids and no muramic acid (Fig. 43). The carbohydrate part of the peptidoglycan of methanogens is the novel carbohydrate *N*-acetyltalosaminuronic acid plus *N*-acetylglucosamine or *N*-acetylgalactosamine. Comparison of X-ray diffraction data for murein and pseudomurein indicates that the two polymers have a similar 3-dimensional architecture (Labischinski *et al.*, 1980). The different cell wall structure of methanogens is reflected in their lack of sensitivity to antibiotics such as penicillin, cycloserine and vancomycin, and to lysozyme (Jones *et al.*, 1977; Kandler and Hippe, 1977; Kandler and Konig, 1978) However, the observation that L-alanine is a product of cell wall synthesis in *Methanobacterium thermoautotrophicum* indicates that cross-linking of the peptide strands of the pseudomurein proceeds by a mechanism very similar to that in eubacterial murein synthesis. The mechanism appears to differ only in that L-alanine, rather than D-alanine, is involved as the terminal amino acid in the transpeptidation reaction. In eubacteria a specific transport system enables the organism to re-utilise the extracellular D-alanine formed during cell wall synthesis. But in *M. thermoautotrophicum* the absence of an analogous transport system leads to the observed accumulation of L-alanine in the medium during growth (Schönheit and Thauer, 1980).

TABLE 53
Summary of cell wall and lipid composition of the methanogens

Order and family	Genus	Gram reaction	Morphology	Cell wall composition	Polar lipids (isopranyl glycerol ethers)
Order I; Methanobacteriales					
Family; *Methanobacteriaceae*	*Methanobacterium*	+	Long rods	Pseudomurein	$C_{20} + C_{40}$
	Methanobrevibacter	+	Short rods	Pseudomurein	$C_{20} + C_{40}$
Order II; Methanococcales					
Family; *Methanococcaceae*	*Methanococcus*	−	Regular to irregular coccus	Protein subunits with trace glucosamine	C_{20}
Order III; Methanomicrobiales					
Family; *Methanomicrobiaceae*	*Methanomicrobium*	−	Short curved rod	Protein subunits	Not determined
	Methanogenium	−	Highly irregular coccus	Protein subunits	Not determined
	Methanospirillum	−	Long curved rod	Protein subunits with external sheath	$C_{20} + C_{40}$
Family; *Methanosarcinaceae*	*Methanosarcina*	+	Irregular coccus in packets	Acid heteropolysaccharide	C_{20}

This table is taken from the data in Balch *et al.* (1979); references are given in Table 52, and the structures of pseudomurein and lipids in Figs 43 and 44 respectively.

```
—NAG—NAM—NAG—                    —NAG—NATM—NAG—
        ↓                                ↓
        Ala                              Glu
        ↓                                ↓γ
        D-Glu                            Ala
        ↓γ                               ↓ε
        Lys⇐ᵋD-Ala                       Lys⇐ᵞGlu
        ↓       ↑                        ↓       ↑
        (D-Ala) Lys                      (Glu)   Lys⇐ᵞGlu
        ↓       ↑γ                       ↓       ↑ε
        (D-Ala) D-Glu                    (Ala)   Ala
                ↑                                ↑γ
                Ala                              Glu
                ↑                                ↑
—NAG—NAM—NAG—                    —NAG —NATM — NAG—

     Murein                            Pseudomurein
```

Fig. 43. The structure of murein of a typical Gram-positive organism (*Gaffkya* sp.) and of pseudomurein from the methanogen *Methanobacterium thermoautotrophicum* (from Hammes et al., 1979).

Some inhibitors (e.g. bacitracin) of murein synthesis in eubacteria also inhibit the synthesis of pseudomurein in methanogens, emphasising that there must be further similar reactions in the synthesis of the two polymers (Hammes et al., 1979).

The lipids of typical bacteria consist predominantly of esters of glycerol and fatty acids. Such ester lipids do not occur in methanogens which, instead, contain the analogous glycerol ethers (Table 52b; Fig. 44). Of the

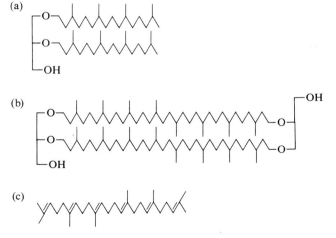

Fig. 44. Principal components of the polar fraction (a and b) and the neutral lipid fraction of methanogens (c) (see Table 52 for references). (a) C_{20} diphytanyl-glycerol diether; (b) C_{40} dibiphytanyl-diglycerol tetraether; (c) $C_{30}H_{50}$ isoprenoid (squalene).

total lipids, about 75% are such polar lipids. Analysis of the acid-hydrolysate of the phospholipid fraction of methanogens has shown that the glycerol is bonded by ether bonds to branched isopranyl side chains thus forming C_{20} phytanyl and/or C_{40} biphytanyl glycerol ethers (Fig. 44). The neutral lipids comprise a wide range of (C_{14} to 3_{30}) isoprenoid hydrocarbons; the principal compounds are the C_{20}, C_{25} and C_{30} components; the C_{30} (squalene) isoprenoid is the predominant neutral lipid.

These findings provide the first evidence for the microbial synthesis of multibranched isoprenoid hydrocarbons comparable to those found in sediments and petroleum, and they may have major implications in the interpretation of biogeochemical evolution (Tornabene and Langworthy, 1978). A possible role of the glycerol ethers in stabilising membrane structure, and of the neutral isoprenoid hydrocarbons, such as squalene, in evolution of membrane systems has been proposed by Rohmer et al. (1979).

IV. The RNAs and DNA of methanogens

The 16S rRNA sequences and the secondary structure of the 5S rRNAs of the methanogens bear no more relationship to those of typical bacteria than they do to those of eucaryotes. The post-transcriptional modification patterns in 16S and 23S rRNAs also differ strikingly from those of typical bacteria, although the modifications tend to occur in the same locales in the two cases. Furthermore, the tRNAs are unique in that they do not contain the otherwise "universal" common arm sequence GTψCG; instead they contain the analogous sequences G$\psi\psi\dot{C}$G or G$\dot{U}\psi$C$\dot{G}$ in which the bases marked with a dot are modified (Balch et al., 1979).

Analysis of the DNA of *Methanobacterium thermoautotrophicum* has shown that its genome is about one third of the size of that of *E. coli* and that satellite DNA is absent (Mitchell, 1978). The percentage (G + C) base ratios vary between 27·5% and 61%, with each of the groups showing, as expected, a more restricted range (Balch et al., 1979).

V. Biosynthesis in methanogenic bacteria

All the major biochemical techniques used for elucidating the assimilation pathways in other methylotrophs (Chapters 2, 3 and 4) have been used in the study of assimilation in methanogens. Key papers in describing this work are listed in Table 52b and their conclusions summarised below.

No methanogen uses the usual pathways for assimilation of C_1 compounds; that is, none has the RuBP, RuMP, DHA or serine pathways and none uses the reductive tricarboxylic acid cycle for CO_2 fixation.

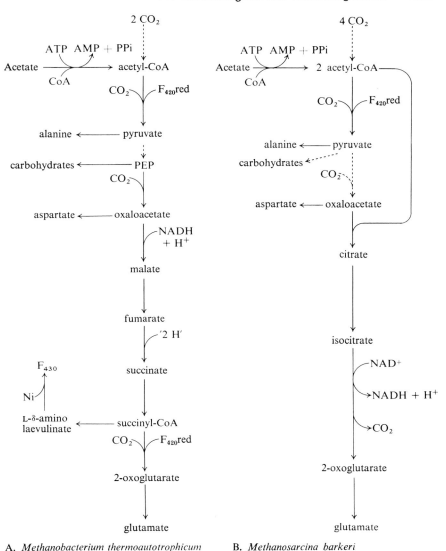

A. *Methanobacterium thermoautotrophicum*   B. *Methanosarcina barkeri*

Fig. 45. Biosynthesis in two methanogens (for references see Table 52b).

Two very different methanogens have been studied with respect to their carbon assimilation pathways; *Methanobacterium thermoautotrophicum* and *Methanosarcina barkeri* (Fig. 45). In both organisms CO_2 is a major carbon source, and acetate can also be assimilated. Although acetate cannot act as sole source of carbon and energy in *M. thermoautotrophicum*, up to 65% of its cell carbon can be derived from acetate, as is also the case with *M. ruminantium* (Fuchs *et al.*, 1978a).

The first part of their assimilation pathways is similar, and involves conversion of CO_2 to acetyl-CoA. Acetate is converted to acetyl-CoA by the action of acetate thiokinase (Oberlies *et al.*, 1980), but the mechanism of formation of acetyl-CoA from CO_2 is completely unknown. A complete reductive tricarboxylic acid cycle has been ruled out (Fuchs and Stupperich, 1978), and a more direct condensation of two C_1 compounds, similar to that found in some clostridia has been proposed (Ljungdahl and Wood, 1969; Taylor *et al.*, 1976; Ljungdahl and Andreeson, 1976). This remains at present a speculation, but there is some evidence that corrinoid compounds may be involved (Kenealy and Zeikus, 1981).

Acetyl-CoA is carboxylated to pyruvate in a reaction that is catalysed by pyruvate synthase; this requires the deazaflavin coenzyme F_{420} as electron donor instead of the ferrodoxin that is usual in some other anaerobic bacteria. Pyruvate is then converted to oxaloacetate. The enzymes involved in the interconversion of pyruvate, phosphoenolpyruvate (PEP) and oxaloacetate remain to be elucidated in the methanogens. PEP carboxylase is present in *Methanobacterium thermoautotrophicum* but was undetected in *Methanosarcina barkeri* (Zeikus *et al.*, 1977).

These two bacteria also differ in their routes for biosynthesis of succinate and 2-oxoglutarate from oxaloacetate (Fig. 45). *M. barkeri* is able to produce 2-oxoglutarate by way of citrate and isocitrate, but the pathway for succinate synthesis remains obscure because of the absence of fumarate reductase and 2-oxoglutarate dehydrogenase (Weimer and Zeikus, 1979). By contrast, *M. thermoautotrophicum* produces 2-oxoglutarate by way of malate, fumarate, fumarate reductase and 2-oxoglutarate synthase (electron donor F_{420}) (see Thauer and Fuchs, 1981).

VI. Energy coupling in methanogens

A. Introduction

The whole purpose of methanogenesis is to make ATP. This process has been studied most extensively as it occurs in methanogens growing on H_2 plus CO_2, and particularly in *Methanobacterium thermoautotrophicum*. All the evidence so far is consistent with the conclusion that methanogens obtain ATP by electron transport phosphorylation coupled to the oxidation of hydrogen, with CO_2 serving as terminal electron acceptor (Kell *et al.*, 1981).

The overall reaction of methanogenesis is:

$$CO_2 + 4\,H_2 \rightarrow CH_4 + 2H_2O \quad \Delta G^{\circ\prime} = -131 \text{ kJ mol}^{-1}$$

In non-standard conditions, particularly with the usual low concentrations of H_2, this free energy change will be more positive, and it is unlikely that more than 1 ATP per mole of CH_4 could be produced. In 1956 Barker proposed a scheme for reduction of CO_2 to CH_4 which has been a model for most subsequent investigations (Fig. 46). Free formate, formaldehyde and methanol are not intermediates; the CO_2 becomes bound to a carrier molecule, and the C_1 unit remains bound during the sequential reduction reactions until it is released as methane. In essence this model remains a suitable model of methanogenesis; the main changes that have been proposed are that the carrier (X) may not be the same at every stage, and that it might be better to represent the process as a closed cycle (Wolfe, 1980; Ellefson and Wolfe, 1981; Figs 52 and 53).

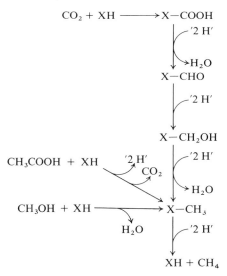

Fig. 46. Barker's scheme for the reduction of CO_2, methanol and acetate to methane. This scheme is from Barker (1956). XH is a carrier molecule. For more recent proposals developed from this model see Figs. 52 and 53.

A critical fact in any description of methanogenesis is that the hydrogen in methane does not come directly from the H_2 energy source but from the protons of water (Daniels *et al.*, 1980; Table 57). Consistent with this important fact is the conclusion that to harness the energy available from methanogenesis, a proton motive force (pmf) must be established, by coupling electron transport between H_2 and CO_2 to the translocation of protons across a closed membrane system. This pmf must then be coupled to ATP synthesis by a proton-translocating ATP synthetase.

This process of electron transport phosphorylation has an extra degree of complexity in methanogens because there are, in effect, four different terminal electron acceptors which must be reduced in sequence (Fig. 46); and hence four electron transport chains, having the same electron donor but different acceptors. Not all of these chains can be coupled to ATP synthesis. The sequential reductions by H_2 of "formaldehyde" to "methanol" and then to methane are exergonic, and can presumably be "arranged" to set up a pmf; whereas both steps in the reduction by H_2 of CO_2 to "formate" and then to "formaldehyde" require energy (they are endergonic). The energy for driving the two earlier reduction steps must come either from ATP or, more likely, from the pmf set up by the two later exergonic reduction steps.

By contrast with the situation in methylotrophs, measurements of growth yields in methanogens have not been important in considering their bioenergetics. This has been partly due to the requirements for growth factors and the earlier unexplained need for complex growth media. This sort of problem is now being overcome, and some studies of growth yields of methanogens are being published (see Table 52b for references).

I have stated here the general problem of energy coupling in methanogens, and the basis of a probable "solution", in very general terms, in order to give some framework to this inchoate, but rapidly developing subject. The facts of the matter will now be considered in three main sections dealing with the reduction of CO_2 to CH_4; these will be followed by a section on the less-understood problem of methanogenesis from formate, methanol, methylated amines and acetate.

B. Novel coenzymes from methanogens

This section is a set of descriptions of the novel coenzymes found in methanogens. Many of these are probably involved in the process of methanogenesis itself, but some may be involved in the carbon assimilation pathways; some of course may be involved in both. Kell has expressed surprise (or even indignation) (Kell *et al.*, 1981) at the "remarkable present emphasis on soluble cofactors in the s udy of a process that must be coupled with membranes". It would, perhaps, be more remarkable if these soluble cofactors were observed, and subsequently ignored. The discovery and characterisation of more than half a dozen novel coenzymes is already a major achievement of biochemistry although their function is, in many cases, uncertain (Table 54).

1. Coenzyme F_{420}, a deazaflavin

Factor F_{420}, now called coenzyme F_{420}, was first described by Cheeseman *et al.* (1972). Its properties have been fully reviewed recently by Wolfe (1980) and references to specific aspects of its biochemistry are listed in Table 54.

TABLE 54
Novel coenzymes from methanogens

1. *Coenzyme* F_{420} (5-deazaflavin) (reviews, Wolfe, 1980; Kell *et al.*, 1981).
   (a) *Structure and spectra:* Cheeseman *et al.* (1972); Eirich *et al.* (1978); Ashton *et al.* (1979); Pol *et al.* (1980); Yamazaki *et al.* (1980).
   (b) *Role as electron carrier:* Cheeseman *et al.* (1972); Tzeng *et al.* (1975a, b); Ferry and Wolfe (1977); Zeikus *et al.* (1977); Walsh (1979); Eirich *et al.* (1979); Schauer and Ferry (1980); Pol *et al.* (1980); Fuchs and Stupperich (1980); Gunsalus and Wolfe (1980); Yamazaki and Tsai (1980); Yamazaki *et al.* (1980).
   (c) *Distribution:* Ferry and Wolfe (1977); Gunsalus and Wolfe (1978a); Eirich *et al.* (1979); Schauer and Ferry (1980): Yamazaki and Tsai (1980).
   (d) *Basis of fluorescence assay for methanogens:* Edwards and McBride (1975); Mink and Dugan (1977); Doddema and Vogels (1978); Vogels *et al.* (1980).

2. *Methanopterin and its derivatives* (B_O, YFC and F_{342}). Gunsalus and Wolfe (1978a); Daniels and Zeikus (1978); Keltjens and Vogels (1981).

3. *Factor* F_{430}; *a nickel-containing tetrapyrrole.* Gunsalus and Wolfe (1978a); Schönheit *et al.* (1979); Diekert *et al.* (1979, 1980a, b, c); Whitman and Wolfe (1980); Thauer and Fuchs (1981).

4. *Coenzyme* M
   (a) *Structure and activity of its derivatives and analogues:* Taylor and Wolfe (1974a); Taylor (1976); Gunsalus *et al.* (1976, 1978); Hutten *et al.* (1981); Romesser and Wolfe (1981).
   (b) *Growth factor for methanogens:* Bryant (1965); McBride and Wolfe (1971a); Taylor *et al.* (1975); Gunsalus *et al.* (1976).
   (c) *Distribution and active transport in methanogens:* Balch and Wolfe (1979a, b).
   (d) *Assay:* Taylor and Wolfe (1974b); Hermans *et al.* (1980); Shapiro and Wolfe (1980).
   (e) *Redox potential:* Kell and Morris (1979).

5. *Compounds* F_A *and* F_C. Keltjens and Vogels (1981).

6. *Component B of methyl Coenzyme M reductase.* Gunsalus and Wolfe (1980).

Coenzyme F_{420} is a 5-deazaflavin analogue of FMN, with a complex side-chain containing glutamylglutamate; and thus reminiscent of folates (Fig. 47). It is a low potential electron carrier and participates in two-electron transfer reactions (midpoint redox potential, E_m7, $-0\cdot340$ V) (Fig. 48). Reduction occurs at positions 1 and 5 of the ring system, and this reduction leads to loss of fluorescence and of the characteristic absorbance at 420 nm, with the concomitant appearance of a new peak at 320 nm. The semiquinone does not appear to be readily formed.

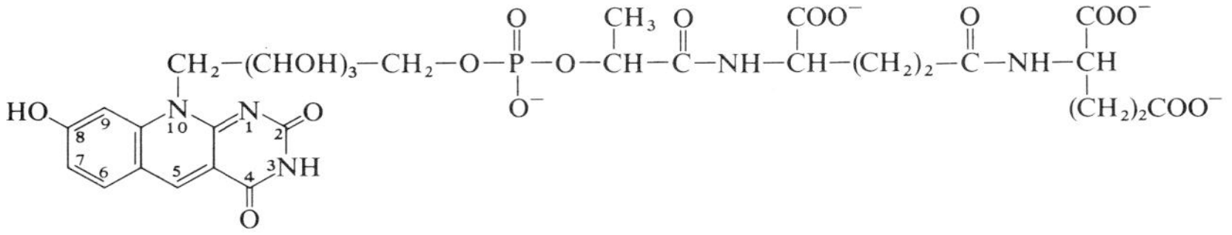

Fig. 47. The structure of Coenzyme F_{420} (proposed by Eirich *et al.*, 1978). This structure is for oxidised Coenzyme F_{420}; the reduced (dihydro) form is reduced at positions 1 and 5. The suggested trivial name is N-(N-L-lactyl-γ-L-glutamyl)-L-glutamic acid phosphodiester of 7,8-didemethyl-8-hydroxy-5-deazariboflavin 5'-phosphate.

A. Hydrogenase (Tzeng *et al.*, 1975a; Eirich *et al.*, 1979; Ellefson and Wolfe, 1980a, b, 1981).

$$H_2 + F_{420} \text{ ox} \longrightarrow F_{420} \text{ red}$$

B. NADP$^+$ reductase (Tzeng *et al.*, 1975b; Ellefson and Wolfe, 1980b; Jones and Stadtman, 1980; Yamazaki and Tsai, 1980; Yamazaki *et al.*, 1980).

$$NADP^+ + F_{420} \text{ red} \longrightarrow NADPH + H^+ + F_{420} \text{ ox}$$

C. Formate dehydrogenase (Tzeng *et al.*, 1975b; Schauer and Ferry, 1980; Jones and Stadtman, 1980, 1981).

$$HCOOH + F_{420} \text{ ox} \longrightarrow CO_2 + F_{420} \text{ red}$$

D. Methyl-Coenzyme-M reductase (Ellefson and Wolfe, 1980a, b, 1981).

$$CH_3\text{-S-CoM} + F_{420} \text{ red} \longrightarrow CH_4 + HS\text{-CoM} + F_{420} \text{ ox}$$

Fig. 48. The two-electron transfer reactions of Coenzyme F_{420} (deazaflavin). These reactions can all occur as written; and extracts containing these proteins plus Coenzyme F_{420} can also catalyse the following reactions: NADP$^+$-linked hydrogenase (A + B); NADP$^+$-linked formate dehydrogenase (B + C); formate hydrogen lyase (A + C); H_2-coupled, methyl-Coenzyme-M reductase (A + D); NADPH-coupled, methyl-Coenzyme-M reductase (B + D). Coenzyme F_{420} is also the electron donor in pyruvate and 2-oxoglutarate synthesis. Although these reactions are written as $2H^+/2e^-$ transfer reactions there is some evidence that Coenzyme F_{420} is a $1H^+/2e^-$ carrier; the second proton during reduction reactions comes from water (see Ashton *et al.*, 1979; Walsh, 1979).

Coenzyme F_{420} occurs in all those methanogens tested, although a slightly modified form may be present in *Methanosarcina barkeri* (Eirich *et al.*, 1979). The only non-methanogen in which it has been observed is *Streptomyces griseus*, in which it participates in the DNA photoreactivation reaction (Eker *et al.*, 1980).

Coenzyme F_{420} is the electron acceptor for hydrogenase and formate dehydrogenase; the electron donor for NADP$^+$ reductase; and it can act as electron donor for methyl-Coenzyme M reductase (Fig. 48). It is also the electron donor for pyruvate and 2-oxoglutarate synthases (Zeikus *et al.*, 1977; Fuchs *et al.*, 1978a; Fuchs and Stupperich, 1980). It has been suggested that coenzyme F_{420} may not be involved directly in methanogenesis (see Kell *et al.*, 1981), and that its main functions are probably in the production of NADPH (from formate, or $CO_2 + H_2$) for biosynthetic reactions, and as the reductant for pyruvate and 2-oxoglutarate synthesis. Ellefson and Wolfe (1980a, 1981), however, suggest that coenzyme F_{420} may be the immediate electron donor in the final step in reduction of methyl-Coenzyme M to methane.

2. Methanopterin and its derivatives (B_0, YFC and F_{342})

In 1978 Gunsalus and Wolfe described a new blue-fluorescent chromophoric factor (F_{342}) from *Methanobacterium thermoautotrophicum*; this factor is the basis of a specific epifluorescent detection method for methanogens (Doddema and Vogels, 1978). Independently, another novel compound (YFC, yellow fluorescent compound) was described by Daniels and Zeikus (1978), who found it to be a major rapidly-labelled metabolite of $^{14}CO_2$ and $^{14}CH_3OH$ in a number of methanogens. The relationship between these compounds (found only in methanogens) has been elucidated by Keltjens and Vogels (1981), who have shown that they are novel pteridine derivatives (Fig. 49). The suggested name for the "parent" compound (previously called B_o) is methanopterin; and YFC is the carboxylated, reduced derivative carboxydihydromethanopterin. Methanopterin has an absorption maximum at 342 nm and a fluorescence emission maximum at about 435 nm (Gunsalus and Wolfe, 1978a). The chromophoric factor F_{342} is a natural product or precursor of methanopterin. There appear to be two forms of methanopterin, one having a single glutamate residue and the other having two. When incubated with dialysed cell extracts of *M. thermoautotrophicum* in the presence of $H_2 + CO_2$, methanopterin is converted to the reduced form (probably 7,8-dihydromethanopterin) and to the carboxydihydromethanopterin. These results, and the stimulation by methanopterin of ATP

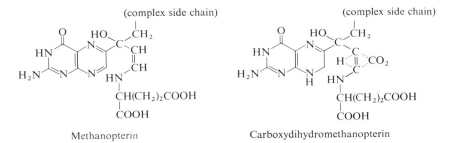

Fig. 49. The structures of methanopterin (factor B_o) and carboxydihydromethanopterin (YFC). These structures were tentatively proposed by Keltjens and Vogels (1981). The nature of the binding of CO_2 is at present unclear. The complex side chain probably contains a 6-carbon ring plus either glucosamine or galactosamine.

synthesis and of methanogenesis from $CO_2 + H_2$, are consistent with it being a C_1 carrier in methanogenesis from CO_2. No derivatives containing the more reduced C_1 units (formyl, methylene or methyl) have, however, been described. A possible alternative role for carboxydihydromethanopterin is as the first carboxylated derivative involved in fixation of CO_2 into cell material (see Section V). The midpoint redox potential (E_m7) of methanopterin is -0.45 V and it is a $2H^+/2e^-$ carrier (Kell et al., 1981; Kell, personal communication). This midpoint potential is consistent with both of the proposed roles of methanopterin.

3. Factor F_{430}: a nickel-containing tetrapyrrole

Factor F_{430} is a low molecular weight, yellow compound isolated initially from *Methanobacterium thermoautotrophicum* by Gunsalus and Wolfe (1978a) (Thauer and Fuchs, 1981). It has recently been shown that nickel is essential for growth of methanogens and that this nickel is mainly required for synthesis of Factor F_{430} (Schönheit et al., 1979; Diekert et al., 1979, 1980a; Whitman and Wolfe, 1980). Incorporation studies with ^{14}C succinate and δ-aminolaevulinate have indicated that F_{430} has a nickel tetrapyrrole structure (Diekert et al., 1980b, c). Evidence supporting this proposal has now been obtained by using F_{430} prepared by a modification of the previously used method (Thauer and Fuchs, 1981). This modification does not produce the usual 3 degradation products, previously thought to be alternative forms of F_{430}. The purified factor has a mass per mole of nickel of 1500 and an extinction coefficient ($\mathscr{E}_{430}$) of 23 000 cm^{-1} litre (mole Ni)$^{-1}$. F_{430} contains the elements C, H, N, O and Ni but no P or S. The nitrogen content is 9 atoms of nitrogen per atom of nickel, consistent with the proposed tetrapyrrole structure and the observed hydrolysis products of 1 mole each of glutamate and aspartate plus 3 moles of NH_3 per mole of nickel.

One of the degradation products of F_{430} is red and has an absorption spectrum resembling that of vitamin B_{12}, having peaks at 478 nm and 560 nm. It has been suggested that the opinion that methanogens contain high concentrations of corrinoids may have been due to the formation of such red degradation products of Factor F_{430} (but see Krzycki and Zeikus, 1980).

No function is yet known for F_{430} but an involvement in methanogenesis might be inferred from the observation that it can be isolated in a form containing coenzyme M derivatives bound to it (Keltjens and Vogels, 1981).

4. Coenzyme M

Coenzyme M was discovered by McBride and Wolfe (1971a) and its structure elucidated by Taylor and Wolfe (1974a) (Table 55). Its structure and function have been reviewed by Wolfe (1979a, 1980), and references to individual aspects of its biochemistry are listed in Table 54.

TABLE 55
The forms of Coenzyme M found in methanogens

Names	Abbreviation	Formula
Coenzyme M; 2-mercaptoethane sulphonic acid	HS-CoM	$HSCH_2CH_2SO_3^-$
2,2'-dithiodiethane sulphonic acid	$(S\text{-CoM})_2$	$^-O_3SCH_2CH_2S\text{-}SCH_2CH_2SO_3^-$
Methyl Coenzyme M; 2-(methylthio)-ethane sulphonic acid	CH_3-S-CoM	$CH_3SCH_2CH_2SO_3^-$
Hydroxymethyl Coenzyme M; 2-(hydroxymethyl)-ethane sulphonic acid	$HOCH_2$-S-CoM	$HOCH_2SCH_2CH_2SO_3^-$

The midpoint redox potential of the $(S\text{-CoM})_2$/HS-CoM redox couple at pH 7·0 and 18°C is -193 mV (Kell and Morris, 1979).

References to the structure and function of these compounds are given in Table 54. Further unidentified forms of Coenzyme M have been found by Daniels and Zeikus (1978) and Balch and Wolfe (1979b).

Coenzyme M, 2-mercaptoethane sulphate, is the smallest of all known coenzymes. It is 45% sulphur, it is strongly acidic, and it is stable at high temperature and low pH. The naturally occurring forms are summarised in Table 55. Coenzyme M occurs in all methanogens (between 0·2 mM and 2·0 mM), and is identical to the growth factor found by Bryant (1965) to be essential for growth of the rumen strain of *Methanobacterium ruminantium*; it thus shows a traditional vitamin-coenzyme relationship. Coenzyme M (HS-CoM) is rapidly oxidised in air to the disulphide $(S\text{-CoM})_2$. This is enzymically reduced to the active HS-CoM, NADPH being the reductant for this. It is not known if the HS-CoM/$(S\text{-CoM})_2$ redox couple is actively involved in methanogenesis; it tends to be assumed that the NADPH – $(S\text{-CoM})_2$ oxidoreductase functions merely as a regenerator of HS-CoM after "accidental" oxidation to the disulphide.

Coenzyme M functions as a methyl- carrier, methyl-CoM being the substrate for methylreductase which catalyses the final step in methane synthesis. The methyl-Coenzyme M reductase is highly specific for $CH_3\text{-CoM}$; modification of the sulphonate, the sulphide, the length of the carbon chain, or the alkyl moiety leads to loss of activity. $CH_3\text{-S-CoM}$ can be replaced only by $C_2H_5\text{-S-CoM}$ which is reduced at a lower rate (20%) to ethane (Gunsalus et al., 1978).

The precursor of $CH_3\text{-S-CoM}$ during the reduction of CO_2 to methane is possibly hydroxymethyl-S-CoM; this has been synthesised by Romesser and shown to be a substrate for methanogenesis in crude extracts of *M. thermoautotrophicum* (Ellefson and Wolfe, 1981).

5. Compounds F_A and F_C

These colourless compounds, isolated from *M. thermoautotrophicum*, stimulate methanopterin-activated methanogenesis from $H_2 + CO_2$ (Keltjens and Vogels, 1981). Factor F_A has a UV absorption peak at 245 nm; and F_C has absorption maxima at 252 nm and 270 nm. Both have a low molecular weight (less than 300) and both contain sulphide. It is suggested that these compounds are possible candidates in the reduction pathway of CO_2 to methane, perhaps at the level of formate or formaldehyde.

6. Component B of methyl-Coenzyme M reductase

Component B is an essential component of the reductase (Gunsalus and Wolfe, 1980). It is oxygen-labile and heat resistant, and has a molecular weight of about 1000. It has no absorption maxima above 250 nm and it is not fluorescent. It occurs in all methanogens tested and cannot be replaced by any known coenzyme.

C. ATP synthesis in methanogens

1. ATP synthetase

As mentioned in the introduction to this section, it has been concluded that ATP is synthesised by way of electron transport phosphorylation in methanogens. For this to occur there must be enzymes (membrane-bound) for the oxidation of H_2; there must be enzymes catalysing the reduction of the C_1 intermediates between CO_2 and CH_4; there must be intervening membrane-bound redox components, coupling these oxidising and reducing reactions; these must be arranged so as to translocate (or pump) protons across a closed membrane system; and there must be a membrane-bound proton-translocating ATP synthetase. All the evidence summarised below (and in Kell et al., 1981) is consistent with these proposals (see Table 56 for references).

TABLE 56
ATP synthesis and the reduction of CO_2 to methane

Hydrogenases of methanogens: Fuchs et al. (1979a); McKellar and Sprott (1979); Gunsalus and Wolfe (1980) Kell et al. (1981).

ATP pool size: Roberton and Wolfe (1970).

Adenine nucleotide translocase: Doddema et al. (1980); Kell et al. (1981).

Proton-motivated ATP synthesis: Roberton and Wolfe (1969, 1970); Mountfort (1978); Doddema et al. (1978, 1979b, 1980); Sauer et al. (1979, 1980a, b); Kell et al. (1981).

Methyltransferase reactions: Taylor and Wolfe (1974b); McBride and Wolfe (1971a); Taylor (1976); Gunsalus et al. (1976, 1978); Ferry and Wolfe (1977).

Methyl-S CoM reductase: McBride and Wolfe (1971a); Taylor and Wolfe (1974a); Gunsalus et al. (1978); Romesser (1978); Gunsalus and Wolfe (1977, 1978b, 1980); Gunsalus et al. (1978); Ellefson and Wolfe (1981); Romesser and Wolfe (1981).

The ATP requirement for methanogenesis in extracts: Wolin et al. (1963); Wood and Wolfe (1966); Roberton and Wolfe (1969); Gunsalus and Wolfe (1978b); Kell et al. (1981); Hutten et al. (1981).

The RPG effect: Gunsalus and Wolfe (1977); Wolfe (1979a, 1980); Sauer et al. (1980b); Kell et al. (1981); Hutten et al. (1981).

The ATP synthetase complex of *M. thermoautotrophicum* is similar to those described in aerobic and anaerobic microorganisms, and it is membrane-bound (Doddema et al., 1978, 1979b). When assayed as an ATPase, the K_m for ATP is 2 mM and activity is dependent on Mg^{++}; other divalent ions can replace the Mg^{++} to some extent. GTP and UTP, but not the diphosphates, are hydrolysed, but not as rapidly as ATP. The ATPase activity is inhibited by the usual ATP synthetase inhibitor N,N'-dicyclohexylcarbodiimide (DCCD).

It has been shown by Doddema et al. (1978, 1979b) that ATP synthesis can be driven in whole cells by a proton motive force (pmf), artificially imposed by a pH shift or by valinomycin. Similar results have been obtained by Sauer et al. (1980a, b) and Mountfort (1978) who have shown that artificially-induced pH gradients can be discharged by either the uncoupler CCCP or by valinomycin (17 μM); methanogenesis in these cells is strongly inhibited when the pH gradient is discharged.

These results are consistent with the conclusions that the ATP synthetase activity is coupled to proton translocation as in other bacteria and in mitochondria. An unexpected observation was, therefore, the failure of DCCD or uncoupling agents to inhibit ATP synthesis in whole cells, particularly as these agents inhibited ATP synthesis by membrane vesicles containing hydrogenase and ATP synthetase (Doddema et al., 1979b). These vesicles supported ATP synthesis coupled to a pmf, set up by either an artificial pH gradient or as the result of hydrogenase activity. In these experiments, when no added acceptor for the hydrogenase was present in the reaction mixture, the rate of ATP synthesis soon diminished; this could be restored by addition of the artificial electron acceptor phenazine ethosulphate or by addition of methanopterin (factor B_o).

Two explanations of the failure of uncouplers and DCCD to inhibit ATP synthesis in whole cells have been suggested. One is that the unusual outer membrane excludes these reagents, and the other is that the energy-generating system is localised internally to the cytoplasmic membrane. This second suggestion is perhaps supported by the observation that ATP synthesis and hydrolysis, and ADP uptake, are inhibited by inhibitors of mitochondrial ADP/ATP translocase (Doddema et al., 1980). Such translocases are extremely rare in bacteria and only occur in those organisms having membrane organelles. It has been suggested, therefore, that energy coupling in methanogens may be located in a primitive organelle, seen in *M. thermoautotrophicum* as invaginations of the cytoplasmic membranes on which the hydrogenase and some of the ATP synthetase is located (Doddema et al., 1979b). The conclusion that the reduction of CO_2 to methane is an essentially membrane-related process is strongly supported by the work of Sauer et al. (1979, 1980a, b). A critical observation in this work is that the methanogenesis catalysed by membrane vesicles is completely prevented by mechanical disruption of the vesicles.

2. The establishment of a proton motive force (pmf) in methanogens

The necessary pmf for ATP synthesis must be established by proton

11. Methanogens and methanogenesis 317

translocation coupled to the oxidation of H_2 by CO_2. Although the reductant is always the same (H_2), the terminal electron acceptor is not; it is a C_1 unit at the level of oxidation of either CO_2, formate, formaldehyde or methanol (Fig. 46). Only the last two reductions are likely to be exergonic and arranged to translocate protons so as to produce a pmf. The reduction of CO_2 to the formaldehyde level is likely to be endergonic. The energy to drive these reductions may be provided as ATP, but it is more likely to be provided by way of the pmf, as suggested by Kell et al. (1981). If this is the case, then these reductive steps must also be coupled to proton translocation; but in the opposite direction to that of the exergonic part of the process of methanogenesis. The exergonic reactions must also result in translocation of *more* protons than are translocated in the opposite direction by the endergonic reductive steps, in order to "drive" the proton-translocating ATP synthetase (see Fig. 50).

Exergonic reactions:

$2 H_2 + X\text{-}CH_2OH \longrightarrow X\text{-}H + H_2O + CH_4$

Plus x protons translocated OUT

Endergonic reactions:

$2 H_2 + X\text{-}COOH \longrightarrow X\text{-}CH_2OH + H_2O$

Plus y protons translocated IN

$ADP + Pi \longrightarrow ATP$ Plus z protons translocated IN

Fig. 50. Possible coupling of exergonic and endergonic reactions by proton translocation across membranes. In principle, the direction of proton translocation is unimportant; but the direction must be opposite for exergonic and endergonic reactions and x must be equal to or greater than $y + z$ (see Fig. 51).

To establish a pmf, the hydrogenase complex must be arranged in the membrane so that it reacts with an appropriate terminal electron acceptor, and so that it liberates protons on the appropriate side of the membrane. By analogy with other electron transport systems, this arrangement will probably involve alternating electron- and H-carrying components—usually flavoproteins, quinones, iron-sulphur centres and cytochromes (see Chapter 8). Such components should be found in either the hydrogenase complex, or in the associated membranes or in both; but as Kell points out there is remarkably little information on this (Kell et al., 1981). Methanogens do not usually contain quinones or cytochromes (Thauer et al., 1977; Wolfe, 1979a, b) although there is a single report of a cytochrome *b* in *Methanosarcina barkeri* growing on methanol or methylated amines (Kuhn et al., 1979). A recent EPR study of *Methanobacterium bryantii* has demonstrated a number of paramagnetic centres including a $[Fe_4 - S_4^*]$ or $[Fe_3 - S_3^*]$ centre, and an unusual type which may be due to a nickel-containing, membrane-bound component (Lancaster, 1980).

318 The biochemistry of methylotrophs

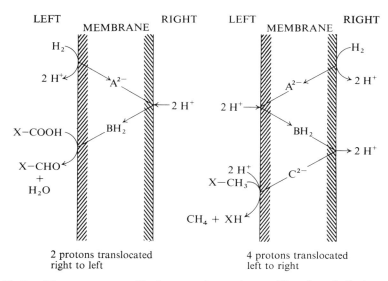

Fig. 51. Possible arrangements of hydrogenase in membranes. These hypothetical arrangements are proposed merely as an illustration of how the oxidation of hydrogen might be coupled to various stoicheiometries and directions of proton translocation. A and C are electron carrying components, analogous in function to iron-sulphur centres and cytochromes. B is a hydrogen carrier, analogous in function to flavoproteins and quinones. The two systems above illustrate how the pmf set up by the reduction of X-CH₃ to methane might be used to drive the endergonic reduction of X-COOH to X-CHO. Such schemes as these do not preclude the possibility of soluble redox mediators (e.g. Coenzyme F_{420}) acting between the proton-translocating membrane and the reduction of the C_1 units (see text).

Nearly all proposals for proton-translocating electron transport systems in bacteria are based on imagination, and any facts come later (if ever). This should be borne in mind in considering schemes in Fig. 51. These give examples of how H_2 may be oxidised in such a way as to produce proton gradients in both directions, and with varying stoicheiometries. Many equally-likely schemes might be drawn; these are included merely as examples; the number of possible schemes is amplified by the fact that substrates and products (CO_2, H_2 and CH_4) can all pass freely across membranes. It should be noted that there may be more than one type of hydrogenase system; these might differ in their solubility, their electron acceptors, or their arrangement in the membrane. The hydrogenase need not always be proton-translocating. If soluble redox carriers such as Coenzyme F_{420} are involved as intermediates between the proton-translocating membrane and the reduction of a C_1 unit, then these carriers will presumably be different for the endergonic reduction reactions (X–COOH to X–CH₂OH) from those required for the more exergonic reduction reactions (X–CH₂OH to CH_4). Although hydrogenase can use the deazaflavin coenzyme F_{420} as a hydrogen acceptor, it has been suggested that it might be involved only in coupling hydrogenase with the reduction of $NADP^+$ to provide NADPH for biosynthesis (Kell *et al.*, 1981) (Figs 45 and 48).

D. The reduction of CO_2 to CH_4

This section is concerned with the nature of the C_1 intermediates between CO_2 and methane during methanogenesis (see Table 56 for references). Barker's scheme (Fig. 46), and modifications of it (Figs 52 and 53), form the basis for this discussion. A key question, of course, concerns the nature of the C_1 carriers in this process. For some time it was thought that vitamin B_{12} derivatives (corrinoids) might be involved. This is almost certainly not the case, and the following section explains briefly how this idea arose. This is followed by a description of the methylreductase complex, and the final section is a consideration of the reduction of CO_2 to the level of formaldehyde, and how this is coupled to the later stages of methanogenesis.

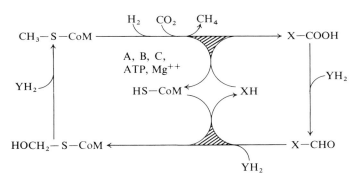

Fig. 52. The reduction of CO_2 to CH_4. This is adapted (merely simplified) from that in Wolfe (1980) and Ellefson and Wolfe (1981). The CDR factor is also involved in the reduction of CO_2 to $HOCH_2$-S-CoM (p. 322). An alternative, more general, scheme is given in Fig. 53.

1. *The role of methyl transfer reactions and vitamin B_{12} derivatives (corrinoids) in methanogenesis*

Before the discovery of Coenzyme *M* by McBride and Wolfe, the chief contender for the role of methyl-carrier (X in Barker's scheme, Fig. 46) was coenzyme B_{12}. Methyl cobalamin is readily reduced to methane by extracts of methanogenic bacteria (Blaylock and Stadtman, 1963, 1966). It has been pointed out, however, that methyl B_{12} is a strong alkylating agent, and probably plays the role of a non-specific methylating agent in methanogens.

Extracts of methanogens catalyse a variety of transmethylation reactions using methyl-B_{12} as methyl donor; these reactions produce methane, dimethylarsine, methyl mercury, methyl selenide, methyltelluride and S-methyl compounds (Wolfe, 1971; Gunsalus et al., 1976). In elucidating the structure of Coenzyme M, Taylor and Wolfe (1974b) employed a methyltransferase (purified 100-fold) that transferred the methyl group from methyl-B_{12} to Coenzyme M. Whether or not methyl-B_{12} is the natural methyl donor for the enzyme has not been determined, and it has been concluded that there is no evidence that corrinoids or the methyl-transferase reactions are involved in methanogenesis from CO_2 and H_2 (Wolfe, 1979a).

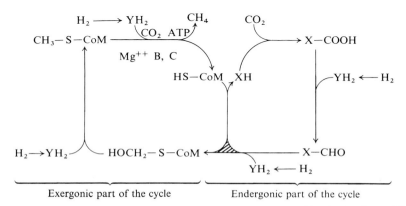

Fig. 53. The reduction of CO_2 to methane. This is an adaptation of Barker's scheme (Fig. 46) to take into account more recent developments; it is a more open, general scheme than that in Fig. 52 (based on Wolfe, 1979a, 1980; Ellefson and Wolfe, 1981). For thermodynamic reasons the reductant (YH_2) cannot be the same in all the reactions. It is produced (directly or indirectly) from H_2 by way of hydrogenase. ATP, Mg^{++}, CO_2 and compound B are "activators" of the reductase (component C).

Recent measurements of corrinoids in methanogens (Krzycki and Zeikus, 1980), and studies of the affects of corrinoid antagonists on the metabolism of methanogens have led, however, to the suggestion that corrinoids may be involved in acetate synthesis and acetate catabolism, although not necessarily involved in methanogenesis itself (Kenealy and Zeikus, 1981).

2. The methyl-coenzyme M reductase complex and the reduction of "XCH_2OH" to CH_4

This enzyme complex catalyses the final step in methanogenesis, which is the reductive demethylation of CH_3-S-CoM (see Table 56 for references and Section B4 for information on Coenzyme M).

Understanding of this reaction has relied heavily on the fact that the proteins from *M. thermoautotrophicum* are stable at room temperature (although O_2-labile), and that the rate of methanogenesis is particularly high at the high temperature optimum of this thermophilic organism (60°C) (Gunsalus and Wolfe, 1980; Ellefson and Wolfe, 1981). The overall reaction with H_2 as reductant is:

$$CH_3\text{-S-CoM} + H_2 \xrightarrow[A, B, C]{ATP\ Mg^{++}} CH_4 + HS\text{-CoM}$$

This reaction is exergonic and the ATP is required as an activator; the mechanism of this activation is not known. Components A and C are protein fractions and component B is a novel O_2-labile, heat resistant, low molecular weight cofactor (Section B6).

Component A is a large protein complex of more than 500 000 daltons that contains hydrogenase as well as other proteins (Gunsalus and Wolfe 1980). The hydrogenase is able to use as electron acceptor, methyl viologen (K_m, 1·5 mM) or the deazaflavin Coenzyme F_{420} (K_m, 2·5 μM), but not NAD(P)$^+$. It is extremely sensitive to oxygen, but insensitive to cyanide and CO. It has a specific activity in *M. thermoautotrophicum* higher than any previously reported from any other bacterial source, and 100 times greater than that in *Methanobacterium* strain G2R (McKellar and Sprott, 1979). Gunsalus and Wolfe (1980) suggest that the large size of the component A fraction might indicate the presence of other enzymes, or electron transport proteins, or both.

Component C, the methylreductase protein, has been purified to homogeneity from *M. thermoautotrophicum* (Ellefson and Wolfe, 1980b). It is an acidic protein with a molecular weight of about 300 000, and is composed of three different subunits (38 500, 48 000, 68 000), which are probably present in an $\alpha_2\beta_2\gamma_2$ arrangement in the native protein. This yellow protein has an absorption maximum at 425 nm and a shoulder at 445 nm; this spectrum is unaffected by addition of oxidants and reductants.

The methylreductase role of component C in the complete methylreductase system has been demonstrated by replacing component A (which includes the hydrogenase) with NADPH—Coenzyme F_{420} oxidoreductase plus NADPH. This combination can itself be replaced by a reduced preparation of Coenzyme F_{420}. This facilitates the analysis of the reductive demethylation reaction but it does not necessarily tell us how CH_3-S-CoM is reduced "*in vivo*". Hydrogen is certainly the origin of the reductant for the process during methanogenesis from $H_2 + CO_2$, but it is not certain that free, soluble Coenzyme F_{420} is the electron donor.

As pointed out by Kell et al. (1981), a high proportion of the energy available from CO_2 reduction to methane must be harnessed in the later stages of reduction ("hydroxymethyl" to "methyl" to "methane"). So the question remains: How is the methylreductase complex coupled to ATP synthesis? By analogy with other electron transport systems, the reduction of CH_3-S-CoM is likely to be arranged in such a way that a proton motive force (pmf) is set up across the membrane (see previous section). In relation to this, it should be noted that although the supernatants after centrifugation are used for studying the methylreductase, these preparations may well contain small membrane vesicles (see Kell et al., 1981 for a discussion of this important point).

In the penultimate step in methanogenesis from CO_2, the C_1 unit at the level of oxidation of formaldehyde is reduced by H_2 to CH_3-S-CoM. The only candidate for the carrier of the hydroxymethyl group is Coenzyme M, and it has been shown by Romesser and Wolfe (1981) that chemically synthesised hydroxymethyl-S-CoM is rapidly reduced to methane by extracts of *M. thermoautotrophicum*. In solution, $HOCH_2$-S-CoM is in equilibrium with HCHO and HS-CoM; and when added to cell extracts, formaldehyde is enzymatically converted to methane at a rate almost equal to that from $HOCH_2$-S-CoM and CH_3-S-CoM. Although highly likely to be an intermediate in reduction of CO_2 to CH_4, neither $HOCH_2$-S-CoM nor free formaldehyde has ever been detected during this process.

3. The RPG effect and the reduction of CO_2 to "X-CH_2OH"

The RPG effect, named after its discoverer R. P. Gunsalus, is the stimulation by CH_3-S-CoM of CO_2 reduction to methane (Gunsalus and Wolfe, 1977). When both CH_3-S-CoM and CO_2 were present in crude extracts of *M. thermoautotrophicum*, the overall rate of CH_4 synthesis increased 30-fold, and for each mole of CH_3-S-CoM added, 12 moles of CH_4 were produced. A heat-stable, dialysable cofactor (Carbon Dioxide Reduction or CDR factor) was found to be essential for reduction of CO_2, but not for reduction of $HOCH_2$-S-CoM or formaldehyde, to methane. So, the CDR factor is thought to function prior to the formaldehyde level of reduction. Even in the absence of CDR, CO_2 stimulates the rate of reduction of CH_3-S-CoM to methane, but in this case only stoicheiometric conversion of CH_3-S-CoM occurs; that is, the RPG effect is abolished. These are obviously critical observations, but their interpretation is not straightforward. This is particularly so in the light of the experiments of Sauer et al. (1980b); these showed that intact vesicles are required for the reduction by H_2 of CO_2 to methane in *M. thermoautotrophicum*. Methyl CH_3-S-CoM stimulated methanogenesis from CO_2, but to nothing like the extent shown by Gunsalus and Wolfe (1977); and, more important perhaps, it was found that the $^{14}CH_3$ group from $^{14}CH_3$-S-CoM did not give rise to $^{14}CH_4$ during methanogenesis from CO_2. Furthermore, in extracts of *Methanosarcina barkeri*, CH_3-S-CoM had no effect, although other HS-CoM derivatives were stimulatory (Hutten *et al.*, 1981).

Gunsalus and Wolfe concluded from their observations that an intermediate, generated in the methylreductase reaction, may be responsible for the activation and subsequent reduction of CO_2 to CH_4. This conclusion, together with the discovery of the CDR factor led to the speculative scheme shown in Fig. 52. This scheme was drawn as a closed cycle in order to emphasise the "possibility of a close connection between the terminal reaction and CO_2 activation". This scheme implies that the carboxylation of the unknown carrier X is intimately coupled to the terminal methylreductase step. This has not yet been proved (in this particular form) and so I have redrawn the scheme in a more general and open form in Fig. 53.

Kell *et al.* (1981) have pointed out that some coupling of the primary and terminal steps in CO_2 reduction to methane is a thermodynamic necessity. That is, the energy available from the reduction of "formaldehyde" to methane must be used, not only to synthesise ATP but also to drive the endergonic reduction of CO_2 to "formaldehyde". Thus, an alternative interpretation of the RPG effect is that the later reactions are coupled to the earlier reaction by way of the "energised membrane", rather than by direct enzyme—substrate interactions (see Section C2).

The nature of the C_1 carrier (X in Figs 46, 52 and 53) is not known. From the little information available on the CDR factor, this might perhaps be the C_1 carrier for the first part of methanogenesis from CO_2. Alternatively, the important observation of Daniels and Zeikus (1978) that ^{14}C-carboxydihydromethanopterin accumulates during incubation of *Methanosarcina barkeri* with $^{14}CH_3OH$ or $^{14}CO_2$, suggests that methanopterin might be the C_1 carrier. In this case, as with all discussions of C_1 carriers in methanogens, there is the difficulty of distinguishing between three potential functions: that of a C_1 carrier in methanogenesis; that of C_1 carrier in biosynthesis from reduced C_1 derivatives and CO_2; and the possibility that different C_1 carriers might sometimes be involved in the oxidation of reduced C_1 compounds to CO_2.

E. Methanogenesis from acetate, methanol, methylated amines and formate

Less is known about biosynthesis, methanogenesis and energy transduction from these substrates than from $H_2 + CO_2$, and most of this information is from studies with *Methanosarcina barkeri* (see Table 57).

All the available evidence on biosynthesis, which is not much, is consistent with assimilation of C_1 substrates (methanol, methylated amines and formate) being by way of a $C_1 + C_1$ condensation to acetate (or acetyl-CoA), and assimilation of this by way of the routes indicated in Fig. 45. Acetate itself is assimilated directly by these routes after conversion to acetyl-CoA (see Section V). Although CH_3-S-CoM is assumed to be an essential intermediate in biosynthesis, as well as in methanogenesis, there is some evidence that other methylated compounds (probably corrinoids) may be involved (Kenealy and Zeikus, 1981; see also Daniels and Zeikus, 1978). Formate is probably oxidised to CO_2 and assimilated as during growth on $CO_2 + H_2$.

TABLE 57
Methanogenesis from acetate, methanol, methylated amines and formate

The metabolism of Methanosarcina barkeri (general papers and reviews): Mah *et al.* (1977, 1978, 1981); Weimer and Zeikus (1978a, b, 1979); Hutten *et al.* (1980, 1981); Walther *et al.* (1981).
Labelling patterns in methane: Stadtman and Barker (1949, 1951); Pine and Barker (1956); Pine and Vishniac (1957); Mah *et al.* (1977, 1981); Walther (1981).
Growth on acetate and on mixed substrates: Smith and Mah (1978, 1980); Baresi *et al.* (1978); Hutten *et al.* (1980).
Formation of methyl-Coenzyme M from methanol: Daniels and Zeikus (1978); Hutten *et al.* (1980, 1981); Hermans *et al.* (1980); Shapiro and Wolfe (1980).
Methanogenesis from methylated amines: Weimer and Zeikus (1978); Hippe *et al.* (1979); Mah *et al.* (1981); Walther *et al.* (1981).
Growth and methanogenesis on formate: Tzeng *et al.* (1975b); Jones and Stadtman (1976, 1977); Jones *et al.* (1977); Schauer and Ferry (1980); Mah *et al.* (1981).
Formate dehydrogenase: Jones and Stadtman (1980, 1981); Jones *et al.* (1979); Schauer and Ferry (1980); Yamazaki and Tsai (1980).

1. Methanogenesis from acetate

All the hydrogen atoms of the methyl group of acetate are incorporated intact into methane; the acetate is not oxidised completely to CO_2, and this CO_2 then reduced to methane (for references see Table 57). CH_3-S-CoM is the likely immediate precursor of methane, and the reductant for this must come from the initial "aceticlastic" reaction of acetate. As pointed out by Mah *et al.* (1981), if every methyl group were to be converted in such a reaction to methane, then no reductant would be available for biosynthesis; this consideration led to their experiments showing that, in fact, about 1% of the methyl groups are oxidised to CO_2 in order (presumably) to provide such reductants.

Although acetate is one of the main sources of methane in non-ruminant environments (about 70% coming from acetate), it is the least "energetic" of the methanogenic substrates.

The overall reaction of methanogenesis from acetate is:

$$CH_3COOH \rightarrow CH_4 + CO_2 \quad \Delta G^{\circ\prime} = -31 \text{kJ (mol } CH_4)^{-1}$$

How the initial aceticlastic reaction occurs, and how ATP is obtained from the overall reaction, is unknown. It has been suggested by Wolfe (1979a) that a free radical mechanism might operate, but it is difficult to see how such a reaction could be involved in establishing the pmf, which is presumably essential to harness the energy of reduction of the methyl group to methane. Wolfe (1979a) has also proposed a speculative model to explain how a methanogen might establish a proton gradient during acetate degradation, and produce methane with the labelling pattern obtained by Pine and Barker (1956). This model appears to be more complex than it need be, and Fig. 54 is an alternative illustration of how the reaction might be arranged so as to translocate protons, and so produce a pmf for use in ATP synthesis.

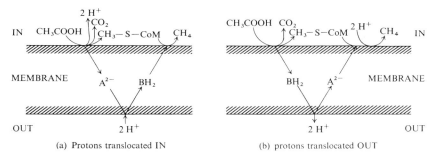

Fig. 54. Possible arrangements of methanogenesis from acetate to produce a pmf across the membrane. Two arrangements are drawn to illustrate that protons might be translocated in either direction; the actual direction will depend on the arrangement of the membrane-bound ATP synthetase. A is a transmembrane electron carrier able to react with the transmembrane hydrogen carrier B.

2. Methanogenesis from methanol

The overall reaction is as follows:

$$4 CH_3OH \rightarrow 3 CH_4 + CO_2 + 2 H_2O \quad \Delta G^{\circ\prime} = -105 \text{ kJ (mol } CH_4)^{-1}$$

The labelling pattern in methane shows that the methyl group of methanol is incorporated intact into methane; the methanol is not all metabolised to $H_2 + CO_2$, and then the CO_2 reduced to methane (Table 57).

The first step in the utilisation of methanol is the formation of CH_3-S-CoM.

This has been shown by radioactive tracer studies (Daniels and Zeikus, 1978), and by the isolation and characterisation of a specific methanol-HS-CoM methyltransferase (Shapiro and Wolfe, 1980):

$$CH_3OH + HS\text{-}CoM \rightarrow CH_3\text{-}S\text{-}CoM + H_2O$$

This transferase is similar to the methylreductase in requiring ATP as an activator. The subsequent reduction of CH_3-S-CoM to methane uses the reductant produced by oxidation of CH_3-S-CoM to CO_2. This oxidation process may occur by way of the same C_1 carriers and reductants as are involved in methanogenesis from $CO_2 + H_2$. The observation that CH_3OH is no longer oxidised to CO_2 when H_2 is also present, is consistent with this suggestion (Hutten et al., 1981). *Methanosarcina barkeri*, the only methanogen able to use methanol, differs from most other methanogens in that CH_3-S-CoM does not stimulate methanogenesis from $CO_2 + H_2$ in cell extracts (the RPG effect), but the significance of this observation is not yet clear (Hutten et al., 1981).

The energy available from the conversion of methanol to methane plus CO_2 is presumably harnessed by way of a pmf, as indicated for the reduction by H_2 of X-CH_3 (see Fig. 51).

3. Methanogenesis from methylated amines

The growth of methanogens on methylated amines has been reviewed by Walther et al. (1981), and other relevant references are listed in Table 57. *Methanosarcina barkeri* is able to grow on methylamine, dimethylamine, trimethylamine and dimethylethylamine and, although unable to grow on choline or betaine, the methyl groups from these compounds can also be converted to methane. The cell yields (per mole of CH_4 produced) are very similar on methylated amines and methanol, as expected from the free energy changes for the overall reactions:

Methylamine:
$$4\,CH_3NH_2 + 2\,H_2O \rightarrow 3\,CH_4 + CO_2 + 4\,NH_3 \quad \Delta G^{\circ\prime} = -75\,\text{kJ}\,(\text{mol}\,CH_4)^{-1}$$

Dimethylamine:
$$2\,(CH_3)_2NH + 2\,H_2O \rightarrow 3\,CH_4 + CO_2 + 2\,NH_3 \,\Delta G^{\circ\prime} = -74\,\text{kJ}\,(\text{mol}\,CH_4)^{-1}$$

Trimethylamine:
$$4\,(CH_3)_3N + 6\,H_2O \rightarrow 9\,CH_4 + 3\,CO_2 + 4\,NH_3 \,\Delta G^{\circ\prime} = -75\,\text{kJ}\,(\text{mol}\,CH_4)^{-1}$$

The methyl groups of the methylated amines are incorporated intact into methane. The cofactors required for methanogenesis from trimethylamine (the best substrate) are the same as required for methanogenesis from methanol, but additional enzyme systems are required and have to be induced in *Methanosarcina barkeri* when trimethylamine serves as substrate (Walther et al., 1981).

ATP synthesis during methanogenesis from methylated amines is presumably dependent on a pmf set up in a manner similar to that operating with methanol as substrate.

4. Methanogenesis from formate

About half the known strains of methanogens are able to use formate (Balch et al., 1979).

The overall reaction for methanogenesis from formate is:

$$4 \text{ HCOOH} \rightarrow \text{CH}_4 + 3 \text{ CO}_2 + 2 \text{ H}_2\text{O} \quad \Delta G^{\circ\prime} = -130.4 \text{ kJ (mol CH}_4)^{-1}$$

Until recently, little was known about growth on formate, except that an unusual formate dehydrogenase is involved. This enzyme is not $NAD(P)^+$-linked. Its electron acceptor is the deazaflavin, Coenzyme F_{420}, and it contains selenium and tungsten, which are also essential for growth on formate (see Table 57 for references).

In a recent important study of formate metabolism in *Methanobacterium formicicum*, Schauer and Ferry (1980) have shown that molar growth yields on formate (per mole CH_4 produced) are very similar to those obtained during growth on $H_2 + CO_2$. This is perhaps to be expected as the free energy changes of the two reactions are almost identical. Both H_2 and formate were used as energy source simultaneously.

It is not known how formate dehydrogenase is coupled to production of methane or to ATP synthesis. Although usually found in the soluble fraction of cells, it appears likely that oxidation of some formate will be coupled directly to proton translocation.

12
The commercial exploitation of methylotrophs

I. The use of methylotrophs for the production of single cell protein (SCP) . 328
  A. Introduction to SCP 328
  B. Single cell protein from methylotrophs 331
    1. Methylotrophic yeasts as a source of single cell protein (SCP) . . 331
    2. Methylotrophic bacteria as a source of SCP 333
II. The use of methylotrophs for the overproduction of metabolites ("fermentation products") 338
  A. Introduction 338
  B. The production of vitamin B_{12} and riboflavin derivatives . . . 339
  C. The production of carboxylic acids and amino acids by methylotrophs . 339
  D. Commercial methanogenesis 342
III. The use of methylotrophs and their enzymes as biocatalysts . . . 342
  A. Enzymic and biological assay systems 342
    1. The bioassay of vitamin B_{12} 342
    2. The enzymic assay of amines and alcohols 343
  B. The oxidation of alcohols 343
  C. The oxidation of hydrocarbons and their derivatives by methane monooxygenase 344
  D. The use of methylotrophs and their enzymes in electro-enzymology and in biofuel cells 348

This short chapter is not intended as a comprehensive, critical review of actual commercial applications of methylotrophs. Rather, it is a brief summary of the biochemical basis of some potential applications together with an introductory guide to the literature.

I. The use of methylotrophs for the production of single cell protein (SCP)

A. Introduction to SCP

One of the major achievements in biotechnology occurred before the word biotechnology was coined. This achievement is the large-scale development of the use of microorganisms as a source of single cell protein (SCP) (Table 58).

TABLE 58
The production of single cell protein (SCP)

(a) *General reviews on SCP* (some include discussion of methylotrophs)
Wilkinson (1971). Hydrocarbons as a source of SCP.
Tannenbaum and Wang (eds) (1975). Single-cell protein.
De Pontanel (ed.) (1975). Proteins from hydrocarbons.
Litchfield (1977). Comparative technical and economic aspects of SCP processes.
Taylor and Senior (1978). Single-cell proteins: a new source of animal feeds.
Mateles (1979). The physiology of SCP production.
Rose (ed.) (1979). Microbial biomass.
Hamer and Harrison (1980). SCP; the technology, economics and future potential.
Hamer and Hamdan (1979). Protein production by microorganisms.

(b) *Reviews of C_1 compounds as sources of SCP*
Cooney and Levine (1972). Microbial utilisation of methanol.
Harrison et al. (1972). Yield and productivity in SCP production from methane and methanol.
Maclennan et al. (1974). Microbial production of protein.
Cooney (1975). Engineering considerations in the production of SCP from methanol.
Barnes et al. (1976). Process considerations and techniques specific to protein production from natural gas.
Cooney and Makiguchi (1977). An assessment of SCP from methanol-grown yeast.
Hamer (1979). Biomass from natural gas.
Krug et al. (1979). SCP from C_1 compounds.
Senior and Windass (1980). The ICI SCP process.
Drozd and Linton (1981). SCP production from methane and methanol in continuous culture.
Smith (1981). Some aspects of ICI's SCP process.
Urakami et al. (1981). Process for producing bacterial cell protein from methanol (the Mitsubishi process).
Faust et al. (1981). SCP production from methanol: production of a high quality product.

330   The biochemistry of methylotrophs

SCP has been mainly considered as a potential replacement for the fish meal and soya bean products used as animal feed protein supplements; but there remains the potential longer-term development of SCP as a protein source for human consumption. The major process routes for SCP production that have been considered by industrial and commercial organisations include:
(a) the growth of yeasts and bacteria on waxy n-alkanes;
(b) the growth of yeasts and bacteria on methanol;
(c) the growth of yeasts on ethanol;
(d) the growth of bacteria on methane and natural gas;
(e) the photosynthetic growth of algae on CO_2;
(f) the growth of bacteria on CO_2 plus H_2;
(g) the growth of fungi on carbohydrates including cellulose;
(h) the growth of yeast and fungi on industrial waste liquors.

In the earlier stages of SCP technology emphasis was almost exclusively on hydrocarbon-derived substrates, but the rising costs of these have shifted the emphasis towards some of the other substrates mentioned above (see Hamer and Hamdan, 1979).

The basic characteristics of a microorganism to be used for production of SCP include the following:
(a) a high yield coefficient on a suitable substrate;
(b) a high growth rate;
(c) the ability to grow to high densities;
(d) a high affinity for the carbon substrate;
(e) a high optimum growth temperature;
(f) stable growth in continuous culture;
(g) no requirement for expensive growth factors;
(h) resistance to contamination;
(i) high nutritional value and absence of toxicity.

Probably the most important of these characteristics is that of high yield; this is important because more than half of the operating cost in SCP production is that of carbon substrate. Besides this obvious point, the yield coefficient has other implications because a lower growth yield on a given substrate leads to a higher oxygen demand. This requires a greater supply of O_2, and the utilisation of this leads to a greater heat output. The cost of aerating and cooling large fermenters constitutes a major proportion of operating costs and must, therefore, be kept to a minimum. It is for this reason that, besides having a high yield coefficient and a low O_2 requirement, it is a great advantage if the organism has a high optimum temperature for growth, so that the cooling requirement is minimal. Because of these interrelated factors it can be shown that a small decrease in yield coefficient can lead to a disproportionately large increase in costs (see Cooney, 1975).

B. Single cell protein from methylotrophs

The two substrates considered here as feedstock for SCP production are methane and methanol (see Table 58b for reviews).

Although methane can be obtained cheaply on a large scale (in some parts of the world), and although it is relatively pure and inexpensive it has some disadvantages. These include its explosive nature, its low solubility, the low yield coefficients on this substrate and the high oxygen requirement for growth. These two last points are both related to the first step in methane oxidation, which requires molecular O_2 plus a reductant (NADH usually), which is produced by the further oxidation of methanol. The amount of O_2 required to oxidise methane to CO_2 is twice that required for methanol oxidation and, in terms of available energy content of the substrate, methane is in effect at the oxidation level of formaldehyde. The biochemistry of this is discussed at some length in Chapter 9.

It is thus concluded that methanol is the substrate of choice for SCP production. It is inexpensive, available in large amounts, very pure, completely miscible with water, easy to store, transport and handle, and any residual methanol after growth can be readily removed from the SCP product. A further advantage is that methanol is less dependent on a single route for its production; although primarily produced from natural gas, it could also be produced from coal, oil shale, cellulose and petroleum. SCP from methanol is thus likely to be influenced less by environmental, climatic and political fluctuations than any other form of SCP.

Many bacteria and yeasts are able to grow well on methanol and so are contenders for the title of most successful commercial methylotroph. The winning contestants at present are bacteria, but yeasts do have some advantages and are still under consideration.

1. Methylotrophic yeasts as a source of single cell protein (SCP)

Interest in the commercial exploitation of yeast as a potential source of single cell protein (SCP) has a considerable history and a number of yeasts have been isolated on methanol and characterised with respect to their potential as SCP (Asthana *et al.*, 1971; Sahm and Wagner, 1972; Levine and Cooney, 1973; Minami *et al.*, 1978). General aspects of the utilisation of yeasts for production of SCP have been discussed in the following valuable reviews: de Pontanel (1975); Cooney and Levine (1972, 1975); Cooney (1975); Tannenbaum and Wang (1975); Harder and van Dijken (1975); Sahm (1977); Litchfield (1977); Cooney and Makiguchi (1977); Tani *et al.* (1978a); Hamer (1979); Levi *et al.* (1979); Mateles (1979); Krug *et al.* (1979); Hamer and Harrison (1980).

Yeasts have a number of general advantages over bacteria as a source of SCP, including their lower nucleic acid content and their greater (psychological) acceptability for human consumption. The lower pH optimum for growth is also an advantage in minimising bacterial contamination. The disadvantages of yeasts include their lower levels of protein, which has a lower methionine content, and their slower growth rates.

With respect to growth on methanol as a source of SCP, yeasts have a number of further disadvantages compared with bacteria. The best bacteria for SCP production have the RuMP pathway of carbon assimilation, and this is not markedly different in terms of energetics from the DHA pathway operating in yeasts. The critical difference between bacteria and yeasts is in the system for initial oxidation of methanol to formaldehyde. In bacteria this is coupled to an electron transport chain, and thus to ATP synthesis, but this is not the case in yeast where a flavoprotein oxidase is responsible for oxidation of methanol. This oxidase has low affinities for both methanol and oxygen (Chapter 10; van Dijken et al., 1976a), and uses oxygen without providing any reductant or ATP for biosynthesis. These characteristics of the oxidase lead to the following generalisations:

(a) The yield of yeast per gram of methanol is lower than in bacteria having the RuMP pathway. This can be predicted on theoretical grounds (Harder and van Dijken, 1975; Anthony, 1980), and has been measured in a range of different yeasts (Levine and Cooney, 1973; Cooney, 1975; van Dijken et al., 1976a; Minami et al., 1978; Egli and Fiechter, 1981); the yields are usually between 0·36 and 0·4 g dry wt (g methanol)$^{-1}$. These compare with yields of up to about 0·54 g (g methanol)$^{-1}$ for bacteria with the RuMP pathway (Goldberg et al., 1976).

(b) The yield per mole of oxygen is likely to be about half that predicted for growth of bacteria by the RuMP pathway (Anthony, 1980); this is because an atom of oxygen is consumed for every molecule of methanol oxidised to formaldehyde, and yet none of the energy from this is harnessed.

(c) The concentration of methanol during methanol-limited continuous culture is high. This is because of the high K_m for methanol of the oxidase (Table 49), which in turn leads to high K_s values for methanol (4 mM to 17 mM) in growing yeasts (Levine and Cooney, 1973; Reuss et al., 1975).

(d) The concentration of oxygen in air-saturated water (about 0·2 mM) is very close to the K_m concentration of oxygen for the oxidase; a high oxygen concentration is therefore required to achieve high growth rates, and indeed this may be the factor limiting the maximum growth rates of yeasts on methanol (van Dijken, 1976; Middlehoven et al., 1976).

These general characteristics lead to the following problems associated with production of yeast SCP from methanol: high concentrations of methanol in the outflowing medium must be retrieved or wasted; oxygen must be provided at a high rate to achieve a high growth rate; extra cooling must be provided to counteract the high heat production due to the high rate of oxygen utilisation (Cooney, 1975).

2. Methylotrophic bacteria as a source of SCP

Methylotrophic bacteria have the advantage over yeasts of higher yield coefficients, growth rates and protein content; and they are the organisms chosen by Imperial Chemical Industries (ICI) (UK), the Mitsubishi Gas Chemical Company (MGC) and Norprotein (Scandinavia) as the basis for investigations and development of methanol-based SCP processes. The production of SCP must be on a large scale to be commercially viable, and the production capacity of a plant must be in the region of 100 000 tonnes per year. The only way of achieving this is by continuous culture of the SCP organisms. This has required specially designed fermenters and the solution of many new problems related to the vast scale of the projects. The story of the development of the ICI "Pruteen" process has been particularly well told by Senior and Windass (1980) and clearly reviewed by Smith (1981). Their pressure-cycle fermenter and simple summaries of the overall process are illustrated in Figs 55 and 56.

The choice of organism for the SCP process is obviously critical, and a wide range of bacteria has been considered (Maclennan *et al.*, 1974; Urakami and Komagata, 1979). Because, as discussed above, the growth yield on methanol must be high, the bacteria of choice are those having the RuMP pathway of carbon assimilation. The relationship between growth yields and the assimilation pathways is discussed fully in Chapter 9 and the reviews in Table 59. The bacterial strain chosen by ICI for their SCP process is *Methylophilus methylotrophus*. It has a high optimum growth temperature (about 40°C), grows rapidly on methanol (μ_{max} is about 0·55 h^{-1}) and it produces a high yield. Because it is to some extent ATP-limited, any increase in ATP yield from methanol oxidation, or decrease in ATP requirement, might lead to an increase in cell yields. Such a valuable increase has been achieved by a remarkable example of "genetic engineering" by the ICI group. They have eliminated the energetically sub-optimal glutamine synthetase/glutamate synthase pathway (the GOGAT pathway) by mutation, and replaced it with the more favourable glutamate dehydrogenase pathway, by introducing the *E. coli* glutamate dehydrogenase gene cloned in a broad host-range plasmid vector (Windass *et al.*, 1980; Holloway, 1981; Chapter 4, page 262; Chapter 9, page 130). The newly incorporated dehydrogenase route presumably leads to a higher K_s for ammonia than the alternative pathway, but this will be irrelevant in a carbon-limited culture, arranged so that the effluent growth medium is recirculated to avoid wastage of the excess ammonia necessary to saturate the ammonia assimilation system.

Fig. 55. The ICI "Pruteen" plant at Billingham, UK. The centre tower is the 1.5 million litre fermenter. A flow-diagram of the process is illustrated in Fig. 56.

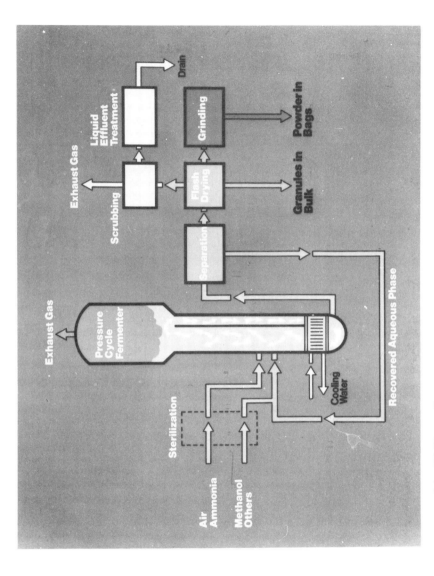

Fig. 56. Flow diagram of the ICI "Pruteen" process.

TABLE 59
The measurement of growth yields of methylotrophs

(a) *Theoretical discussions of growth yields on C_1 compounds* (see Chapter 9)

Van Dijken and Harder (1975). Growth yields of microorganisms on methanol and methane. A theoretical study.
Harder and van Dijken (1975). A theoretical study of growth yields of yeasts on methanol.
Harder and van Dijken (1976). Theoretical considerations on the relation between energy production and growth of methane-utilising bacteria.
Stouthamer (1977b). Theoretical calculations on the influence of the inorganic nitrogen source on parameters for aerobic growth in microorganisms.
Anthony (1978b). The prediction of growth yields in methylotrophs.
Anthony (1980). Methanol as substrate; theoretical aspects.
Krug et al. (1979). SCP from C_1 compounds.
Drozd and Linton (1981). SCP production from methane and methanol in continuous culture.
Harder et al. (1981). Utilisation of energy in methylotrophs.
Tempest and Neijssel (1981). Comparative aspects of microbial growth yields with specific reference to C_1-utilisers.

(b) *Investigations of growth rates and yields on C_1 compounds, and physiological factors affecting them*

Maclennan et al. (1971). The influence of dissolved O_2 on *Pseudomonas* AM1 grown on methanol in continuous culture.
Harrison (1973). Studies on the affinity of methanol- and methane-utilising bacteria for their C_1 substrates.
Goldberg et al. (1976). Bacterial yields on methanol, methylamine, formaldehyde and formate.
Linton and Buckee (1977). Interactions in a methane-utilising mixed bacterial culture in a chemostat.
Brooks and Meers (1973). The effect of discontinuous methanol addition on the growth of a carbon-limited culture of *Pseudomonas* (*Methylophilus methylotrophus*).

Rokem et al. (1978a). Maintenance requirements for bacteria growing on C_1 compounds.

Linton and Vokes (1978). Growth of methane-utilising bacteria in mineral salts medium with methanol as sole source of carbon.

Minami et al. (1978). A new methanol-assimilating, high productivity, thermophilic yeast.

Hirt et al. (1978). Formaldehyde incorporation by a new methylotroph (L3).

van Verseveld and Stouthamer (1978). Growth yields and the efficiency of oxidative phosphorylation during autotrophic growth of *Paracoccus denitrificans* on methanol and formate.

Drozd et al. (1978). An *in situ* assessment of the specific lysis rate in continuous culture of *Methylococcus* sp. grown on methane.

Keevil and Anthony (1979b). Effect of growth conditions on the involvement of cytochrome *c* in electron transport, proton translocation and ATP synthesis in the facultative methylotroph *Pseudomonas* AM1.

Drozd and Wren (1980). Growth energetics in the production of bacterial SCP from methanol.

Tsuchiya et al. (1980). Medium optimisation for a methanol-utilising bacterium based on chemostat theory.

Rokem et al. (1980). Growth of mixed cultures of bacteria on methanol.

Swartz and Cooney (1981). Methanol inhibition in continuous culture of *Hansenula polymorpha*.

Egli and Fiechter (1981). Theoretical analysis of media used in the growth of yeasts on methanol.

Drozd and Linton (1981). SCP production from methane and methanol in continuous culture.

As a high yield is essential for high productivity and profitability, any physiological and environmental factors affecting yields must be identified and controlled. Although this topic is not covered in detail here, Table 59 gives references to a range of important papers on physiological factors affecting growth yields in methylotrophic microorganisms. It is, of course, necessary to grow the organism under carbon-limitation. On a large scale, however, true carbon-limitation is not readily achieved because of the relatively long mixing times compared with lab-scale fermenters. In the ICI process, the lower yields arising from local high concentrations of methanol within the fermenter are avoided by introducing the methanol through many (about 3000) small nozzles throughout the fermenter which is 1500 M^3 (1·5 million litres) in volume (Brooks and Meers, 1973; Senior and Windass, 1980).

A final point that should be mentioned is that, whereas the use of a single organism for SCP production has many advantages, mixed cultures are sometimes more stable and may therefore be better for some purposes (see Wilkinson and Harrison, 1973; Wilkinson et al., 1974; Snedecor and Cooney, 1974; Linton and Buckee, 1977; Harrison, 1978; Bull and Brown, 1979).

II. The use of methylotrophs for the overproduction of metabolites ("fermentation products")

A. Introduction

The same factors that make methanol attractive as a substrate for production of SCP make it attractive as a potential source of metabolites such as keto acids, carboxylic acids, amino acids, purines and vitamins. Production by microbial metabolism of methanol of large amounts of solvents, such as butanol or acetone, is not possible because such processes depend on the use of anaerobic fermentation. Metabolite overproduction on methanol is more akin to the process used for amino acid production by corynebacteria growing on carbohydrates (for reviews of such processes see Hughes and Rose, 1971; Rose, 1978; Bull et al., 1979; Harrison et al., 1980).

Perhaps in a separate class is the possibility of production of the storage compound poly 3-hydroxybutyrate as a precursor for the chemical synthesis of polymers for use in the plastics industry. Many methylotrophs produce poly 3-hydroxybutyrate, and its content in cells can be dramatically increased by growth in excess-carbon (usually low-nitrogen) conditions. Analogous to this is the possibility of using methylotrophs for polysaccharide production. Hou et al. (1978) have shown that the obligate methanotroph *Methylocystis parvus* can be adapted to grow rapidly on methanol (μ_{max}, 0·65 h^{-1}).

and that during this growth it produces 62% of its dry weight as polysaccharide.

B. The production of vitamin B_{12} and riboflavin derivatives

Vitamin B_{12} is usually produced during fermentation of carbohydrates, using bacteria such as *Propionibacterium shermanii*. The possibility of producing a cheaper source of this vitamin led Tanaka *et al.* (1974) to investigate its production from methanol by methylotrophs. They found that a pink facultative methylotroph, isolated by them (organism FM02T), produced vitamin B_{12} at a concentration of 2·6 mg l^{-1} under suitable growth conditions (Toraya *et al.*, 1976). This is about 20 times the concentration produced by another methylotroph (Klebsiella sp. 101) (Nishio *et al.*, 1975a, b). It has been suggested that the role of vitamin B_{12} derivatives in the pink facultative methylotrophs such as *Protaminobacter ruber* may be in a B_{12}-dependent methionine synthetase system or in a methylmalonyl-CoA mutase reaction (Sato *et al.*, 1976, 1977; Ueda *et al.*, 1981).

The high concentration of flavoprotein alcohol oxidase produced during growth of yeasts on methanol has drawn attention to the possibility of using methylotrophic yeasts as a source of flavin derivatives. During induction of the alcohol oxidase the enzymes for conversion of riboflavin to FAD are also produced at a higher level (see Chapter 10), and these high levels can be used to catalyse production of up to 45 μg ml^{-1} of FAD in the presence of added FMN and AMP (Shimizu *et al.*, 1977c).

C. The production of carboxylic acids and amino acids by methylotrophs

So far, there is little published work on the use of methanol to replace the more conventional carbohydrate substrates for the microbial production of carboxylic acids and amino acids. Production of some of these by methylotrophic yeasts have been described, but the concentrations achieved were relatively low (0·3–1·5 gl^{-1}) (cited in Sahm, 1977). Likewise, the production of citrate by *Candida boidinii* growing on methanol gave a low yield (1 g citrate l^{-1}) compared with the amounts produced by yeasts growing on carbohydrate or alkane substrates (Miall, 1980), and it was also necessary to add fluoroacetate for citrate to accumulate during growth with methanol (cited in Sahm, 1977).

There is relatively more information on the potential production of amino acids by methylotrophic bacteria, but the yields are not yet up to those achieved from the more conventional substrates.

Methylotrophic microorganisms with appropriate characteristics for production of amino acids may be freshly isolated from natural sources, or mutants may be isolated by procedures designed to select mutants with appropriate lesions that might lead to metabolite overproduction. Three main types of mutant may be isolated in this way: mutants with enzymes missing from catabolic pathways; auxotrophic mutants with growth factor requirements; mutants with altered regulatory enzymes. Examples of all four approaches are given below. After these initial selection procedures, it is also necessary to explore further properties of the isolated bacteria and mutants to achieve maximum production of metabolites. Furthermore, once optimum conditions are established it might be possible to use non-growing immobilised bacteria for the production process.

Although bacteria assimilating methanol by way of the RuMP pathway are the best choice for production of single cell protein, for purposes of metabolite overproduction the serine pathway bacteria may sometimes be more suitable. This is because they are usually facultative methylotrophs, and hence have the complement of TCA cycle enzymes in the high concentrations required for growth on multicarbon compounds. Intermediates of this cycle are themselves sometimes valuable (such as fumarate and citrate), and they also provide the carbon skeletons for production of many amino acids.

Mutants of the serine-pathway bacterium *Pseudomonas* AM1 are readily isolated with lesions in 2-oxoglutarate dehydrogenase; because a complete TCA cycle is not essential for growth on methanol such mutants are still able to grow well on this substrate and have high levels of other TCA cycle enzymes (Taylor and Anthony, 1976a; Bolbot and Anthony, 1980b). During growth on methanol small amounts of amino acids are released into the growth medium (up to 6 mg l^{-1}). It is likely that further lesions would have to be produced in regulatory proteins to achieve greater amounts of these products. The same logic that applies to the use of TCA cycle mutants also applies to the use of mutants lacking pyruvate dehydrogenase; such mutants of *Pseudomonas* AM1 have been isolated (Bolbot and Anthony, 1980b) and these have been shown to accumulate alanine and valine during growth with methanol.

Having suggested why bacteria with the serine pathway might be better than others for overproduction of amino acids, it is obvious that pathways for the biosynthesis of all amino acids must be present in the obligate methylotrophs having the alternative carbon assimilation pathway, and the first reports of production of L-glutamate by methanol-utilising bacteria considered using strains of the obligate methylotroph *Methanomonas methylovora* (M12–4 and M8–5) which produced up to 1 g l^{-1} of L-glutamate (Kouno *et al.*, 1972; Oki *et al.*, 1973). More recent reports have described a number of different mutant strains of methylotrophs which support production of higher yields of 7–12 g l^{-1} (Nakayama et al., 1976). These strains were derived from methanol-utilisers that were already able to produce small amounts of L-glutamate (at least 100 mg l^{-1}). L-glutamate-producing mutants had at least one of the following characteristics: (a) a requirement for L-methionine; (b) a requirement for L-isoleucine; (c) a requirement for L-phenylalanine; (d) resistance to DL-lysine. L-Glutamate is used as a seasoning agent in the food industry and at present is produced in vast amounts from carbohydrates using corynebacteria (Demain, 1971).

An interesting potential use of serine-pathway bacteria relies on the fact that the first step in the assimilation of methanol in these bacteria is catalysed by serine transhydroxymethylase which adds formaldehyde to glycine to give serine. This was exploited by Keune et al. (1976), using the pink facultative methylotroph *Pseudomonas* 3ab, who showed that yields of up to 4·7 g l^{-1} could be achieved. Because glycine was inhibitory to growth, it had to be added at the end of the exponential phase of growth. This problem was overcome by Tani et al. (1978a, b) who selected a glycine-resistant strain from soil (*Arthrobacter globiformis* SK-200) as a suitable strain for serine overproduction. A problem in attempting such overproduction is that the regulatory systems of bacteria are designed to prevent excessive production. In this case the prevention of accumulation of serine by the internal inhibitor methionine was avoided by selecting a methionine-requiring auxotroph from the glycine-resistant strain: hence the internal methionine concentration could be diminished by controlling the amounts of methionine added to the growth medium. By this method L-serine could be produced at 5·2 g l^{-1}.

The overproduction of serine described here is an unusual case in that serine is an intermediate in the main carbon assimilation pathway. Alternative types of mutant can be used for production of amino acids that are end-products rather than metabolic intermediates. The amount of amino acid produced is regulated by end-product inhibition. This regulatory system must somehow be modified to achieve overproduction of amino acids. Isolation of mutants in which this has been achieved has depended on the fact that analogues of the end-product amino acid, while unable to satisfy the cell's requirements, are able to inhibit the regulatory enzyme and thus act as growth inhibitors. Analogue-resistant mutants often have altered regulatory enzymes, which are insensitive to both the analogue and its natural end-product. Such mutants have been isolated from species of *Methylomonas* (an obligate methanol-utiliser). Some of these are resistant to the valine analogue valine hydroxamate, and these overproduce L-leucine (800 mg l^{-1}) and L-valine (2·2 g l^{-1}) (Ogata et al., 1977; Izumi et al., 1977). Others were resistant to aromatic amino acid analogues, and overproduce phenylalanine (4 g l^{-1}), *L*-tyrosine (1·1 g l^{-1}) and L-tryptophan (200 mg l^{-1}) (Suzuki et al., 1977b). The fact that methylotrophs do not degrade these aromatic amino acids may well be an advantage in using them for production of these compounds.

Although I know of no published work on microbial production of glycine, it is possible that methylotrophs having the serine pathway might be used as a means of overproduction of this compound which is used as a flavour enhancer, and is being used on an increasingly large scale in soft drinks to mask the bitter components of the taste of the sweetener saccharine.

D. Commercial methanogenesis

As discussed in Chapter 11, some methanogers are honorary methylotrophs, and so this subject has to be mentioned even if rather cursorily. Methanogenesis is the last phase in the anaerobic degradation of organic matter, and so methane is the final product of most sewage treatment processes. Conversion of carbohydrate to methane in such systems results in loss of only about 10% of the calorific value of the starting material. In modern plants, the methane is used to generate electricity. The main potential for methane production lies in the degradation of waste material, where the costs of building and operating the system are largely offset by the need to dispose of the waste. Although mainly done on a large scale by industrial or civic corporations, there is considerable potential for development of smaller-scale domestic digesters. Reviews of some commercial aspects of methanogenesis have been published (Hobson *et al.*, 1981; Hungate, 1977; Ghosh and Klass, 1978).

III. The use of methylotrophs and their enzymes as biocatalysts

As a glance at the main headings in this book will indicate, the diversity of the biochemistry of methylotrophs is most impressive, embracing not only novel pathways but many novel enzymes. It is this fact. together with the availability of their inexpensive substrates, that make methylotrophs attractive as potential commercial or industrial biocatalysts.

A. Enzymic and biological assay systems

1. The bioassay of vitamin B_{12}

Relatively few methylotrophs are auxotrophs and so their use in bioassay systems is very limited. One particularly interesting example, however, is the use of a marine methanol-utilising bacterium for bioassay of vitamin B_{12}. Coastal fisheries are susceptible to the harmful effects of abnormal propagations of marine algae called "red tides". As these algae require vitamin B_{12} for growth, the presence of substantial amounts of vitamin B_{12} in some seawater was suspected to be an essential factor in their occurrence. The more conventional microbial assays are unsuitable for use with low concentrations of B_{12} in saline solutions and procedures with alternative organisms are slow and complex. This problem has been solved by Yamamoto et al. (1979) who have developed a method for assaying B_{12} using a marine methanol-utilising bacterium (strain YK4042). The bioassay is rapid (less than 24 h) and the effective range is 0–3 ngl^{-1}. This is ten times the sensitivity of the usual bioassay using *Lactobacillus leichmanii*, and the assay is also more specific.

2. The enzymic assay of amines and alcohols

The first of these methods is for the specific microestimation of methylamine, ethylamine and *n*-propylamine using the primary amine dehydrogenase from methylamine-grown *Pseudomonas* AM1 (Chapter 7, Section V.A) (Large et al., 1969). The method is suitable for 10–140 nmoles of amine. The method depends on coupling the oxidation of amine to the enzymic reduction of 2,6-dichlorophenolindophenol. We have found this assay system is rapid and sensitive even in the presence of a wide range of nitrogen-containing metabolites (Cook and Anthony, 1978).

The second of these assays is for the microestimation of trimethylamine (Large and McDougall, 1975). The method is preferable to some of the alternatives which are unspecific, insensitive or involve unpleasant reagents or extraction into organic solvents. Although the GLC method is probably preferable, this enzymic method is very sensitive (10–140 nmoles) and can be used in the presence of methylamine or trimethylamine *N*-oxide. The method is based on the trimethylamine dehydrogenase isolated from *Hyphomicrobium vulgare* (see Chapter 7, Section III.A). Like that for the primary amines, this method depends on coupling the oxidation of trimethylamine to the reduction of 2,6-dichlorophenolindophenol. The enzyme is stable and easily prepared, and calibration curves and standards are not essential.

A method for the detection and assay of primary alcohols bas been developed by Guilbault (1970). This is based on yeast alcohol oxidase and is specific for primary alcohols.

B. The oxidation of alcohols

Although chemical methods for the large-scale oxidation of alcohols are available, these tend to be unspecific and require extreme physical conditions. For some applications, more gentle specific methods may be more suitable.

One obvious candidate forming the basis for such methods is the methanol dehydrogenase present in all methanol-grown bacteria (Chapter 6, Section III) The enzyme oxidises only primary alcohols but it oxidises a very large. though well-defined, range of these at high rates. The enzyme is easy to prepare in large amounts and is stable. Most of the alcohols that are oxidised by the isolated enzyme are also oxidised at high rates by whole cells (Anthony and Zatman, 1965), and these may be a more suitable catalyst for such oxidations.

A second possibility is that of using the alcohol oxidase of yeasts for industrial purposes. This enzyme oxidises a small range of primary aliphatic alcohols using molecular O_2 and producing H_2O_2 (Chapter 10, Section II.A). It is produced in large amounts in methanol-grown yeasts and is located, together with catalase, in organelles called peroxisomes. The enzyme bas been immobilised on solid phase materials (Baratti et al., 1978), as have isolated peroxisomes (Tanaka et al., 1977, 1978), and whole yeast cells containing the oxidase (Couderc and Baratti, 1980b). Such immobilised preparations may be used for the oxidation of alcohols or, in the absence of catalase activity (by removal or inhibition), for production of H_2O_2.

The third enzyme, whose activity in oxidising alcohols might be exploited, is the secondary alcohol dehydrogenase from methanotrophic bacteria. The methane monooxygenase of these bacteria oxidise n-alkanes to a mixture of the corresponding primary and secondary alcohols (substituted at the C-2 position) (Chapter 6, Section II.F; Patel et al., 1980a; Hou et al., 1979a, 1980b). In some methanotrophs the secondary alcohol may be oxidised to the corresponding methyl ketone by an NAD^+-linked secondary alcohol dehydrogenase. This enzyme is also found in other (non-methanotrophic) methylotrophs, and has been purified from bacteria (Hou et al., 1979c), and from methylotrophic yeasts (Patel et al., 1979d, e). The oxidation of the best substrate, 2-butanol, to 2-butanone has been most extensively investigated, and is reviewed by Hou et al. (1980a).

C. The oxidation of hydrocarbons and their derivatives by methane monooxygenase

All methanotrophic bacteria oxidise their growth substrate by way of methane monooxygenase. The nature of the reaction catalysed by this enzyme and its substrate specificity is discussed extensively in Chapter 6 (Section II.F). The broad specificity of the monooxygenase is the basis of a potential industrial exploitation of this enzyme for transformation of hydrocarbons produced in the petroleum industry; and this potential is the subject of a number of patent applications (e.g. Dalton, 1977b; Higgins, 1978, 1979b; Hou et al., 1979d). In order to insert an oxygen atom into a substrate,

12. Commercial exploitation 345

a reductant (usually NADH) is also required, and the cost and instability of this precludes the utilisation of the oxygenase in an isolated system. Whole cells must therefore be used, in which NADH is regenerated by further metabolism of the initial hydroxylation product, or by metabolism of a second oxidisable substrate (e.g. methanol or formaldehyde), or by oxidation of internal storage polymers. The range of substrates oxidised by whole cells is more limited than for the isolated oxygenase, and some bacterial strains (e.g. *Methylosinus* sp.) have a wider specificity than others (e.g. *Methylococcus* sp.) (see Table 60). The problem of substrate range and choice of organism has been extensively discussed by Dalton (1980a, b, 1981a) and by Higgins *et al.* (1979, 1981a, b), and the general advantages and disadvantages of using whole cells for biological transformations clearly reviewed by Drozd (1980). In some of the examples shown in Table 60 (especially for *M. trichosporium*), the transformations include more than simple, single-step oxidations. The formation of aldehydes and acids must involve other oxidoreductases, which in some cases will be the wide-specificity methanol dehydrogenase. In some of the transformations shown, the recently-discovered secondary alcohol dehydrogenase is probably involved (see above). It is suggested by Higgins *et al.* (1981a) that dechlorination of substituted aromatic compounds, and shifts in the positions of ring substituents, may be manifestations of the mechanism of the methane monooxygenase, or that they may involve other enzyme activities. In some cases dechlorination is associated with covalent attachment of a methyl or alkanoate group to an aromatic ring, and it is suggested that this may be consistent with a free radical mechanism for the oxygenase as proposed by Hutchinson *et al.* (1976).

A transformation of particular potential importance to industry which is catalysed by methane monooxygenase in whole cells of methylotrophs is the epoxidation of propylene (propene) to the epoxide (1,2-epoxypropane) which is excreted:

$$CH_3\text{-}CH=CH_2 + O_2 + NADH + H^+ \rightarrow CH_3\text{-}\underset{\underset{O}{\diagdown\diagup}}{CH\text{-}CH_2} + H_2O + NAD^+$$

In relation to this possibility, many new methanotrophs have been isolated by Hou *et al.* (1979b, c) and shown to have the ability to catalyse the epoxidation of propylene. The longer-chain terminal alkenes are also oxidised to the corresponding 1,2-epoxides. These are industrially important because of their ability to polymerise under thermal, ionic and free radical catalysis, to form epoxy homopolymers and copolymers. The use of methanotrophs to catalyse the production of epoxides from alkenes is potentially very important, because most epoxide production depends at present on processes

TABLE 60
The oxidation of hydrocarbons by whole cells of methanotrophs

Substrate	Products from *Methylococcus capsulatus*	Products from *Methylosinus trichosporium*
Methane	Methanol	Methanol
Carbon monoxide	Carbon dioxide	
Ethane	Ethanol	Acetaldehyde, acetate, acetone
Propane	Propan-1-ol	
Butane	Butan-1-ol	
Hexane	Hexan-1-ol, hexan-2-ol	
Tetradecane	Tetradecan-1-ol	
Hexadecane		Hexadecan-1-ol
Ethene	Epoxyethane	Epoxyethane
Propene	Epoxypropane	Epoxypropane
But-1-ene	1,2-epoxybutane	
Styrene	Styrene epoxide	Styrene epoxide
Pyridine		Pyridine *N*-oxide
1-Phenylbutane		Benzoic acid, 4-phenyl butanoic acid, other neutral products
1-Phenylheptane		1-Hydroxy-1-phenylheptane, 1-phenylheptan-7-al, 1-oxo-1-phenylheptane, cinnamic acid, benzoic acid
m-Chlorotoluene		Benzyl alcohol, benzyl epoxide, *m*- or *p*-methyl benzyl alcohol
m-Chlorophenol		Probably hydroxyphenylheptanoic acid
m-Cresol		*m*- and *p*-hydroxybenzaldehyde
o-Cresol		5-Methyl-1, 3-benzene diol
cis-But-2-ene	*cis*-2,3-Epoxybutane; *cis*-2-Buten-1-ol	2,3-Epoxybutane

trans-But-2-ene	trans-2,3-Epoxybutane; trans-2-Buten-1-ol
Butadiene	3,4-Epoxy-1-butene
Cyclohexane	Cyclohexanol, 3-hydroxycyclohexanone
Cyclohexanol	3-Hydroxycyclohexanone
Benzene	Phenol
Phenol	Catechol, quinol
Toluene	Benzyl alcohol, cresol	Benzoic acid, p-cresol
Ethylbenzene	Benzoic acid, 2-phenylethanol, p-hydroxyethylbenzene; phenylacetic acid
Isopropylbenzene	p-Hydroxy isopropylbenzene

This information is taken from the reviews of Dalton (1980a, b) and Higgins et al. (1981a).

requiring multiple steps, and which depend heavily on the market price of by-products such as styrene (for a review of the microbial epoxidation of gaseous 1-alkenes see Hou et al., 1980a).

D. The use of methylotrophs and their enzymes in electro-enzymology and in biofuel cells

Redox reactions involve electron transfer from one molecule to another, and are thus made up of two half reactions which either donate or accept electrons. These two half reactions may be kept physically separate, but coupled together by way of electrodes and an external circuit. Assuming that electrodes can be found that will react with the redox couples at sufficiently high rates, and assuming that the difference in redox potentials between the two couples is sufficiently great, then the external circuit can carry a load and we have a fuel cell. If the oxidising and reducing reactions are catalysed by enzymes or by whole microorganisms, then the fuel cell is an enzymic or microbial fuel cell (see Fig. 57).

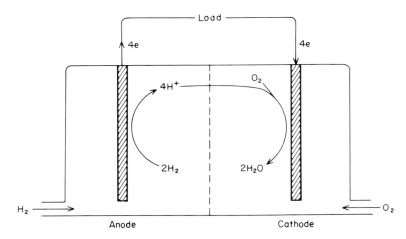

Fig. 57. A simple hydrogen-oxygen fuel cell. Oxidation occurs at the anode and reduction at the cathode, the two electrodes being separated by a membrane. As the hydrogen passes over the anode surface it is electrochemically oxidised to H^+ which migrate to the cathode where oxygen is reduced to water. The electrons released at the anode flow through the external circuit. If the anode compartment contains hydrogenase able to react at the electrode, and the cathode compartment contains cytochrome oxidase (for example) then this becomes an enzymic fuel cell.

If the enzymically-catalysed reduction reaction can use electrons provided by a cathode, or if the oxidation reaction can donate its electrons to an anode, then we have an electroenzymological process. If microorganisms are used instead of enzymes for such a process, driven by an electrical current, then we have a microbial "reactor". All these possibilities are being vigorously investigated and indeed some biofuel cells have been known and used for many years. For reviews of this topic see Bockris and Srinivasan (1969), Allen, (1972), Higgins and Hill (1979) and Higgins et al. (1980a).

The limiting factor in developing electroenzymology, and microbial and enzymic fuel cells, is the problem of the interaction between protein catalysts and the electrodes. This is overcome (to some extent) by using low molecular weight intermediary electron carriers, such as methylene blue, phenazine methosulphate or viologen dyes, or by modifying the electrodes by binding molecules (e.g. 4,4'-bipyridyl) which promote electron transfer (Higgins et al., 1980a). The same problems of mediation have led to the conclusion that biofuel cells and reactors using whole microbes are not practicable, but recent developments in the use of mediators such as 2,6-dichlorophenolindophenol between oxidising reactions in the bacteria and the platinum electrode of the fuel cells have been more encouraging (Bennetto et al., 1980).

Methylotrophs and their enzymes have the same potential importance in electroenzymology and biofuel cells as they have for other processes mentioned previously. One obvious possibility is to use methanol and methanol dehydrogenase in an enzymic fuel cell, and this has been achieved by Plotkin et al. (1981). In their methanol dehydrogenase bioelectrochemical cell, the anode contains the enzyme in buffer at its optimum pH (9·5), and the mediator between enzyme and the platinum electrode is phenazine ethosulphate. Oxygen-free nitrogen is passed through the anode compartment which is separated from the cathode compartment by a cation exchange membrane. The cathode is also platinum, over which O_2 is passed. Besides acting as an enzymic fuel cell, the system can also be used for the assay of tiny amounts of any primary alcohols that can be oxidised by the dehydrogenase; the usable range is about 100 pmoles to 100 nmoles. If suitable electrodes and mediators could be found, then perhaps whole cells oxidising methanol could be used in the anode compartment. Even more complex biofuel cells have been suggested by Higgins and Hill (1979); in these the oxidation of methane to methanol in the cathode compartment is coupled to the oxidation of methanol to formate in the anode compartment.

An example of an enzymic reactor, also proposed by Higgins et al. (1980a), is a reactor in which methane monooxygenase catalyses the oxidation of alkanes or alkenes by O_2, the reductant for the reaction being provided as electrons directly at a cathode, instead of the usual, expensive NADH.

Clearly the possibilities in this field are many and exciting, provided that the fundamental problems of mediation between electrodes and enzymes or microbial cells are overcome.

References

Akiba, M. M., Veyama, H., Seki, M. and Fukimbara, T. (1970). *J. Ferment. Technol.* **48**, 323.
Alefounder, P. R. and Ferguson, S. J. (1981). *Biochem. Biophys. Res. Comm.* **98**, 778–784.
Allen, M. J. (1972). *In* "Methods in Microbiology" (J. R. Norris and D. W. Ribbons, eds), Vol. 6B, pp. 247–283. Academic Press, London and New York.
Amano, Y., Sarvada, M., Takada, N. and Terui, G. (1975). *J. Ferment. Technol.* **53**, 315–326.
Ameyama, M., Matshita, K., Ohno, Y., Shinagawa, E. and Adachi, O. (1981). *FEBS Lett.* **130**, 179–183.
Andrews, T. J. and Lorimer, G. H. (1978). *FEBS Lett.* **90**, 1–9.
Andrews, T. J., Abel, K. M., Menzel, D. and Badger, M. R. (1981). *Arch. Microbiol.* **130**, 344–348.
Anthony, C. (1975a). *Sci. Progr. (London)* **62**, 167–206.
Anthony, C. (1975b). *Biochem. J.* **146**, 289–298.
Anthony, C. (1978a). *Proc. Soc. Gen. Microbiol.* **5**, 67.
Anthony, C. (1978b). *J. Gen. Microbiol.* **104**, 91–104.
Anthony, C. (1980). *In* "Hydrocarbons in Biotechnology" (D. E. F. Harrison, I. J. Higgins and R. Watkinson, eds), pp. 35–57. Heyden, London.
Anthony, C. (1981). *In* "Microbial growth on C_1 compounds" (H. Dalton, ed.), pp. 220–230. Heyden, London.
Anthony, C. and Taylor, I. J. (1975). *Proc. Soc. Gen. Microbiol.* **4**, 67.
Anthony, C. and Zatman, L. J. (1964a). *Biochem. J.* **92**, 609–614.
Anthony, C. and Zatman, L. J. (1964b). *Biochem. J.* **92**, 614–627.
Anthony, C. and Zatman, L. J. (1965). *Biochem. J.* **96**, 808–812.
Anthony, C. and Zatman, L. J. (1967a). *Biochem. J.* **104**, 953–959.
Anthony, C. and Zatman, L. J. (1967b). *Biochem. J.* **104**, 960–969.
Aperghis, P. N. G. and Quayle, J. R. (1981). Abstract of the Third International Symposium on Microbial growth on C_1 compounds.
Asai, Y., Makiguchi, N., Shimada, M. and Kurimura, Y. (1976). *J. Gen. Appl. Microbiol.* **22**, 197–202.
Ashton, W. T., Brown, R. D., Jacobson, F. and Walsh, C. (1979). *J. Amer. Chem. Soc.* **101**, 4419–4420.
Asthana, H., Humphrey, A. E. and Moritz, V. (1971). *Biotechnol. Bioeng.* **13**, 923–928.
Attwood, M. M. and Harder, W. (1972). *Antonie van Leeuwenhoek* **38**, 369–378.
Attwood, M. M. and Harder, W. (1974). *J. Gen. Microbiol.* **84**, 350–356.
Attwood, M. M. and Harder, W. (1977). *FEMS Microbiol. Lett.* **1**, 25–30.
Attwood, M. M. and Harder, W. (1978). *FEMS Microbiol. Lett.* **3**, 111–114.
Austin, B. and Goodfellow, M. (1979). *Int. J. Syst. Bacteriol.* **29**, 373–378.
Austin, B., Goodfellow, M. and Dickinson, C. H. (1978). *J. Gen. Microbiol.* **104**, 139–155.
Babel, W. (1979). *Z. Allg. Mikrobiol.* **19**, 671–677.
Babel, W. and Hoffmann, K. (1975). *Z. Allg. Mikrobiol.* **15**, 53–57.

Babel, W. and Loffhagen, N. (1977). *Z. Allg. Mikrobiol.* **17**, 75–79.
Babel, W. and Loffhagen, N. (1979). *Z. Allg. Mikrobiol.* **19**, 299–302.
Babel, W. and Miethe, D. (1974). *Z. Allg. Mikrobiol.* **14**, 153–156.
Babel, W. and Mothes, G. (1978). *Z. Allg. Mikrobiol.* **18**, 17–26.
Babel, W. and Müller, R. (1977). *Z. Allg. Mikrobiol.* **17**, 175–182.
Babel, W. and Müller-Kraft, G. (1979). *Z. Allg. Mikrobiol.* **19**, 687–693.
Babel, W. and Steudel, A. (1977). *Z. Allg. Mikrobiol.* **17**, 267–275.
Baker, F. D., Papiska, H. R. and Campbell, L. L. (1962). *J. Bacteriol.* **84**, 973–978.
Balch, W. E. and Wolfe, R. S. (1979a). *J. Bacteriol.* **137**, 256–263.
Balch, W. E. and Wolfe, R. S. (1979b). *J. Bacteriol.* **137**, 264–273.
Balch, W. E., Schoberth, S., Tanner, R. S. and Wolfe, R. S. (1977). *Int. J. Syst. Bacteriol.* **27**, 355–361.
Balch, W. E., Fox, G. E., Magrum, L. J., Woese, C. R. and Wolfe, R. S. (1979). *Microbiol. Revs.* **43**, 260–296.
Bamforth, C. W. and Large, P. J. (1977a). *Biochem. J.* **161**, 357–370.
Bamforth, C. W. and Large, P. J. (1977b). *Biochem. J.* **167**, 509–512.
Bamforth, C. W. and O'Connor, M. L. (1979). *J. Gen. Microbiol.* **110**, 143–149.
Bamforth, C. W. and Quayle, J. R. (1977). *J. Gen. Microbiol.* **101**, 259–267.
Bamforth, C. W. and Quayle, J. R. (1978a). *Arch. Microbiol.* **119**, 91–97.
Bamforth, C. W. and Quayle, J. R. (1978b). *Biochem. J.* **169**, 677–686.
Bamforth, C. W. and Quayle, J. R. (1979). *Biochem. J.* **181**, 517–524.
Baratti, J., Couderc, R., Cooney, C. L. and Wang, D. I. C. (1978). *Biotechnol. Bioeng.* **20**, 333–348.
Baresi, L., Mah, R. A., Ward, D. M. and Kaplan, I. R. (1978). *Appl. Environ. Microbiol.* **36**, 186–197.
Barker, H. A. (1956). Bacterial fermentations. *In* "Biological Formation of Methane," pp. 1–27. Wiley, New York.
Barnes, R. O. and Goldberg, E. D. (1976). *Geology* **4**, 297–300.
Barnes, L. J., Drozd, J. W., Harrison, D. E. F. and Hamer, G. (1976). *In* "Microbial Production and Utilisation of Gases (H_2, CH_4, CO)" (H. G. Schlegel, G. Gottschalk and N. Pfennig, eds), pp. 389–402. Erich Goltze KG, Göttingen.
Bassham, J. A. (1963). *Adv. Enzymol.* **25**, 39–117.
Bassham, J. A. (1971a). *Proc. Nat. Acad. Sci. U.S.A.* **68**, 2877–2882.
Bassham, J. A. (1971b). *Science* **172**, 526–534.
Bassham, J. A., Benson, A. A., Kay, L. D., Harris, A. Z., Wilson, A. T. and Calvin, M. (1954). *J. Amer. Chem. Soc.* **76**, 1760–1770.
Baudhuin, P., Beaufay, H. and de Duve, C. (1965). *J. Cell. Biol.* **26**, 219–243.
Beardmore-Gray, M. and Anthony, C. (1981). Unpublished observations.
Beardsmore, A. J., Aperghis, P. N. G. and Quayle, J. R. (1982). *J. Gen. Microbiol.* **128**. In Press.
de Beer, R., van Ormondt, D., van Ast, M. A., Banen, R., Frank, J. A. and Duine, J. A. (1979). *J. Chem. Phys.* **70**, 4491–4495.
de Beer, R., Duine, J. A., Frank, J. and Large, P. J. (1980). *Biochim. Biophys. Acta.* **622**, 370–374.
Bellion, E. and Hersh, L. B. (1972). *Arch. Biochem. Biophys.* **153**, 368–374.
Bellion, E. and Kelley, R. L. (1979). *J. Bacteriol.* **138**, 519–522.
Bellion, E. and Kim, Y. S. (1978). *Biochim. Biophys. Acta.* **541**, 425–434.
Bellion, E. and Kim, Y. S. (1979). *Current Microbiol.* **2**, 31–34.
Bellion, E. and Spain, J. C. (1976). *Can. J. Microbiol.* **22**, 404–408.
Bellion, E. and Woodson, J. (1975). *J. Bacteriol.* **122**, 557–564.

Bellion, E. and Wu, G. T. S. (1978). *J. Bacteriol.* **135**, 251–258.
Bellion, E., Bolbot, J. A. and Lash, T. D. (1981). *Curr. Microbiol.* **6**, 367–372.
Ben-Bassat, A. and Goldberg, I. (1977). *Biochim. Biophys. Acta.* **497**, 586–597.
Ben-Bassat, A. and Goldberg, I. (1980). *Biochim. Biophys. Acta.* **611**, 1–10.
Ben-Bassat, A., Goldberg, I. and Mateles, R. I. (1980). *J. Gen. Microbiol.* **116**, 213–223.
Bennetto, H. P., Stirling, J. L., Tanaka, K. and Vega, C. A. (1980). *Soc. Gen. Microbiol. Quart.* **8**, 37.
Beringer, J. E., Bedrion, J. L., Buchanan-Wollaston, A. V. and Johnston, A. W. B. (1978). *Nature* **276**, 633–634.
Best, D. J. and Higgins, I. J. (1981). *J. Gen. Microbiol.* **125**, 73–84.
Bicknell, B. and Owens, J. D. (1980). *J. Gen. Microbiol.* **117**, 89–96.
Bird, C. W., Lynch, J. M., Pirt, S. J., Reid, W. W., Brooks, C. J. W. and Middleditch, B. S. (1971). *Nature* **229**, 473–474.
Blackmore, M. A. and Quayle, J. R. (1968). *Biochem. J.* **107**, 705–713.
Blackmore, M. A. and Quayle, J. R. (1970). *Biochem. J.* **118**, 53–59.
Blaylock, B. A. (1968). *Arch. Biochem. Biophys.* **124**, 314–324.
Blaylock, B. A. and Stadtman, T. C. (1963). *Biochem. Biophys. Res. Comm.* **11**, 34–38.
Blaylock, B. A. and Stadtman, T. C. (1966). *Arch. Biochem. Biophys.* **116**, 138–152.
Bockris, J. O'M. and Srinivasan, S. (1969). "Fuel Cells, their Electrochemistry." McGraw Hill, New York.
de Boer, W. E. and Hazeu, W. (1972). *Anton. van Leeuwenhoek* **38**, 33–47.
Bolbot, J. A. (1979). Ph.D. thesis, University of Southampton.
Bolbot, J. A. and Anthony, C. (1980a). *J. Gen. Microbiol.* **120**, 245–254.
Bolbot, J. A. and Anthony, C. (1980b). *J. Gen. Microbiol.* **120**, 233–244.
de Bont, J. A. M., van Dijken, J. P. and Harder, W. (1981). *J. Gen. Microbiol.* **127**, 315–323.
Boogerd, F. C., van Verseveld, H. W. and Stouthamer, A. H. (1980). *FEBS Lett.* **113**, 279–284.
Borman, C. and Sahm, H. (1978). *Arch. Microbiol.* **117**, 67–72.
Boulton, C. A. and Large, P. J. (1975). *FEBS Lett.* **55**, 286–290.
Boulton, C. A. and Large, P. J. (1977). *J. Gen. Microbiol.* **101**, 151–156.
Boulton, C. A. and Large, P. J. (1979a). *Biochim. Biophys. Acta.* **570**, 22–30.
Boulton, C. A. and Large, P. J. (1979b). *FEMS Microbiol. Lett.* **5**, 159–162.
Boulton, C. A., Crabbe, M. J. C. and Large, P. J. (1974). *Biochem. J.* **140**, 253–263.
Boulton, C. A., Haywood, G. W. and Large, P. J. (1980). *J. Gen. Microbiol.* **117**, 293–304.
Brady, R. S. and Flatmark, T. (1971). *J. Mol. Biol.* **57**, 529–539.
Brammar, W. J. (1981). In "Microbial Growth on C_1 Compounds" (H. Dalton, ed.), pp. 312–316. Heyden, London.
Braun, M., Mayer, F. and Gottschalk, G. (1981). *Arch. Microbiol.* **128**, 288–293.
Bringer, S., Sprey, B. and Sahm, H. (1979). *Eur. J. Biochem.* **101**, 563–570.
Brook, D. F. and Large, P. J. (1975). *Eur. J. Biochem.* **55**, 601–609.
Brook, D. F. and Large, P. J. (1976). *Biochem. J.* **157**, 197–205.
Brooks, J. O. and Meers, J. L. (1973). *J. Gen. Microbiol.* **77**, 513–519.
Brown, L. R., Strawinsky, R. J. and McCleskey, C. S. (1964). *Can. J. Microbiol.* **10**, 791–799.
Brown, C. M., Ellwood, D. C. and Hunter, J. R. (1977). *FEMS Microbiol. Lett.* **1**, 163–166.

Bryant, M. P. (1965). *In* "Physiology of Digestion in the Ruminant" (R. W. Dougherty, ed.), pp. 411–418. Butterworths, Washington, D. C.
Bryant, M. P., Wolin, E. A., Wolin, M. J. and Wolfe, R. S. (1967). *Arch. Mikrobiol.* **59**, 20–31.
Budohoski, L., Michalik, J. and Raczynska-Bojanowska, K. (1978). *Acta. Microbiol. Polonica* **27**, 257–266.
Bull, A. T. and Brown, C. M. (1979). *In* "Microbial Biochemistry". International review of biochemistry (R. J. Quayle, ed.), Vol. 21, pp. 177–266. University Park Press, Baltimore.
Bull, A. T., Ellwood, D. C. and Ratledge, C. (1979). "Microbial Technology: Current State, Future Prospects". Cambridge University Press, Cambridge.
Bykovskaya, S. V. and Voronkov, V. V. (1977). *Mikrobiologiya* **46**, 36–41.
Byrom, D. (1981). *In* "Microbial Growth on C_1 Compounds" (H. Dalton, ed.), pp. 278–284. Heyden, London.
Byrom, D. and Ousby, J. C. (1975). "Microbial Growth on C_1 Compounds", pp. 23–27. Society of Fermentation Technology, Tokyo.
Bystrykh, L. V., Sokolov, A. P. and Trotsenko, Y. A. (1981). *FEBS Lett.* **132**, 324–328.
Chalfan, Y. and Mateles, R. I. (1972). *Appl. Microbiol.* **23**, 135–140.
Chandra, T. S. and Shethna, Y. I. (1977). *J. Bacteriol.* **131**, 389–398.
Cheeseman, P., Toms-Wood, A. and Wolfe, R. S. (1972). *J. Bacteriol.* **112**, 527–531.
Chen, B. J., Hirt, W., Lim, H. C. and Tsao, G. T. (1977). *Appl. Environ. Microbiol.* **33**, 269–274.
Colby, J. and Dalton, H. (1976). *Biochem. J.* **157**, 495–497.
Colby, J. and Dalton, H. (1978). *Biochem. J.* **171**, 461–468.
Colby, J. and Dalton, H. (1979). *Biochem. J.* **177**, 903–908.
Colby, J. and Zatman, L. J. (1972). *Biochem. J.* **128**, 1373–1376.
Colby, J. and Zatman, L. J. (1973). *Biochem. J.* **132**, 101–112.
Colby, J. and Zatman, L. J. (1974). *Biochem. J.* **143**, 555–567.
Colby, J. and Zatman, L. J. (1975a). *Biochem. J.* **148**, 505–511.
Colby, J. and Zatman, L. J. (1975b). *Biochem. J.* **150**, 141–144.
Colby, J. and Zatman, L. J. (1975c). *Biochem. J.* **148**, 513–520.
Colby, J. and Zatman, L. J. (1975d). *J. Gen. Microbiol.* **90**, 169.
Colby, J., Dalton, H. and Whittenbury, R. (1975). *Biochem. J.* **151**, 459–462.
Colby, J., Stirling, D. I. and Dalton, H. (1977). *Biochem. J.* **165**, 395–402
Colby, J., Dalton, H. and Whittenbury, R. (1979). *Ann. Rev. Microbiol.* **33**, 481–517.
Cole, J. A. (1976). *Adv. Microbiol. Physiol.* **14**, 1–92.
Conti, S. F. and Hirsch, P. (1965). *J. Bacteriol.* **89**, 503–512.
Cook, R. J. and Anthony, C. (1978). *J. Gen. Microbiol.* **109**, 265–274.
Cooney, C. L. (1975). *In* "Microbial Growth on C_1 Compounds", pp. 183–197. Society of Fermentation Technology, Tokyo.
Cooney, C. L. and Levine, D. W. (1972). *Adv. Appl. Microbiol.* **15**, 337–365.
Cooney, C. L. and Levine, D. W. (1975). *In* "Single-cell protein II" (S. R. Tannenbaum and D. I. C. Wang, eds), pp. 402–423. MIT Press, Cambridge, Massachusetts.
Cooney, C. L. and Makiguchi, N. (1977). "Biotechnology and Bioengineering Symposium," No 7, pp. 65–76. Wiley, New York.
Cooper, R. A. and Kornberg, H. L. (1967). *Proc. Roy. Soc. B.* **168**, 263–280.
Couderc, R. and Baratti, J. (1980a). *Agric. Biol. Chem.* **44**, 2279–2289.
Couderc, R. and Baratti, J. (1980b). *Biotechnol. Bioeng.* **22**, 1155–1173.

Cowan, S. T. (1970). *J. Gen. Microbiol.* **61**, 145–154.
Cox, R. B. and Quayle, J. R. (1975). *Biochem. J.* **150**, 569–571.
Cox, R. B. and Quayle, J. R. (1976a). *J. Gen. Microbiol.* **95**, 121–133.
Cox, R. B. and Quayle, J. R. (1976b). *J. Gen. Microbiol.* **97**, 137–139.
Cox, R. B. and Zatman, L. J. (1973). *Biochem. Soc. Trans.* **1**, 669–671.
Cox, R. B. and Zatman, L. J. (1974). *Biochem. J.* **141**, 605–608.
Cox, R. B. and Zatman, L. J. (1976). *J. Gen. Microbiol.* **93**, 397–400.
Cross, A. R. (1979). Ph.D. thesis, University of Southampton.
Cross, A. R. and Anthony, C. (1978). *Proc. Soc. Gen. Microbiol.* **5**, 42.
Cross, A. R. and Anthony, C. (1980a). *Biochem. J.* **192**, 421–427.
Cross, A. R. and Anthony, C. (1980b). *Biochem. J.* **192**, 429–439.
Cypionka, H., Meyer, O. and Schlegel, H. G. (1980). *Arch. Microbiol.* **127**, 301–307.
Dahl, J. S., Mehta, R. J. and Hoare, D. S. (1972). *J. Bacteriol.* **109**, 916–921.
Dalton, H. (1977a). *Arch. Microbiol.* **114**, 273–279.
Dalton, H. (1977b). U.K. Patent No. 27886.
Dalton, H. (1980a). In "Hydrocarbons in Biotechnology" (D. E. F. Harrison, I. J. Higgins and R. Watkinson, eds), pp. 85–97. Heyden, London.
Dalton, H. (1980b). *Adv. Appl. Microbiol.* **26**, 71–87.
Dalton, H. (1981a). In "Microbial Growth on C_1 Compounds" (H. Dalton, ed.), pp. 1–10. Heyden, London.
Dalton, H. (1981b) (ed.). "Microbial Growth on C_1 Compounds." Heyden, London.
Dalton, H. and Whittenbury, R. (1976). *Adv. Microbiol.* **109**, 147–151.
Daniels, L. and Zeikus, J. G. (1978). *J. Bacteriol.* **136**, 75–84.
Daniels, L., Fuchs, G., Thauer, R. K. and Zeikus, J. G. (1977). *J. Bacteriol.* **132**, 118–126.
Daniels, L., Fulton, G., Spencer, R. W. and Orme-Johnson, W. H. (1980). *J. Bacteriol.* **141**, 694–698.
Davey, J. F. and Mitton, J. R. (1973). *FEBS Lett.* **37**, 335–337.
Davey, J. F., Whittenbury, R. and Wilkinson, J. F. (1972). *Arch. Microbiol.* **87**, 359–366.
Davies, S. L. and Whittenbury, R. (1970). *J. Gen. Microbiol.* **61**, 227–232.
Davis, J. B. and Yarborough, H. F. (1966). *Chem. Geol.* **1**, 137–144.
Davis, J. B., Coty, V. F. and Stanley, J. P. (1964). *J. Bacteriol.* **88**, 468–472.
Davis, D. H., Doudoroff, M. and Stanier, R. Y. (1969). *Int. J. Syst. Bact.* **19**, 375–390.
Dawson, M. J. and Jones, C. W. (1981a). In "Microbial Metabolism of C_1 Compounds" (H. Dalton, ed.), pp. 251–257. Heyden, London.
Dawson, M. J. and Jones C. W. (1981b). *Biochem. J.* **194**, 915–924.
Dawson, M. J. and Jones, C. W. (1981c). *Eur. J. Biochem.* **118**, 113–118.
Demain, A. L. (1971). *Symp. Soc. Gen. Microbiol.* **21**, 77–101.
Dickens, F. and Williamson, D. H. (1958). *Nature* **181**, 1790.
Diekert, G., Klee, B. and Thauer, R. K. (1979). *Arch. Microbiol.* **124**, 103–106.
Diekert, G., Weber, B. and Thauer, R. K. (1980a). *Arch. Microbiol.* **127**, 273–278.
Diekert, G., Gilles, H-H., Jaenchen, R. and Thauer, R. K. (1980b). *Arch. Microbiol.* **128**, 256–262.
Diekert, G., Jaenchen, R. and Thauer, R. K. (1980c). *FEBS Lett.* **119**, 118–120.
Diel, F., Held, W., Schlanderer, G. and Dellweg, H. (1974). *FEBS Lett.* **38**, 274–276.
Dietrich, J. and Hennig, U. (1970). *Eur. J. Biochem.* **14**, 258–269.
van Dijken, J. P. (1976). Ph.D. thesis, University of Groningen, Holland.
van Dijken, J. P. and Bos, P. (1981). *Arch. Microbiol.* **128**, 320–324.

van Dijken, J. P. and Harder, W. (1974). *J. Gen. Microbiol.* **84**, 409–411.
van Dijken, J. P. and Harder, W. (1975). *Biotechnol. Bioeng.* **17**, 15–30.
van Dijken, J. P. and Quayle, J. R. (1977). *Arch. Microbiol.* **114**, 281–286.
van Dijken, J. P., Otto, R. and Harder, W. (1975a). *Arch Microbiol.* **106**, 221–226.
van Dijken, J. P., Veenhuis, M., Krejer-van Rij, N. J. W. and Harder, W. (1975b). *Arch. Microbiol.* **102**, 41–44.
van Dijken, J. P., Veenhuis, M., Vermeulen, C. A. and Harder, W. (1975c). *Arch. Microbiol.* **105**, 261–267.
van Dijken, J. P., Otto, R. and Harder, W. (1976a). *Arch. Microbiol.* **111**, 137–144.
van Dijken, J. P., Oostra-demkes, G. J., Otto, R. and Harder, W. (1976b). *Arch. Microbiol.* **111**, 77–83.
van Dijken, J. P., Harder, W., Beardsmore, A. J. and Quayle, J. R. (1978). *FEMS Microbiol. Lett.* **4**, 97–102.
van Dijken, J. P., Harder, W. and Quayle, J. R. (1981a). In "Microbial Growth on C_1 Compounds" (H. Dalton, ed.), pp. 191–210. Heyden, London.
van Dijken, J. P., Duine, J. A., Frank, J. and Large, P. J. (1981b). Unpublished. Cited in Large (1981).
Dijkhuizen, L. and Harder, W. (1979a). *Arch. Microbiol.* **123**, 47–53.
Dijkhuizen, L. and Harder, W. (1979b). *Arch. Microbiol.* **123**, 55–63.
Dijkhuizen, L., Knight, M. and Harder, W. (1978). *Arch. Microbiol.* **116**, 77–83.
Dijkhuizen, L., Timmerman, J. W. C. and Harder, W. (1979). *FEMS Microbiol. Lett.* **6**, 53–56.
Doddema, H. J. and Vogels, G. D. (1978). *Appl. Environ. Microbiol.* **36**, 752–754.
Doddema, H. J., Hutten, T. J., van der Drift, C. and Vogels, G. D. (1978). *J. Bacteriol.* **136**, 19–23.
Doddema, H. J., Derksen, J. W. M. and Vogels, G. D. (1979a). *FEMS Microbiol. Lett.* **5**, 135–138.
Doddema, H. J., van der Drift, C., Vogels, G. D. and Veenhuis, M. (1979b). *J. Bacteriol.* **140**, 1081–1089.
Doddema, H. J., Claesen, C. A., Kell, D. B., van der Drift, C. and Vogels, G. D. (1980). *Biochem. Biophys. Res. Comm.* **95**, 1288–1293.
Donnelly, M. I. and Dagley, S. (1980). *J. Bacteriol.* **142**, 916–924.
Douthit, H. A. and Pfennig, N. (1976). *Arch. Microbiol.* **107**, 233–234.
Dow, C. S. and Whittenbury, R. (1980). In "Contemporary Microbial Ecology". (D. C. Ellwood, J. N. Hedger, M. J. Lothan, J. M. Lynch and J. H. Slater, eds), pp. 391–417. Academic Press, London and New York.
Drabikowska, A. W. (1977). *Biochem. J.* **168**, 171–178.
Drabikowska, A. K. (1981). In "Microbial Growth on C_1 Compounds" (H. Dalton, ed.), pp. 240–250. Heyden, London.
Drozd, J. W. (1980). In "Hydrocarbons in Biotechnology", (D. E. F. Harrison, I. J. Higgins and R. Watkinson, eds), pp. 75–83. Heyden, London.
Drozd, J. W. and Linton, J. D. (1981). In "Continuous Culture of Cells" (P. H. Calcott, ed.). CRC Uniscience Publications.
Drozd, J. W. and Wren, S. J. (1980). *Biotechnol. Bioeng.* **22**, 353–362.
Drozd, J. W., Linton, J. D., Downs, J. and Stephenson, R. J. (1978). *FEMS Microbiol. Lett.* **4**, 311–314.
Dudina, L. P., Sukoratova, L. V. and Eroshin, V. K. (1977). *Abstract of International Symposium on Microbial Growth on C_1 Compounds*, pp. 89–90. Pushchino.
Duine, J. A. and Frank, J. (1980a). *Biochem. J.* **187**, 213–219.
Duine, J. A. and Frank, J. (1980b). *Biochem. J.* **187**, 221–226.

Duine, J. A. and Frank, J. (1981a). *In* "Microbial Growth on C_1 Compounds" (H. Dalton, ed.), pp. 31–41. Heyden, London.
Duine, J. A. and Frank, J. (1981b). *J. Gen. Microbiol.* **122**, 201–209.
Duine, J. A., Frank, J. and Westerling, J. (1978). *Biochim. Biophys. Acta.* **524**, 277–287.
Duine, J. A., Frank, J. and de Ruiter, L. G. (1979a). *J. Gen. Microbiol.* **115**, 523–526.
Duine, J. A., Frank, J. and van Zeeland, J. K. (1979b). *FEBS Lett.* **108**, 443–446.
Duine, J. A., Frank, J. and Verwiel, P. E. J. (1980). *Eur. J. Biochem.* **108**, 187–192.
Duine, J. A., Frank, J. and Verwiel, P. E. J. (1981). *Eur. J. Biochem.* **118**, 395–399.
Dunstan, P. M. and Anthony, C. (1973). *Biochem. J.* **132**, 797–801.
Dunstan, P. M., Drabble, W. T. and Anthony, C. (1972a). *Biochem. J.* **128**, 99–106.
Dunstan, P. M., Drabble, W. T. and Anthony, C. (1972b). *Biochem. J.* **128**, 107–115.
Durham, D. R. and Perry, J. J. (1978a). *J. Bacteriol.* **134**, 837–843.
Durham, D. R. and Perry, J. J. (1978b). *J. Bacteriol.* **135**, 981–986.
Durham, D. R. and Perry, J. J. (1978c). *J. Gen. Microbiol.* **105**, 39–44.
Dworkin, M. and Foster, J. W. (1956). *J. Bact.* **72**, 646–659.
Eady, R. R. and Large, P. J. (1968). *Biochem. J.* **106**, 245–255.
Eady, R. R. and Large, P. J. (1971). *Biochem. J.* **123**, 757–771.
Eady, R. R., Jarman, T. R. and Large, P. J. (1971). *Biochem. J.* **125**, 449–459.
Edwards, T. and McBride, B. C. (1975). *Appl. Microbiol.* **29**, 540–545.
Eggeling, L. and Sahm, H. (1978). *Eur. J. Appl. Microbiol. Biotechnol.* **5**, 197–202.
Eggeling, L. and Sahm, H. (1980). *Appl. Environ. Microbiol.* **39**, 268–269.
Eggeling, L., Sahm, H. and Wagner, F. (1977). *FEMS Lett.* **1**, 205–211.
Egli, T. and Fiechter, A. (1981). *J. Gen. Microbiol.* **123**, 365–369.
Egli, T., van Dijken, J. P., Veenhuis, M., Harder, W. and Fiechter, A. (1980). *Arch. Microbiol.* **124**, 115–121.
Egorov, A. M., Avilova, T. V., Dikov, M. M., Popov, U. O., Rodinov, Y. V. and Berezin, I. V. (1979). *Eur. J. Biochem.* **99**, 569–576.
Eirich, L. D., Vogels, G. D. and Wolfe, R. S. (1978). *Biochemistry* **17**, 4583–4593.
Eirich, L. D., Vogels, G. D. and Wolfe, R. S. (1979). *J. Bacteriol.* **140**, 20–27.
Eker, A. P. M., Pol, A., van der Meyden, P. and Vogels, G. D. (1980). *FEMS Microbiol. Lett.* **8**, 161–165.
Ellefson, W. E. and Wolfe, R. S. (1980a). *J. Biol. Chem.* **255**, 8388–8389.
Ellefson, W. E. and Wolfe, R. S. (1980b). *Fed. Proc.* **39**, 1773.
Ellefson, W. E. and Wolfe, R. S. (1981). *In* "Microbial Metabolism of C_1 Compounds" (H. Dalton, ed.), pp. 171–180. Heyden, London.
Elwell, M. and Hersh, L. B. (1979). *J. Biol. Chem.* **254**, 2434–2438.
Evans, H. J. and Wood, H. G. (1971). *Biochemistry* **10**, 721–728.
Evans, M. C. W., Buchanan, B. B. and Amon, D. E. (1966). *Proc. Nat. Acad. Sci. U.S.A.* **55**, 928–934.
Faust, U., Präve, P. and Schlingmann, M. (1981). *In* "Microbial Growth on C_1 Compounds" (H. Dalton, ed.), pp. 335–341. Heyden, London.
Fennewald, M. A. and Shapiro, J. A. (1979). *J. Bact.* **139**, 264–269.
Ferenci, T. (1974). *FEBS Lett.* **41**, 94–98.
Ferenci, T. (1976a). *Arch. Microbiol.* **108**, 217–219.
Ferenci, T. (1976b). *In* "Microbial Production and Utilisation of Gases (H_2, CH_4, CO)" (H. G. Schlegel, G. Gottschalk and N. Pfennig, eds), pp. 389–402. Erich Goltze KG, Göttingen.
Ferenci, T., Strøm, T. and Quayle, J. R. (1974). *Biochem. J.* **144**, 477–486.
Ferenci, T., Strøm, T. and Quayle, J. R. (1975). *J. Gen. Microbiol.* **91**, 79–91.

Ferry, J. G. and Wolfe, R. S. (1977). *Appl. Env. Microbiol.* **34**, 371–376.
Ferry, J. G., Smith, P. H. and Wolfe, R. S. (1974). *Int. J. Syst. Bacteriol.* **24**, 465–469.
Flechtner, V. R. and Hanson, R. S. (1970). *Biochim. Biophys. Acta.* **222**, 253–264.
Forrest, W. W. and Walker, D. J. (1971). *Adv. Microbiol. Physiol.* **5**, 213–274.
Forrest, H. S., Salisbury, S. A. and Kitty, C. G. (1980). *Biochem. Biophys. Res. Comm.* **97**, 248–251.
Foster, J. W. and Davis, R. H. (1966). *J. Bacteriol.* **91**, 1924–1931.
Friedrich, C. G., Bowien, B. and Friedrich, B. (1979). *J. Gen. Microbiol.* **115**, 185–192.
Fuchs, G. and Stupperich, E. (1978). *Arch. Microbiol.* **118**, 121–125.
Fuchs, G. and Stupperich, E. (1980). *Arch. Microbiol.* **127**, 267–272.
Fuchs, G., Stupperich, E. and Thauer, R. K. (1978a). *Arch. Microbiol.* **117**, 61–66.
Fuchs, G., Stupperich, E. and Thauer, R. K. (1978b). *Arch. Microbiol.* **119**, 215–218.
Fuchs, G., Moll, J., Scherer, P. A. and Thauer, R. K. (1979a). *In* "Hydrogenases: Their Catalytic Structure and Function" (H. G. Schlegel, ed.), pp. 83–92. Erich Goltze KG, Göttingen.
Fuchs, G., Thauer, R., Ziegler, H. and Stichler, W. (1979b). *Arch. Microbiol.* **120**, 135–139.
Fujuii, T. and Tonomura, K. (1972). *Agric. Biol. Chem.* **36**, 2297–2306.
Fujuii, T. and Tonomura, K. (1973). *Agric. Biol. Chem.* **37**, 447–449.
Fujuii, T. and Tonomura, K. (1974). *Agric. Biol. Chem.* **38**, 1763–1765.
Fujuii, T. and Tonomura, K. (1975a). *Agric. Biol. Chem.* **39**, 2325–2330.
Fujuii, T. and Tonomura, K. (1975b). *Agric. Biol. Chem.* **39**, 1891–1892.
Fujuii, T., Asada, Y. and Tonomura, K. (1974). *Agric. Biol. Chem.* **38**, 1121.
Fukui, S. and Tanaka, A. (1979a). *J. Appl. Bacteriol.* **1**, 171–201.
Fukui, S. and Tanaka, A. (1979b). *TIBS* **4**, 246–249.
Fukui, S., Tanaka, A., Kawamoto, S., Yasuhara, S., Teranıshi, Y. and Osumi, M. (1975a). *J. Bacteriol.* **123**, 317–328.
Fukui, S., Kawamoto, S., Yasuhara, S., Tanaka, A., Osumi, M. and Imaizumi, F. (1975b). *Eur. J. Biochem.* **59**, 561–566.
Fukui, S., Tanaka, A. and Osumi, M. (1975c). *In* "Microbial Growth on C_1 Compounds" (H. Dalton, ed.), pp. 139–148. Heyden, London.
Galchenko, V. F., Shishkina, V. N., Suzina, N. E. and Trotsenko, Y. A. (1978). *Microbiology* **46**, 723–728.
Gautier, F. and Bonewald, R. (1980). *Mol. Gen. Genet.* **178**, 375–380.
Ghosh, R. (1980). *Biochem. Soc. Trans.* **8**, 639–640.
Ghosh, S. and Klass, D. L. (1978). *Proc. Biochem.* **13**, 15–24.
Ghosh, R. and Quayle, J. R. (1979). *Anal. Biochem.* **99**, 112–117.
Ghosh, R. and Quayle, J. R. (1981). *Biochem. J.* **199**, 245–250
Goldberg, I. (1976). *Eur. J. Biochem.* **63**, 233–240.
Goldberg, I. and Mateles, R. A. (1975). *J. Bacteriol.* **124**, 1028–1029,
Goldberg, I. and Schechter, I. (1978). *J. Bacteriol.* **135**, 717–720.
Goldberg, I., Rock, J. S., Ben-Bassat, A. and Mateles, R. I. (1976). *Biotechnol. Bioeng.* **18**, 1657–1668.
Goodwin, P. M. (1980). Abstract of the Third International Symposium on Microbial Growth on C_1 Compounds (Sheffield).
Gottschal, J. C. and Kuenen, J. G. (1981). *In* "Microbial Growth on C_1 Compounds" (H. Dalton, ed.), pp. 92–104. Heyden, London.
Green, P. N. and Bousfield, I. J. (1981). *In* "Microbial Growth on C_1 Compounds" (H. Dalton, ed.), pp. 285–293. Heyden, London.
Guest, J. R. (1978). *Adv. Neurol.* **21**, 219–244.

Guest, J. R. and Creaghan, I. T. (1973). *J. Gen. Microbiol.* **75**, 197–210.
Guilbault, G. G. (1970). "Enzymatic Methods of Analysis". Pergamon, Oxford.
Gunsalus, R. P. and Wolfe, R. S. (1977). *Biochem. Biophys. Res. Comm.* **76**, 790–795.
Gunsalus, R. P. and Wolfe, R. S. (1978a). *FEMS Microbiol. Lett.* **3**, 191–193.
Gunsalus, R. P. and Wolfe, R. S. (1978b). *J. Bacteriol.* **135**, 851–857.
Gunsalus, R. P. and Wolfe, R. S. (1980). *J. Biol. Chem.* **255**, 1891–1895.
Gunsalus, R. P., Eirich, D., Romesser, J., Balch, W., Shapiro, S. and Wolfe, R. S. (1976). *In* "Microbial Production and Utilisation of Gases" (H. G. Schlegel, G. Gottschalk and N. Pfennig, eds), pp. 191–198. Erich Goltze KG, Göttingen.
Gunsalus, R. P., Romesser, J. A. and Wolfe, R. S. (1978). *Biochemistry* **17**, 2374–2377.
Hacking, A. J. and Quayle, J. R. (1974). *Biochem. J.* **139**, 399–405.
Haddock, B. A. and Jones, C. W. (1977). *Bacteriol. Rev.* **41**, 47–99.
Hamer, G. (1979). *In* "Microbial Biomass" (A. H. Rose, ed.), pp. 315–360. Academic Press, London and New York.
Hamer, G. and Hamdan, I. Y. (1979). *Chem. Soc. Rev.* **8**, 143–170.
Hamer, G. and Harrison, D. E. F. (1980). *In* "Hydrocarbons in Biotechnology" (D. E. F. Harrison, I. J. Higgins and R. Watkinson, eds), pp. 59–73. Heyden, London.
Hammes, W. P., Winter, J. and Kandler, O. (1979). *Arch. Microbiol.* **123**, 275–279.
Hammond, R. C. and Higgins, I. J. (1978). *Proc. Soc. Gen. Microbiol.* **5**, 43.
Hammond, R. C., Taylor, F. and Higgins, I. J. (1979). *Proc. Soc. Gen. Microbiol.* **6**, 89.
Hammond, R. C., Rees, B. and Higgins, I. J. (1981). *Biochim. Biophys. Acta* **638**, 22–29.
Hampton, D. and Zatman, L. J. (1973). *Biochem. Soc. Trans.* **1**, 667–668.
Hanson, R. S. (1980). *Adv. Appl. Microbiol.* **26**, 3–39.
Harder, W. and Attwood, M. M. (1975). *Antonie van Leeuwenhoek* **41**, 421–429.
Harder, W. and Attwood, M. M. (1978). *Adv. Microbial. Physiol.* **17**, 303–359.
Harder, W. and van Dijken, J. P. (1975). *In* "Microbial Growth on C_1 Compounds", pp. 155–161. The Society of Fermentation Technology, Japan.
Harder, W. and van Dijken, J. P. (1976). *In* "Microbial Production and Utilisation of Gases' (H. G. Schlegel, G. Gottschalk, and N. Pfennig eds), pp. 403–418. Erich Goltze KG, Göttingen.
Harder, W. and Quayle, J. R. (1971a). *Biochem. J.* **121**, 753–762.
Harder, W. and Quayle, J. R. (1971b). *Biochem. J.* **121**, 763–769.
Harder, W., Attwood, M. M. and Quayle, J. R. (1973). *J. Gen. Microbiol.* **78**, 155–163.
Harder, W., Martin, A. and Attwood, M. M. (1975). *J. Gen. Microbiol.* **86**, 319–326.
Harder, W., van Dijken, J. P. and Roels, J. A. (1981). *In* "Microbial Growth on C_1 Compounds" (H. Dalton, ed.), pp. 258–269. Heyden, London.
Harrison, D. E. F. (1973). *J. Appl. Bact.* **36**, 301–308.
Harrison, D. E. F. (1976). *Adv. Microbiol. Physiol.* **14**, 243–313.
Harrison, D. E. F. (1978). *Adv. Appl. Microbiol.* **24**, 129–164.
Harrison, D. E. F., Topiwala, H. H. and Hamer, G. (1972). *Proc. IVth Int. Ferment. Soc.* Fermentation Technology Today (G. Terui, ed.), pp. 491–495. Society of Fermentation Technology, Japan.
Harrison, D. E. F., Higgins, I. J. and Watkinson, R. (eds) (1980). "Hydrocarbons in Biotechnology". Heyden, London.

Harwood, J. H., Williams, E. and Bainbridge, B. W. (1972). *J. Appl. Bact.* **35**, 99–108.
Hatch, M. D. and Slack, C. R. (1968). *Biochem. J.* **106**, 141–146.
Hayward, H. R. and Stadtman, E. R. (1959). *J. Bacteriol.* **78**, 557–561.
Hazeu, W. (1975). *Antonie van Leeuwenhoek* **41**, 121–134.
Hazeu, W. and Steenis, P. J. (1970). *Antonie van Leeuwenhoek* **36**, 67–72.
Hazeu, W., de Bruyn, J. C. and Box, P. (1972). *Arch. Mikrobiol.* **87**, 185–188.
Hazeu, W., Batenburg-van der Vegte, W. H. and Nieuwdorp, P. J. (1975). *Experientia* **31**, 926–927.
Hazeu, W., Batenburg-van der Vegte, W. H. and deBruyn, J. C. (1980). *Arch. Microbiol.* **124**, 211–220.
Henninger, W. and Windisch. S. (1975). *Arch. Microbiol.* **105**, 47–48.
Heptinstall, J. and Quayle, J. R. (1969). *J. Gen. Microbiol.* **55**, xvi–xvii.
Heptinstall, J. and Quayle, J. R. (1970). *Biochem. J.* **117**, 563–572.
Hermans, J. M. H., Hutten, T. J., van der Drift, C. and Vogels, G. D. (1980). *Anal. Biochem.* **106**, 363–366.
Hersh, L. B. (1973). *J. Biol. Chem.* **248**, 7295–7303.
Hersh, L. B. (1974a). *J. Biol. Chem.* **249**, 5208–5212.
Hersh, L. B. (1974b). *J. Biol. Chem.* **249**, 6264–6271.
Hersh, L. B. and Bellion, E. (1972). *Biochem. Biophys. Res. Comm.* **48**, 712–719.
Hersh, L. B., Tsai, L. and Stadtman, E. R. (1969). *J. Biol. Chem.* **244**, 4677–4683.
Hersh, L. B., Perterson, J. A. and Thompson, A. A. (1971). *Arch. Biochem. Biophys.* **145**, 115–120.
Hersh, L. B., Stark, M. J., Wortham, S. and Fiero, M. K. (1972). *Arch. Biochem. Biophys.* **150**, 219–226.
Higgins, I. J. (1978). U.K. Patents Nos 19712, 2024205A, 35123.
Higgins, I. J. (1979a). Microbial biochemistry. "International Review of Biochemistry" (J. R. Quayle, ed.), Vol. 21, pp. 300–353. University Park Press, Baltimore.
Higgins, I. J. (1979b). U.K. Patent No. 8005577.
Higgins, I. J. (1980). In "Diversity of Bacterial Respiratory Systems" (C. J. Knowles, ed.), Vol. I, pp. 187–221. CRC Press, Florida.
Higgins, I. J. and Hill, H. A. O. (1979). *Soc. Gen. Microbiol. Symp.* **29**, 359–377.
Higgins, I. J. and Quayle, J. R. (1970). *Biochem. J.* **118**, 201–208.
Higgins, I. J., Knowles, C. J. and Tonge, G. M. (1976a). In "Microbial Production and Utilisation of Gases (H_2, CH_4, CO)" (H. G. Schlegel, G. Gottschalk and N. Pfennig, eds), pp. 389–402. Erich Goltze KG, Göttingen.
Higgins, I. J., Taylor, S. C. and Tonge, G. M. (1976b). *Proc. Soc. Gen. Microbiol.* **3**, 179–180.
Higgins, I. J., Hammond, R. C., Sariaslani, F. S., Best, D., Davies, M. M., Tryhorn, S. E. and Taylor, F. (1979). *Biochem. Biophys. Res. Comm.* **89**, 671–677.
Higgins, I. J., Hammond, R. C., Plotkin, E., Hill, H. A. O., Vosaki, K., Eddowes, M. J. and Cass, A. E. G. (1980a). In "Hydrocarbons in Biotechnology" (D. E. F. Harrison, I. J. Higgins and R. Watkinson, ed), pp. 181–193. Heyden, London.
Higgins, I. J., Best, D. J. and Hammond, R. C. (1980b). *Nature* **286**, 561–564.
Higgins, I. J., Best, D. J. and Scott, D. (1981a). In "Microbial Growth on C_1 Compounds" (H. Dalton, ed.), pp. 11–20. Heyden, London.
Higgins, I. J., Best, D. J. and Hammond, R. C. (1981b). *Nature* **291**, 169–170.
Hill, B. and Attwood, M. M. (1974). *J. Gen. Microbiol.* **83**, 187–190.
Hill, B. and Attwood, M. M. (1976a). *J. Gen. Microbiol.* **96**, 185–193.

Hill, B. and Attwood, M. M. (1976b). *J. Gen. Microbiol.* **97**, 335–338.
Hill, C. L.., Steenkamp, D. J., Holm, R. H. and Singer, T. P. (1977). *PNAS* **74**, 547–551
Hippe, H., Caspari, D., Fiebig, K. and Gottschalk, G. (1979). *Proc. Nat. Acad. Sci.* **78**, 494.
Hirsch, P. (1974). *Ann. Rev. Microbiol.* **28**, 391–444.
Hirsch, P. and Conti, S. F. (1964a). *Arch. Mikrobiol.* **48**, 339–357.
Hirsch, P. and Conti, S. F. (1964b). *Arch. Mikrobiol.* **48**, 358–367.
Hirsch, P., Morita, S. and Conti, S. F. (1963). *Bacteriol. Proc.* 97.
Hirt, W., Papoutsakis, E., Krug, E., Lim, H. C. and Tsao, G. T. (1978). *Appl. Environ. Microbiol.* **36**, 56–62.
Ho, K. P. and Payne, W. J. (1979). *Biotechnol. Bioeng.* **21**, 787–802.
Hobson, P. N., Bousefield, S. and Summers, R. (1981). "Methane Production from Agricultural and Domestic Wastes." Applied Science Publishers, Barking, U.K.
Hoffman, K. H. and Babel, W. (1980). *Z. Allg. Mikrobiol.* **20**, 389–398.
Hollinshead, J. A. (1966). *Biochem. J.* **99**, 389–395.
Holloway, B. W. (1981). *In* "Microbial Growth on C_1 Compounds" (H. Dalton, ed.), pp. 317–324. Heyden, London.
Höpfnier, T. and Trautwein, A. (1971). *Arch. Mikrobiol.* **77**, 26–35.
Horecker, B. L., Tsolas, O. and Lai, C. Y. (1972). *Enzymes* **6**, 213–258.
Hornig, D. and Wagner, C. (1968). *Bacteriol. Proc.* 115.
Hou, C. T., Laskin, A. I. and Patel, R. N .(1978). *Appl. Environ. Microbiol.* **37**, 800–804.
Hou, C. T., Patel, R. N., Laskin, A. I., Barnabe, N. and Marezak, I. (1979a). *Appl. Environ. Microbiol.* **38**, 135–142.
Hou, C. T., Patel, R., Laskin, A. I. and Barnabe, N. (1979b). *Appl. Environ. Microbiol.* **38**, 127–134.
Hou, C. T., Patel. R. N., Laskin, A. I. Barnabe, N. and Marezak, I. (1979c). *FEBS Lett.* **101**, 179–183.
Hou, C. T., Patel, R. N. and Laskin, A. I. (1979d). U.K. Patent Nos 7913054, 7913062, 7913063.
Hou, C. T., Patel, R. N. and Laskin, A. I. (1980a). *Adv. Appl. Microbiol.* **26**, 41–69.
Hou, C. T., Patel, R. N., Laskin, A. I. and Barnabe, N. (1980b). *FEMS Microbiol. Lett.* **9**, 267–270.
Hou, C. T., Patel, R., Laskin, A. I. Barnabe, N. and Marezak, I. (1981). *Appl. Environ. Microbiol.* **41**, 829–832.
Hubley, J. H., Mitton, J. R. and Wilkinson, J. F. (1974). *Arch. Microbiol.* **95**, 365–368.
Hubley, J. H., Thompson, A. W. and Wilkinson, J. F. (1975). *Arch. Microbiol.* **102**, 199–202.
Hughes, D. E. and Rose, A. H. (1971). "Microbes and Biological Productivity." Cambridge University Press, Cambridge, U.K.
Hungate, R. E. (1966). "The Rumen and its Microbes." Academic Press, New York and London.
Hungate, R. E. (1976). *In* "Microbial Production and Utilisation of Gases" (H. G. Schlegel, G. Gottschalk and N. Pfennig, eds), pp. 119–124. Erich Goltze KG, Göttingen.
Hungate, R. E. (1977). *In* "Microbial Energy Conservation" (H. G. Schlegel and J. Barnea, eds), pp. 339–346. Pergamon Press, Oxford.
Hutchinson, D. W., Whittenbury, R. and Dalton, H. (1976). *J. Theor. Biol.* **58**, 325–335.

Hutten, T. J., Bongaerts, H. C. M., van der Drift, C. and Vogels, G. D. (1980). *Antonie van Leeuwenhoeck* **46**, 601–610.
Hutten, T. J., de Jong, M. H., Peters, B. P. H., van der Drift, C. and Vogels, G. D. (1981). *J. Bacteriol.* **145**, 27–34.
Hutton, W. E. and Zobell, C. E. (1949). *J. Bacteriol.* **58**, 463–473.
Ianotti, E. L., Kafkawitz, D., Wolin, M. J. and Bryant, M. P. (1973). *J. Bacteriol.* **114**, 1231–1240.
Ida, S. and Alexander, M. (1965). *J. Bacteriol.* **90**, 151–156.
Ishimoto, M. and Shimokawa, O. (1978). *Z. Allg. Mikrobiol.* **18**, 173–181.
Ivanovsky, R. N., Zacharova, E. V., Netrusov, A. I., Rodionov, Y. V. and Kondratieva, E. N. (1980). *FEMS Microbiol. Lett.* **8**, 139–142.
Izumi, Y., Asano, Y., Tani, Y. and Ogata, K. (1977). *J. Ferment. Technol.* **55**, 452–458.
Janssen, F. W. and Ruelius, H. W. (1968). *Biochim. Biophys. Acta.* **151**, 330–342.
Jarman, T. R. and Large, P. J. (1972). *J. Gen. Microbiol.* **73**, 205–208.
Jayaseelan, K. and Guest, J. R. (1979). *FEMS Microbiol. Lett.* **6**, 87–88.
Jensen, R. G. and Bahr, J. T. (1977). *Ann. Rev. Plant. Physiol.* **28**, 379–400.
Johanson, R. A., Hill, J. M. and McFadden, B. A. (1974a). *Biochim. Biophys. Acta.* **364**, 327–340.
Johanson, R. A., Hill, J. M. and McFadden, B. A. (1974b). *Biochim. Biophys. Acta.* **364**, 341–352.
John, P. and Whatley, F. R. (1977). *Biochim. Biophys. Acta.* **463**, 129–153.
Johnson, P. A. and Quayle, J. R. (1964). *Biochem. J.* **93**, 281–290.
Johnson, P. A. and Quayle, J. R. (1965). *Biochem. J.* **95**, 859–867.
Jones, C. W. (1977). *Symp. Soc. Gen. Microbiol.* **27**, 23–59.
Jones, J. B. and Stadtman, T. C. (1976). In "Microbial Production and Utilisation of Gases" (H. G. Schlegel, G. Gottschalk and N. Pfennig, eds), pp. 199–205. Erich Goltze KG, Göttingen.
Jones, J. B. and Stadtman, T. C. (1977). *J. Bacteriol.* **130**, 1404–1406.
Jones, J. B. and Stadtman, T. C. (1980). *J. Biol. Chem.* **255**, 1049–1053.
Jones, J. B. and Stadtman, T. C. (1981). *J. Biol. Chem.* **256**, 656–663.
Jones, J. B., Bowers, B. and Stadtman, T. C. (1977a). *J. Bacteriol.* **130**, 1357–1363.
Jones, C. W., Brice, J. M. and Edwards, C. (1977b). *Arch. Microbiol.* **115**, 85–93.
Jones, J. B., Dilworth, G. L. and Stadtman, T. C. (1979). *Arch. Biochem. Biophys.* **195**, 255–260.
Jordan, P. M. and Akhtar, M. (1970). *Biochem. J.* **116**, 277–286.
Kamanaka, K. (1981). In "Microbial Growth on C_1 Compounds" (H. Dalton, ed.), pp. 21–30. Heyden, London.
Kandler, O. (1979). *Naturwissenschaften* **66**, 95–105.
Kandler, O. and Hippe, H. (1977). *Arch. Microbiol.* **113**, 57–60.
Kandler, O. and Konig, H. (1978). *Arch. Microbiol.* **118**, 141–152.
Kaneda, T. and Roxburgh, J. M. (1959a). *Can. J. Microbiol.* **5**, 87–98.
Kaneda, T. and Roxburgh, J. M. (1959b). *Biochim. Biophys. Acta.* **33**, 106–110.
Kaneda, T. and Roxburgh, J. M. (1959c). *Can. J. Microbiol.* **5**, 187–195.
Kato, N., Tamaoki, T., Tani, Y. and Ogata, K. (1972). *Agr. Biol. Chem.* **36**, 2411–2419.
Kato, N., Kano, M., Tani, Y and Ogata, K. (1974a). *Agr. Biol. Chem.* **38**, 111–116.
Kato, N., Tani, Y. and Ogata, K. (1974b). *Agr. Biol. Chem.* **38**, 675–677.
Kato, K., Kurimura, Y., Makaguchi, N. and Asai, Y. (1974c). *J. Gen. Appl. Microbiol.* **20**, 123–127.

Kato, N., Kazako, T., Tani, Y. and Ogata, K. (1974d). *J. Ferment. Technol.* **52**, 917–920.
Kato, N., Omuri, Y., Tani, Y. and Ogata, K. (1976). *Eur. J. Biochem.* **64**, 341–350.
Kato, N., Ohashi, H., Hori, T., Tani, Y. and Ogata, K. (1977a). *Agr. Biol. Chem.* **41**, 1133–1140.
Kato, N., Tsuji, K., Ohashi, H., Tani, Y. and Ogata, K. (1977b). *Agr. Biol. Chem.* **41**, 29–34.
Kato, N., Ohashi, H., Tani, Y. and Ogata, K. (1978). *Biochim. Biophys. Acta.* **523**, 236–244.
Kato, N., Nishizawa, T., Sakazawa, C., Tani, Y. and Yamada, H. (1979a). *Agr. Biol. Chem.* **43**, 2013–2015.
Kato, N., Sahm, H., Schütte, H. and Wagner, F. (1979b). *Biochim. Biophys. Acta.* **566**, 1–11.
Kato, N., Sahm, H. and Wagner, F. (1979c). *Biochim. Biophys. Acta.* **566**, 12–20.
Kawamoto, S., Yamada, T., Tanaka, A. and Fukui, S. (1979). *FEBS Lett.* **97**, 253–256.
Kawamoto, S., Yamada, T., Tanaka, A. and Fukui, S. (1980). *Arch. Microbiol.* **128**, 145–151.
Kawashima, N. and Wildman, S. G. (1970). *Am. Rev. Plant Physiol.* **21**, 325–358.
Keevil, C. W. and Anthony, C. (1979a). *Biochem. J.* **180**, 237–239.
Keevil, C. W. and Anthony, C. (1979b). *Biochem. J.* **182**, 71–79.
Keilin, D. and Hartree, E. F. (1945). *Biochem. J.* **39**, 293–301.
Kell, D. B. and Morris, J. D. (1979). *FEBS Lett.* **108**, 481–484.
Kell, D. B., John, P. and Ferguson, S. J. (1978). *Biochem. J.* **174**, 257–266.
Kell, D. B., Doddema, H. J., Morris, J. G. and Vogels, G. D. (1981). *In* "Microbial Metabolism of C_1 Compounds" (H. Dalton, ed.), pp. 159–170. Heyden, London.
Kelly, D. P. (1971). *Ann. Rev. Microbiol.* **25**, 177–210.
Kelly, D. P., Wood, A. P., Gottschal, J. C. and Kuenen, J. G. (1979). *J. Gen. Microbiol.* **114**, 1–13.
Keltjens, J. T. and Vogels, G. D. (1981). *In* "Microbial Metabolism of C_1 Compounds" (H. Dalton, ed.), pp. 152–158. Heyden, London.
Kenealy, W. and Zeikus, J. G. (1981). *J. Bacteriol.* **146**, 133–140.
Kemp. M. B. (1972). *Biochem. J.* **127**, 64P–65P.
Kemp, M. B. (1974). *Biochem. J.* **139**, 129–134.
Kemp. M. B. and Quayle, J. R. (1965). *Biochim. Biophys. Acta.* **107**, 174–196.
Kemp, M. B. and Quayle, J. R. (1966). *Biochem. J.* **99**, 41–48.
Kemp, M. B. and Quayle, J. R. (1967). *Biochem. J.* **102**, 94–102.
Kenue, H., Sahm, H. and Wagner, F. (1976). *Eur. J. Appl. Microbiol.* **2**, 175–184.
Khambata, S. R. and Bhat, J. V. (1953). *J. Bacteriol.* **66**, 505–507.
Kihlberg, R. (1972). *Ann. Rev. Microbiol.* **26**, 427–466.
Kim, K. E. and Chang, G. W. (1974). *Can. J. Gen. Microbiol.* **20**, 1745–1748.
Kirikova, N. N. (1970). *Microbiologiya* **39**, 18–31.
Knobloch, K., Ishagne, M. and Aleem, M. I. H. (1971). *Arch. Mikrobiol.* **76**, 114–124.
Komagata, K. (1981). *In* "Microbial Growth on C_1 Compounds" (H. Dalton, ed.), pp. 301–311. Heyden, London.
Konig, H. and Kandler, O. (1979a). *Arch. Microbiol.* **121**, 271–275.
Konig, H. and Kandler, O. (1979b). *Arch. Microbiol.* **123**, 295–299.

Kornberg, H. L. (1966). *In* "Essays in Biochemistry" (P. N. Campbell and G. D. Greville, eds), Vol. 1, pp. 1–31. Academic Press, London and New York.
Kortstee, G. J. J. (1980). *FEMS Microbiol. Lett.* **8,** 59–65.
Kortstee, G. J. J. (1981). *In* "Microbial Growth on C_1 Compounds" (H. Dalton, ed.), pp. 211–219. Heyden, London.
Kouno, K. and Ozaki, A. (1975). *In* "Microbial Growth on C_1 Compounds" pp. 11–21. Society of Fermentation Technology, Tokyo, Japan.
Kouno, K., Oki, T., Kitai, A. and Ozaki, A. (1972). United States Patent No. 3,663,370.
Kouno, K., Oki, T., Nomura, T. and Ozaki, A. (1973). *J. Gen. Appl. Microbiol.* **19,** 11–21.
Krug, E. L. R., Lim, H. C. and Tsao, G. T. (1979). *Ann. Rep. Ferment. Proc.* **3,** 141–195.
Krzycki, S. and Zeikus, J. G. (1980). *Curr. Microbiol.* **3,** 243–245.
Kuhn, W., Fiebig, K., Watther, R. and Gottschalk, G. (1979). *FEBS Lett.* **105,** 271–274.
Kung, H. F. and Wagner, C. (1969). *J. Biol. Chem.* **244,** 4136–4140.
Kung, H. F. and Wagner, C. (1970a). *Biochem. J.* **116,** 357–365.
Kung, H. F. and Wagner, C. (1970b). *Biochim. Biophys. Acta.* **201,** 513–516.
Labischinski, H., Barnickel, G., Leps, B., Bradaczek, H. and Giesbrecht, P. (1980). *Arch. Microbiol.* **127,** 195–201.
Ladner, A. and Zatman, L. J. (1969). *J. Gen. Microbiol.* **55,** xvi.
Lancaster, J. R. (1980). *FEBS Lett.* **115,** 285–288.
Langenberg, K. F., Bryant, M. P. and Wolfe, R. S. (1968). *J. Bacteriol.* **95,** 1124–1129.
Large, P. J. (1971). *FEBS Lett.* **18,** 297–300.
Large, P. J. (1981). *In* "Microbial Growth on C_1 Compounds" (H. Dalton, ed.), pp. 55–69. Heyden, London.
Large, P. J. and Carter, R. H. (1973). *Trans. Biochem. Soc.* **1,** 1291–1293.
Large, P. J. and Haywood, G. W. (1981). *FEMS Microbiol. Lett.* **11,** 207–209.
Large, P. J. and McDougall, H. (1975). *Anal. Biochem.* **64,** 304–310.
Large, P. J. and Quayle, J. R. (1963). *Biochem. J.* **87,** 386–396.
Large, P. J., Peel, D. and Quayle, J. R. (1961). *Biochem. J.* **81,** 470–480.
Large, P. J., Peel, D. and Quayle, J. R. (1962a). *Biochem. J.* **82,** 483–488.
Large, P. J., Peel, D. and Quayle, J. R. (1962b). *Biochem. J.* **85,** 243–250.
Large, P. J., Eady, R. R. and Murden, D. J. (1969). *Anal. Biochem.* **32,** 402–407.
Large, P. J., Meiberg, J. B. M. and Harder, W. (1979). *FEMS Microbiol. Lett.* **5,** 281–286.
Large, P. J., Haywood, G. W., van Dijken, J. P. and Harder, W. (1980). *Soc. Gen. Microbiol. Quart.* **7,** 96–97.
Laskin, A. I. and Lechevalier, H. A. (1973). "Handbook of Microbiology." Vol. 1. CRC Press, The Chemical Rubber Company, Ohio, USA.
Lawford, H. G., Cox, J. C., Garland, P. B. and Haddock, B. A. (1976). *FEBS Lett.* **64,** 369–374.
Lawlis, V. B., Gordon, G. L. R. and McFadden, B. A. (1979). *J. Bacteriol.* **139,** 287–298.
Lawrence, A. J. and Quayle, J. R. (1970). *J. Gen. Microbiol.* **63,** 371–374.
Lawrence, A. J., Kemp, M. B. and Quayle, J. R. (1970). *Biochem. J.* **116,** 631–639.
Lawton, S. A. and Anthony, C. (1981). Unpublished observations.

Leadbetter, E. R. and Foster, J. W. (1958). *Arch. Mikrobiol.* **30**, 91–118.
Leadbetter, E. R. and Foster, J. W. (1959). *Nature* **184**, 1428–1429.
Leadbetter, E. R. and Gottlieb. J. A. (1967). *Arch. Mikrobiol.* **59**, 211–217.
Levering, P. R., van Dijken, J. P., Veenhuis, M. and Harder, W. (1981) *Arch. Microbiol.* **129**, 72–80.
Levi, J. D., Shennan, J. L. and Ebbon, G. P. (1979). *In* "Microbial Biomass" (A. H. Rose, ed.), pp. 362–419. Academic Press, London and New York.
Levine, D. W. and Cooney, C. L. (1973). *Appl. Microbiol.* **26**, 982–990.
Levitch, M. E. (1977a). *Biochem. Biophys. Res. Comm.* **76**, 609–614.
Levitch, M. E. (1977b). *Anal. Biochem.* **82**, 463–467.
Lin, M. C. M. and Wagner, C. (1975). *J. Biol. Chem.* **250**, 3746–3751.
Lindley, N. D., Waites, M. J. and Quayle, J. R. (1980). *FEMS Microbiol. Lett.* **8**, 13–16.
Lindley, N. D., Waites, M. J. and Quayle, J. R. (1981). *J. Gen. Microbiol.* **126**, 253–259.
Linton, J. D. and Buckee, J. C. (1977). *J. Gen. Microbiol.* **101**, 219–225.
Linton, J. D. and Stephenson, R. J. (1978). *FEMS Microbiol. Lett.* **3**, 95–98.
Linton, J. D. and Vokes, J. (1978). *FEMS Lett.* **4**, 125–128.
Litchfield, J. H. (1977). *Adv. Appl. Microbiol.* **22**, 267–305.
Ljungdahl, L. G. and Andreeson, J. R. (1976). *In* "Microbial Production and Utilisation of Gases" (H. G. Schlegel, G. Gottschalk and N. Pfennig, eds), pp. 163–172. Erich Goltze KG, Göttingen.
Ljungdahl, L. G. and Wood, H. G. (1969). *Ann. Rev. Microbiol.* **23**, 515–538.
Lodder, J. (ed.) (1970). "The Yeasts—A Taxonomic Study." North Holland, Amsterdam.
Loffhagen, N. and Babel, W. (1978). *Wiss. Z. Karl-Marx-Univ. Leipzig, Math. Nat. R.* **27**, 81–91.
Loffhagen, N., Bley, T. and Babel, W. (1979). *Z. Allg. Mikrobiol.* **19**, 367–372.
Loginova, N. V. and Trotsenko, Y. A. (1974). *Microbiology* **43**, 831–836.
Loginova, N. V. and Trotsenko, Y. A. (1976a). *Microbiology* **44**, 892–896.
Loginova, N. V. and Trotsenko, Y. A. (1976b). *Microbiology* **45**, 217–223.
Loginova, N. V. and Trotsenko, Y. A. (1977a). *Microbiology* **46**, 170–175.
Loginova, N. V. and Trotsenko, Y. A. (1977b). Abstract of the Second International Conference on Microbial Growth on C_1 Compounds, pp. 37–39 (Putschino).
Loginova, N. V. and Trotsenko, Y. A. (1979a). *Microbiology* **47**, 765–770.
Loginova, N. V. and Trotsenko, Y. A. (1979b). *Mikrobiologya* **48**, 785–791.
Loginova, N. V. and Trotsenko, Y. A. (1979c). *FEMS Microbiol. Lett.* **5**, 239–243.
Loginova, N. V. and Trotsenko, Y. A. (1979d). *Microbiology* **48**, 158–162.
Loginova, N. V., Shishkina, V. N. and Trotsenko, Y. A. (1976). *Microbiology* **45**, 34–40.
Loginova, N. V., Shishkina, V. N., Fillipova, T. M. and Trotsenko, Y. A. (1977). *Izr. Akad. Nauk. SSSR, Ser. Biol. (Eng. trans.)* **2**, 235–243.
Loginova, N. V., Namsaraev, B. B. and Trotsenko, Y. A. (1978). *Microbiology* **47**, 134–135.
Lynch, M., Wopat, A. E. and O'Connor, M. L. (1980). *Appl. Environ. Microbiol.* **40**, 400–407.
Maclennan, D. G., Ousby, J. C., Vasey, R. B. and Cotton, N. T. (1971). *J. Gen. Microbiol.* **69**, 395–404.
Maclennan, D. G., Ousby, J. C., Owen, T. R. and Steer, D. C. (1974). UK Patent No. 1370892.

Madigan, M. T. and Gest, H. (1978). *Arch. Microbiol.* **117**, 119–122.
Mah, R. A., Ward, D. M., Baresi, L. and Glass, T. L. (1977). *Ann. Rev. Microbiol.* **31**, 309–342.
Mah, R. A., Smith, M. R. and Baresi, L. (1978). *Appl. Env. Microbiol.* **35**, 1174–1184.
Mah, R. A., Smith, M. R., Ferguson, T. and Zinder, S. (1981). *In* "Microbial Metabolism of C_1 Compounds" (H. Dalton, ed.), pp. 131–142. Heyden, London.
Makula, R. A. (1978). *J. Bacteriol.* **134**, 771–777.
Makula, R. A. and Singer, M. E. (1978). *Biochem. Biophys. Res. Comm.* **82**, 716–722.
Malashenko, Y. R. (1976). *In* "Microbial Production and Utilisation of Gases (H_2, CH_4, CO)." (H. G. Schlegel, G. Gottschalk and N. Pfennig, eds), pp. 293–300. Erich Goltze KG, Göttingen.
Marison, I. W. (1980). Ph.D. thesis, University of Sheffield.
Marison, I. W. and Attwood, M. M. (1980). *J. Gen. Microbiol.* **117**, 305–313.
Martens, C. S. and Berner, R. A. (1977). *Limnol. Oceanogr.* **22**, 10–25.
Martin, A. (1978). *Ann. Rev. Microbiol.* **32**, 433–468.
Matsumoto, T. (1978). *Biochim. Biophys. Acta.* **522**, 291–302.
Matsumoto, T. and Tobari, J. (1978a). *J. Biochem.* **83**, 1591–1597.
Matsumoto, T. and Tobari, J. (1978b). *J. Biochem.* **84**, 461–465.
Matsumoto, T., Hiraoka, B. Y. and Tobari, J. (1978). *Biochim. Biophys. Acta.* **522**, 303–310.
Matsumoto, T., Shirai, S., Ishii, Y. and Tobari, J. (1980). *J. Biochem.* **88**, 1097–1102.
Mateles, R. I. (1979). *Symp. Soc. Gen. Microbiol.* **29**, 29–52.
McBride, B. C. and Wolfe, R. S. (1971a). *Biochemistry* **10**, 2317–2324.
McBride, B. C. and Wolfe, R. S. (1971b). *Biochemistry* **10**, 4312–4317.
McCarthy, J. E. G., Ferguson, S. J. and Kell, D. B. (1981). *Biochem. J.* **196**, 311–321.
McFadden, B. A. (1969). *In* "Methods in Enzymology" (J. M. Lowenstein, ed.), Vol. 13, pp. 163–170.
McFadden, B. A. (1973). *Bacteriol. Rev.* **37**, 289–319.
McFadden, B. A. (1978). *In* "The Bacteria" (L. N. Ornston and J. R. Sokatch, eds), Vol. 6, pp. 219–304. Academic Press, London and New York.
McFadden, B. A. and Purohit, S. (1977). *J. Bacteriol.* **131**, 136–144.
McFadden, B. A., Rao, G. R., Cohen, A. L. and Roche, T. E. (1968). *Biochemistry* **7**, 3574–3582.
McKellar, R. C. and Sprott, G. D. (1979). *J. Bacteriol.* **139**, 231–238.
McNerney, T. and O'Connor, M. L. (1980). *Appl. Environ. Microbiol.* **40**, 370–375.
Mehta, R. J. (1973a). *Antonie van Leeuwenhoek* **39**, 295–302.
Mehta, R. J. (1973b). *Antonie van Leeuwenhoek* **39**, 303–312.
Mehta, R. J. (1975a). *J. Bacteriol.* **124**, 1165–1167.
Mehta, R. J. (1975b). *Experientia* **31**, 407–408.
Mehta, R. J. (1975c). *Antonie van Leeuwenhoek* **41**, 89–95.
Mehta, R. J. (1976). *J. Ferment. Technol.* **54**, 596–602.
Mehta, R. J. (1977). *Can. J. Microbiol.* **23**, 402–406.
Meiberg, J. B. M. (1979). Ph.D. thesis, University of Groningen, The Netherlands.
Meiberg, J. B. M. and Harder, W. (1978). *J. Gen. Microbiol.* **106**, 265–276.
Meiberg, J. B. M. and Harder, W. (1979). *J. Gen. Microbiol.* **115**, 49–58.
Meiberg, J. B. M., Bruinenberg, P. M. and Harder, W. (1980). *J. Gen. Microbiol.* **120**, 453–463.
Miall, L. M. (1980). *In* "Hydrocarbons in Biotechnology" (D. E. F. Harrison, I. J. Higgins and R. Watkinson, eds), pp. 25–34. Heyden, London.

Michalik, J. and Raczynska-Bojanowska, K. (1976). *Acta. Biochim. Polon.* **23**, 375–386.
Michalik, J., Budohoski, L. and Raczynska-Bojanowska, K. (1979). *Acta. Biochim. Polon.* **26**, 397–406.
Michels, P. A. M. and Haddock, B. A. (1980). *FEMS Microbiol. Lett.* **7**, 327–331.
Middlehoven, W. J., Berends, J., van Aert, A. J. M. and Bruinsma, J. (1976). *J. Gen. Microbiol.* **93**, 185–188.
Miethe, D. and Babel, W. (1976). *Z. Allg. Mikrobiol.* **16**, 289–299.
Millay, R. H. and Hersh, L. B. (1976). *J. Biol. Chem.* **251**, 2754–2760.
Millay, R. H., Schilling, H. and Hersh, L. B. (1978). *J. Biol. Chem.* **253**, 1371–1377.
Minami, K., Yanamura, M., Shimizu, S., Ogawa, K., and Sekine, N. (1978). *J. Ferment. Technol.* **56**, 1–7.
Mink, R. W. and Dugan, P. R. (1977). *Appl. Env. Microbiol.* **33**, 713–717.
Mitchell, R. M., Loeblich, L. A., Klotz, L. C. and Loeblich, A. R. (1978). *Science* **204**, 1082–1084.
Monosov, E. Z. and Netrusov. A. I. (1975). *Microbiology* **45**, 518–523.
Moore, G. R., Pettigrew, G. W., Pitt, R. C. and Williams, R. J. P. (1980). *Biochim. Biophys. Acta.* **590**, 261–271.
Mountfort, D. O. (1978). *Biochem. Biophys. Res. Comm.* **85**, 1346–1351.
Mue, S., Tubori, S. and Kikuchi, G. (1964). *J. Biochem.* **56**, 545–551.
Müller, R. and Babel, W. (1980). *Z. Allg. Mikrobiol.* **20**, 325–333.
Müller, R. and Sokolov, A. P. (1979). *Z. Allg. Mikrobiol.* **19**, 261–267.
Müller, U., Willnow, P., Ruschig, U. and Höpfner, T. (1978). *Eur. J. Biochem.* **83**, 485–498.
Myers, P. A. and Zatman, L. J. (1971). *Biochem. J.* **121**, 10P.
Nakayama, K., Kobata, M., Tanaka, Y., Nomura, T. and Katsumata, R. (1976). United States Patent No. 3,939,042.
Namsarev, B. B., Nozhevnikova, A. N. and Zavarzin, G. A. (1971). *Microbiology* **40**, 675–678.
Natori, Y., Nagasaki, T., Kobayashi, A. and Fukawa H. (1978). *Agr. Biol. Chem.* **42**, 1799–1800.
Neben, I., Sahm, H. and Kula, M-R. (1980). *Biochim. Biophys. Acta.* **614**, 81–91.
Neill, A. R., Grime, D. W. and Dawson, R. M. C. (1978). *Biochem. J.* **170**, 529–535.
Netrusov, A. I. (1975). *Microbiology* **44**, 487–488.
Netrusov, A. I. (1981). *In* "Microbial Growth on C_1 Compounds" (H. Dalton, ed.), pp. 231–239. Heyden, London.
Netrusov, A. I. and Anthony, C. (1979). *Biochem. J.* **178**, 353–360
Netrusov, A. I., Rodionov, Y. V. and Kondratieva, E. N. (1977). *FEBS Lett.* **76**, 56–58.
Newaz, S. S. and Hersh, L. B. (1975). *J. Bacteriol.* **124**, 825–833.
Newsholme, E. A. and Start, C. (1973). "Regulation in Metabolism." Wiley, London.
Nishio, N., Yano, T. and Kamikubo, T. (1975a). *Agr. Biol. Chem.* **39**, 21–27.
Nishio, N., Yano, T. and Kamikubo, T. (1975b). *Agr. Biol. Chem.* **39**, 207–213.
Oberlies, G., Fuchs, G. and Thauer, R. K. (1980). *Arch. Microbiol.* **128**, 248–252.
O'Connor, M. L. (1981). *In* "Microbial Growth on C_1 Compounds" (H. Dalton, ed.), pp. 293–300. Heyden, London.
O'Connor, M. L. and Hanson, R. S. (1975). *J. Bacteriol.* **124**, 985–996.
O'Connor, M. L. and Hanson, R. S. (1977). *J. Gen. Microbiol.* **101**, 327–332.
O'Connor, M. L. and Hanson, R. S. (1978). *J. Gen. Microbiol.* **104**, 105–111.

O'Connor, M. L. and Quayle, J. R. (1979). *J. Gen. Microbiol.* **113**, 203–208.
O'Connor, M. L. and Quayle, J. R. (1980). *J. Gen. Microbiol.* **120**, 219–225.
O'Connor, M. L., Wopat, A. and Hanson, R. S. (1977). *J. Gen. Microbiol.* **98**, 265–272.
Ogata, K., Nishikawa, H. and Ohsugi, M. (1969). *Agr. Biol. Chem.* **33**, 1519–1520.
Ogata, K., Tani, Y. and Kato, N. (1975). "Microbial Growth on C_1 Compounds." pp. 99–119. The Society of Fermentation Technology, Japan.
Ogata, K., Izumi, Y., Kawamori, M., Asano, Y. and Tani, Y. (1977). *J. Ferment. Technol.* **55**, 444–451.
O'Keeffe, D. T. and Anthony, C. (1978). *Biochem. J.* **170**, 561–567.
O'Keeffe, D. T. and Anthony, C. (1980a). *Biochem. J.* **190**, 481–484.
O'Keeffe, D. T. and Anthony, C. (1980b). *Biochem. J.* **192**, 411–419.
O'Keeffe, D. T. and Anthony, C. (1981). Unpublished observations.
Oki, T. and Kitai, A. (1974). *Process Biochem.* **9** (9), 31–32.
Oki, T., Kuono, K., Kitai, A. and Ozaki, A. (1972). *J. Gen. Appl. Microbiol.* **18**, 295–305.
Oki, T., Kitai, A., Kouno, K. and Ozaki, A. (1973). *J. Gen. Appl. Microbiol.* **19**, 79–83.
Okomura, S., Yamanoi, A., Tsugawa, R. and Nakase, T. (1969). Patent specification 1210770, London, U.K.
Ooyama, J. and Foster, J. W. (1965). *Antonie van Leeuwenhoek* **31**, 45–65.
Osumi, M. and Sato, M. (1978). *J. Electron Microsc.* **27**, 127–136.
Panganiban, A. T., Patt, T. E., Hart, W. and Hanson, R. S. (1979). *Appl. Env. Microbiol.* **37**, 303–309.
Patel, R. N. and Felix, A. (1976). *J. Bacteriol.* **128**, 413–424.
Patel, R. N. and Hoare, D. S. (1971). *J. Bacteriol.* **107**, 187–192.
Patel, R. N., Hoare, D. S. and Taylor, B. F. (1969). *Bacteriol. Proc.* **69**, 128.
Patel, R. N., Bose, H. R., Mandy, W. J. and Hoare, D. S. (1972). *J. Bacteriol.* **110**, 570–577.
Patel, R. N., Mandy, W. J. and Hoare, D. S. (1973). *J. Bacteriol.* **113**, 937–945.
Patel, R., Hoare, S. L., Hoare, D. S. and Taylor, B. F. (1975). *J. Bacteriol.* **123**, 382–384.
Patel, R., Hou, C. T. and Felix, A. (1976). *J. Bacteriol.* **126**, 1017–1019.
Patel, R. N., Hoare, S. L., Hoare, D. S. and Taylor, B. F. (1977). *Appl. Environ. Microbiol.* **34**, 607–610.
Patel, R. N., Hou, C. T. and Felix, A. (1978a). *J. Bacteriol.* **133**, 641–649.
Patel, R. N., Hou, C. T. and Felix, A. (1978b). *J. Bacteriol.* **136**, 352–358.
Patel, R. N., Hou, C. T. and Felix, A. (1979a). *J. Gen. Appl. Microbiol.* **25**, 197–204.
Patel, R. N., Hou, C. T. and Felix, A. (1979b). *Arch. Microbiol.* **122**, 241–248.
Patel, R. N., Hou, C. T., Laskin, A. I., Felix, A. and Derelanko, P. (1979c). *J. Bacteriol.* **139**, 675–679.
Patel, R. N., Hou, C. T., Laskin, A. I., Derelanko, P. and Felix, A. (1979d). *Appl. Environ. Microbiol.* **38**, 209–223.
Patel, R. N., Hou, C. T., Laskin, A. I., Derelanko, P. and Felix, A. (1979e). *Eur. J. Biochem.* **101**, 401–406.
Patel, R. N., Hou, C. T., Laskin, A. I., Felix, A. and Derelanko, P. (1980a). *Appl. Environ. Microbiol.* **39**, 720–726.
Patel, R. N., Hou, C. T., Derelanko, P. and Felix, A. (1980b). *Arch. Biochem. Biophys.* **203**, 654–662.
Patt, T. E. and Hanson, R. S. (1978). *J. Bacteriol.* **134**, 636–644.

Patt, T. E., Cole, G. C., Bland, J. and Hanson, R. S. (1974). *J. Bact.* **120**, 955–964.
Patt, T. E., Cole, G. C. and Hanson, R. S. (1976). *Int. J. Syst. Bact.* **26**, 226–229.
Payne, W. J. (1970). *Ann. Rev. Microbiol.* **24**, 17–52.
Payne, W. J. and Wiebe, W. J. (1978). *Ann. Rev. Microbiol.* **32**, 155–184.
Peel, D. and Quayle, J. R. (1961). *Biochem. J.* **81**, 465–469.
Penley, M. and Wood, J. (1972). *Biochim. Biophys. Acta.* **273**, 265–274.
Perry, J. J. (1968). *Antonie van Leeuwenhoek* **34**, 27–36.
Pine, M. J. and Barker, H. A. (1956). *J. Bacteriol.* **71**, 644–648.
Pine, M. J. and Vishniac, W. (1957). *J. Bacteriol.* **73**, 736–742.
Plotkin, E. V., Higgins, I. J. and Hill, H. A. O. (1981). *Biotechnol. Lett.* **3**, 187–192.
Pol. A., van der Drift, C., Vogels, G. D., Cuppen, T. J. H. M. and Laarkhoven, W. H. (1980). *Biochem. Biophys. Res. Comm.* **92**, 255–260.
Pollock, R. J. and Hersh, L. B. (1971). *J. Biol. Chem.* **246**, 4737–4743.
Pollock, R. J. and Hersh, L. B. (1973). *J. Biol. Chem.* **248**, 6724–6733.
de Pontanel, H. G. (1975). (ed.) "Proteins from Hydrocarbons". Academic Press, New York and London.
Porte, F. and Vignais, P. M. (1980). *Arch. Microbiol.* **127**, 1–10.
Proctor, H. M., Norris, J. R. and Ribbons, D. W. (1969). *J. Appl. Bacteriol.* **32**, 118–121.
Quadri, S. M. H. and Hoare, D. S. (1969). *J. Bacteriol.* **95**, 2344–2357.
Quayle, J. R. (1961). *Ann. Rev. Microbiol.* **15**, 119–152.
Quayle, J. R. (1969). *Meth. Enzymol.* **13**, 292–296.
Quayle, J. R. (1972). *Adv. Microb. Physiol.* **7**, 119–203.
Quayle, J. R. (1975). *In* "Microbial Growth on C_1 Compounds" (G. Terui, ed.), pp. 59–65. Society of Fermentation Technology, Japan.
Quayle, J. R. (1980). *Proc. Biochem. Soc.* **8**, 1–10.
Quayle, J. R. and Ferenci, T. (1978). *Microbiol. Rev.* **42**, 251–273.
Quayle, J. R. and Keech, D. B. (1958). *Biochim. Biophys. Acta.* **29**, 223–225.
Quayle, J. R. and Keech, D. B. (1959a). *Biochem. J.* **72**, 623–630.
Quayle, J. R. and Keech, D. B. (1959b). *Biochem. J.* **72**, 631–637.
Quayle, J. R. and Keech, D. B. (1959c). *Biochim. Biophys. Acta.* **31**, 587–588.
Quayle, J. R. and Keech, D. B. (1960). *Biochem. J.* **75**, 515–523.
Quayle, J. R. and Pfennig, N. (1975). *Arch. Microbiol.* **102**, 193–198.
Quayle, J. R., Fuller, R. C., Benson, A. A. and Calvin, M. (1954). *J. Am. Chem. Soc.* **76**, 3610.
Quayle, J. R., Keech, D. B. and Taylor, G. A. (1961). *Biochem. J.* **78**, 225–236.
Reeburgh, W. S. (1976). *Earth Planet. Sci. Lett.* **28**, 337–344.
Reeburgh, W. S. (1980). *Earth Planet. Sci. Lett.* **47**, 345–352.
Reeburgh, W. S. (1981). *In* "The Dynamic Environment of the Ocean Floor" (K. Fanning and F. T. Manheim, eds), pp. 203–217. D. C. Heath, Lexington, U.S.A.
Reeburgh, W. S. and Heggie, D. T. (1977). *Limnol. Oceanogr.* **22**, 1–9.
Reed, W. M., Titus, J. A., Dugan, P. R. and Pfistor, R. M. (1980). *J. Bacteriol.* **141**, 908–913.
Reuss, M., Gneiser, J., Reng, H. G. and Wagner, F. (1975). *Eur. J. Appl. Microbiol.* **1**, 295–305.
Ribbons, D. W. (1975). *J. Bacteriol.* **122**, 1351–1363.
Ribbons, D. W. and Michalover, J. L. (1970). *FEBS Lett.* **11**, 41–44.
Ribbons, D. W. and Wadzinski, A. M. (1976). *In* "Microbial Production and Utilisation of Gases (H_2, CH_4, CO)". (H. G. Schlegel, G. Gottschalk and N. Pfennig, eds), pp. 359–369. Erich Goltze KG, Göttingen.

Ribbons, D. W., Harrison, J. E. and Wadzinski, A. M. (1970). *Ann. Rev. Microbiol.* **24**, 135–150.
Roberton, A. M. and Wolfe, R. S. (1969). *Biochim. Biophys. Acta.* **192**, 420–429.
Roberton, A. M. and Wolfe, R. S. (1970). *J. Bacteriol.* **102**, 43–51.
Rock, J. S., Goldberg, I., Ben-Bassat, A. and Mateles, R. I. (1976). *Agr. Biol. Chem.* **40**, 2129–2135.
Rodionov, Y. V. and Zakharova, E. V. (1980). *Biochemistry (USSR)* **45**, 654–661.
Rodionov, Y. V., Avilova, T. V., Zakharova, E. V., Platonenkova, L. S., Egerov, A. M. and Berezin, I. V. (1977a). *Biokhimiya* **42**, 1896–1904.
Rodionov, Y. V., Avilova, T. A. and Popov, V. O. (1977b). *Biochemistry (USSR)* **42**, 1594–1598.
Rodionov, Y. V., Avilova, T. V. and Popov, V. O. (1977c). *Biokhimiya* **42**, 2020–2026.
Roels, J. A. (1980). *Biotechnol. Bioeng.* **22**, 33–53.
Roggenkamp, R., Sahm, H. and Wagner, F. (1974). *FEBS Lett.* **41**, 283–286.
Roggenkamp, R., Sahm, H., Hinkelmann, W. and Wagner, F. (1975). *Eur. J. Biochem.* **59**, 231–236.
Rohmer, M., Bouvier, P. and Ourisson, G. (1979). *Proc. Nat. Acad. Sci.* **76**, 847–851.
Rokem, J. S., Goldberg, I. and Mateles, R. I. (1978a). *Biotechnol. Bioeng.* **20**, 1557–1564.
Rokem, J. S., Reichler, J. and Goldberg, I. (1978b) *Antonie van Leeuwenhoek* **44**, 123–127.
Rokem, J. S., Goldberg, I. and Mateles, R. I. (1980). *J. Gen. Microbiol.* **116**, 225–232.
Romanova, A. K. and Nozhevnikova, A. N. (1977). Abstract of the Second International Symposium on Microbial Growth on C_1 Compounds, pp. 109–110. Pushchino, USSR.
Romanovskaya, V. A., Malashenko, Y. R. and Bogachenko, V. N. (1978). *Microbiology* **47**, 120–130.
Romesser, J. A. (1978). Ph.D. thesis, University of Illinois, Urbana.
Romesser, J. A. and Wolfe, R. S. (1981). *Biochem. J.* **197**, 565–571.
Romesser, J. A., Wolfe, R. S., Mayer, F., Spiess, E. and Walther-Mauruschat, A. (1979). *Arch. Microbiol.* **121**, 147–153.
Rose, A. H. (ed.) (1978). "Economic Microbiology", Vol. 2. Academic Press, London and New York.
Rose, A. H. (ed.) (1979). "Economic Microbiology," Vol. 4. Academic Press, London and New York.
Rottenberg, H. (1979). *Biochim. Biophys. Acta.* **549**, 225–253.
Sahm, H. (1977). *Adv. Biochem. Eng.* **6**, 77–103.
Sahm, H. and Wagner, F. (1972). *Arch. Mikrobiol.* **84**, 29–42.
Sahm, H. and Wagner, F. (1973a). *Eur. J. Biochem.* **36**, 250–256.
Sahm, H. and Wagner, F. (1973b). *Arch. Mikrobiol.* **90**, 263–268.
Sahm, H. and Wagner, F. (1974). *Arch. Microbiol.* **97**, 163–168.
Sahm, H. and Wagner, F. (1975). *Eur. J. Appl. Microbiol.* **1**, 147–158.
Sahm, H., Roggenkamp. R., Wagner, F. and Hinkelman, W. (1975). *J. Gen. Microbiol.* **88**, 218–222.
Sahm, H., Cox, R. B. and Quayle, J. R. (1976a). *J. Gen. Microbiol.* **94**, 313–322.
Sahm, H., Schütte, H. and Kula, M-R. (1976b). *Eur. J. Biochem.* **66**, 591–596.
Sakaguchi, K., Kurane, R. and Murata, M. (1975). *Agr. Biol. Chem.* **39**, 1695–1702
Salem, A. R. and Quayle, J. R. (1971). *Biochem. J.* **124**, 74P.
Salem, A. R., Large, P. J. and Quayle, J. R. (1972). *Biochem. J.* **128**, 1203–1211.

References

Salem, A. R., Wagner, C., Hacking, A. J. and Quayle, J. R. (1973a). *J. Gen. Microbiol.* **76**, 375–388.
Salem, A. R., Hacking, A. J. and Quayle, J. R. (1973b). *Biochem. J.* **136**, 89–96.
Salem, A. R., Hacking, A. J. and Quayle, J. R. (1974). *J. Gen. Microbiol.* **81**, 525–527.
Salisbury, S. A., Forrest, H. S., Cruse, W. B. T. and Kennard, O. (1979). *Nature* **280**, 843–844.
Sato, K. (1978). *FEBS Lett.* **85**, 207–210.
Sato, K., Veda, S. and Shimizu, S. (1976). *FEBS Lett.* **71**, 248–250.
Sato, K., Veda, S. and Shimizu, S. (1977). *Appl. Environ. Microbiol.* **33**, 515–521.
Sauer, F. D., Erfle, J. D. and Mahadevan, S. (1979). *Biochem. J.* **178**, 165–172.
Sauer, F. D., Mahadevan, S. and Erfle, J. D. (1980a). *Biochem. Biophys. Res. Comm.* **95**, 715–721.
Sauer, F. D., Erfle, J. D. and Mahadevan, S. (1980b). *Biochem. J.* **190**, 177–182.
Schauer, N. L. and Ferry, J. G. (1980). *J. Bacteriol.* **142**, 800–807.
Schlegel, H. G. and Meyer, O. (1981). *In* "Microbial Growth on C_1 Compounds" (H. Dalton, ed.), pp. 105–115. Heyden, London.
Schlegel, H. G., Gottschalk, G. and Pfennig, N. (eds) (1976). "Symposium on Microbial Production and Utilisation of Gases (H_2, CH_4, CO)". Erich Goltze KG, Göttingen.
Schönheit, P. and Thauer, R. K. (1980). *FEMS Lett.* **9**, 77–80.
Schönheit, P., Moll, J. and Thauer, R. K. (1979). *Arch. Microbiol.* **123**, 105–107.
Schönheit, P., Moll, J. and Thauer, R. K. (1980). *Arch. Microbiol.* **127**, 59–65.
Schütte, H., Flossdorf, J., Sahm, H. and Kula, M-R. (1976). *Eur. J. Biochem.* **62**, 151–160.
Scott, D., Brannan, J. and Higgins, I. J. (1981). *J. Gen. Microbiol.* **125**, 63–72.
Seifert, E. and Pfennig, N. (1979). *Arch. Microbiol.* **122**, 177–182.
Seigel, M. I., Wishnick, M. and Lane, M. E. (1972). *In* "The Enzymes" (P. D. Boyer, ed.), 3rd edn, Vol. VI, pp. 169–192. Academic Press, New York and London.
Senior, P. J. and Windass, J. (1980). *Biotechnol. Lett.* **2**, 205–210.
Seto, N., Sakayanagi, S. and Lizuka, H. (1975). *In* "Microbial Growth on C_1 Compounds", pp. 35–44. Society of Fermentation Technology, Tokyo, Japan.
Shapiro, S. and Wolfe, R. S. (1980). *J. Bacteriol.* **141**, 728–734.
Shaw, W. V. and Stadtman, E. R. (1970). *Meth. Enzymol.* **17**, 868–873.
Shaw, W. V., Tsai, L. and Stadtman, E. R. (1966). *J. Biol. Chem.* **241**, 935–945.
Shiio, I. and Ozaki, H. (1968). *J. Biochem.* **64**, 45–53.
Shimizu, S., Ishida, M., Tani, Y. and Ogata, K. (1977a). *Agr. Biol. Chem.* **41**, 423–424.
Shimizu, S., Ishida, M., Kato, N., Tani, Y. and Ogata, K. (1977b). *Agr. Biol. Chem.* **41**, 2215–2220.
Shimizu, S., Ishida, M., Tani, Y. and Ogata, K. (1977c). *J. Ferment. Technol.* **55**, 630–632.
Shirai, S., Matsumoto, T. and Tobari, J. (1978). *J. Biochem.* **83**, 1599–1607.
Shishkina, V. N. and Trotsenko, Y. A. (1979). *FEMS Microbiol. Lett.* **5**, 187–191.
Shishkina, V. N., Yurchenko, V. V., Romanovskaya, V. A., Malashenko, Y. R. and Trotsenko, Y. A. (1976). *Microbiology* **45**, 359–361.
Shively, J. M., Saluja, A. and McFadden, B. A. (1978). *J. Bacteriol.* **134**, 1123–1132.
Sjölin, B. and Vestermark, A. (1973). *Biochim. Biophys. Acta.* **297**, 165–173.
Smith, S. R. L. (1981). *In* "Microbial Growth on C_1 Compounds" (H. Dalton, ed.), pp. 342–348. Heyden, London.
Smith, A. J. and Hoare, D. S. (1977). *Bact. Rev.* **41**, 419–448.

Smith, M. R. and Mah, R. A. (1978). *Appl. Environ. Microbiol.* **36**, 870–879.
Smith, M. R. and Mah, R. A. (1980). *Appl. Environ. Microbiol.* **39**, 993–999.
Smith, V. and Ribbons, D. W. (1970). *Arch. Mikrobiol.* **74**, 116–122.
Smith, A. L., Kelly, D. P. and Wood, A. P. (1980). *J. Gen. Microbiol.* **121**, 127–138.
Snedecor, B. and Cooney, C. L. (1974). *Appl. Microbiol.* **27**, 1112–1117.
Söhngen, N. L. (1906). *Zentbl. Bakt. ParasitKde.* (*Abt. II*) **15**, 513.
Sokolov, A. P. and Trotsenko, Y. A. (1977). *Mikrobiologiya* **46**, 1119–1121.
Sokolov, A. P. and Trotsenko, Y. A. (1978a). *Microbiology* **46**, 902–904.
Sokolov, A. P. and Trotsenko, Y. A. (1978b). *Biochemistry* **43**, 620–625.
Sokolov, A. P., Luchin, S. V. and Trotsenko, Y. A. (1980). *Biokhimiya* **45**, 1371–1378.
Spector, L. B. (1972). In "The Enzymes" (P. D. Boyer, ed.), Vol. 7, pp. 357–389. Academic Press, London and New York.
Sperl, G. T. and Hoare, D. S. (1971). *J. Bacteriol.* **108**, 733–736.
Sperl, G. T., Forrest, H. S. and Gibson, D. T. (1974). *J. Bacteriol.* **118**, 541–550.
Stadtman, T. C. (1967). *Ann. Rev. Microbiol.* **21**, 121–142.
Stadtman, T. C. and Barker, H. A. (1949). *Arch. Biochem.* **21**, 255–264.
Stadtman, T. C. and Barker, H. A. (1951). *J. Bacteriol.* **61**, 81–86.
Stanier, R. Y., Palleroni, N. J. and Doudoroff, M. (1966). *J. Gen. Microbiol.* **43**, 159–271.
Steenkamp, D. J. (1979). *Biochem. Biophys. Res. Comm.* **88**, 244–250.
Steenkamp, D. J. and Gallup, M. (1978). *J. Biol. Chem.* **253**, 4086–4089.
Steenkamp, D. J. and Mallinson, J. (1976). *Biochim. Biophys. Acta.* **429**, 705–719.
Steenkamp, D. J. and Singer, T. P. (1976). *Biochem. Biophys. Res. Comm.* **71**, 1289–1295.
Steenkamp, D. J., Singer, T. P. and Beinert, H. (1978a). *Biochem. J.* **169**, 361–369.
Steenkamp, D. J., Kenney, W. C. and Singer, T. P. (1978b). *J. Biol. Chem.* **253**, 2812–2817.
Steenkamp. D. J., McIntire, W. and Kenney, W. C. (1978c). *J. Biol. Chem.* **253**, 2818–2824.
Steenkamp, D. J., Beinert, H., McIntire, W. and Singer T. P. (1978d). In "Mechanisms of Oxidising Enzymes" (T. P. Singer and R. N. Oudarza, eds), pp. 127–141. Elsevier, Amsterdam.
Steinbach, R. A., Sahm, H. and Schütte, H. (1978). *Eur. J. Biochem.* **87**, 409–415.
Stieglitz, D. and Mateles, R. I. (1973). *J. Bacteriol.* **114**, 390–398.
Stirling, D. I. and Dalton, H. (1977). *Arch. Microbiol.* **114**, 71–76.
Stirling, D. I. and Dalton, H. (1978). *J. Gen. Microbiol.* **107**, 19–29.
Stirling, D. I. and Dalton, H. (1979a). *Eur. J. Biochem.* **96**, 205–212.
Stirling, D. I. and Dalton, H. (1979b). *FEMS Microbiol. Lett.* **5**, 315–318.
Stirling, D. I. and Dalton, H. (1980). *J. Gen. Microbiol.* **116**, 277–283.
Stirling, D. I. and Dalton, H. (1981). *Nature* **291**, 169.
Stirling, D. I., Colby, J. and Dalton, H. (1979). *Biochem. J.* **177**, 361–366.
Stocks, P. K. and McCleskey, C. S. (1964a). *J. Bacteriol.* **88**, 1065–1070.
Stocks, P. K. and McCleskey, C. S. (1964b). *J. Bacteriol.* **88**, 1071–1077.
Stokes, J. E. and Hoare, D. S. (1969). *J. Bacteriol.* **100**, 890–894.
Stouthamer, A. H. (1976). "Yield Studies in Microorganisms." Meadowfield Press, Durham.
Stouthamer, A. H. (1977a). *Symp. Soc. Gen. Microbiol.* **27**, 285–315.
Stouthamer, A. H. (1977b). *Antonie van Leeuwenhoek* **43**, 351–367.
Stouthamer, A. H. (1978). In "The Bacteria" (L. N. Ornston and J. R. Sokatch, eds), Vol. 6, pp. 219–304. Academic Press, London and New York.

Stouthamer, A. H. (1979). *In* "International Review of Biochemistry" (J. R. Quayle, ed.), Vol. 21, pp. 300–353. University Park Press, Baltimore.
Strøm, T., Ferenci, T. and Quayle, J. R. (1974). *Biochem. J.* **144**, 465–476.
Stucki, J. W. (1980). *Eur. J. Biochem.* **109**, 269–283.
Stucki, G., Gälli, R., Ebersold, H. R. and Leisinger, T. (1981). *Arch. Microbiol.* **130**, 366–371.
Suzuki, M., Kühn, I., Berglund, A., Unden, A. and Heden, C. G. (1977a). *J. Ferment. Technol.* **55**, 459–465.
Suzuki, M., Berglund, A., Unden, A. and Heden, C. G. (1977b). *J. Ferment. Technol.* **55**, 466–475.
Swartz, J. R. and Cooney, C. L. (1981). *Appl. Environ. Microbiol.* **41**, 1206–1213.
Tanaka, A., Ohya, Y., Shimizu, S. and Fukui, S. (1974). *J. Ferment. Technol.* **52**, 921–924.
Tanaka, A., Yasuhara, S., Kawamoto, S., Fukui, S. and Osumi, M. (1976). *J. Bacteriol.* **126**, 919–927.
Tanaka, A., Yasuhara, S., Osumi, M. and Fukui, S. (1977). *Eur. J. Biochem.* **80**, 193–197.
Tanaka, A., Yasuhara, S., Gellf, G., Osumi, M. and Fukui, S. (1978). *Eur. J. Appl. Microbiol. Technol.* **5**, 17–27.
Tani, Y., Miya, T., Nishikawa, H. and Ogata, K. (1972a). *Agr. Biol. Chem.* **36**, 68–75.
Tani, Y., Miya, T. and Ogata, K. (1972b). *Agr. Biol. Chem.* **36**, 76–83.
Tani, Y., Kato, N. and Yamada, H. (1978a). *Adv. Appl. Microbiol.* **24**, 165–186.
Tani, Y., Kanagawa, T., Hanpongkittikun, A., Ogata, K. and Yamada, H. (1978b). *Agr. Biol. Chem.* **42**, 2275–2279.
Tannenbaum, S. R. and Wang, D. I. C. (eds) (1975). "Single-cell protein II." MIT Press, Cambridge, Massachusetts.
Tate, R. L. and Alexander, M. (1976). *Appl. Environ. Microbiol.* **31**, 399–403.
Taylor, C. D. (1976). *In* "Microbial Production and Utilisation of Gases" (H. G. Schlegel, G. Gottschalk and N. Pfennig, eds), pp. 181–190. Erich Goltze KG, Göttingen.
Taylor, I. J. (1977). "Microbial Growth on C_1 Compounds," pp. 52–54. Pushchino, USSR.
Taylor, I. J. and Anthony, C. (1976a). *J. Gen. Microbiol.* **93**, 259–265.
Taylor, I. J. and Anthony, C. (1976b). *J. Gen. Microbiol.* **95**, 134–143.
Taylor, S. C. and Dow, C. S. (1980). *J. Gen. Microbiol.* **116**, 81–87.
Taylor, G. T. and Pirt, S. J. (1977). *Arch. Microbiol.* **113**, 17–22.
Taylor, I. J. and Senior, P. J. (1978). *Endeavour* **2**, 31–34.
Taylor, C. D. and Wolfe, R. S. (1974a). *J. Biol. Chem.* **249**, 4879–4885.
Taylor, C. D. and Wolfe, R. S. (1974b). *J. Biol. Chem.* **249**, 4886–4890.
Taylor, C. D., McBride, B. C., Wolfe, R. S. and Bryant, M. P. (1975). *J. Bacteriol.* **120**, 974–975.
Taylor, G. T., Kelly, D. P. and Pirt, S. J. (1976). *In* "Microbial Production and Utilisation of Gases" (H. G. Schlegel, G. Gottschalk and N. Pfennig, eds), pp. 173–179. Erich Goltze KG, Göttingen.
Taylor, S. C., Dalton, H. and Dow, C. (1980). *FEMS Microbiol. Lett.* **8**, 157–160.
Taylor, S. C., Dalton, H. and Dow, C. S. (1981). *J. Gen. Microbiol.* **122**, 89–94.
Tempest, D. W. and Neijssel, O. M. (1980). *In* "Diversity of Bacterial Respiration" (C. J. Knowles, ed.), Vol. I, pp. 1–31. CRC. Press, Florida.
Tempest, D. W. and Neijssel, O. M. (1981). *In* "Microbial Growth on C_1 Compounds" (H. Dalton, ed.), pp. 325–334. Heyden, London.

Tezuka, H., Nakahara, T., Minoda, Y. and Yamada, K. (1975). *Agr. Biol. Chem.* **39**, 285–286.
Thauer, R. and Fuchs, G. (1981). In "Microbial Growth on C_1 Compounds" (H. Dalton, ed.), pp. 143–145. Heyden, London.
Thauer, R. K., Jungermann, K. and Decker, K. (1977). *Bact. Rev.* **41**, 100–180.
Thomson, A. W., O'Neill, J. G. and Wilkinson, J. F. (1976). *Arch. Microbiol.* **109**, 243–246.
Tonge, G. M., Knowles, C. J., Harrison, D. E. F. and Higgins, I. J. (1974). *FEBS Lett.* **44**, 106–110.
Tonge, G. M., Harrison, D. E. F., Knowles, C. J. and Higgins, I. J. (1975). *FEBS Lett.* **58**, 293–299.
Tonge, G. M., Harrison, D. E. F. and Higgins, I. J. (1977a). *Biochem. J.* **161**, 333–344.
Tonge, G. M., Drozd, J. W. and Higgins, I. J. (1977b). *J. Gen. Microbiol.* **99**, 229–232.
Tonomura, K., Kansaki, F. and Kambayashi, A. (1972). *Report Ferment. Res. Inst.* **41**, 49–55.
Toraya, R., Yongsmith, B., Tanaka, A. and Fukui, S. (1975). *Appl. Microbiol.* **30**, 477–479.
Toraya, R., Yongsmith, B., Honda, S., Tanaka, A. and Fukui, S. (1976). *J. Ferment. Technol.* **54**, 102–108.
Tornabene, T. G. and Langworthy, T. A. (1978). *Science* **203**, 51–53.
Tornabene, T. G., Wolfe, R. S., Balch, W. E., Hölzer, G., Fox, G. E. and Oró, J. (1978). *J. Mol. Evol.* **11**, 259–266.
Tornabene, T. G., Langworthy, T. A., Holzer, G. and Oró, J. (1979). *J. Mol. Evol.* **13**, 1–8.
Trotsenko, Y. A. (1976). In "Microbial Production and Utilisation of Gases" (H. G. Schlegel, G. Gottschalk and N. Pfennig, eds), pp. 329–336. Erich Goltze KG, Göttingen.
Trotsenko, Y. A. and Loginova, N. V. (1974). *Microbiology* **42**, 695–700.
Trotsenko, Y. A. and Loginova, N. V. (1978). *Usp. Mikrobiol.* **14**, 28–55.
Trotsenko, Y. A., Bykovskaya, S. V. and Kirikova, N. N. (1973). *Dokl. Akad. NAUK SSSR. (Proc. Acad. Sci. USSR)* **211**, 1230–1232.
Trotsenko, Y. A., Loginova, N. V. and Shishkina, V. N. (1974). *Proc. Acad. Sci. USSR* **216**, 1413–1415.
Trotsenko, Y. A., Kirikova, N. N. and Bykovskaya, S. V. (1975). *Microbiology* **43**, 915–917.
Trudinger, P. A. (1955). *Biochim. Biophys. Acta.* **18**, 581–582.
Trudinger, P. A. (1956). *Biochem. J.* **64**, 274–286.
Tsuchiya, Y., Nishio, N. and Nagai, S. (1980). *Eur. J. Appl. Microbiol. Biotechnol.* **9**, 121–127.
Tuboi, S. and Kikuchi, G. (1962). *Biochim. Biophys. Acta.* **62**, 188–190.
Tuboi, S. and Kikuchi, G. (1963). *J. Biochem.* **53**, 364–373.
Tuboi, S. and Kikuchi, G. (1965). *Biochim. Biophys. Acta.* **96**, 148–153.
Tye, R. and Willets, A. (1973). *J. Gen. Microbiol.* **77**, 1P.
Tzeng, S. F., Wolfe, R. S. and Bryant, M. P. (1975a). *J. Bacteriol.* **121**, 184–191.
Tzeng, S. F., Bryant, M. P. and Wolfe, R. S. (1975b). *J. Bacteriol.* **121**, 192–196.
Ueda, S., Sato, K. and Shimizu, S. (1978). *J. Nutr. Sci. Vitaminol.* **24**, 477–489.
Ueda, S., Sato, K. and Shimizu, S. (1981). *Agric. Biol. Chem.* **45**, 823–830.
Uotila, L. and Koivusala, M. (1974). *J. Biol. Chem.* **249**, 7653–7663.

Urakami, T. and Komagata, K. (1979). *J. Gen. Appl. Microbiol.* **25**, 343–360.
Urakami, T., Terao, I. and Nagai, I. (1981). In "Microbial Growth on C_1 Compounds" (H. Dalton, ed.), pp. 349–359. Heyden, London.
Urushibara, T., Forrest, H. S., Hoare, D. S. and Patel, R. N. (1971). *Biochem. J.* **125**, 141–146.
Utter, M. F. and Kolenbrander, H. M. (1972). In "The Enzymes" (P. D. Boger ed.), Vol. 6, pp. 117–168. Academic Press, London and New York.
Veenhuis, M., van Dijken, J. P. and Harder, W. (1976). *Arch. Microbiol.* **111**, 123–135.
Veenhuis, M., van Dijken, J. P., Pilon, S. A. F. and Harder, W. (1978a). *Arch. Microbiol.* **117**, 153–163.
Veenhuis, M., Zwart, K. and Harder, W. (1978b). *FEMS Microbiol. Lett.* **3**, 21–28.
Veenhuis, M., Keizer, I. and Harder, W. (1979). *Arch. Microbiol.* **120**, 167–176.
Veenhuis, M., Zwart, K. B. and Harder, W. (1981). *Arch. Microbiol.* **129**, 35–41.
van Verseveld, H. W. and Stouthamer, A. H. (1978a). *Arch. Microbiol.* **118**, 13–20.
van Verseveld, H. W. and Stouthamer, A. H. (1978b). *Arch. Microbiol.* **118**, 21–26.
van Verseveld, H. W., Boon, J. P. and Stouthamer, A. H. (1979). *Arch. Microbiol.* **121**, 213–223.
van Verseveld, H. W., Krab, K. and Stouthamer, A. H. (1981). *Biochim. Biophys. Acta.* **635**, 525–534.
Vishniac, W., Horecker, B. L. and Ochoa, S. (1957). *Adv. Enzymol.* **19**, 1–77.
van Vliet-Smits, M., Harder, W. and van Dijken, J. P. (1981). *FEMS Microbiol. Lett.* **11**, 31–35.
Vogels, G. D. (1979). *Antonie van Leeuwenhoek* **45**, 347–352.
Vogels, G. D., Hoppe, W. F. and Stumm, C. K. (1980). *Appl. Env. Microbiol.* **40**, 608–612.
Wadzinski, A. M. and Ribbons, D. W. (1975a). *J. Bacteriol.* **122**, 1364–1374.
Wadzinski, A. M. and Ribbons, D. W. (1975b). *J. Bacteriol.* **123**, 380–381.
Wagner, C. (1964). *Bacteriol. Proc.* 91.
Wagner, C. and Levitch, M. E. (1975). *J. Bacteriol.* **122**, 905–910.
Wagner, C. and Quayle, J. R. (1972). *J. Gen. Microbiol.* **72**, 485–491.
Wagner, C., Lusty, S., Kung, H. F. and Rogers, N. L. (1967). *J. Biol. Chem.* **242**, 1287–1293.
Waites, M. J. and Quayle, J. R. (1980). *J. Gen. Microbiol.* **118**, 321–327.
Waites, M. J. and Quayle, J. R. (1981). *J. Gen. Microbiol.* **124**, 309–316.
Waites, M. J., Lindley, N. D. and Quayle, J. R. (1981). *J. Gen. Microbiol.* **122**, 193–199.
Walker, D. A. (1977). *Curr. Top. Cell Regul.* **11**, 203–241.
Walsh, C. (1979). In "Enzyme Reaction Mechanisms", pp. 865–866. Freeman, San Francisco.
Walther, R., Fiebig, K., Fahlbusch, K., Caspari, D., Hippe, H. and Gottschalk, G. (1981). In "Microbial Growth on C_1 Compounds" (H. Dalton, ed.), pp. 146–151. Heyden, London.
Ward, T. E. and Frea, J. I. (1980). *Appl. Environ. Microbiol.* **39**, 597–603.
Warner, P. J., Higgins, I. J. and Drozd, J. W. (1977). *FEMS Microbiol. Lett.* **1**, 339.
Warner, P. J., Higgins, I. J. and Drozd, J. W. (1980). *FEMS Microbiol. Lett.* **7**, 181–185.
Weaver, T. L. and Duggan, P. R. (1975). *J. Bacteriol.* **122**, 433–436.
Weimer, P. J. and Zeikus, J. G. (1978a). *Arch. Microbiol.* **119**, 43–57.
Weimer, P. J. and Zeikus, J. G. (1978b). *Arch. Microbiol.* **119**, 175–182.

Weimer, P. J. and Zeikus, J. G. (1979). *J. Bacteriol.* **137**, 332–339.
Weissbach, A., Smyrniotis, P. Z. and Horecker, B. L. (1954). *J. Am. Chem. Soc.* **76**, 3611.
Weitzmann, P. D. J. and Danson, M. J. (1976). *Curr. Top. Cell. Regul.* **10**, 161–205.
Weitzmann, P. D. J. and Dunmore, P. (1969a). *Biochem. Biophys. Acta.* **171**, 198–200.
Weitzmann, P. D. J. and Dunmore, P. (1969b). *FEBS Lett.* **3**, 265–267.
Weitzmann, P. D. J. and Jones, D. (1968). *Nature* **219**, 270–272.
Westerling, J., Frank, J. and Duine, J. A. (1979). *Biochem. Biophys. Res. Comm.* **87**, 719–722.
Whitman, W. B. and Wolfe, R. S. (1980). *Biochem. Biophys. Res. Comm.* **92**, 1196–1201.
Whittenbury, R. and Kelly, D. P. (1977). *Symp. Soc. Gen. Microbiol.* **27**, 119–149.
Whittenbury, R., Phillips, K. C. and Wilkinson, J. F. (1970a). *J. Gen. Microbiol.* **61**, 205–218.
Whittenbury, R., Davies, S. L. and Davey, J. F. (1970b). *J. Gen. Microbiol.* **61**, 219–226.
Whittenbury, R., Dalton, H., Eccleston, M. and Reed, H. L. (1975). In "Microbial Growth on C_1 Compounds", pp. 1–9. Society of Fermentation Technology, Japan.
Whittenbury, R., Colby, J., Dalton, H. and Reed, H. L. (1976). In "Microbial Production and Utilisation of Gases" (H. G. Schlegel, G. Gottschalk and N. Pfennig, eds), pp. 281–292. Erich Goltze KG, Göttingen.
Widdowson, D. and Anthony, C. (1975). *Biochem. J.* **152**, 349–356.
Wiken, T. O., Hazeu, W., van Leengaed, L. and Snoo, T. J. J. (1977). Alcohol, Industry and Research, pp. 220–226. Helsinki, Finland.
Wilkinson, J. F. (1971). In "Microbes and Biological Productivity" (D. E. Hughes and A. H. Rose, eds), SGM Symposium **21**, pp. 15–46. Cambridge University Press, Cambridge.
Wilkinson, J. F. (1975). In "Microbial Growth on C_1 Compounds" (G. Terui, ed.), pp. 45–57. Society of Fermentation Technology, Japan.
Wilkinson, J. F. and Harrison, D. E. F. (1973). *J. Appl. Bact.* **36**, 309–313.
Wilkinson, T. G., Topiwala, H. H. and Hamer, G. (1974). *Biotechnol. Bioeng.* **16**, 41–59.
Williams, E. and Shimmin, M. A. (1978). *FEMS Microbiol. Lett.* **4**, 137–141.
Williams, J. B., Roche, T. E. and McFadden, B. A. (1971). *Biochemistry* **10**, 1384–1390.
Williams, E., Shimmin, M. A. and Bainbridge, B. W. (1977). *FEMS Microbiol. Lett.* **2**, 293–296.
Willison, J. C. and Haddock, B. A. (1981). *FEMS Microbiol. Lett.* **10**, 53–57.
Willison, J. C. and John, P. (1979). *J. Gen. Microbiol.* **115**, 443–450.
Willison, J. C., Haddock, B. A. and Boxer, D. H. (1981a). *FEMS Microbiol. Lett.* **10**, 249–255.
Willison, J. C., Ingledew, W. J. and Haddock, B. A. (1981b). *FEMS Microbiol. Lett.* **10**, 363–368.
Windass, J. D., Worsey, M. J., Pioli, E. M., Pioli, D., Barth, P. T., Atherton, K. T., Dart, E. C., Byrom, D., Powell, K. and Senior, P. J. (1980). *Nature* **287**, 396–401.
Woese, C. R., Magrum, L. J. and Fox, G. E. (1978). *J. Mol. Evol.* **11**, 245–252.
Wolf, H. J. (1981). In "Microbial Growth on C_1 Compounds" (H. Dalton, ed.), pp. 202–210. Heyden, London.
Wolf, H. J. and Hanson, R. S. (1978). *Appl. Environ. Microbiol.* **36**, 105–114.

References

Wolf, H. J. and Hanson, R. S. (1979). *J. Gen. Microbiol.* **114**, 187–194.
Wolf, H. J., Christiansen, M. and Hanson, R. S. (1980). *J. Bacteriol.* **141**, 1340–1349.
Wolfe, R. S. (1971). *Adv. Microbiol. Physiol.* **6**, 107–146.
Wolfe, R. S. (1979a). *Microbial Biochemistry, Int. Rev. Biochem.* **21**, 270–300.
Wolfe, R. S. (1979b). *Antonie van Leeuwenhoek* **45**, 353–364.
Wolfe, R. S. (1980). *In* "Diversity of Bacterial Respiratory Systems", (C. J. Knowles ed.), Vol. I, pp. 161–186. CRC Press, Florida.
Wolin, M. J. (1976). *In* "Microbial Production and Utilisation of Gases" (H. G. Schlegel, G. Gottschalk and N. Pfennig, eds), pp. 141–150. Erich Goltze KG, Göttingen.
Wolin, M. J., Wolin, E. A. and Wolfe, R. S. (1963). *Biochem. Biophys. Res. Comm.* **12**, 464–468.
Wood, W. A. (ed.) (1966). "Methods in Enzymology", Vol. 9.
Wood, W. A. (ed.) (1975a). "Methods in Enzymology", Vol. 41.
Wood, W. A. (ed.) (1975b). "Methods in Enzymology", Vol. 42.
Wood, A. P. and Kelly, D. P. (1981). *J. Gen. Microbiol.* **125**, 55–62.
Wood, J. M. and Wolfe, R. S. (1966). *J. Bacteriol.* **92**, 696–700.
Wood, H. G., O'Brien, W. E. and Michaels, G. (1977). *Adv. Enzymol.* **45**, 88–155.
Yamada, H., Kishimoto, N. and Kumagai, H. (1976). *J. Ferment. Technol.* **54**, 726–737.
Yamada, H., Shin, K. C., Kato, N., Shimizu, S. and Tani, Y. (1979). *Agr. Biol. Chem.* **43**, 877–878.
Yamamoto, M., Seriu, Y., Kouno, K., Okamoto, R. and Inui, T. (1978a). *J. Ferment. Technol.* **56**, 451–458.
Yamamoto, M., Seriu, Y., Goto, S., Okamoto, R. and Inui, T. (1978b). *J. Ferment. Technol.* **56**, 459–466.
Yamamoto, M., Okamoto, R. and Inui, T. (1979). *J. Ferment. Technol.* **57**, 400–407.
Yamamoto, M., Iwaki, H., Kouno, K. and Inui, T. (1980). *J. Ferment. Technol.* **58**, 99–106.
Yamanaka, K. (1981). *In* "Microbial Growth on C_1 Compounds" (H. Dalton, ed.), pp. 21–30. Heyden, London.
Yamanaka, K. and Matsumoto, K. (1977a). Abstracts of the Second International Symposium of Microbial Growth on C_1 Compounds", pp. 72–74. Pushchino.
Yamanaka, K. and Matsumoto, K. (1977b). *Agr. Biol. Chem.* **41**, 467–475.
Yamanaka, K. and Matsumoto, K. (1979). *Agr. Biol. Chem.* **43**, 1–7.
Yamazaki, S. and Tsai, L. (1980). *J. Biol. Chem.* **255**, 6462–6465.
Yamazaki, S., Tsai, L., Stadtman, T. C., Jacobson, F. S. and Walsh, C. (1980). *J. Biol. Chem.* **255**, 9025–9027.
Yasuhara, S., Kawamoto, S., Tanaka, A., Osumi, M. and Fukui, S. (1976). *Agr. Biol. Chem.* **40**, 1771–1780.
Yoch, D. C. and Lindstrom, E. S. (1967). *Biochem. Biophys. Res. Comm.* **28**, 65–69.
Yokote, J., Sugimoto, M. and Abe, S. (1974). *J. Ferment. Technol.* **52**, 201.
Zatman, L. J. (1981). *In* "Microbial Growth on C_1 Compounds" (H. Dalton, ed.), pp. 42–54. Heyden, London.
Zavarzin, G. S. and Nozhevnikova, A. N. (1976). *In* "Microbial Production and Utilisation of Gases" (H. G. Schlegel, G. Gottschalk and N. Pfennig, eds), pp. 207–214. Erich Goltze KG, Göttingen.
Zehnder, A. J. B. and Brock, T. D. (1979). *J. Bacteriol.* **137**, 420–432.

Zehnder, A. J. B., Huser, B. A., Brock, T. D. and Wuhrmann, K. (1980). *Arch. Microbiol.* **124**, 1–11.
Zeikus, J. G. (1977). *Bacteriol. Rev.* **41**, 514–541.
Zeikus, J. G. (1980). *Ann. Rev. Microbiol.* **34**, 423–464.
Zeikus, J. G. and Bowen, V. G. (1975a). *Can. J. Microbiol.* **21**, 121–129.
Zeikus, J. G. and Bowen, V. G. (1975b). *J. Bacteriol.* **121**, 373–380.
Zeikus, J. G. and Wolfe, R. S. (1973). *J. Bacteriol.* **113**, 461–467.
Zeikus, J. G., Fuchs, G., Kenealy, W. and Thauer, R. K. (1977). *J. Bacteriol.* **132**, 604–613.
Zeikus, J. G., Ben-Bassat, A. and Hegge, P. W. (1980). *J. Bacteriol.* **143**, 432–440.
Zhilina, T. (1971). *Microbiology* **40**, 587–591.
Zinder, S. H. and Mah, R. A. (1979). *Appl. Environ. Microbiol.* **38**, 996–1008.
Zwart, K. B., Veenhuis, M., van Dijken, J. P. and Harder, W. (1980). *Arch. Microbiol.* **126**, 117–126.

List of figures

Fig. 1. *Methylomonas methanica* (a) and *Methylococcus capsulatus* (b) showing Type 1 membranes. 17
Fig. 2. *Methanomonas methano-oxidans* showing Type II membranes. . . 18
Fig. 3. Photomicrographs and electron micrographs of *Hyphomicrobium* sp. . 37
Fig. 4. The ribulose bisphosphate cycle of CO_2 fixation (sedoheptulose bisphosphatase variant). 44
Fig. 5. The ribulose bisphosphate cycle of CO_2 assimilation (transaldolase variant). 45
Fig. 6. The ribulose monophosphate (RuMP) cycle of formaldehyde assimilation (KDPG aldolase/transaldolase variant). 63
Fig. 7. The ribulose monophosphate (RuMP) cycle of formaldehyde assimilation (fructose bisphosphate aldolase/sedoheptulose bisphosphatase variant). 64
Fig. 8. The ribulose monophosphate (RuMP) cycle of formaldehyde assimilation (fructose bisphosphate aldolase/transaldolase variant). . . 65
Fig. 9. The ribulose monophosphate (RuMP) cycle of formaldehyde assimilation (KDPG aldolase/sedoheptulose bisphosphatase variant). . 66
Fig. 10. The biosynthesis from formaldehyde of triose, pentose and hexose. . 68
Fig. 11. The biosynthesis of erythrose 4-phosphate from formaldehyde. . 69
Fig. 12. The oxidation of formaldehyde to CO_2 by a dissimilatory RuMP cycle. 69
Fig. 13. The interconversion of phosphoenolpyruvate and oxaloacetate. . . 71
Fig. 14. The serine pathway of formaldehyde assimilation (icl+ variant). . 97
Fig. 15. The serine pathway operating to synthesise oxaloacetate (A) and succinate (B). 98
Fig. 16. Potential routes for assimilation of radioactive (^{14}C) methanol and CO_2 into glycine, serine and malate in *Pseudomonas* AM1. . . 100
Fig. 17. The homoisocitrate glyoxylate cycle for oxidation of acetyl-CoA to glyoxylate in icl⁻-serine pathway bacteria. 103
Fig. 18. Model for arrangement of genes coding for C_1 enzymes in *Methylobacterium organophilum* XX. 132
Fig. 19. Pathways for the assimilation of ethanol, malonate, pyruvate, lactate, propane 1,2-diol and 3-hydroxybutyrate in *Pseudomonas* AM1 . 149
Fig. 20. Pathway of electron transfer between the components of the soluble methane monooxygenase complex during the oxidation of methane to methanol. 158
Fig. 21. (a) Absorption spectra of different forms of methanol dehydrogenase. (b) The absorption spectrum of the quinol and quinone forms of the prosthetic group of methanol dehydrogenase. 179
Fig. 22. The prosthetic group of methanol dehydrogenase and derivatives of it. 181
Fig. 23. A tentative reaction cycle for *in vitro* activity of methanol dehydrogenase. 185
Fig. 24. A mechanism proposed for the involvement of PQQ in catalysis by methanol dehydrogenase. 186

List of figures

Fig. 25. The cyclic route for oxidation of formaldehyde. . . . 192
Fig. 26. The oxidation of methylated amines to formaldehyde. . . . 196
Fig. 27. The prosthetic group of trimethylamine dehydrogenase. . . 199
Fig. 28. A speculative mechanism for methylamine dehydrogenase. . . 207
Fig. 29. The possible involvement of N-methylated amino acids in methylamine oxidation. 209
Fig. 30. Proposed reaction sequence for N-methylglutamate synthase . . 211
Fig. 31. A speculative mechanism for the autoreduction of cytochrome c and its involvement in reaction with methanol dehydrogenase. . . . 228
Fig. 32. Electron transport and proton translocation in *Pseudomonas* AM1 . 235
Fig. 33. Electron transport in *Methylophilus methylotrophus*. . . . 237
Fig. 34. The respiratory chain of *Paracoccus denitrificans*. . . 239
Fig. 35. Tentative scheme for electron transport in *Methylosinus trichosporium* and *Pseudomonas extorquens*. 241
Fig. 36. Possible arrangements of methanol dehydrogenase and cytochrome c in the bacterial membrane. 243
Fig. 37. Metabolism of carbon substrate for provision of cell material, NADH and ATP. 254
Fig. 38. Assimilation of methanol and CO_2 into *Rhodopseudomonas acidophila* during anaerobic growth in the light.. 263
Fig. 39. (a) The oxidation of methanol to CO_2 in methylotrophic yeasts. (b) and (c) Peroxisomes and crystalloids in *Hansenula polymorpha* . 276
Fig. 40. The dihydroxyacetone cycle of formaldehyde assimilation in yeast . 286
Fig. 41. The dissimilatory DHA cycle of formaldehyde oxidation. . . 287
Fig. 42. The production of methane in natural habitats. 299
Fig. 43. The structure of murein of a typical Gram-positive organism (*Gaffkya* sp.) and of pseudomurein from the methanogen *Methanobacterium thermoautotrophicum*. 303
Fig. 44. Principal components of the polar fraction (a and b) and the neutral lipid fraction of methanogens (c). 303
Fig. 45. Biosynthesis in two methanogens. 305
Fig. 46. Barker's scheme for the reduction of CO_2, methanol and acetate to methane. 307
Fig. 47. The structure of Coenzyme F_{420}. 310
Fig. 48. The two-electron transfer reactions of Coenzyme F_{420} (deazaflavin). 310
Fig. 49. The structures of methanopterin (factor B_0) and carboxydihydromethanopterin (YFC). 311
Fig. 50. Possible coupling of exergonic and endergonic reactions by proton translocation across membranes. 317
Fig. 51. Possible arrangements of hydrogenase in membranes. . . . 318
Fig. 52. The reduction of CO_2 to CH_4. 319
Fig. 53. The reduction of CO_2 to methane. 320
Fig. 54 Possible arrangements of methanogenesis from acetate to produce a pmf across the membrane. 325
Fig. 55. The ICI "Pruteen" plant at Billingham, UK. 334
Fig. 56. Flow diagram of the ICI "Pruteen" process.. 335
Fig. 57. A simple hydrogen-oxygen fuel cell. 348

List of tables

Table 1. Substrates used for methylotrophic growth	3
Table 2. Properties of the obligate methanotrophs in Professor Whittenbury's collection.	6
Table 3. Further properties of methanotrophic bacteria.	6
Table 4. Properties characterising the main groups (Genera) of methanotrophs.	7
Table 5. Properties of species (sub-groups) of methanotrophs (taken from Whittenbury et al., 1970a)	8
Table 6. Summary of some properties of *Methylococcus* species as defined by Romanovskaya et al. (1978).	11
Table 7. The facultative methanotrophs (*Methylobacterium* sp.) . . .	12
Table 8. Distribution of carbon assimilation pathways in methylotrophic bacteria	23
Table 9. Obligate methylotrophs unable to use methane but able to use methanol or methylamine.	25
Table 10. Properties common to the obligate methylotrophs unable to use methane.	26
Table 11. The pink facultative methylotrophs.	28
Table 12. The non-pigmented "pseudomonads".	30
Table 13. Gram-negative (or variable), non-motile rods.	31
Table 14. Gram-positive facultative methylotrophs.	32
Table 15. Facultative autotrophs or phototrophs growing on methanol or formate	34
Table 16. Methylotrophs able to use the ribulose bisphosphate pathway. .	48
Table 17. Summary of variants of the ribulose monophosphate cycle of formaldehyde fixation.	67
Table 18. The occurrence and distribution of the RuMP pathway of formaldehyde fixation.	72
Table 19. Distribution of cleavage and rearrangement variants of the RuMP cycle.	75
Table 20. Hexulose phosphate synthase.	79
Table 21. The glucose 6-phosphate dehydrogenases of methylotrophs. . .	82
Table 22. Evidence for the serine pathway from ^{14}C-labelling experiments .	99
Table 23. Evidence for the serine pathway from enzyme studies. . .	102
Table 24. Mutants of *Pseudomonas* AM1.	106
Table 25. Mutants of *Methylobacterium organophilum*, *Pseudomonas aminovorans* and *Pseudomonas* MS.	108
Table 26. Methylotrophs using the serine pathway as their main assimilation pathway.	110
Table 27. Distribution of the icl$^-$ and icl$^+$ variants of the serine pathway .	112
Table 28. Regulation of synthesis of C_1 enzymes in facultative methylotrophs.	128
Table 29. The distribution of the TCA cycle in methylotrophs. . .	138
Table 30. Thermodynamic constants for reactions involved in the oxidation of C_1 compounds.	154

List of tables

Table 31. Studies on methane oxidation in bacteria. 156
Table 32. The substrate specificity of the soluble methane monooxygenase from *Methylococcus capsulatus* (Bath). 164
Table 33. Methanol dehydrogenase. 168
Table 34. Summary of properties of the methanol dehydrogenases listed in Table 33. 170
Table 35. The PQQ prosthetic group of methanol dehydrogenase. . . 180
Table 36. Dehydrogenases having a PQQ prosthetic group (quinoproteins). 183
Table 37. The oxidation of formaldehyde. 188
Table 38. The properties of N-methylglutamate dehydrogenase. . . . 212
Table 39. Distribution of enzymes involved in methylamine oxidation in facultative methylotrophs 214
Table 40. The oxidation of methylated amines by obligate methylotrophs. . 216
Table 41. Cytochromes, electron transport systems and proton-translocation in methylotrophs. 222
Table 42. Properties of the cytochromes c of methylotrophs. . . . 225
Table 43. The relationship between microbial physiology and cell yields . 247
Table 44. Summary of assimilation equations. 252
Table 45. The limitation of cell yields by ATP, NADH and carbon supply. . 256
Table 46. Predicted yields on methanol and methane 261
Table 47. Electron flow from each dehydrogenase expressed as a proportion of total electron transport. 267
Table 48. Eucaryotic methylotrophs. 270
Table 49. The alcohol oxidase of yeast. 275
Table 50. Structure and function of peroxisomes in methanol-utilising yeasts. 282
Table 51. The assimilation of methanol in yeasts; the DHA pathway. . . 289
Table 52 (a) Reviews of methanogens and methanogenesis. . . .
(b). The methanogens 297
Table 53. Summary of cell wall and lipid composition of the methanogens. . 302
Table 54. Novel coenzymes from methanogens. 309
Table 55. The forms of Coenzyme M found in methanogens. . . . 313
Table 56. ATP synthesis and the reduction of CO_2 to methane. . . . 315
Table 57. Methanogenesis from acetate, methanol, methylated amines and formate. 324
Table 58. The production of single cell protein (SCP). 329
Table 59. The measurement of growth yields of methylotrophs. . . . 336
Table 60. The oxidation of hydrocarbons by whole cells of methanotrophs. 346

Index

Page numbers marked f denote a figure and those marked t denote a table.

A

Acetaldehyde (ethanal)
  formation of adduct with PQQ, 180t, 181f
  oxidation by methanol dehydrogenase, 173
  product of methane monooxygenase, 165t, 346t
Acetate
  assimilation in methanogens, 305, 306, 320, 324
  metabolism by methylotrophs, 27, 35, 101, 104, 147–150
  methanogenesis from, 298, 299f, 307f, 320, 323–325
  oxidation in methanogens, 323, 324
Acetate, radioactive, in study of serine pathway, 101, 104
"Acetate organisms", 298
Acetate thiokinase, *see* Acetyl-CoA synthetase
Aceticlastic reaction of acetate, 324, 325
Acetoacetyl-CoA, intermediate in assimilation of multicarbon compounds, 149f
Acetobacter pasteurianum, alcohol dehydrogenase of, 182, 183t
Acetobacterium sp., in methanogenesis, 299f
Acetone
  adduct formation with PQQ, 179, 180t, 181f
  product of methane monooxygenase, 346t
Acetyl-Coenzyme A
  as activator of PEP carboxylase, 119
  as activator of pyruvate carboxylase, 92
  in assimilation of multicarbon compounds, 147–150

Acetyl-Coenzyme A (*continued*)
  in formaldehyde assimilation (serine pathway), 97f, 101, 103–105, 122–125
  in formaldehyde assimilation (RuMP pathway), 70, 71f
  as inhibitor of glucose 6-phosphate dehydrogenase, 82t
  in methanogen biosynthesis, 305f, 306, 324
  oxidation to glyoxylate, 97f, 103, 107t, 124, 144, 148, 149f
Acetyl-CoA synthetase (acetate thiokinase)
  in assimilation of multicarbon compounds, 149f
  of methanogens, 305f, 306
  mutants lacking, 107t, 149f
Acetylene (ethyne)
  as inhibitor of methane monooxygenase, 22, 161–163, 164t
Acetylene-reduction test for nitrogenase, 21, 163
N-Acetylgalactosamine, 301, 303f
N-Acetylglucosamine, 301, 303f
N-Acetyltalosaminuronic acid, 301, 303f
Achromobacter (genus), 29
Achromobacter parvulus
  formate dehydrogenase of, 194
Achromobacter 1L, 34t, 48t
  RuBP pathway in, 48t
  TCA cycle in, 140t
Acid-labile iron
  in dimethylamine monooxygenase, 202
Acid-labile sulphide
  in formate dehydrogenase, 194
  in methane monooxygenase, 157, 158f
  in trimethylamine dehydrogenase, 198, 199
Acinetobacter (genus), 29, 31t

383

Acinetobacter calcoaceticus
  alcohol dehydrogenase of, 173, 174, 182, 183*t*
  glucose dehydrogenase of, 182, 183*t*
Acinetobacter S50, see Strain S50
Aconitase (aconitate hydratase)
  in assimilation of formaldehyde (serine pathway), 97*f*, 114
  inhibition by fluoroacetate, 114
  low levels in obligate methylotrophs, 138*t*
Active transport
  of Coenzyme M, 309*t*
  effect on growth yields, 249
Acyl-CoA oxidation in yeasts, 281
Adenine nucleotide translocase
  in methanogens, 315*t*, 316
ADP
  as inhibitor of PEP carboxylase, 120, 121, 135
  in regulation of enzyme activity, 132–136
Affinity chromatography
  of malate thiokinase, 122
Affinity of methylotrophs for methanol, 40, 332, 336*t*
Alanine
  accumulation by pyruvate dehydrogenase mutant, 340
  growth of restricted facultative methylotrophs on, 33
  as intermediate in oxidation of methylated amines, 209*f*
L-Alanine
  accumulation during growth of methanogens, 301
  in methanogen cell walls, 301, 303*f*
β-Alanine esters, as activators of methanol dehydrogenase, 174
Alcaligenes (genus), 29, 31*t*
Alcaligenes eutrophus, see *Alkaligenes eutrophus*
Alcaligenes sp., RuBP pathway, in 48*t*
Alcaligenes FOR$_1$, 48*t*
Alcohols, see also individual alcohols
  commercial oxidation of, 343, 344
  products of methane monooxygenase, 164*t*, 165*t*, 346*t*
Alcohols (long chain)
  carbon limitation during growth on, 256*t*, 258

Alcohols (primary)
  enzymic assay of, 343, 349
  oxidation, see Alcohol oxidase, Methanol dehydrogenase, Methane monooxygenase
Alcohols (secondary), see Secondary alcohols
Alcohol dehydrogenases of bacteria, see also Methanol dehydrogenase, 182, 183*t*
Alcohol dehydrogenase of methylotrophic yeasts, 279
Alcohol oxidase (yeast), 274–279
  in assay of alcohols, 343
  as commercial source of flavins, 339
  effect on growth yields, 255, 261*t*, 262, 279
  immobilisation, 344
  in large scale oxidation of alcohols, 344
  localisation, 276*f*, 277*f*, 282*t*, 283
  mutant lacking, 278, 283, 294
  regulation of, 278, 282*t*, 283, 292–294
  relevance to SCP production, 332
Aldehydes
  oxidation, see Formaldehyde dehydrogenase, Methanol dehydrogenase
  oxidation by methanol dehydrogenase, 172, 173
Aldehyde dehydrogenase, see Formaldehyde dehydrogenase (bacterial dye-linked)
Aldehyde dehydrogenase (non-specific), of bacteria, 190
Aldehyde dehydrogenase of *Gluconobacter*, 183*t*
Aldolase, see also Fructose bisphosphate aldolase
  in assimilation of CO_2, 44*f*, 45*f*, 54
Aldol condensation, see also Aldolase, Fructose bisphosphate aldolase, Hexulose phosphate synthase, Serine transhydroxymethylase, 62, 77, 115
Alkaligenes eutrophus, 34*t*
Alkanes and derivatives
  assimilation equations for, 252*t*
  carbon-limitation of growth on, 256*t*, 258, 281
  oxidation by methane monooxygenase, 163–167, 346*t*

Index

Alkane-utilisers
  limitation of growth yields of, 256t, 258
Alkane-utilising yeasts
  peroxisomes in, 258, 281, 283
Alkenes
  epoxidation by methane monooxygenase, 163–167, 345, 346t
Alkylamines
  as activators of methanol dehydrogenase, 174
Allulose phosphate, 60–61
Alteromonas thalassomethanolica, 38
Amines, *see also* Methylated amines and individual amines
  as activators of methanol dehydrogenase, 173, 174, 185, 186f
  enzymic assay of, 343
Amine oxidase (bacteria), *see* Methylamine oxidase
Amine oxidase of *Candida boidinii*, 208, 273
Amine oxidase-containing peroxisomes, 273
Amino acid composition
  of alcohol oxidase (yeast), 278
  of cytochromes c, 226
  of methanol dehydrogenases, 175
  of methylamine dehydrogenases, 207
Amino acids
  production by methylotrophs, 339–342
δ-Aminolaevulinate in methanogens, 305f, 312
o-Aminophenol, as inhibitor of methane monooxygenase, 162
Aminotriazole as catalase inhibitor, 208
Ammonia
  as activator of methanol dehydrogenase, 173, 185–187
  altered assimilation route in *M. methylotrophus*, 262, 333
  oxidation by methane monooxygenase, 163
Ammonia assimilation, effect on growth yield of pathway for, 262, 333

AMP
  in regulation of enzyme activity, 59, 132, 133, 144, 145
  in relief of inhibition of citrate synthase, 144, 145
Amytal, as inhibitor of electron transport, 241
Anaerobic methylotrophs, *see also* Methanol, methanogenesis from, Methanol, anaerobic oxidation of
  Hyphomicrobium, 33, 35, 38–41, 48t, 50, 263, 296, 299f, 323–327
Anaerobic oxidation of methane, 38, 167
Anaerobic oxidation of methanol, 39, 50, 239f, 240, 263
Anaerobic oxidation of methylated amines, 196f, 201, 203
Anaerobic preparation of methanol dehydrogenase, 174, 187, 231
Analogue-resistant mutants
  in overproduction of metabolites, 341
Antibiotics
  in isolation of yeasts, 273
  sensitivity of methanogens to, 301, 303
Antibiotics-resistance, transfer of, 130
Antimycin A, as inhibitor of electron transport, 230, 235f, 237f, 239f, 241
D-Arabino-hexulose 6-phosphate, *see* hexulose 6-phosphate
D-Arabino-3-hexulose 6-phosphate formaldehyde lyase, *see* 3-Hexulose phosphate synthase
D-Arabino 3-hexulose 6-phosphate 3,2-ketoisomerase, *see* 3-hexulose phosphate isomerase
Archaebacteria, 300
Aromatic compounds, oxidation by methane monooxygenase, 163, 165t, 345–348
Arthrobacter (genus), 29, 33, 23t, 31t
Arthrobacter globiformis, *see also* *Arthrobacter* P1, 31t
  methylamine oxidation in, 196f, 215t
  overproduction of serine by, 341
  pyruvate carboxylase, PEP carboxylase and PEP carboxykinase of, 74
  RuMP pathway in, 73t, 75t
  TCA cycle in, 140t

Arthrobacter rufescens, 31*t*
Arthrobacter sp.
  methylamine oxidase of, 214*t*, 218
  RuMP pathway in, 73*t*
  serine pathway in, 110*t*
Arthrobacter 1A1 and 1A2, 31*t*
Arthrobacter 2B2, 31*t*
  formaldehyde fixation reaction in, 288
  radioisotope evidence for TCA cycle, 141
  RuMP pathway in, 73*t*
Arthrobacter P1, 31*t*, 33
  catalase in, 208
  growth yields, *see* Methylamine, growth yields on
  limitation of growth yields in, 255, 256*t*
  methylamine oxidase of, 208, 215*t*
  RuMP pathway in, 73*t*, 75*t*
  TCA cycle in, 140*t*
Ascorbate
  as reductant for methane monooxygenase, 159, 160
  stimulation of trimethylamine *N*-oxide demethylase by, 201
Ascorbate/TMPD, oxidation of, 230, 234, 236, 241, 242, 243
Ascospores of methylotrophic yeasts, 273
Aspartate, radioactive labelling during elucidation of serine pathway, 98–100
Asporogenous yeasts, 273
Assay of amines and alcohols, 343, 349
Assimilation of carbon, *see* Carbon assimilation pathways, distribution; Dihydroxyacetone cycle, Glyoxylate cycle, Methanogens, biosynthesis in, Ribulose bisphosphate pathway, Ribulose monophosphate pathway, Serine pathway
Assimilation equations, *see* Summary equations
Assimilation equations (theoretical)
  assumptions and methods for derivations, 249–251
  summary of equations, 252*t*, 253*t*, 254*f*, 263

ATP, *see also* Y_{ATP}
  as activator of methyltransferase, 326
  as inhibitor of citrate synthase, 144
  as inhibitor of formate dehydrogenase, 194
  as inhibitor of formate and formaldehyde dehydrogenase (yeast), 284, 295
  as inhibitor of glucose 6-phosphate dehydrogenase, 82*t*, 83, 94, 193
  as inhibitor of phosphofructokinase, 93
  as inhibitor of 6-phosphogluconate dehydrogenase, 89, 94
  limitation of growth yields by, *see* Growth yields, limitation by ATP
  pool size in methanogens, 315*t*
  proportion of substrate oxidised to provide, 251–257*t*
  in regulation of enzyme activity, 93, 132–136
  requirement for methanogenesis, 315*t*, 319*f*–321, 326
  synthesis, *see* ATP synthesis
ATP malate lyase (malate ATP lyase)
  in assimilation of formaldehyde (serine pathway), 101, 103, 112*t*, 113, 123
  mutant lacking, 108*t*, 113
  relationship with isocitrate lyase, 113
ATP synthesis, *see also* Y_{ATP}
  coupled to methane hydroxylation, 265
  during methanol oxidation, 230, 234, 235, 240, 242–244, 266
  in growth of methanotrophs on methanol, 266
  in methanogens, 306–308, 311, 315–319, 322, 325–327
ATP synthesis in methylotrophs, references to, 223*t*
ATP synthetase
  of methanogens, 307, 315, 316, 317*f*, 325*f*
Autoreduction of cytochromes *c*, 186, 227–229, 231
Autotrophs, *see also* Chemolithotrophs, Methanogens, Ribulose bisphosphate pathway
  carbon assimilation in, 42–59

Autotrophs (*continued*)
definition of, 42
energy-limitation during growth on, 255
regulation in, 59
Autoxidation of cytochromes *c*, 225*t*, 230, 241
Azide, as inhibitor of cytochrome oxidases, 236, 237*f*
"*Azotobacter*"-type cysts, 7*t*, 9, 15, 20, 39

B

Bacillus (genus), 23*t*, 24, 32*t*
Bacillus cereus, 32*t*
Bacillus methanicus, 2
Bacillus PM6, 24, 29, 32*t*, 33
citrate synthase of, 144
glucose 6-phosphate dehydrogenase in, 76, 83, 90
incomplete TCA cycle in, 139*t*, 141
NADPH oxidase in, 90
oxidation of methylamine in, 215*t*
oxidation of methylated amines in, 215*t*, 217
6-phosphogluconate dehydrogenase in, 76, 83, 90
RuMP pathway in, 73*t*, 75*t*
trimethylamine *N*-oxide demethylase of, 200
Bacillus S2A1, 24, 29, 32*t*, 33
glucose 6-phosphate dehydrogenase in, 83, 90
incomplete TCA cycle in, 139*t*
NADPH oxidase in, 90
oxidation of methylamine in, 215*t*
oxidation of methylated amines in, 215*t*, 217
6-phosphogluconate dehydrogenase of, 76, 90
RuMP pathway in, 73*t*, 75*t*
Bacitracin, effect on methanogens, 303
Bacterium formoxidans
RuBP pathway in, 48*t*
Bacterium 7d
methylamine dehydrogenase in, 215*t*
RuBP pathway in, 48*t*
Bacterium 2B2, *see Arthrobacter* 2B2
Bacterium 5B1, *see* Organism 5B1
Bacterium C2A1, *see* Organism C2A1
Bacterium 5H2, *see* Organism 5H2

Bacterium MB58, 6-hexulose phosphate synthase of, 78
Barker's scheme for methanogenesis, 307, 319
Basidiomyces, alcohol oxidase from, 275*t*, 278
Benzene, oxidation by methane monooxygenase, 165*t*, 347*t*
Betaine
growth of restricted facultative methylotrophs on, 33
methanogenesis from, 326
Bicarbonate, assimilation of, *see* Carbon dioxide
Bioassay of Vitamin B_{12}, 342
Biocatalysis, by methylotrophs (commercial exploitation), 342–349
Bioenergetics, *see also* Electron transport and energy transduction in methylotrophic bacteria, and growth yields of methylotrophs, 245–268
of methanogenesis, 306, 315–319, 325–327
Biofuel cells, 348, 349
Biomass, *see* Single cell protein (SCP)
Biosynthesis, *see* Carbon assimilation pathways
Biosynthesis in methanogens, 297*t*, 304–306, 323
Biotechnology, 328–349
Biotin requirement of yeast, 273
Biphytanyl glycerol ethers, 303*f*, 304
α,α′Bipyridine
as inhibitor of methanol dehydrogenase, 174
4,4′-Bipyridyl
in biofuel cells, 348
1,3-Bisphosphoglycerate, *see* 1,3-Diphosphoglycerate
Blastobacter viscosus, 34*t*
RuBP pathway in, 48*t*
Brevibacterium fuscum, 32*t*
RuMP pathway in, 73*t*, 75*t*
Brevibacterium 24, *N*-methylglutamate dehydrogenase in, 215*t*
Brilliant cresyl blue
as electron acceptor for methylamine dehydrogenase, 205
as electron acceptor for trimethylamine dehydrogenase, 197

Bromomethane oxidation by methane monooxygenase, 155, 163, 164*t*, 166
Budding in methylotrophic yeasts, 273, 283
Butadiene oxidation by methane monooxygenase, 347*t*
Butane, oxidation by methane monooxygenase, 164*t*, 166, 346*t*
n-Butanol, *see* Alcohols
2-Butanol, oxidation by methylotrophs, 344
But-1-ene (1-butene), oxidation by methane monooxygenase, 165*t*, 166, 346*t*
But-2-ene (2-butene), oxidation by methane monooxygenase, 165*t*, 166, 346*t*

C

Calvin cycle, *see* Ribulose bisphosphate pathway
Candida boidinii, 270*t*
  alcohol oxidase of, 275*t*, 278
  amine oxidase of, 208
  apparent alcohol dehydrogenase in, 279
  DHA synthase of, 289*t*, 290, 291
  evidence for DHA cycle, 288, 289*t*
  formaldehyde dehydrogenase of, 284, 295
  formate dehydrogenase of, 285, 295
  S-formylglutathione hydrolase of, 284
  glucose 6-phosphate dehydrogenase of, 82*t*
  localisation of oxidase and catalase in, 282*t*
  mutant isolation and characterisation, 289*t*, 290
  mutant lacking alcohol oxidase, 278, 283
  peroxisomes and crystalloids in, 282*t*, 283
  production of citrate by, 339
  utilisation of methylamine, 273
  rearrangement enzymes of, 289*t*
  regulation of oxidising enzymes in, 292–295
  regulation of peroxisome synthesis and degradation in, 282*t*, 292, 293
  triokinase of, 289*t*, 292

Candida methylica, triokinase of, 289*t*, 292
Candida sp., 270*t*, 273
  alcohol oxidase of, 275*t*
  evidence for DHA cycle, 288, 289*t*
  formate dehydrogenase of, 285
Capsules, 8*t*, 13
Carbon assimilation in methanogens, *see* Methanogens, biosynthesis
Carbon assimilation pathways, *see* Dihydroxyacetone cycle, Ribulose bisphosphate pathway, Ribulose monophosphate pathway, Serine pathway
Carbon assimilation pathways (distribution), 7*t*, 20, 23*t*, 46, 48*t*, 67*t*, 72*t*, 73*t*, 75*t*, 110*t*, 111*t*, 112*t*, 286*f*, 304, 305*f*
Carbon conversion efficiency (CCE), 251, 256*t*, 261*t*
Carbon dioxide, *see also* carboxylases
  as activator of methyl-Coenzyme M reductase, 320*f*
  assimilation, *see* Ribulose bisphosphate pathway
  assimilation in methanogens, 305*f*, 306, 311, 312
  assimilation by serine pathway, 97*f*, 98*f*, 99, 100, 119–121
  binding to methanopterin, 311, 312, 323
  methanogenesis from, 298, 299*f*, 306, 307*f*, 311, 312, 315*t*, 316, 319–323
  production during assimilation of alkanes etc., 249, 253*t*, 254, 256*t*, 258
  reduction to methane, 307*f*, 316, 319, 320*f*
Carbon dioxide reduction (CDR) factor in methanogenesis, 319*f*, 322, 323
Carbon monoxide
  binding to cytochrome a_3, 221
  binding to cytochromes *b* and *o*, 202, 221, 233, 236, 239
  methanogenesis from, 298
  oxidation by methane monooxygenase, 155, 163–166, 346*t*
  reaction with cytochromes *c*, 221, 226, 241

Carbon monoxide (*continued*)
  reaction with dimethylamine
    monooxygenase, 202
  reaction with methane
    monooxygenase, 159, 162
  as substrate for methylotrophs, 3,
    48*t*
Carbon supply, limitation of growth
  yields by, *see* Growth yields
Carbon-limiting growth conditions,
  effect on electron transport
  systems, 234–238
Carbonyl reagents
  reaction with amine oxidase of
    C. boidinii, 209
  reaction with methylamine
    dehydrogenase, 205
  reaction with *N*-methylglutamate
    synthase, 211
Carboxydihydromethanopterin, 311,
  323
Carboxydismutase, *see* Ribulose
  bisphosphate carboxylase
Carboxylation
  in biosynthesis of methanogens,
    305*f*, 306, 311
Carboxylation of C_3 compounds
  in RuMP pathway, 51, 70–74, 91, 92
  in serine pathway, 97*f*, 98*f*, 99, 100,
    119–121
Carboxylic acids, production by
  methylotrophs, 339
Carnitine degradation, 40
Carotenoids, *see also* Pigmentation,
  19, 27, 28*t*
  mutants lacking, 106*t*, 107*t*
Catabolic inactivation of enzymes in
  yeast, 294
Catabolite repression
  of serine pathway enzymes, 126–130
  in yeasts, 293
Catalase in amine-oxidising bacteria,
  208
Catalase in methylotrophic yeasts,
  276*f*, 279–281, 282*t*, 283
  localisation of, 276*f*, 277*f*, 282*t*, 283
  peroxidative activity of, 280
  regulation of, 279, 280, 282*t*, 283,
    292, 293
CDR factor, *see* Carbon dioxide
  reduction factor

Cell material (general formula),
  249–254*f*, 263
Cell walls of methanogens, 297*t*,
  301–303
Chelating agents, *see* Metal chelating
  agents
Chemolithotrophs, *see also*
  Methanogens (introduction), 46,
  298
Chlorobium thiosulphatophilum, 43, 52
Chloromethane, oxidation by methane
  monooxygenase, 163, 164*t*, 166
Chlorotoluene, oxidation by methane
  monooxygenase, 346*t*
Chlorophenol, oxidation by methane
  monoxygenase, 346*t*
Choline
  degradation, 40
  methanogenesis from, 326
Chromatium, ribulose bisphosphate
  carboxylase of, 53
Chromophoric factor F_{342}, *see*
  Methanopterin
Chromosome mobilisation in
  methylotrophs, 130, 131
Citramalyl-CoA lyase, 124
Citrate
  growth of restricted facultative
    methylotrophs on, 33
  inhibition of hydroxypyruvate
    reductase, 117
  inhibition of phosphofructokinase, 93
  intermediate in assimilation of
    formaldehyde (serine pathway),
    97*f*
  production by yeast from methanol,
    339
Citrate synthase, 144–146
  in assimilation of formaldehyde
    (serine pathway), 97*f*
  hydrolysis of malyl-CoA by, 123, 148
  regulation of, 136, 144–146
Cleavage reactions of the RuMP
  pathway, *see also* individual
  reactions, 62, 67*t*, 81–85
Cleavage reaction of the serine
  pathway, *see also* malyl-CoA
  lyase, 100–103
Cloning of genes of methylotrophs, 131
Clostridium aceticum, in
  methanogenesis, 299*f*

390 The biochemistry of methylotrophs

Coenzyme A, in regulation of enzyme activity, 132
Coenzyme F_{420}, a deazaflavin, 306, 308–311, 318, 319, 321, 327
Coenzyme M, 309t, 310f, 312–314, 319–323, 325f
  binding to factor F_{430}, 312
Coenzyme Q, *see also* Electron transport in methylotrophs, 219
  types found in methylotrophs, 26t, 27, 35, 219
Coenzymes, novel, from methanogens, 308, 314
Co-metabolism in methanotrophs, 166, 167
Commercial exploitation of methylotrophs, 328–349
Component B of methyl-CoM reductase, 309t, 314, 320f, 321
Compounds F_A and F_C, 309t, 314
Continuous culture, and SCP production, 329t, 332–358
Coordinate regulation of serine pathway enzymes, 101, 127–132
Co-oxidation in methanotrophs, 6t, 166, 167
Copper
  in cytochrome c, 225t
  in methylamine oxidase, 208
  in methane monooxygenase, 158, 161
Corrinoids
  in methanogen biosynthesis, 306
  in methanogenesis, 312, 319
Corynebacteria, use for overproduction of metabolites, 338, 341
Coupling constant and Y_{ATP}, 246, 250
Cresol, oxidation by methane monooxygenase, 346t
Crystalloids in yeast peroxisomes, 276f, 277f, 282t, 283
Cyanide
  as inhibitor of cytochrome oxidases, 233, 235f, 237f, 239f, 241
  as inhibitor of dimethylamine monooxygenase, 202
  as inhibitor of formate dehydrogenase, 193, 285
  as inhibitor of methane monooxygenase, 159, 162, 241f
  as inhibitor of methanol dehydrogenase, 174, 184, 185f

Cyanide (*continued*)
  as inhibitor of trimethylamine N-oxide demethylase, 201
Cyanomethane, oxidation by methane monooxygenase, 164t
Cyclohexane, oxidation by methane monooxygenase, 165t, 347t
Cyclohexanol, oxidation by methane monooxygenase, 347t
6 S-Cysteinyl-FMN
  prosthetic group of dimethylamine dehydrogenase, 199f, 203
  prosthetic group of trimethylamine dehydrogenase, 199
Cytochemical determination of alcohol oxidase and catalase in yeast, 282t, 283
Cytochemistry
  of amine oxidase, 208
Cytochromes in methylotrophs, *see also* Electron transport chains
  general summary, 221–224
  references to, 222t
Cytochrome a/a_3
  effect of growth conditions on, 224, 236–238, 239f, 240
  in electron transport in methylotrophs, 221, 229, 232–241
Cytochrome b, *see also* Dimethylamine dehydrogenase, Dimethylamine monooxygenase
  binding to CO, *see also* Dimethylamine dehydrogenase, Cytochrome o, 221
  effect of growth conditions on, 224, 236
  in electron transport in methylotrophs, 221–224, 229, 230, 232–241
  interaction with dimethylamine dehydrogenase ETF, 204
  interaction with N-methylglutamate dehydrogenase, 213
  interaction with trimethylamine dehydrogenase ETF, 200
  in a methanogen, 317
Cytochrome c, *see also* Electron transport in methylotrophs, Proton translocation in methylotrophs, Electron transport in individual methylotrophs, 221, 224–232, 236

Cytochrome c (continued)
  autoreduction of, 186, 227–229, 231
  autoxidation, 225t, 230, 241
  binding of methane to, 159
  effect of growth conditions on, 221
  effect of pH on midpoint potential of, 228
  as electron donor to methane monooxygenase, 155, 159–161, 241f, 264, 265
  in electron transport chains of methylotrophs, 221–242
  interaction with methylamine dehydrogenase, 206
  interaction with N-methylglutamate dehydrogenase, 213
  lack of interaction with dimethylamine and trimethylamine dehydrogenases, 204
  and methane monooxygenase, 158–161, 241f, 259, 265
  midpoint redox potential of, 224, 225t, 226, 227, 236, 237f, 238, 265
  mutants lacking, 106t, 108t, 127, 131, 132f, 149f, 191, 226, 229, 233–235, 240
  periplasmic location, 243
  proposed oxidase or oxygenase function, 159, 226, 227, 236, 241
  reaction with carbon monoxide, 221, 226, 241
  reaction with methanol dehydrogenase, 176, 186, 187, 228t, 229–232, 242–244
  in regulation of metabolism, 133
  regulation of synthesis, 131, 132f
  release into culture medium, 226, 236
  in reversed electron transport from methanol, 263–265
  and spectrum of aldehyde dehydrogenase, 190
Cytochrome c_{CO}, see also Cytochromes c of methylotrophs, Carbon monoxide, Cytochrome c, proposed oxidase function, 226, 227, 241f
Cytochrome o
  as oxidase in methylotrophs, 221–224, 236–240
Cytochrome oxidase, see Oxidase (cytochrome)

Cytochrome P-420, 202
Cytochrome P-450, 202

D

Deazaflavin, see Coenzyme F_{420}
Decarboxylation reactions, effect on growth yields, 249, 253t, 254, 256t, 258
Dechlorination, catalysed by methane monooxygenase, 164t, 345, 346t
Definition of autotroph, 42
Definition of methylotroph, 2, 42
2,3-Dehydroglutarate, in assimilation of formaldehyde (serine pathway), 103–105
Dehalogenation catalysed by methane monooxygenase, 164t, 345, 346t
Denitrification by methylotrophs, 35, 38–41
Diamine oxidation by methylamine dehydrogenase, 205
Dichloromethane, growth of Hyphomicrobium on, 35
Dichloromethane, oxidation by methane monooxygenase, 164t
2,6-Dichlorophenolindophenol
  in biofuel cells, 349
  as electron acceptor for N-methylglutamate dehydrogenase, 211, 212t
  as electron acceptor for methylamine dehydrogenase, 205
N,N'-Dicyclohexylcarbodiimide (DCCD), inhibitor of ATPase of methanogens, 316
Diethylamine, substrate for trimethylamine dehydrogenase, 198
Diethyldithiocarbamate, as inhibitor of methane monooxygenase, 161, 162
Diethylether, oxidation by methane monooxygenase, 165t, 166
Dihydroxyacetone
  intermediate in assimilation of formaldehyde (DHA cycle), 286f, 291, 292
  intermediate in oxidation of formaldehyde (DHA cycle), 287f
  radioactivity from methanol into (yeast), 289t, 290

Dihydroxyacetone cycle in yeasts
  description, 285–287
  evidence for, 288–290
  and growth yields, 247t, 252t, 256t, 261t, 262
  reactions of, *see also* individual enzymes, 286f, 287f, 289t, 290–292
  regulation of, 289t, 291, 292, 295
  relevance to SCP production, 247t, 332
  summary equation, 252t, 285
Dihydroxyacetone kinase, *see* Triokinase
Dihydroxyacetone phosphate
  intermediate in assimilation of formaldehyde (DHA cycle), 286f, 292
  intermediate in assimilation of formaldehyde (RuMP pathway), 62, 64f, 65f, 66f, 85, 88, 91
  intermediate in CO_2 assimilation (RuBP pathway), 44f, 45f, 54
  intermediate in oxidation of formaldehyde (DHA cycle), 287f
  and NADH oxidation in yeasts, 281
Dihydroxyacetone synthase
  in formaldehyde assimilation (DHA cycle), 286f, 288, 289t, 291, 292
  in formaldehyde oxidation (yeasts), 287f
Dimethylamine
  assimilation equations for, 252t
  growth of methanogens on, 326
  growth yields on, 257t, 258, 259
  as intermediate in oxidation of other methylated amines, 196f
  oxidation of, 196f, 200, 201–204, 216t, 217
  oxidation by trimethylamine dehydrogenase, 203
  oxidation by trimethylamine monooxygenase, 200
  utilisation by yeasts, 273
Dimethylamine dehydrogenase, 196f, 203, 204, 217
  limitation of growth yields and, 252t, 257t, 258, 259
  proportion of electron transport from 266, 268

Dimethylamine monooxygenase, 196f, 201–203, 216t, 217
  limitation of growth yields and, 252t, 257t, 258, 259
Dimethylarsine, 320
Dimethylcarbonate as carbon source, 6t
Dimethylether
  as intermediate in methane oxidation, 155
  oxidation by methane monooxygenase, 163, 165t, 166
Dimethylethylamine, growth of methanogens on, 326
Dimethylsulphide, production by *Pseudomonas* MS, 151
2,3-Diphosphoglycerate, activator of phosphoglyceromutase, 118
1,3-Diphosphoglycerate
  in assimilation of CO_2 (RuBP pathway), 44f, 45f, 53, 54
  in assimilation of formaldehyde (DHA cycle), 286f
  in assimilation of formaldehyde (RuMP pathway), 64f, 65f, 90
Diplococcus PAR, 31t
  absence of homoisocitrate lyase in, 104
  elucidation of serine pathway in, 98, 99t, 111t, 112t
  growth inhibition by itaconate, 114
  methanol dehydrogenase of, 168t, 170t
  TCA cycle in, 139t
α,α'-Dipyridyl, as inhibitor of methane monooxygenase, 161
Dissimilatory (DHA) cycle of formaldehyde oxidation in yeasts, 287, 295
Dissimilatory (RuMP) cycle of formaldehyde oxidation
  description, 68–70
  distribution, 76, 118t, 192
  evidence for and against, 76
  key enzyme (6-phosphogluconate dehydrogenase), 89
  regulation, 89, 90, 93, 193
Distribution of carbon assimilation pathways, *see* Carbon assimilation pathways, distribution of and individual pathways and enzymes

Distribution of pathways for oxidation of methylated amines, 196t, 214t–218
2,2′-Dithiodiethane sulphonic acid, *see* Coenzyme M
DNA base ratios
  of methanogens, 304
  of methanotrophs 7t, 9, 10, 11t, 12t, 22
  of other methylotrophs, 24, 26t, 27, 35, 37
DNA photoreactivation and Coenzyme F_{420}, 310
DNA repair in methanotrophs, 130
Dye-linked aldehyde dehydrogenases, 188t–191
Dye-linked methanol dehydrogenase, *see* methanol dehydrogenase

E

Ecology of methylotrophs, 38–41
Economic aspects of SCP production, 329t, 330, 333, 338
Ectothiorhodospira
  ribulose bisphosphate carboxylase of, 53
EDTA, inhibitor of methanol oxidation, 172
Efficiency of carbon utilisation, what use is it? *see also* Growth yields, P/O ratios, 244
Electroenzymology, 348–349
Electron micrographs
  of methylotrophs, 13, 16–18f, 24, 277f
  of yeasts and their organelles, 277f
Electron transferring flavoprotein (ETF)
  electron acceptor from dimethylamine dehydrogenase, 204
  electron acceptor from trimethylamine dehydrogenase, 199
Electron transport
  percent from each dehydrogenase, 266–268
  in *Methylophilus methylotrophus*, 236–238
  in *Methylosinus trichosporium* and other methanotrophs, 240–242
  in methylotrophs, introduction and summary, 219–224

Electron transport (*continued*)
  in methylotrophs, references to, 222t
  in *Paracoccus denitrificans*, 239, 240
  in *Pseudomonas* AM1, 233–236
Electron transport chains
  effect of growth conditions on, 221–224, 234, 235, 240, 236–239
  in methanotrophs, 308, 317–319, 325f
  in methylotrophs, *see also* individual organisms, 222t, 232–242, 235f, 237f, 239f, 241f
  in regulation, 131
Endergonic reactions, *see* Thermodynamics
End-product inhibition in overproduction of metabolites, 341
"Energised membrane", *see* Protonmotive force
Energy transduction in methanogens, 306–327
Energy transduction in methylotrophs, *see also* ATP synthesis, Electron transport, Proton translocation, 219–224, 242–244
Energy-limitation of growth yields, *see also* Growth yields, limitation by ATP and NAD(P)H 248, 255
Enolase (PEP hydratase)
  in assimilation of formaldehyde (RuMP pathway), 63f, 66f, 91
  in assimilation of formaldehyde (serine pathway), 97f, 118
Entner/Doudoroff enzymes, *see also* Glucose 6-phosphate dehydrogenase, 6-Phosphogluconate dehydrase, KDPG aldolase, 62, 67t, 68, 188t
Enzymic fuel cell, 348, 349
Epifluorescent detection method for methanogens, 309t, 311
Epoxides, products of methane monooxygenase, 163–166, 345–348
Epoxy polymers, 345
Equilibrium position, effect of concentration on, 78, 123
D-Erythro-L-glycero 3-hexulose 6-phosphate, *see* Hexulose 6-phosphate

Erythrose 4-phosphate
  in assimilation of CO_2 (RuBP pathway), 44f, 45f, 55, 56, 58
  in assimilation of formaldehyde (DHA cycle), 286f
  in assimilation of formaldehyde (RuMP pathway), 63f–66f, 68, 69f, 85, 87, 88
Escherichia coli
  cloning of methanol dehydrogenase gene in, 131
  gene transfer into methylotrophs, 130
  pyruvate dehydrogenase of, 146
Ethane, oxidation by methane monooxygenase, 163, 164t, 166, 346t
Ethanoic acid, *see* Acetate
Ethanol
  growth of yeast on, 278, 285
  metabolism by methylotrophs, 27, 35, 147–150
Ethanolamine, oxidation by methylamine dehydrogenase, 205
Ethene, oxidation by methane monooxygenase, 163, 165t, 346t
Ethers
  of glycerol in lipids of methanogens, 302–304
  oxidation by methane monooxygenase, 163, 165t, 166
Ethylamine, enzymic assay of, 343
Ethylbenzene, oxidation by methane monooxygenase, 347t
Ethyl glycine, as activator of methanol dehydrogenase, 174
Ethyne, *see* Acetylene
Eucaryotic methylotrophs, *see also* Yeasts, 269–274
Eucaryotic methylotrophs other than yeasts, 110t, 270t
Euglena gracilis, RuBP carboxylase of, 53
Evolution
  of carbon assimilation pathways, 51, 52, 61
  of membrane systems, 304
  of methanol oxidising systems, 244
Exergonic reactions, *see* Thermodynamics
Exospores of methanotrophs, *see* Spores

F
Facultative autotrophs, 33, 34t, 141
  electron transport in, 239
  oxidative routes for methylated amines in, 215t, 218
Facultative methanotrophs, *see* Methylobacterium
Facultative methylotrophs
  cytochromes and electron transport in, *see also* individual organisms, 222t
  hexulose phosphate synthase of, 79t
  methanol dehydrogenases of, 168t
  RuMP pathway in, 67t, 72t, 74, 76
  serine pathway in, 109, 110t, 112t
  TCA cycle in, 138t, 139t, 141, 143
Facultative methylotrophs unable to use methane, 27–35
Facultative phototrophs, *see also* individual organisms, 33, 34t
Factor F_{430}, nickel-containing tetrapyrrole, 309t, 312
Factor B_0, *see* Methanopterin
Factor F_{342}, *see* Methanopterin
Factor F_{420}, *see* Coenzyme F_{420}
Factor YFC, *see* Methanopterin
FAD
  commercial production of, 339
  in ETF of trimethylamine dehydrogenase, 200
  in methane monooxygenase, 157, 158f
  prosthetic group of yeast alcohol oxidase, 275t, 278, 294
  prosthetic group of N-methylglutamate dehydrogenase, 212t
  synthesis in yeasts, 294
Fatty acids, *see also* Acetate, Formate
  carbon-limitation of growth on, 258
  role in methanogenesis, 299f
Fatty acid composition of methylotrophs, 6t, 22, 24, 26t, 27, 35
FBP, *see* Fructose bisphosphate
Fermenter (pressure cycle), *see* Pressure cycle fermenter
"Fermentation products" from methylotrophs, 338–342
Ferron, as inhibitor of methane monooxygenase, 161, 162

Flagella, 7t, 8t, 9, 10, 11t, 13, 19, 26t, 27, 28t, 29, 30t, 35, 37
Flavin
  component of dimethylamine monooxygenase, 202
  prosthetic group of N-methylglutamate dehydrogenase, 212t, 213
Flavoprotein, *see* FAD, FMN, Flavin, Riboflavin
Flavoprotein component of methane monooxygenase, 157, 158f
Flavoprotein dehydrogenases and growth yields, 251, 252t, 254f, 256t, 261t, 266, 267t
Fluoroacetate, inhibitor of aconitate hydratase, 114, 339
Fluorescence
  of Coenzyme F_{420}, 309
  of methanogens, 309t, 311
  of methanopterin, 311
  of methylamine dehydrogenase and its prosthetic group, 205
  of PQQ from methanol dehydrogenase, 177, 180t, 184
FMN
  deazaflavin analogue, *see* Coenzyme F_{420}
  prosthetic group of formate dehydrogenase, 194
  prosthetic group of N-methylglutamate synthase, 211
FMN adenyltransferase in methylotrophic yeasts, regulation of, 294
FMN-derivative, prosthetic group of amine dehydrogenases, 199, 203
Formaldehyde
  assimilation of, *see* Serine pathway and Ribulose monophosphate pathway
  cyclic route for oxidation (bacteria), *see* Dissimilatory (RuMP) cycle of formaldehyde oxidation
  cyclic route for oxidation (yeast), 287, 295
  effect of oxidation route on growth yields, 252t, 255, 256t, 259, 261t, 266
  growth of methylotrophs on, 6t, 26t, 270t

Formaldehyde (*continued*)
  growth yields on, 247t, 336t
  importance in regulation, 93, 133, 244, 294
  as intermediate in oxidation of methylated amines, 195, 196f, 209f
  methanogenesis from, 322, 323
  oxidation by alcohol oxidase (yeast), 275t, 278
  oxidation by bacteria to CO_2, 68–70, 187–194
  oxidation by catalase (yeast), 280
  oxidation by methanol dehydrogenase, 133, 172, 173, 192
  oxidation by serine pathway and TCA cycle, 136, 193
  production in methazotrophs, 208
  as reductant for methane monooxygenase, 163, 167, 345
  substrate for hexulose phosphate synthase, *see also* Hexulose phosphate synthase, 77–79t
  substrate for DHA synthase, 286f, 287f, 291
  substrate for serine transhydroxymethylase, 115
  transport, 244
Formaldehyde oxidation
  in methanotrophs, 76
  in a mutant lacking cytochrome c, 191
  limitation of growth yields and, 255, 256t, 258, 259
Formaldehyde oxidation (cyclic route) in yeasts, 287, 288
Formaldehyde dehydrogenase
  bacterial dye-linked, 188t, 190, 191
  bacterial NAD^+-dependent, 187–190
  basis of alcohol dehydrogenase of yeast, 279
  limitation of growth yields and, 252t, 255, 256t, 259, 261t, 266
  location of, 244
  mutant lacking, 108t
  proportion of electron transport from, 266, 267t
  requirement for NAD^+-linked enzyme in methanotrophs, 159
Formaldehyde dehydrogenase of methanotrophs, 76, 159, 188t

Formaldehyde dehydrogenase of
methylotrophic yeasts, 273, 276f,
283, 284
regulation of, 292–295
Formamide, growth of facultative
autotrophs on, 48t
Formate
assimilation equations for, 253t
assimilation by methanogens, see
also Biosynthesis in methanogens,
324
assimilation by serine pathway, 99
energy-limitation during growth on,
255
growth of facultative autotrophs on,
48t
growth of methylotrophs on, 26t,
27, 29, 30t, 33, 34t, 35, 39, 40, 43,
47–48t, 58, 270t
growth yields on, 256t, 336t, 337t
growth yields of methanogens on,
297t, 327
methanogenesis from, 298, 299f, 323,
324, 327
microbial production and
utilisation of, see also Formate,
methanogenesis from, 39
NADH dehydrogenase during
growth on, 267t, 268
oxidation by bacteria, 154t, 194
oxidation by catalase (yeast), 280
oxidation by yeast, 276f, 284, 285
as product of methanol
dehydrogenase activity, 172, 173,
185, 192
regulation of oxidation, 194
transport of, 238
Formate dehydrogenase
bacterial, 194
low levels in methylotrophs with
RuMP pathway, 192, 193
of methanogens, 310, 324t, 327
in methanotrophs, 76
mutant lacking, 108t
regulation, 131, 194
Formate dehydrogenase (yeast), 273,
276f, 284, 285
regulation of, 292–295
Formate hydrogen lyase
of methanogens, 310f

S-Formylglutathione, product of yeast
formaldehyde dehydrogenase,
276f, 284
S-Formylglutathione hydrolase in
yeasts, 276f, 284
Formyltetrahydrofolate synthetase, 191
Fortuitous metabolism in
methanotrophs, 166, 167
Free energy changes (standard) for C_1
oxidation reactions, see also
Thermodynamics, Midpoint redox
potentials, 154t
Free radicals
as electron acceptor from methanol
dehydrogenase, 175
in methanogenesis from acetate, 325
in oxidation of methane to
methanol, 153, 345
possible involvement in cytochrome
c activity, 228t, 231
in reduction of PQQ, 181f, 184, 185,
231
Fructose
growth of marine bacteria on, 37
growth of methylotrophs on, 23, 25t,
26t, 37
Fructose 1,6-bisphosphate
in assimilation of CO_2 (RuBP
pathway), 44f, 45f, 54, 55
in assimilation of formaldehyde
(DHA cycle), 286f
in assimilation of formaldehyde
(RuMP pathway), 62, 64f, 66f,
84, 85, 88
in oxidation of formaldehyde
(DHA cycle), 287f
Fructose bisphosphate aldolase
variants of the RuMP pathway,
see Ribulose monophosphate
pathway, variants
Fructose bisphosphate aldolase
see also Aldolase, in assimilation of
CO_2
in assimilation of formaldehyde,
64f, 65f, 67t, 75t, 85, 88
Fructose bisphosphate aldolase (yeast)
in formaldehyde assimilation, 286f
in formaldehyde oxidation, 287f
Fructose bisphosphatase
in CO_2 assimilation, 44f, 45f, 55, 59
inhibition by AMP, 59

Fructose bisphosphatase (yeast)
  in formaldehyde assimilation (DHA cycle), 286f, 290
  in formaldehyde oxidation, 287f
  mutants lacking, 290
Fructose 6-phosphate
  in assimilation of formaldehyde (DHA cycle), 286f
  in assimilation of CO_2 (RuBP pathway), 44f, 45f, 55, 56, 58
  in assimilation of formaldehyde (RuMP pathway), 62, 63f–66f, 68f, 69f, 80, 81, 84, 85, 87
  in dissimilatory cycles of formaldehyde oxidation, 69f, 287f
  in oxidation of formaldehyde (DHA cycle), 287f
Fuel cells, 348, 349
Fungi, methylotrophic, *see* Eucaryotic methylotrophs
Fumarate, in assimilation of formaldehyde (serine pathway), 97f
Fumarase (fumarate hydratase), in assimilation of formaldehyde (serine pathway), 97f
Fumarate reductase of methanogens, 305f, 306

G

Gaffkya, sp., cell wall of, 303f
Galactosamine in structure of methanopterin, 311f
Genetics of methylotrophs
  gene mapping in *Methylobacterium organophilum*, 131–132
  preliminary experiments on gene transfer, 130, 131
Genetic engineering, in the ICI "Pruteen" organism, 131, 333
Gliocladium deliquescens, 270t
  serine pathway in, 111t
Gluconate
  assimilation equation for, 252t
  growth of restricted facultative methylotrophs on, 33
Gluconeogenesis, 97, 118
Gluconeogenic enzymes
  in growth of methylotrophic yeasts, 287

Gluconobacter oxydans
  glucose dehydrogenase of, 182, 183t
  polyol dehydrogenase of, 182, 183t
Gluconobacter suboxydans
  alcohol dehydrogenase of, 183t
  aldehyde dehydrogenase of, 183t
  glucose dehydrogenase of, 183t
Glucosamine in structure of methanopterin, 311f
Glucose
  growth of restricted facultative methylotrophs on, 33
  growth of yeasts on, 278, 283, 293, 294
  as radioactive tracer of phosphogluconate metabolism, 76
  theoretical assimilation equation for, 251, 252t
Glucose dehydrogenase of *Acinetobacter* spp., a quinoprotein, 182, 183t
Glucose 6-phosphate
  in dissimilatory cycle of formaldehyde oxidation, 69f, 81–83
  in assimilation of formaldehyde (RuMP pathway), 63f, 66f, 69f, 81–83
  in oxidation of formaldehyde (DHA cycle), 82t, 287f, 288
  radioactivity from methanol into (yeast), 289t, 290
Glucose 6-phosphate dehydrogenase, 81–83
  in formaldehyde assimilation, 63f, 66f, 81–83, 295
  in formaldehyde oxidation, 69f, 74, 82t, 83, 288
  in methanotrophs, 6t, 20
  regulation of, 82t, 83, 93, 94, 288, 295
Glucose phosphate isomerase
  in assimilation of formaldehyde, 63f, 66f, 81
  in oxidation of formaldehyde, 69f
  in oxidation of formaldehyde (yeasts), 287f
L-Glutamate
  overproduction by methylotrophs, 340
  in oxidation of methylated amines, 196f, 209f, 210

398 The biochemistry of methylotrophs

L-Glutamate (*continued*)
  as substrate for N-methylglutamate synthase, 209f, 210
Glutamate dehydrogenase, gene transfer on plasmid, 131, 333
Glutamate synthase, 333
Glutaryl-enzyme intermediate, *see* N-methylglutamate synthase
Glutamine synthetase, and SCP production, 333
Glutamylglutamate side chain of Coenzyme F_{420}, 309, 310f
γ-Glutamylmethylamide, in oxidation of methylated amines, 209f, 210
γ-Glutamylmethylamide synthetase, 209f, 211
  mutant lacking, 108t
Glutarate, in assimilation of formaldehyde (serine pathway), 103–105
Glutathione
  in formaldehyde oxidation by bacteria, 187, 188t
  regulatory role in yeasts, 294
  requirement for formaldehyde dehydrogenase of yeasts, 276f, 279, 284
  stimulation of trimethylamine N-oxide demethylase by, 201
Glyceraldehyde 3-phosphate
  in assimilation of formaldehyde (DHA cycle), 286f, 291
  in assimilation of formaldehyde (RuMP pathway), 62, 63f–66f, 67t, 68, 69f, 84–88, 90
  in CO_2 assimilation (RuBP pathway), 44f, 45f, 54, 56
  inhibitor of phosphoribulokinase, 59
  as inhibitor of glucose 6-phosphate dehydrogenase, 82t
  in oxidation of formaldehyde (DHA cycle), 287f
Glyceraldehyde phosphate dehydrogenase
  in CO_2 assimilation, 44f, 45f, 53
  in formaldehyde assimilation, 64f, 65f, 90, 286f
Glycerate
  in assimilation of formaldehyde (serine pathway), 97f, 117

Glycerate kinase
  in assimilation of formaldehyde (serine pathway), 97f, 101, 102t, 117
  mutants lacking, 107t, 108t, 132f
  regulation of synthesis, 126, 128t, 132f
Glycerol
  assimilation equation for, 252t
  growth of yeasts on, 278, 285, 292, 293
Glycerol ethers, in methanogens, 302t–304
Glycine
  in assimilation of formaldehyde (serine pathway), 97f, 98–101, 115, 116
  inhibition of methylotrophs by, 341
  overproduction by methylotrophs, 342
Glycine esters, as activators of methanol dehydrogenase, 174
Glycine-resistance
  in overproduction of serine, 341
Glycine synthesis by serine pathway, 99–100
Glycolaldehyde, oxidation by formaldehyde dehydrogenase, 190
Glycollate
  inhibitor of malyl-CoA lyase, 124
  metabolism by *Pseudomonas* AM1, 149f, 151
  production by hydroxypyruvate reductase, 117
Glycolytic enzymes, *see also* individual enzymes
  in assimilation of formaldehyde, 62, 67t, 85, 93, 287
  in growth of methylotrophic yeasts, 287
Glyoxal, oxidation by formaldehyde dehydrogenase, 190
Glyoxylate
  as activator of serine transhydroxymethylase, 115, 116, 134
  in assimilation of multicarbon compounds, 148–150
  inhibition of isocitrate dehydrogenase by, 146

Glyoxylate (*continued*)
  in assimilation of formaldehyde (serine pathway), 97*f*, 100–104, 116, 121–125
  reduction to glycollate by hydroxypyruvate reductase, 117
Glyoxylate cycle (or by-pass), *see also* Isocitrate lyase, Acetyl-Coenzyme A, in formaldehyde assimilation, 103, 125, 147
  relevance to growth yields, 249, 258
Gram-negative methylotrophs, 6*t*, 26*t*, 28*t*, 30*t*, 31*t*, 34*t*, 35, 37
Gram-positive facultative methylotrophs, 32*t*, 33
Gram-positive methanotrophs, 10
Gram-positive methylotrophs
  oxidative routes for methylated amines in, 215*t*
Gram-negative (or variable) non-motile rods, 29, 31*t*
  oxidative routes for methylated amines in, 215*t*
Gram reaction, of methanogens, 301, 302*t*
Growth factors for methanogens, *see also* Nickel, Coenzyme M, 309*t*, 312, 314, 327
Growth inhibitors in the study of methylotrophs, 113, 114
Growth rates
  of methylotrophic bacteria, *see also* individual organisms, 26*t*, 49*t*, 333, 336*t*, 338
  and SCP production, 332, 333
  of yeast on methanol, 332
Growth yields, *see also* individual organisms, Y_{ATP}
  effect of nitrogen source on, 247*t*, 249, 336*t*
  on alkanes, 256*t*, 258, 281
  and bioenergetics of methylotrophs, 234, 240, 244, 245–268, 331–338
  of cytochrome *c*-deficient mutant, 234
  on dimethylamine, 257*t*, 258, 259
  on formate, 247*t*, 256*t*, 337*t*
  limitation by ATP, 251–257*t*, 260, 262, 264, 265, 333
  limitation by carbon supply, 251–259, 262, 265, 333

Growth yields (*continued*)
  limitation by NAD(P)H, 251–260, 262, 264, 265
  on methane, 247*t*, 256*t*, 258, 259, 261*t*, 264–266, 336*t*
  of methanogens, 297*t*, 308, 326, 327
  on methanol, 50, 247*t*, 258, 259, 260–263, 265, 329*t*, 332–338
  of methanotrophs on methanol, 265
  on methylamine, 247*t*, 256*t*, 258, 336*t*
  on oxygen (Y_{O_2}), 246, 248, 251, 255, 261*t*, 262, 279, 332, 336*t*
  physiological factors affecting, 247*t*, 333, 336*t*
  predictions of, 247*t*, 259, 260–266
  prediction of (methods and assumptions), 245–253*t*, 259
  and production of single cell protein, 247*t*, 329*t*, 330–338
  reviews of literature, 247*t*, 329*t*, 336*t*
  on trimethylamine, 257*t*
  of yeasts on methanol, *see* Dihydroxyacetone cycle in yeasts, and growth yields
  of yeasts on alkanes, 258, 281

H

Halobacteria, 300
Halogenated alkanes, oxidation by methane monooxygenase, 163, 164*t*, 345, 346*t*
Halophila, ribulose bisphosphate carboxylase of, 53
Hansenula polymorpha, 270*t*
  alcohol oxidase of, 275*t*, 278
  apparent alcohol dehydrogenase in, 279
  catalase of, 280
  DHA synthase of, 289*t*, 290, 291
  electron micrographs of sections of, 277*f*
  evidence for DHA cycle in, 289*f*, 290
  formaldehyde dehydrogenase of, 284
  formate dehydrogenase of, 284
  hydrolysis of *S*-formylglutathione in, 284
  inhibition by methanol, 337*t*
  localisation of oxidase and catalase in, 282*t*

400 The biochemistry of methylotrophs

Hansenula polymorpha (*continued*)
  mutant isolation and characterisation, 289*t*, 290
  peroxisomes and crystalloids of, 277*f*, 282*t*, 283
  rearrangement enzymes of, 289*t*
  regulation of oxidising enzymes in, 293, 294
  regulation of peroxisome synthesis and degradation in, 282*t*, 293
  thermotolerance and isolation, 274
  triokinase of, 289, 292
  utilisation of methylamine, 273
Hansenula sp., 270*t*, 273
Haem
  ligation in cytochrome *c*, 228, 229
  reaction with carbon monoxide, 221
  spectrum in aldehyde dehydrogenases, 190, 191
  unusual environment in cytochrome *c*, 227
Haem *c*, prosthetic group of primary amine dehydrogenase from *Ps. putida*, 204
Heats of combustion and growth yields, 248, 252*t*, 255, 264
Heptane, oxidation by methane monooxygenase, 164*t*
n-Heptyl quinoline *N*-oxide (HQNO), as inhibitor of electron transport, 230, 236, 237*f*
Heretical taxonomy for bacteriologists, 4
Heterocyclic compounds, oxidation by methane monooxygenase, 163, 165*t*, 345, 346*t*
Hexadecane, oxidation by methane monooxygenase, 346*t*
Hexane, oxidation by methane monooxygenase, 164*t*, 346*t*
3-Hexulose 6-phosphate
  in assimilation of formaldehyde (RuMP pathway), 61, 62, 63*f*–66*f*, 69*f*, 77–80
  in dissimilatory cycle of formaldehyde oxidation, 69*f*
  in assimilation of formaldehyde, 63*f*–66*f*, 80, 93
  in oxidation of formaldehyde, 69*f*, 93

3-Hexulose phosphate synthase, 77–79*t*
  in assimilation of formaldehyde, 63*f*–66*f*, 72*t*
  in oxidation of formaldehyde, 69*f*
  regulation of, 78, 93
  unusual kinetics of enzyme from methanol-utilisers, 78
Histamine, oxidation by methylamine dehydrogenase, 205
Homocitrate, in assimilation of formaldehyde (serine pathway), 103–105
Homocitrate synthase, in assimilation of formaldehyde (serine pathway), 104
Homoisocitrate, in assimilation of formaldehyde (serine pathway), 103–105
Homoisocitrate-glyoxylate cycle, 103–105, 150
Homoisocitrate lyase, in assimilation of formaldehyde (serine pathway), 102*t*, 103*f*, 104
Hydration of aldehydes, 173
Hydrazines, inhibition of trimethylamine dehydrogenase by, 198
Hydrocarbons
  production by methanogens, 304
  as source of SCP, 329*t*, 330
Hydrocarbons and their derivatives, *see also* Alkanes and their derivatives
  commercial oxidation of, 329*t*, 344–348
Hydrogen
  energy source for aerobic methylotrophs, 33, 34*t*, 47, 48*t*, 50
  reductant in methanogenesis, *see also* Hydrogenase, 298, 299*f*, 306–308, 315–323
Hydrogenase
  and Coenzyme F_{420}, 310, 318, 321
  in enzymic fuel cell, 348*f*
  in methanogenesis, 310, 315*t*–318, 320*f*, 321
Hydrogen carriers, in methanol oxidation, 242
Hydrogenomonas eutropha
  ribulose bisphosphate carboxylase of, 53
  RuBP pathway in, 48*t*

Hydrogen/oxygen fuel cell, 348f
Hydrogen peroxide
  commercial production by alcohol oxidase, 344
  production and utilisation in yeasts, 274–281
3-Hydroxybutyrate, metabolism by methylotrophs, 27, 35, 147–150
4-Hydroxybutyrate, metabolism by methylotrophs, 150, 172
2-Hydroxyglutarate, in assimilation of formaldehyde (serine pathway), 103–105
Hydroxylamine, as inhibitor of methylamine dehydrogenase, 205
Hydroxylation reactions, *see also* Dimethylamine monooxygenase, Methane monooxygenase, Trimethylamine monooxygenase, 163–167, 252t, 256t, 267t, 346t, 347t
Hydroxymethyl-Coenzyme M, 313t, 314, 319f, 320f, 322
2-(Hydroxymethyl)-ethane sulphonic acid, *see* Hydroxymethyl-Coenzyme M
3-Hydroxy 3-methylglutaryl-CoA lyase, 124
S-Hydroxymethylglutathione, substrate for yeast formaldehyde dehydrogenase, 276f, 284, 294
5-Hydroxy N-methylpyroglutamate, 211
Hydroxypyruvate
  in assimilation of formaldehyde (serine pathway), 97f, 116, 117
  as alternative substrate for DHA synthase, 291
Hydroxypyruvate reductase
  in assimilation of formaldehyde (serine pathway), 97f, 101, 102t, 110t, 117
  as "indicator" enzyme for the serine pathway, 20, 50, 109, 110t, 117
  in *Methylococcus capsulatus*, 51, 117
  mutants lacking, 106t, 108t, 132f, 149f
  in *Paracoccus denitrificans*, 50, 117
  regulation of activity of, 117
  regulation of synthesis, 126, 128t, 132f

8-Hydroxyquinoline, as inhibitor of methane monooxygenase, 161, 162
3-Hydroxysuccinyl-Coenzyme A, *see* Malyl-CoA
Hyphomicrobia, *see Hyphomicrobium*
Hyphomicrobium (genus), 23t, 24, 35–37, 39–41
Hyphomicrobium sp.
  alternative route for synthesis of N-methylglutamate, 209f, 211
  anaerobic growth on trimethylamine, 203
  ATP malate lyase activity in, 103
  Coenzyme Q of, 219
  2 cytochromes c in, 226
  cytochrome c reduction by methanol dehydrogenase, 187, 231
  cytochromes and electron transport, 223t
  dimethylamine dehydrogenase of, 196f, 203
  dye-linked aldehyde dehydrogenase of, 189t, 190, 191
  elucidation of serine pathway in, 98, 99, 110t, 112t
  formaldehyde oxidation in, 189t–191
  formate dehydrogenase of, 194
  glycerate kinase of, 118
  growth yields, *see* Methanol, growth yields on
  limitation of growth yields in, 257t
  methanol dehydrogenase of, 168t, 170t, 171, 172, 174, 176, 178f–182, 184–187
  N-methylglutamate dehydrogenase of, 213, 214t
  oxidative routes for methylated amines in, 214t–218
  PEP carboxylase in, 119
  phosphoglycerate mutase of, 118
  regulation of serine pathway in, 118, 127, 128t, 135
  a restricted facultative methylotroph, 139t, 143
  TCA cycle enzymes of, 139t, 141
  trimethylamine dehydrogenase of, 198, 203, 343

I

ICI "Pruteen" process, *see* "Pruteen"

Imidazole, inhibitor of methane monooxygenase, 161
Immobilisation
  of bacteria for overproduction of metabolites, 340
  of enzymes, 343
  of methylamine dehydrogenase, 207
  of peroxisomes, 344
  of whole cells, 344
Incorrect assimilation pathways, see Malate synthase pathway, Homoisocitrate-glyoxylate cycle
Indicator enzymes, 20, 50, 109, 117
Interspecies hydrogen transfer, 298
Iron
  as component of tetrathylammonium monooxygenase, see also Non-haem iron, 197
  ligation in cytochrome c, 228, 229
  stimulation of trimethylamine N-oxide demethylase by, 201
Iron-sulphur centre
  in dimethylamine dehydrogenase, 203
  in methane mono oxygenase, 157, 158f
  in methanogens, 317
  in trimethylamine dehydrogenase, 198, 199
Isocitrate, in assimilation of formaldehyde (serine pathway), 97f, 125
Isocitrate dehydrogenase, 145
  leading to false isocitrate lyase results, 113
  low levels in restricted facultative methylotrophs, 139t, 141
  in methanotrophs, 6t, 20
  mutants lacking, 108t
Isocitrate lyase
  absence in *Pseudomonas* AM1 during growth on C_2 compounds, 147
  in assimilation of formaldehyde (serine pathway), 97f, 102t, 103, 105, 112t, 113, 114, 125
  inhibition by itaconate, 114
  interpretation of assays for, 113
  mutant lacking, 108t, 113, 129
  relationship with ATP malate lyase, 113
  regulation of synthesis, 129
  in yeasts, 281

Isoelectric point
  of cytochromes c, 225t, 226, 231
  of methanol dehydrogenase, 170t, 175
Isoenzymes, see also Cytochromes c of methylotrophs, general properties
  of isocitrate lyase, 125
  of serine transhydroxymethylase, 115, 135
L-Isoleucine, in overproduction of glutamate, 341
Isoniazid, as inhibitor of methylamine dehydrogenase, 205
Isopranyl glycerol ethers, in methanogen lipids, 302–304
Isoprenoids, in methanogens, 303f, 304
Isopropylbenzene, oxidation by methane monooxygenase, 347t
Itaconate, as inhibitor of isocitrate lyase, 114

K
KDPG (2-keto, 3-deoxy, 6-phosphogluconate), in assimilation of formaldehyde (RuMP pathway), 62, 63f, 66f, 83, 84
KDPG aldolase, in assimilation of formaldehyde, 63f, 66f, 67t, 84
KDPG aldolase variants of the RuMP pathway, see Ribulose monophosphate pathway, variants
Kinetic mechanism
  of dimethylamine monooxygenase, 202
  of malyl-CoA lyase, 124
  of methylamine dehydrogenase, 206
  of N-methylglutamate synthase, 210, 211f
  of trimethylamine dehydrogenase, 198
  of yeast formaldehyde dehydrogenase, 284
  of yeast formate dehydrogenase, 285
Klebsiella 101, 31t
  vitamin B_{12} production by, 339
Kloeckera sp., 272t, 273
  formaldehyde dehydrogenase of, 284
  formate dehydrogenase of, 285
  localisation of oxidase and catalase in, 282t

Kloeckera sp. 2201, (*continued*)
  peroxisomes and crystalloids in, 282*t*, 283
  regulation of oxidising enzymes in, 293, 294
  regulation of peroxisome synthesis and degradation in, 282*t*
Kloeckera sp. 2201, 271*t*, 272*t*, 273
  alcohol oxidase of, 275*t*, 278, 279
  apparent alcohol dehydrogenase in, 279
  DHA synthase of, 288, 289*t*, 291
  evidence for DHA cycle in, 289*t*
  probable strain of *C. boidinii*, 272*t*
  regulation of oxidising enzymes in, 292–294
Krebs cycle, *see* Tricarboxylic acid cycle

L

Lactaldehyde, potential intermediate in oxidation of propanediol, 150
Lactate
  assimilation equation for, 252*t*
  metabolism by methylotrophs, 147–150
Lactate dehydrogenase of *Propionibacterium*, 183*t*
Lecithin degradation, 40
Less restricted facultative methylotrophs, 24, 32*t*, 33, 35, 139*t*, 141, 215*t*, 217
L-Leucine, overproduction by methylotrophs, 341
Lignin, 39
Limitation of growth yields, *see* Growth yields, limitation by
Lipid, *see also* Fatty acid composition of methylotrophs, Squalene, Poly 3-hydroxybutyrate
  content of methylotrophs, 24
Lipid cysts, 7*t*, 14, 20, 39
Lipid composition of methanogens, 297*t*, 302–304
Lipid droplets in yeast, 277*f*
Lipoamide dehydrogenase, 146
Localisation
  of alcohol oxidase and catalase in yeasts, 276*f*, 277*f*, 282*t*, 283
  of methane monooxygenase, 155, 159–161

Localisation (*continued*)
  of methanol dehydrogenases, 176, 177
  of methylamine oxidase (amine oxidase), 208
Lumazine, possible prosthetic group of trimethylamine dehydrogenase, 198
Lysine esters, as activators of methanol dehydrogenase, 174
Lysine-resistance, in overproduction of glutamate, 341
Lysis of bacteria, during continuous culture, 337*t*

M

Maintenance energy, 246, 247*t*, 250, 337*t*
Malate
  assimilation equation for, 252*t*
  in assimilation of formaldehyde (serine pathway), 97*f*, 98–100, 121–123
Malate ATP lyase, *see* ATP malate lyase
Malate dehydrogenase, 145
  in assimilation of formaldehyde (serine pathway), 97*f*, 121
  in *Hyphomicrobia*, 139*t*
  in methanotrophs, 6*t*, 20, 138*t*
Malate synthase
  mutant lacking activity, 107*t*, 148–150
  in yeasts, 281
Malate synthase pathway, 148–150
Malate thiokinase (Malyl-CoA synthetase)
  absence in bacteria with icl⁻ serine pathway, 103, 113
  in assimilation of formaldehyde (serine pathway), 97*f*, 101, 102*t*, 103, 112*t*, 113, 121, 122
Malonate, metabolism by *Pseudomonas* AM1, 149*f*
Malyl-Coenzyme A (Malyl-CoA)
  in assimilation of formaldehyde (RuMP pathway), 70
  in assimilation of formaldehyde (serine pathway), 97*f*, 121–124
  hydrolysis by citrate synthase, 123, 148
  intermediate in assimilation of multicarbon compounds, 148–150

404 The biochemistry of methylotrophs

Malyl-CoA hydrolase, 148, 149f
  mutants lacking, 107t, 148–150
Malyl-CoA lyase
  in assimilation of formaldehyde
    (serine pathway), 97f, 101, 102t,
    112t, 122–124
  in assimilation of formaldehyde
    (RuMP pathway), 70
  in assimilation of multicarbon
    compounds, 148–150
  mutants lacking, 106t, 132f, 149f, 150
  regulation of synthesis, 127, 128t,
    132f
  as source of acetyl-CoA in bacteria
    lacking pyruvate dehydrogenase,
    70, 98
Malyl-Coenzyme A synthetase, see
  Malate thiokinase
Mannitol, assimilation equation for,
  252t
Marine algae, 343
Marine methylotrophs, 37, 342
Mediation
  in biofuel cells, 349
Membranes, see also Lipid
  composition of methanogens
  arrangement of "methanol oxidase"
    in, 242–244
  binding of methanol dehydrogenase
    to, 176, 177, 242–244
  cytochromes of *Methylophilus
    methylotrophus* in, 236
  and energy coupling in methanogens,
    307, 315, 316, 322, 323, 325
  in methanotrophs, 9, 10, 12t, 15–19,
    16–18f, 22, 160, 161, 176, 177
  in other methylotrophs, 24, 35
  relevance to methane oxidation,
    160, 161
Membranes, internal, and difficulty in
  measuring proton translocation,
  242
Membrane vesicles
  electron transport and ATP
    synthesis in (methylotrophs), 229,
    230, 234
  methanogenesis in, 316, 322
2-Mercaptoethane sulphonic acid, see
  Coenzyme M
Mercaptoethanol, as reductant of
  methanol dehydrogenase 184

Metabolites, see Overproduction of
  metabolites, and individual
  compounds
Metal-chelating agents, as inhibitors
  of methane monooxygenase, 159,
  161, 162
Methane
  affinity of methylotrophs for, 336t
  anaerobic oxidation of, 38, 167
  assimilation equations for, 252t
  binding to cytochrome c, 159
  growth yields on, see Growth yields,
    on methane
  hydroxylation see Methane
    monooxygenase
  microbial production and utilisation
    of, see also Methanogenesis, 38
  oxidation by bacteria to methanol,
    153–167
  production of, see Methanogenesis
  production of single cell protein
    from, 247t, 329t, 331, 336t
  utilisation by yeasts, 272t, 273, 283
Methane monooxygenase, 153–167
  activity in whole cells, 156t, 345–348
  commercial oxidation of
    hydrocarbons and their
    derivatives by, 344–348
  dehalogenation reactions catalysed
    by, 164t, 345, 346t
  description of component proteins,
    157–160
  effect on growth yields of, 258, 264–266
  electron donor, 155, 158–161, 163,
    166, 167, 241f, 252t, 256t, 258,
    259, 264–266, 345
  electron transport and, 240–242,
    266–268
  in enzymic fuel cells, 349
  epoxidation reactions catalysed by,
    165t, 345–348
  erroneous mechanism based on
    growth yields, 264–266
  inhibitors of, 159, 161–163
  inhibition by acetylene, 163
  limitation of growth yields and,
    256t, 258–260
  from *Methylococcus capsulatus* (Bath),
    155–158f, 160, 161, 346t, 347t
  from *Methylosinus trichosporium*,
    158–161, 346t, 347t

Methane monooxygenase (*continued*)
  oxidation of methanol by (effect on growth yields), 266
  possible free radical mechanism, 153, 345
  possibility of two types, 155, 159–161
  products of oxidation, 164*t*, 165*t*, 346*t*, 347*t*
  proportion of oxygen consumed by, 266–268
  substrate specificity, 163–167, 344–348
  solubility of, 155, 160, 161
Methanethiol, oxidation by methane monooxygenase, 164*t*
Methanobacillus omelianski, 300
Methanobacteriaceae, 302*t*
Methanobacteriales, 302*t*
Methanobacterium (genus), 301, 302*t*
Methanobacterium bryantii, iron sulphur centres and nickel components in membranes, 317
Methanobacterium formicicum, growth and methanogenesis from formate, 327
Methanobacterium ruminantium, Coenzyme M growth factor for, 314
Methanobacterium thermoautotrophicum
  ATP synthetase of, 315
  biosynthesis in, 297*t*, 305, 306
  cell wall structure and synthesis, 301, 303*f*
  energy coupling in, 306, 315, 316
  Factor F_{430} investigations in, 312
  genome size, 304
  hydroxymethyl-Coenzyme M reduction in, 314, 322
  membrane vesicles and methanogenesis, 322
  methanogenic organelles in, 316
  methanopterin (discovery) in, 311
  methanopterin, reductive carboxylation of, 311
  methyl-Coenzyme M reductase of, 321
  PEP carboxylase in, 306
  production of L-alanine by, 301
  reduction of CO_2 to methane in, 314, 319, 320*f*, 322, 323
Methanobacterium strain M.o.H., 300

Methanobrevibacter (genus), 301, 302*t*
Methanococcaceae, 302*t*
Methanococcales, 302*t*
Methanococcus (genus), 302*t*
Methanogenium (genus), 302*t*
Methanogenesis, *see also* Methanogens, 296–300
  from acetate, 298, 299*f*, 307*f*, 320, 323–325
  commercial aspects, 342
  from formate, 298, 299*f*, 323, 324, 327
  from methanol, 298, 307*f*, 323–326
  from methylated amines, 298, 317, 326
  pH optimum for, 298
  methyl transfer reactions and corrinoids in, 319
  reviews of, 297*t*
Methanogens, *see also* individual organisms, Methanogenesis, 296–302*t*
  ATP synthesis in, 306–308, 311, 315–319, 322, 325–327
  biosynthesis, 297*t*, 304–306, 323
  cell walls, 297*t*, 301–303
  energy coupling in, 306–327
  epifluorescent detection method for, 309*t*, 311
  growth yields, 297*t*, 308, 326, 327
  habitat of, 38, 296–300
  incorporation of radioactive CO_2 and methanol, 311, 323–325
  lipid composition, 297*t*, 302–304
  novel coenzymes from, 308–314
  nucleic acids of, 304
  structure and ultrastructure, 297*t*, 301–302*t*
  reviews of, 297*t*
  taxonomy, 300–302*t*
Methanol
  adduct formation with PQQ, 180*t*, 181*f*
  affinity of methylotrophs for, 40, 332, 336*t*
  anaerobic growth on (photosynthetic), *see also* Anaerobic methylotrophs, 48*t*, 50, 263
  anaerobic oxidation of, *see also* Anaerobic methylotrophs, 39, 50, 239*f*, 240, 263

Methanol (*continued*)
  assimilation equations for, 252*t*, 263
  assimilation by serine pathway,
    see also Serine pathway, 99, 100,
    110*t*, 112*t*
  assimilation, *see also* Ribulose
    bisphosphate pathway, Ribulose
    monophosphate pathway
  in enzymic fuel cells, 349
  growth of autotrophs on, *see also*
    Ribulose bisphosphate pathway,
    33, 34*t*, 48*t*, 50
  growth of methanotrophs on, 6*t*, 8*t*,
    15, 21
  growth yields on, *see* Growth
    yields on methanol
  growth yields of methanotrophs on,
    265, 266
  growth yields of methanogens on,
    297*t*, 326
  as inducer of serine pathway
    enzymes, 126–129
  metabolism in yeasts, 269–295
  methanogenesis from, 298, 307*f*,
    323–326
  microbial production and
    utilisation of, 39
  for overproduction of metabolites,
    338–342
  oxidation by bacteria to
    formaldehyde, 154*t*, 167–187
  oxidation by catalase (yeast), 280
  oxidation coupled to ATP synthesis,
    242–244
  oxidation by methane
    monooxygenase, 163–166, 266
  oxidation by methanogens, 326
  oxidation by whole bacteria, 172,
    344, 345
  production from carbon dioxide,
    see Reduction of carbon dioxide
    to methane
  production of single cell protein
    from, 247*t*, 329*t*, 331–338
Methanol-Coenzyme M
  methyltransferase, 326
Methanol dehydrogenase, 167–186,
  192
  absorption spectrum, 177, 178*f*,
    180*t*, 184
  activity in whole cells, 166, 172, 345

Methanol dehydrogenase (*continued*)
  activators and inhibitors, 173, 174,
    185, 186*f*
  amino acid composition, 175
  anaerobically prepared enzyme,
    174, 187, 231
  in assay of alcohols by fuel cells,
    349
  cloning of gene for, 131
  effect on growth yields of, 258, 261*t*,
    263
  electron acceptor (artificial), 174,
    175, 185*f*
  in enzymic fuel cells, 349
  general review, 167–171
  interaction with cytochrome system,
    186, 187, 228–244
  isoelectric point, 170*t*, 175
  in large-scale oxidation of alcohols,
    344
  limitation of growth yields and,
    252*t*, 256*t*, 258, 261*t*
  localisation, 176, 177
  mechanism, 180*t*, 184–186, 228*t*,
    230–232
  molecular weight, 170*t*, 175
  mutant lacking, 106*t*, 108*t*, 127,
    131, 132, 149*f*, 150, 171, 192
  oxidation of formaldehyde by,
    133, 185, 188*t*, 192
  oxidation of propane 1,2-diol and
    4-hydroxybutyrate by, 149*f*, 150
  periplasmic location, 243
  proportion of electron transport
    from, 266, 267*f*
  prosthetic group, *see also*
    Pyrrolo-quinoline quinone,
    177–186
  reaction with cytochrome *c*, 186,
    187, 228*t*, 230–232, 242–244, 265
  as reductant for NAD^+ in light, 263
  regulation of activity of, 133
  regulation of synthesis, 127, 128*t*,
    131, 132*f*
  reversed electron transport from,
    260, 263–265
  serological relationships, 176
  specific activity, 171
  substrate specificity, 171–173
"Methanol oxidase", 177
  arrangement in membranes, 242–244

Index   407

Methanomicrobiaceae, 302*t*
Methanomicrobiales, 302*t*
Methanomicrobium (genus), 302*t*
Methanomonas methano-oxidans, 9, 18*f*
  elucidation of serine pathway in, 98, 99*t*, 110*t*
  methane monooxygenase of, 155, 156*t*
  TCA cycle in, 138*t*, 141
"*Methanomonas*" *methylovora*,
  glutamate production from methanol, 340
Methanopterin and derivatives, 309*t*, 311, 312, 314, 316, 323
Methanosarcina (genus), 302*t*
Methanosarcina barkeri
  biosynthesis in, 297*t*, 305, 306, 323, 324*t*
  carboxylation of methanopterin in, 323
  Coenzyme F_{420} (modified form) in, 310
  cytochrome *b* in, 317
  growth on methanol, methylated amines, acetate or CO, 298, 323–327
  reduction of carbon dioxide to methane in, 323
Methanosarcina sp.
  biosynthesis in, 297*t*
  methanogenesis in, 297*t*
Methanosarcinaceae, 302*t*
Methanospirillum (genus), 302*t*
Methanotrophs, *see also* Carbon assimilation pathways,
  distribution of
  basis for obligate methylotrophy in, 142, 143
  description and taxonomy, 4–23
  cytochromes and electron transport in, 6*t*, 222*t*, 240–242
  electron micrographs of, 16
  electron transport and proton translocation in, 222*t*, 240–242
  growth yields, *see* Methane and Methanol, growth yields on
  limitation of growth yields in, 256*t*, 259, 264–266
  low growth yields on methanol, 265, 266
  methanol dehydrogenases of, 169*t*

Methanotrophs (*continued*)
  oxidations catalysed by whole cells of, 156*t*, 345–348
  secondary alcohol dehydrogenase of, 344
  TCA cycle in, 6, 20, 138*t*, 141, 143
Methanotrophic yeasts, 272*t*, 273, 283
Methazotrophs, 208
Methazotrophic yeasts, 196*f*, 208, 273
Methionine
  ligation to iron in cytochrome *c*, 227, 228
  low content in SCP yeast, 332
  in overproduction of glutamate, 341
  in overproduction of serine, 341
Methionine synthetase, 339
Methoxatin, *see also* Pyrrolo-quinoline quinone, 179, 181*f*
N-Methylalanine, as intermediate in oxidation of methylated amines, 209*f*, 210
N-Methylalanine dehydrogenase, 209*f*, 210
Methylamine
  as activator of methanol dehydrogenase, 173, 174, 186*f*
  assimilation equations for, 252*t*
  assimilation by serine pathway, *see also* Serine pathway, 99, 110*t*, 112*t*
  electron transport from, 229, 233
  enzymic assay of, 343
  growth of methanogens on, 326
  growth of mutants lacking methanol dehydrogenase on, 192
  growth of *Trichoderma lignorum* on, 270*t*
  growth substrate for obligate methylotrophs, 26*t*
  growth of yeasts on, 273
  growth of facultative autotrophs on, 48*t*
  growth yields on, *see* Growth yields on methylamine
  as inducer of isocitrate lyase, 130
  oxidation of, 204–218
  in oxidation of other methylated amines, 196*f*
  as substrate for *N*-methylglutamate synthase, 196*f*, 209*f*, 210

408   The biochemistry of methylotrophs

Methylamine assimilation, *see*
  Ribulose bisphosphate pathway,
  Ribulose monophosphate pathway,
  Serine pathway
Methylamine dehydrogenase (primary
  amine dehydrogenase), 196*f*,
  204–208, 214*t*–216*t*, 218
  assay, substrates and inhibitors, 205
  in assay of amines, 343
  catalytic mechanism, 205–207*f*
  distribution, 214*t*–218
  limitation of growth yields and,
    252*t*, 256*t*, 258
  mutant lacking, 106*t*
  proportion of electron transport
    from, 266–268
  a quinoprotein, 182, 183*t*, 206, 207*f*
  regulation of synthesis, 127, 128*t*
  subunit structure, 206
Methylamine oxidase (amine oxidase),
  196*f*, 208, 214*t*, 218
  limitation of growth yields and,
    253*t*, 257*t*
Methylated amines, *see also* individual
    amines
  anaerobic oxidation of, 203, 217
  assimilation equations for, 252*t*
  bacterial oxidation of, 195–218
  distribution of oxidative pathways
    for, 196*f*, 214*t*–218
  electron transport from, 266–268
  growth of methylotrophs on, 3*t*,
    26*t*, 27, 29, 30*t*, 31*t*, 33, 34*t*, 35,
    37, 40, 48*t*, 110*t*, 112*t*, 195
  growth substrate for obligate
    methylotrophs, 26*t*
  growth of yeasts on, 273
  growth yields on, *see* Growth yields
    on methylamine, dimethylamine
    and trimethylamine
  growth yields of methanogens on,
    297*t*, 326
  isocitrate lyase during growth on, 103
  limitation of growth yields on, *see*
    Methylamine, Dimethylamine,
    Trimethylamine
  methanogenesis from, 298, 317, 324*t*,
    326
  microbial production and utilisation
    of, *see also* Methylated amines,
    methanogenesis from, 40

Methylated amines (*continued*)
  oxidation of (summary), 195, 196*t*
Methylated amines as nitrogen sources,
  see Methazotrophs
Methylated amino acids, *see also*
  individual amino acids, and
  Methylamine oxidation of, 209*f*–
    218
  substrates for *N*-methylglutamate
    dehydrogenase, 212*t*
Methyl-cobalamin, reduction to
  methane, 319
Methyl-Coenzyme M, *see also*
  Coenzyme M
Methyl-Coenzyme M, 309*t*, 313*t*, 314,
  319*f*–326
  oxidation of, 326
Methyl-Coenzyme M reductase, 314,
  315*t*, 320–323, 326
  component B of, 309*t*, 314
  and Coenzyme F_{420}, 310
S-Methyl compounds *see also*
  Dimethyl sulphide, Dimethyl
  sulphoxide, Trimethylsulphonium
  salts, 320
Methylene blue
  in biofuel cells, 349
  as electron acceptor for
    dimethylamine dehydrogenase,
    203
  as electron acceptor for
    trimethylamine dehydrogenase, 197
Methylenetetrahydrofolate, in
  regulation of serine pathway, 133
Methylenetetrahydrofolate
  dehydrogenase, 102*t*, 133, 188*t*,
  191
  in assimilation of formaldehyde
    (serine pathway), 102*t*
Methylformate
  oxidation by methane
    monooxygenase, 163
  utilisation by yeasts, 273
N-Methylglutamate
  as intermediate in oxidation of
    methylated amines, 196*f*, 209*f*–213
  oxidation of, *see also*
    N-Methylglutamate
    dehydrogenase, 196*f*, 209*f*,
    211–213
  synthesis of, 196*f*, 209*f*–211

N-Methylglutamate dehydrogenase (NAD$^+$-dependent), 196f, 213, 214t–218, 253t
N-Methylglutamate dehydrogenase (NAD$^+$-independent), 196f, 211–218, 253t
N-Methylglutamate synthase, 196f, 209f–211
  mutant lacking, 108t
Methylglyoxal, substrate for formaldehyde dehydrogenase, 188t, 284
N-Methylhydroxylamine, product of dimethylamine oxidation, 200
Methylmalonyl-CoA mutase, 339
Methyl mercury, 320
Methylobacter (genus), 6t, 7t, 9, 15, 19, 21, 22, 24
Methylobacter bovis, 8t, 11t
  RuMP pathway in, 72t
Methylobacter capsulatus, 8t, 11t
  RuMP pathway in, 72t
Methylobacter chroococcum, 8t, 11t
  RuMP pathway in, 72t
Methylobacter vinelandii, 8t, 9, 11t
  RuMP pathway in, 72t
Methylobacter sp.
  incomplete TCA cycle in, 6t, 138t
  methanol dehydrogenase of, 177
Methylobacterium (genus), 6t, 7t, 10–13, 15, 20, 21, 29
Methylobacterium organophilum, 10–13, 27, 28t
  cytochromes and electron transport, 222t, 229
  formaldehyde oxidation in, 188t
  gene for cytochrome c, 149f, 229
  gene transfer in, 130–132
  membrane formation in, 15
  methanol dehydrogenase of, 169t, 170t, 172, 174, 176
  mutants of, 108t, 132f
  regulation of serine pathway in, 128t, 129, 131, 132
  serine pathway in, 110t, 112t
  serine transhydroxymethylases of, 115, 134
  TCA cycle in, 20, 138t
Methylobacterium organophilum R6, see *Methylobacterium* R6

Methylobacterium R6, 12t, 13
  formaldehyde oxidation in, 188t
  methanol dehydrogenase of, 169t
  serine pathway in, 110t
  TCA cycle in, 138t
Methylobacterium ethanolicum, 12t
  regulation of serine pathway in, 128t, 129
  serine pathway in, 110t, 112t
  TCA cycle in, 138t
Methylobacterium hypolimneticum, 12t
  serine pathway in, 110t, 112t
  TCA cycle in, 138t
Methylococcus (genus), 7t, 10, 15, 20, 22
Methylococcus (genus), according to Romanovskaya, 10, 11t, 22
Methylococcus bovis, 11t
Methylococcus capsulatus, 8t, 9, 11t, 17f, 20, 51
  commercial exploitation of, 345
  cytochromes and electron transport, 222t, 267t
  elucidation of RuMP pathway in, 61, 72t, 74, 75t, 78
  formaldehyde dehydrogenase of, 187, 188t
  growth yields, see Methane, growth yields on
  3-hexulose phosphate isomerase of, 80
  3-hexulose phosphate synthase of, 78, 79t
  hydroxypyruvate reductase in, 20, 51, 117
  incomplete TCA cycle in, 21, 138t, 142
  isocitrate dehydrogenase of, 21
  malate dehydrogenase of, 21
  malyl-CoA lyase in, 123
  methane monooxygenase of, 155–158, 160–167
  methane monooxygenase activity in whole cells of, 163–167, 344–348
  methanol dehydrogenase of, 169t, 170t, 176, 177
  phosphoribulokinase of, 20, 50
  proton translocation in, 242
  ribulose bisphosphate carboxylase of, 20, 50, 53
  RuBP pathway in, 51

Methylococcus chroococcus, 11t
Methylococcus luteus, 11t
Methylococcus minimus, 8t, 11t, 21
   RuMP pathway in, 72t
Methylococcus mobilis, 9
Methylococcus thermophilus, 9, 11t
   RuMP pathway in, 72t
Methylococcus ucrainicus, 11t
   RuMP pathway in, 72t
Methylococcus vinelandii, 11t
Methylococcus whittenburii, 11t
Methylococcus sp.
   incomplete TCA cycle in, 6t, 20, 138t
   lysis ("in situ"), 337t
   methanol dehydrogenase of, 177
Methylocystis (genus), 6t, 7t, 15, 20
Methylocystis parvus, 8t, 13
   growth rate on methanol, 338
   polysaccharide production by, 338
   serine pathway in, 110t
Methylocystis sp.
   methanol dehydrogenase of, 177
   TCA cycle in, 6t, 138t
Methylomonas (genus), 6t, 7t, 15, 20, 24
Methylomonas agile, 8t
   cytochromes and electron transport, 222t
   RuMP pathway in, 72t
Methylomonas albus, 8t
   citrate synthase of, 144
   cytochromes and electron transport, 222t
Methylomonas aminofaciens, 25t
   3-hexulose phosphate isomerase of, 80
   3-hexulose phosphate synthase of, 79t
   RuMP pathway in, 72t
Methylomonas clara, 25t
Methylomonas methanica, 8t, 9, 16f, 19
   cytochromes and electron transport, 222t
   cytochrome c of, 226
   elucidation of RuMP pathway in, 60–61, 72t, 74, 75t
   formaldehyde oxidation in, 188t
   malyl-CoA lyase in, 123
   methane monooxygenase of, 155, 156t, 162

Methylomonas methanica (*continued*)
   methanol dehydrogenase of, 169t, 170t, 175
   phosphofructokinase in, 74
Methylomonas methanolica, see Pseudomonas methanolica
Methylomonas methylovora, 25t, 26t
   dye-linked aldehyde dehydrogenase of, 190
   formaldehyde oxidation in, 188t, 190
   6-hexulose phosphate synthase of, 78
   methylamine dehydrogenase of, 204
   overproduction of amino acids by, see *"Methanomonas" methylovora*
   oxidative routes for methylated amines in, 216t
Methylomonas rosaceous, 8t, 19
   RuMP pathway in, 72t
Methylomonas rubrum, 8t, 19
   RuMP pathway in, 72t
Methylomonas streptobacterium, 8t
Methylomonas sp.
   incomplete TCA cycle in, 6t, 138t
   methanol oxidation in, 156t
   methanol dehydrogenase of, 177
   mutants of, 341
   overproduction of amino acids by, 341
Methylomonas GB3 and GB8
   6-hexulose phosphate synthase of, 78
   RuMP pathway in, 72t
Methylomonas M15, 25t
   glucose 6-phosphate dehydrogenase of, 82t, 83
   3-hexulose phosphate synthase of, 79t
   RuMP pathway in, 72t
Methylomonas P11, 25t
   cytochromes and electron transport, 222t
   methanol dehydrogenase of, 168t, 173
Methylomonas thalassica, 38
Methylophilus (genus), 24
Methylophilus methylotrophus, 24, 25t, 26t
   ammonia assimilation and cell yields, 262, 333
   cytochrome c of, 225t–230, 236–238

Index   411

Methylophilus methylotrophus (continued)
  cytochrome *o* of, 236–238, 240
  cytochromes and electron transport, 222*t*, 236–238, 267*t*
  cyclic route for oxidation of formaldehyde in, 192*f*
  effect of growth conditions on cytochromes, 224 236–238
  effect on growth of discontinuous methanol addition, 336*t*
  formaldehyde oxidation in, 188*t*, 192*f*
  genetic engineering in, 131, 333
  glucose 6-phosphate dehydrogenase of, 82*t*, 83
  growth rate, 333
  growth yield, *see* Methanol, growth yields on
  3-hexulose phosphate isomerase in, 80
  3-hexulose phosphate synthase in, 79*t*
  high optimum growth temperature, 333
  incomplete TCA cycle in, 138*t*
  limitation of growth yields in, 255, 256*t*, 258, 259
  methanol dehydrogenase of, 168*t*, 170*t*, 176
  outer membranes and lipid content of, 24
  oxidation of methanol by extracts, 230
  oxidation of methylated amines by, 216*t*
  oxidation of NAD(P)H in, 94, 236, 237
  PEP carboxykinase of, 70
  permeability to NAD(P)H, 236
  6-phosphogluconate dehydrogenase of, 89
  potassium transport in, 238
  proton translocation and ATP synthesis, 222*t*, 236–238,
  pyruvate carboxylase of, 70
  RuMP pathway in, 72*t*, 75*t*
  single cell protein, 333–338
Methylosinus (genus), 6*t*, 7*t*, 13, 15, 20
Methylosinus sporium, 8*t*, 13, 18
  methanol dehydrogenase of, 169*t*, 170*t*, 175
  serine pathway in, 110*t*

Methylosinus trichosporium, 8*t*, 13
  citrate synthase of, 144
  commercial exploitation of, 345–348
  cytochrome *c* of, 158–161, 224–227, 241
  cytochromes and electron transport, 222*t*, 225*t*, 230, 240–242, 267*t*
  dye-linked aldehyde dehydrogenase of, 188*t*, 191
  formaldehyde oxidation in, 188*t*, 191
  gene transfer in, 130
  growth yields, *see* Methane, growth yields on
  intracellular membranes of, 19, 161
  methane monooxygenase of, 155, 156*t*, 158–163, 241*f*
  methane monooxygenase activity in whole cells of, 163–167, 345–348
  methanol dehydrogenase of, 177
  oxidation of methanol by extracts, 230
  PEP carboxylase in, 119
  proton translocation and ATP synthesis, 222*t*, 240–242
  serine pathway in, 110*t*
Methylosinus sp.
  cytochromes and electron transport, 222*t*
  methanol dehydrogenase of, 177
  TCA cycle in, 6*t*, 138*t*
Methylovibrio sohngenii, 9
Micrococcus denitrificans, *see Paracoccus denitrificans*
N-Methyl, *N'*-nitro, *N*-nitrosoguanidine, use as mutagen for methylotrophs, 105
Methylotroph, definition, 2
Methylotrophs
  as commercial biocatalysts, 342–349
  commercial exploitation of, 328–350
  cytochromes, electron transport, proton translocation and ATP synthesis in, 219–244
  factors affecting growth rates, 336*t*
  growth yields and bioenergetics, 245–268, 331–338
  growth yields (review of literature), 247*t*, 329*t*
  overproduction of metabolites by, 338–342
  place in nature, 38–41

Methylotrophs (*continued*)
  single cell protein from 247*t*, 328–338
  substrates used by, 3*t*
Methylotrophs unable to use methane, description and taxonomy, 22–38
Methylotrophic yeasts, *see also* Yeasts, 269–274
Methyl selenide, 320
2-(Methylthio)-ethane sulphonic acid, *see* Methyl-Coenzyme M
Methyltelluride, 320
Methyltransferase of methanogens, 315*t*, 320, 326
Methyl transfer reactions in methanogens, 315*t*, 319
Methyl viologen, electron acceptor for hydrogenase, 321
Microbial fuel cell, 348
Microbial reactors, 349
Microbodies in yeasts, *see* Peroxisomes
Microscopic colonies of *Methanomonas methano-oxidans*, 9
Microcyclus aquaticus, 34*t*
  RuBP pathway in, 48*t*
Microcyclus ebrunous, 34*t*
Microcyclus sp., Coenzyme Q of, 220
Midpoint redox potential
  of Coenzyme F_{420}, 309
  of Coenzyme M, 309*t*, 313*t*
  of cytochrome *b*, 236, 237*f*
  of cytochromes *c*, 224, 225*t*, 226, 227, 236, 237*f*, 238, 265
  of cytochrome *o*, 236–238
  effect of pH on cytochrome *c*, 227
  of formaldehyde/methanol couple, 154*t*, 265
  for methanol oxidation, 154*t*, 244, 265
  of methanopterin, 312
  of $NAD^+/NADH$ couple, 154*t*, 265
  for oxidation of C_1 compounds, 154*t*
  of PQQ, 181*f*
Mitochondria of yeasts, 274, 277*f*, 281
Mitsubishi process of SCP production, 329*t*, 333
Mixed cultures involving methylotrophs, 40
Mixed cultures, in production of SCP, 337*t*, 338

Monoamine oxidase inhibitors, inhibition of trimethylamine dehydrogenase by, 198
Monooxygenases, *see* Oxygenases and individual enzymes
More restricted facultative methylotrophs, 23, 139*t*, 141, 142, 216*t*, 217
Motility of methylotrophs, 7*t*, 8*t*, 9, 10, 11*t*, 12*t*, 19, 26*t*, 27, 28*t*, 29, 30*t*, 35, 37
Multicarbon compounds, growth of methylotrophs on, 137–151
Mutagenesis in methanotrophs, 130
Mutagens and multiple lesions, 105
Mutant
  lacking acetyl-CoA synthetase, 107*t*, 149*f*
  lacking alcohol oxidase (yeast), 278, 283, 294
  lacking ATP malate lyase, 108*t*, 113
  lacking carotenoids, 106*t*, 107*t*
  lacking cytochrome *c*, 106*t*, 108*t*, 127, 131, 132*f*, 149*f*, 191, 226, 229, 232–235, 240
  lacking formaldehyde dehydrogenase, 108*t*
  lacking formate dehydrogenase, 108*t*
  lacking fructose bisphosphatase (yeast), 290
  lacking γ-glutamylmethylamide synthetase, 108*t*
  lacking glycerate kinase, 107*t*, 108*t*, 132*f*
  lacking hydroxypyruvate reductase, 106*t*, 108*t*, 132*f*, 149*f*
  lacking isocitrate dehydrogenase, 108*t*
  lacking isocitrate lyase, 108*t*, 113, 129
  lacking malate synthase activity, 107*t*, 148–150
  "lacking" malyl-CoA hydrolase, 107*t*, 148–150
  lacking malyl-CoA lyase, 106*t*, 132*f*, 149*f*, 150
  lacking methanol dehydrogenase, 106*t*, 108*t*, 127, 131, 132, 149*f*, 150, 171, 192
  lacking methylamine dehydrogenase, 106*t*

Index 413

Mutant (*continued*)
  lacking *N*-methylglutamate dehydrogenase, 108*t*
  lacking *N*-methylglutamate synthase, 108*t*
  lacking 2-oxoglutarate dehydrogenase, 107*t*, 143, 146, 340
  lacking PEP carboxylase, 108*t*
  lacking phosphoserine phosphatase 100, 106*t*, 107*t*
  lacking pyruvate dehydrogenase, 106*t*, 143, 146, 149*f*, 340
  lacking serine-glyoxylate aminotransferase, 106*t*, 108*t*, 132*f*
  lacking serine transhydroxymethylase, 106*t*, 108*t*, 132*f*
  lacking triokinase, 290
  operator, 108*t*
  regulatory, 108*t*, 129
  unable to oxidise acetyl CoA to glyoxylate, 103, 105, 107*t*, 148–150
  in elucidation of serine pathway, 100, 105–109, 148–150
  in overproduction of metabolites, 340–342
Muramic acid, 301, 303*f*
Murein, 301, 303
Mycobacterium vaccae, 32*t*
  RuMP pathway in, 73*t*, 75*t*
Mycobacterium 10, methylamine dehydrogenase in, 215*t*
Mycobacterium 50, 34*t*
  RuBP pathway in, 48*t*
  TCA cycle in, 140*t*
Mycobacterium sp., growth on methane, 10

N

NAD$^+$-independent alcohol dehydrogenase, *see* Alcohol dehydrogenase, Methanol dehydrogenase
NADH, *see also* NAD(P)H
  as activator of PEP carboxylase, 120, 121, 135, 136
  as activator of phosphoribulokinase, 59
  anaerobic oxidation by nitrite and nitrate, 239*f*, 240

NADH (*continued*)
  as electron donor for methane monooxygenase, 155–161, 163, 252*t*, 258, 264–266
  as inhibitor of citrate synthase, 136, 144
  as inhibitor of formaldehyde dehydrogenase (yeast), 284, 295
  as inhibitor of formate dehydrogenase, 194, 284
  oxidation in alkane-utilising yeasts, 281
  oxidation by electron transport chains, *see also* Electron transport in methylotrophs, Electron transport in individual organisms, 219, 230, 234, 235*f*, 236–242, 266, 268
  oxidation in methylotrophic yeasts, 274
NADH-acceptor reductase activity of methane monooxygenase, 157, 158*f*
NADH dehydrogenase
  function during oxidation of methanol and methane, 266
  proportion of electron transport from, 266–268
NADPH, *see also* NAD(P)H
  as inhibitor of glucose 6-phosphate dehydrogenase in yeast, 82*t*, 288
  oxidation by electron transport chains, 70, 90, 94, 237
  production for biosynthesis, 69, 70, 89, 93, 132, 146, 252*t*, 254*f*, 287*f*, 288, 295
NADPH-Coenzyme F_{420} oxidoreductase, 310, 321
NADPH-Coenzyme M oxidoreductase, 314
NADPH oxidase activity in *Bacillus* sp., 90
NADPH oxidase activity in *Methylophilus methylotrophus*, 94
NAD(P)H
  as electron donor for dimethylamine monooxygenase, 201, 252*t*, 258, 259
  as electron donor for trimethylamine monooxygenase, 200, 252*t*, 258, 259

414   The biochemistry of methylotrophs

NAD(P)H (*continued*)
  as inhibitor of glucose 6-phosphate dehydrogenase, 82t, 83, 93, 193
  as inhibitor of 6-phosphogluconate dehydrogenase, 89, 93
  limitation of growth yield by, *see* Growth yields
  permeability of whole cells to, 236
  proportion of substrate oxidised to provide, 251–257t
  production by cyclic oxidation route for formaldehyde, *see* Formaldehyde, cyclic route for oxidation of
  production during formaldehyde oxidation, 187, 188t, 191–194, 287f, 288
  in regulation of enzyme activity *see also* individual enzymes, 59, 94, 132–136, 193, 295
Natural gas, *see* Methane
Neocuproine, as inhibitor of methane monooxygenase, 161
Nickel, requirement for growth of methanogens, 312
Nickel-containing tetrapyrrole (Factor F_{430}), 309t, 312
Nickel-containing paramagnetic centre in a methanogen, 317
Nitrate as terminal electron acceptor, 33, 35, 38–41, 203, 239f, 240
Nitrite as terminal electron acceptor, 239f, 240
Nitrobacter agilis, 34t
Nitrogen fixation, 6t, 9, 10, 12t, 21, 26t, 24, 28t, 29, 39
  effect on growth yields, 249, 262
Nitrogen source, effect on growth yields, 247t, 249, 336t
Nitrogen-limiting growth conditions, effect on electron transport systems, 234–238
Nitrogenase assay, in methanotrophs, 163
Nitromethane, oxidation by methane monooxygenase, 164t
Nocardia sp., growth on methane, 10
Nomenclature of methylotrophs, 3, 29
"Non-equilibrium" enzymes, importance in regulation, 134

Non-haem iron, *see also* Iron-sulphur centres
  in formate dehydrogenase, 194
  in methane monooxygenase, 157, 158
  in trimethylamine dehydrogenase, 198, 199
Non-pigmented pseudomonads, 23t, 29, 30t
  limitation of growth yields in, 257t
  oxidative routes for methylated amines in, 214t, 218
Norprotein (Scandinavia) process of SCP production, 333
Novel coenzymes from methanogens, 308–314
Nucleic acids of methanogens, 304
Nucleus of yeasts, 277f

O

Obligate autotrophy, metabolic basis of, 142
Obligate methylotrophs
  Coenzyme Q of, 219
  cytochromes and electron transport in, 222t
  glucose 6-phosphate dehydrogenase in, 83
  hexulose phosphate synthase of, 79t
  limitation of growth yields on trimethylamine, 257t
  methanol dehydrogenases of, 168t
  overproduction of metabolites by, 340, 341
  oxidative routes for methylated amines, in, 216t, 217
  RuMP pathway in, 67t, 72t, 74, 76
  serine pathway in, 109, 110t
  TCA cycle and, 138t, 141–143
Obligate methylotrophs unable to use methane, 23–26, 38
Obligate methylotrophy, 2
Obligate methylotrophy, metabolic basis, 141–143
Obligate proton-reducers, 299f
Octane oxidation by methane monooxygenase, 164t
Oligocarbophilic growth, 35, 39
Operator mutants, 108t, 132f
Operons, *see* Serine pathway, regulation, 126–129, 131–132

Index 415

Organelles, see Mitochondria, Nucleus, Peroxisomes
Organelles in methanogenesis, 316
Organism 3A2, see Pseudomonas 3A2
Organism 4B6, 25t, 26t
  citrate synthase of, 144
  formaldehyde oxidation in, 188t
  glucose 6-phosphate dehydrogenase in, 83
  incomplete TCA cycle in, 138t, 142
  oxidative routes for methylated amines in, 216t
  RuMP pathway in, 72t, 75t
  trimethylamine dehydrogenase of, 197
Organism 5B1, 31t
  citrate synthase of, 144
  growth inhibition by fluoroacetate, 114
  isocitrate lyase in, 112t, 113
  oxidation of methylamine in, 215t
  serine pathway in, 110t, 112t
  TCA cycle in, 139t, 141
Organism BC3, 25t, 26t
Organism C2A1, 25t, 26t
  citrate synthase of, 144
  glucose 6-phosphate dehydrogenase in, 83
  incomplete TCA cycle in, 138t, 142
  limitation of growth yields in, 258
  oxidative routes for methylated amines in, 216t
  RuMP pathway in, 72t, 75t
Organism FM02T, 28t
  vitamin B_{12} production by, 339
Organism 5H2, 31t
  ATP malate lyase activity in, 103
  formaldehyde oxidation in, 189t
  growth inhibition, 114
  isocitrate lyase in, 103, 112t
  oxidation of methylamine in, 215t
  serine pathway in, 111t, 112t
  tetramethylammonium monooxygenase of, 197
Organism L3, 25t, 26t
  RuMP pathway in, 72t
Organisms MB 53 and MB 55 to 60, RuMP pathway in, 73t
Organism PAR, see "*Diplococcus*" PAR
Organism PM6, see *Bacillus* PM6
Organism R6, see *Methylobacterium* R6
Organism S2A1, see *Bacillus* S2A1
Organism S50, see Strain S50
Organism W1, see *Pseudomonas* W1
Organism W3A1, 24, 25t, 26t
  citrate synthase of, 144
  glucose 6-phosphate dehydrogenase in, 83,
  incomplete TCA cycle in, 138t, 142
  oxidation of methylated amines, 216t
  RuMP pathway in, 72t, 75t
  trimethylamine dehydrogenase of, 198–200
Organism W6A, 24, 25t, 26t
  glucose 6-phosphate dehydrogenase in, 83
  incomplete TCA cycle in, 139t
  oxidation of methylated amines, 216t
  RuMP pathway in, 72t, 75t
Overproduction of metabolites, 338–342
Oxalate
  assimilation by *Pseudomonas oxalaticus*, 47, 58
  growth substrate for methylotrophs, 27, 28, 47
  inhibitor of malyl-CoA lyase, 124
Oxaloacetate
  inhibition of isocitrate dehydrogenase by, 146
  in assimilation of formaldehyde (RuMP pathway), 70–71f, 91, 92
  in assimilation of formaldehyde (serine pathway), 97f, 98f, 101, 119–121
Oxenoid-type mechanism of methane monooxygenase, 158
Oxidase (alcohol), see Alcohol oxidase (yeast)
Oxidase (cytochrome), see also Cytochrome a_3, Cytochrome c, proposed oxidase function, Cytochrome o, 228t
  in enzymic fuel cell, 348f
  proportion of oxygen used by, 266–268
  as proton pump, 240
  regulation of synthesis, 224, 236–238, 239f, 240

Oxidase-negative (or variable) methylotrophs, 29, 31t
Oxidation of acyl-CoA in alkane-utilising yeasts, 281
Oxidation (commercial) of alcohols, 343, 344
Oxidation (commercial) of hydrocarbons, 344–348
2-Oxoglutamate, 211
2-Oxoglutarate
  in assimilation of formaldehyde (serine pathway), 103–105
  as inhibitor of citrate synthase, 144
  reaction with phenylhydrazine, 113
  synthesis in methanogens, 305f, 306
2-Oxoglutarate dehydrogenase, 20, 146, 147
  in methanotrophs, 20, 143
  mutants lacking, 107t, 143, 146, 147, 340
  and obligate methylotrophy, 137–143
  and overproduction of metabolites, 340
2-Oxoglutarate synthase of methanogens, 305f, 306, 310
Oxygen, *see also* Growth yields with respect to O_2
  affinity for alcohol oxidase of yeast, 275t, 279, 332,
  affinity for cytochrome oxidases, 237
  affinity for trimethylamine monooxygenase, 200
  "damaging" effect on methanol dehydrogenase, 174, 187
  effect on cytochrome oxidase synthesis, 237
  as inhibitor of methanol dehydrogenase, 172
  incorporation into methanol during methane oxidation, 155
  sensitivity of methanotrophs to, 6t, 10, 39
Oxygenases, *see also* Cytochrome P-420, Cytochrome P-450, Dimethylamine monooxygenase, Methane monooxygenase, Tetramethylammonium monooxygenase, Trimethylamine monooxygenase
  limitation of growth yields and, 256t, 258, 259, 264–266

Oxygenases (*continued*)
  proportion of oxygen consumed by, 266–268
Oxygenase activity of ribulose bisphosphate carboxylase, 50–52
Oxygen demand, in production of SCP, 330–333
Oxygen-limiting growth conditions, effect on electron transport systems, 234, 236–238
Oxygen tension
  effect on growth in continuous culture, 336t
  effect on growth of yeast 279, 332
  effect on membrane formation, 18, 160, 161

P

Paecilomyces varioti, 270t
  serine pathway in, 111t
Paracoccus denitrificans, 33, 34t, 41, 47, 48t
  anaerobic electron transport to nitrate and nitrite from methanol, 239
  2 cytochromes c in, 226, 239f
  cytochromes and electron transport, 222t, 229, 230, 239, 267t
  effect of growth conditions on cytochromes, 224, 239
  efficiency of oxidative phosphorylation and growth yields in, 337t
  formaldehyde oxidation in, 189t
  growth yields, *see also* Methanol, growth yields on, 240, 256t
  hydroxypyruvate reductase in, 50, 117
  methanol dehydrogenase of, 168t, 170t, 182
  methylamine dehydrogenase in, 215t
  mutant lacking cytochrome c, 226, 229, 240
  periplasmic location of methanol dehydrogenase and cytochrome c, 243
  proton-pumping cytochrome oxidase in, 240
  proton translocation and ATP synthesis, 222t, 239
  ribulose bisphosphate carboxylase of, 53, 59
  RuBP pathway in, 48t, 50

Pectin, 39
Pentane, oxidation by methane monooxygenase, 164t
Pentose phosphate epimerase
  in CO_2 assimilation, 44f, 45f, 57
  in formaldehyde assimilation, 63f–66f, 86
  in formaldehyde assimilation (yeast), 286f
  in formaldehyde oxidation (yeasts), 287f
  in CO_2 assimilation, 44f, 45f, 57
  in formaldehyde assimilation, 63f–66f, 87
  in formaldehyde assimilation (yeast), 286f
  in methanotrophs, 6t
PEP, see Phosphoenolpyruvate
PEP hydratase, see Enolase
Peptide cofactor of formaldehyde dehydrogenase, 188t, 190
Peptidoglycan, in walls of methanogens, 301–303
Periplasm, location of methanol dehydrogenase and cytochrome c in, 243
Peroxidative role of catalase in yeast, 276f, 280
Peroxisomes
  in alkane-utilising yeasts, 281–283
  containing amine oxidase and catalase, 208, 273
  immobilisation, 344
  in methanol-utilising yeasts, 274, 276f, 277f, 282t, 283
  regulation of synthesis and degradation, 282t, 283, 293
1,10-Phenanthroline, as inhibitor of methane monooxygenase, 161
Phenazine ethosulphate
  in biofuel cells, 349
  electron acceptor for dimethylamine dehydrogenase, 203
  electron acceptor for methanol dehydrogenase, 175
  in methanogenesis, 316
Phenazine methosulphate
  in biofuel cells, 349
  electron acceptor for dimethylamine dehydrogenase, 203

Phenazine methosulphate (*continued*)
  electron acceptor for formaldehyde dehydrogenases, 190
  electron acceptor for methanol dehydrogenase, 174, 185
  electron acceptor for methylamine dehydrogenase, 205–207f
  electron acceptor for trimethylamine dehydrogenase, 197–199
  electron acceptor for N-methylglutamate dehydrogenase, 211, 212t
  inactivation of methanol dehydrogenase by, 184
Phenol
  oxidation by methane monooxygenase, 347t
  product of methane monooxygenase, 165t, 347t
Phenylalanine
  in overproduction of glutamate, 341
  overproduction by methylotrophs, 341
Phenylbutane, oxidation by methane monooxygenase, 346t
Phenylheptane, oxidation by methane monooxygenase, 346t
Phenylhydrazines, inhibitors of methanol oxidation, 172
Phenylhydrazine assay for glyoxylate, 113
Phenylmethylsulphonyfluoride, as stabiliser of methane monooxygenase, 157
Phosphate
  inhibition of growth by, 26t
  inhibitor of methanol oxidation, 172
Phosphoenolpyruvate (PEP)
  interconversion with triose phosphates, 70, 71f, 90–92
  in formaldehyde assimilation (RuMP pathway), 63f, 66f, 70, 71f, 90–92
  in formaldehyde assimilation (serine pathway), 97f, 100f, 118–121
  inhibitor of fructose bisphosphatase and phosphoribulokinase, 59
Phosphoenolpyruvate carboxylase
  in assimilation of formaldehyde (RuMP pathway), 70, 71, 74, 91

Phosphoenolpyruvate carboxylase (*continued*)
  in assimilation of formaldehyde (serine pathway), 97*f*, 101, 102*t*, 119–121
  of methanogens, 305*f*, 306
  mutant lacking, 108*t*
  regulation of activity of, 119–121, 134–136
  regulation of synthesis, 127, 128*t*, 132*f*
Phosphoenolpyruvate carboxykinase
  in assimilation of formaldehyde (RuMP pathway), 70, 71, 74, 92
  in assimilation of formaldehyde (serine pathway), 118
  in assimilation of multicarbon compounds, 147
Phosphoenolpyruvate synthetase, in assimilation of formaldehyde, 63*f*, 66*f*, 91
Phosphofructokinase, in assimilation of formaldehyde, 62, 64*f*–66*f*, 67*t*, 74, 75*t*, 84, 93
6-Phosphogluconate
  in assimilation of formaldehyde (RuMP pathway), 63*f*, 66*f*, 81–83, 93
  in dissimilatory cycles of formaldehyde oxidation, 69*f*, 76, 81–83, 93, 287*f*
  as inhibitor of ribulose bisphosphate carboxylase, 53, 59
  radioactive labelling in RuMP pathway, 76
6-Phosphogluconate dehydrase, in assimilation of formaldehyde, 63*f*, 66*f*, 83
6-Phosphogluconate dehydrogenase, 89, 90
  in formaldehyde oxidation, 69, 76, 77, 188*t*, 192, 193
  in formaldehyde oxidation (yeasts), 287*f*, 288, 295
  inhibition by NAD(P)H and ATP, 89, 90, 93, 94, 193, 295
  in methanotrophs, 6*t*, 20
2-Phosphoglycerate, in assimilation of formaldehyde (serine pathway), 97*f*, 117–119

3-Phosphoglycerate
  in assimilation of CO_2 (RuBP pathway), 44*f*, 45*f*, 47, 51–53
  in assimilation of formaldehyde (RuMP pathway), 63*f*–66*f*, 90, 91
  "end-product" of assimilation pathways, *see also* individual pathways (below), 249, 250
  as "end product" of the serine pathway, 96, 97*f*, 118
  as "end product" of RuMP pathway, 63, 64–66*f*, 67*t*
  as "end product" of DHA cycle in yeasts, 285, 286*f*
Phosphoglycerate kinase
  in assimilation of CO_2, 44*f*, 45*f*, 53, 59
  in formaldehyde assimilation, 64*f*, 65*f*, 90
  in formaldehyde assimilation (DHA cycle), 286*f*
  regulation, 59
Phosphoglyceromutase (phosphoglycerate mutase)
  in assimilation of formaldehyde (RuMP pathway), 63*f*, 66*f*, 91
  in assimilation of formaldehyde (serine pathway), 97*f*, 118
Phosphoglycollate, 50–52, 117
Phosphoribulokinase
  in CO_2 assimilation, 44*f*, 45*f*, 53
  in *Methylococcus capsulatus*, 6*t*, 20, 21, 51
  occurrence in methanotrophs, 6*t*, 20, 21, 51
  regulation by AMP and NADH, 58–59
Phosphorylation, of malate thiokinase, 122
Phosphoserine phosphatase, mutants lacking, 100, 106*t*, 107*t*
Photophosphorylation, 263*f*
Photosynthetic bacteria, growth yields on methanol, 50, 263
Phototrophic methylotrophs, 33–34*t*, 48*t*, 50
Photorespiration, *see also* Phosphoglycollate, 52
Phytanyl glycerol ethers, 303*f*, 304
Pichia pastoris, 271*t*
  alcohol oxidase of, 275*t*, 278

Pichia pinus, 271t
  apparent alcohol dehydrogenase in, 279
Pichia sp., 271t, 273
  localisation of oxidase and catalase in 282t
  peroxisomes and crystalloids in, 282t, 283
  regulation of peroxisome synthesis and degradation in, 282t
Pigmentation of methylotrophs, 11, 12t, 15, 19, 27, 28t, 29, 31t, 33, 37
Pink facultative methylotrophs, 23t, 27–29
  coenzyme Q of, 219
  growth yields, *see* Methanol, growth yields on
  limitation of growth yields in, 257t
  oxidative routes for methylated amines in, 214t, 218
  vitamin B_{12} production by, 339
Plasmids, *see also* Genetic engineering
  in genetics of methylotrophs, 130, 131, 333
  for growth on methane, 11, 22
Pleomorphism, 12t, 27, 29
Poly 3-hydroxybutyrate, 9, 10, 12t, 14, 21, 27, 35
  production for plastics industry, 338
Polyol dehydrogenases of bacteria, quinoproteins, 182, 183t
Polysaccharide
  commercial production by methylotrophs, 338
  in methanogen cell walls, 301–303
P/O ratios, *see also* Proton translocation, ATP synthesis, Y_{ATP}, 230, 232–234, 242, 244, 245–248
  effect on activity of dehydrogenases and oxygenases, 266–268
  effect on growth yields, 245–248, 251, 254–259, 261t, 265, 266
  in yeasts, 274, 281
Potassium translocation in methylotrophs, 238
Predictions of growth yields, *see* Growth yields, prediction(s) of
Pressure-cycle fermenter, 333, 334f, 335f, 338

Primary alcohol dehydrogenase, *see* Methanol dehydrogenase
Primary amine dehydrogenase, *see* Methylamine dehydrogenase
Propane, oxidation by methane monooxygenase, 164t, 346t
1,2-Propanediol (propane 1,2-diol)
  growth on, 27
  metabolism by methylotrophs, 149f, 150, 172
Propene (propylene), epoxidation by methane monooxygenase, 163, 165t, 166, 345, 346t
Propionate, on haem of cytochrome c, 227
Propionibacterium pentosaceum, lactate dehydrogenase of, 183t
Propionyl-Coenzyme A, substrate for malyl-CoA lyase, 123, 124
n-Propylamine, enzymic assay of, 343
Prosthecate cells, 33, 35, 36f, 37f, 39
Prosthetic group
  of alcohol oxidase (yeast), 275t, 278
  of aldehyde dehydrogenase, 190
  of dimethylamine dehydrogenase, 199f, 203
  of methanol dehydrogenase (PQQ), 177–186
  of methylamine dehydrogenase, *see also* Pyrrolo-quinoline quinone (PQQ), 205–207f
  of *N*-methylglutamate dehydrogenase, 212f, 213
  of *N*-methylglutamate synthase, 211
  of quinoproteins, *see* Pyrrolo-quinoline quinone (PQQ)
  of trimethylamine dehydrogenase, 198, 199
Protaminobacter ruber, 28t
  formaldehyde oxidation in, 189t
  methanol dehydrogenase of, 168t
  methionine synthetase in, 339
  methylmalonyl-CoA mutase in, 339
  serine pathway in, 112t
Protein feed supplements, *see* Single cell protein
Proton motive force, *see also* ATP synthesis, Proton translocation, 232, 260
  and active transport, 249

420   The biochemistry of methylotrophs

Proton motive force (*continued*)
  in methanogens, 307, 315–318, 322, 325–327
  in reduction of NAD^+, 260, 263
Proton translocation
  in methanogens, 307, 315–318, 325, 327
  in *Methylophilus methylotrophus*, 236–238
  in *Methylosinus trichosporium* and other methanotrophs, 240–242
  in methylotrophs, *see also* individual organisms, 143, 222*t*, 232–244
  in *Paracoccus denitrificans*, 239, 240
  in *Pseudomonas* AM1, 233–236
"Pruteen" (ICI SCP from methanol), 329*t*, 333, 334*f*, 335*f*, 338
Pseudomethylotroph, 43
Pseudomonas aeruginosa
  cytochrome *c* of, 227
  promiscuous plasmids from, 130
  quinoprotein nature of alcohol dehydrogenase of, 183*t*
  glucose dehydrogenase of, 183*t*
Pseudomonas aminovorans, 30*t*
  ATP malate lyase activity in, 103
  dimethylamine monooxygenase of, 201–203
  formaldehyde oxidation in, 189*t*
  hydroxypyruvate reductase of, 117
  isocitrate lyase in, 103, 112*t*
  mutants of, 108*t*
  mutant unable to grow on amines, 210
  N-methylglutamate dehydrogenase of, 212*t*, 213, 214*t*
  PEP carboxylase in, 119
  regulation of serine pathway in, 129
  serine pathway in, 111*t*, 112*t*
  trimethylamine monooxygenase of, 200
  trimethylamine *N*-oxide demethylase of, 201
Pseudomonas extorquens, 28*t*
  cytochrome *c* of, 225*t*–227, 235
  cytochromes and electron transport, 222*t*, 230, 235, 241*f*
  effect of growth conditions on cytochromes in, 224
  formaldehyde oxidation in, 189*t*
  gene transfer in, 130

Pseudomonas extorquens (*continued*)
  methanol dehydrogenase of, 168*t*
  methylamine dehydrogenase in, 214*t*
  proton translocation and ATP synthesis, 222*t*
  serine pathway in, 110*t*
Pseudomonas fluorescens, glucose dehydrogenase of, 183*t*
Pseudomonas gazotropha, 34*t*
  RuBP pathway in, 48*t*
Pseudomonas mesophilica, 29
Pseudomonas methanica, *see also Methylomonas methanica*, 4, 9
Pseudomonas methanitrificans, 9
Pseudomonas methanolica, 24, 25*t*, 26*t*
Pseudomonas methylica, 28*t*
  cytochromes and electron transport, 223*t*
  formaldehyde oxidation in, 189*t*
  N-methylglutamate dehydrogenase of, 213, 214*t*
  PEP carboxylase in, 119
  serine pathway in, 110*t*
  TCA cycle in, 139*t*
Pseudomonas oleovorans, 23*t*, 29, 30*t*
  citrate synthases of, 145
  cyclic route for oxidation of formaldehyde in, 192
  exception to many generalisations, 23*t*, 29, 30*t*, 75*t*, 76, 145, 218
  glucose 6-phosphate dehydrogenase in, 83
  6-hexulose phosphate synthase of, 78, 79*t*
  isocitrate dehydrogenase of, 145
  oxidative routes for methylated amines in, 215*t*, 217, 218
  pyruvate carboxylase, PEP carboxylase and PEP carboxykinase of, 70, 71
  RuMP pathway in, 73*t*, 75*t*
  TCA cycle in, 140*t*
Pseudomonas oxalaticus, 23*t*, 34*t*, 40, 43, 48*t*
  assimilation of oxalate by, 47
  formate dehydrogenases of, 194
  regulation in, 58–59
  ribulose bisphosphate carboxylase of, 53, 59
  RuBP pathway in, 47, 48*t*, 99

Index   421

Pseudomonas putida, primary amine dehydrogenase of, 204
Pseudomonas rosea, 28*t*
Pseudomonas 1A3, 1B1, 7B1, 8B1, 30*t*
Pseudomonas 2A3, 30*t*
Pseudomonas 3A2, 28*t*
  citrate synthase of, 144
  methylamine dehydrogenase in, 214*t*
  regulation of serine pathway in, 127, 128*t*, 135
  serine pathway in, 110*t*, 112*t*
  TCA cycle in, 139*t*, 141
Pseudomonas 3ab, overproduction of serine by, 341
Pseudomonas AM1, 28*t*
  absence of homoisocitrate lyase and synthase in, 104
  assimilation of acetyl-CoA during growth on multicarbon compounds, 147–150
  citrate synthase of, 145
  cytochrome *c* of, 191, 225*t*–231, 233–235, 242
  cytochrome *c*-deficient mutant of, 106*t*, 127, 226, 229, 233, 234, 240
  cytochromes and electron transport, 222*t*, 229, 230, 233–235, 267*t*
  dye-linked aldehyde dehydrogenase of, 188*t*, 190
  effect of growth conditions on cytochromes in, 224
  elucidation of serine pathway in, 98–110*t*, 112*t*
  formaldehyde oxidation in, 188*t*, 190, 191
  formate dehydrogenase of, 194
  formate transport in, 238
  gene transfer in, 130, 131
  glycerate kinase of, 118
  growth inhibition by sulphanilamide, fluoroacetate and itaconate, 114
  growth yields on methanol, *see* Methanol, growth yields on
  growth yields on succinate, 235
  hydroxypyruvate reductase of, 117
  isocitrate dehydrogenase of, 146
  limitation of growth yields in, 256*t*, 259
  malate synthase activity in, 148–150
  malyl-CoA hydrolase of, 148–150

Pseudomonas AM1 (*continued*)
  malyl-CoA lyase of, 123, 124, 148–150
  metabolism of 1,2-propanediol and 4-hydroxybutyrate, 150, 172
  metabolism of ethanol, lactate, pyruvate and 3-hydroxybutyrate, 147–150
  membrane vesicles (oxidation and ATP synthesis), 229, 230, 234
  methanol dehydrogenase of, 131, 168*t*, 170*t*, 174, 176, 187
  methylamine dehydrogenase of, 204–208, 214*t*
  methylenetetrahydrofolate dehydrogenase in, 191
  mutants of, *see also* Mutants of each enzyme, 100, 105–109, 127, 143, 145, 146, 148–150, 171, 191, 192, 229, 233, 340
  obligate methylotroph derived from, 143, 145
  2-oxoglutarate dehydrogenase of, 146
  overproduction of amino acids by, 340
  PEP carboxykinase of, 147
  PEP carboxylase of, 119, 120
  phosphoglycerate mutase of, 118
  phosphorylation of pyruvate in, 148
  proposed homoisocitrate-glyoxylate cycle in, 103*f*, 104
  proton translocation and ATP synthesis, 222*t*, 230, 233–235, 242
  pyruvate dehydrogenase of, 146
  quinoprotein nature of primary amine dehydrogenase, 182, 183*t*
  regulation of serine pathway in, 101, 126–129, 134
  restricted facultative methylotroph derived from, 143
  serine transhydroxymethylase of, 115
  TCA cycle in, 139*t*–143, 340
Pseudomonas AM2, 28*t*
Pseudomonas AT2, 28*t*
  N-methylglutamate dehydrogenase of, 211–213, 214*t*
  serine pathway in, 110*t*
Pseudomonas C, 24, 25*t*, 26*t*
  cyclic route for oxidation of formaldehyde in, 76, 77, 192
  experiments on dissimilatory cycle of formaldehyde oxidation, 76, 77

Pseudomonas C (*continued*)
  formaldehyde oxidation in, 76, 77, 188*t*, 192
  glucose 6-phosphate dehydrogenase of, 82*t*, 83
  growth yields, *see* Methanol, growth yields on
  incomplete TCA cycle in, 138*t*
  limitation of growth yields in, 258
  membranes and lipid content of, 24
  methanol dehydrogenase of, 168*t*, 170*t*, 172, 175
  6-phosphogluconate dehydrogenase of, 89
  RuMP pathway in, 72*t*, 75*t*
Pseudomonas EN, proton translocation and ATP synthesis, 223*t*
Pseudomonas J, 25*t*, 26*t*
  methylamine dehydrogenase of, 204, 206–208
  oxidative routes for methylated amines in, 216*t*
Pseudomonas J26
  methanol dehydrogenase of, 168*t*
Pseudomonas JB1, 28*t*
  serine pathway in, 110*t*, 112*t*
Pseudomonas MA, 30*t*
  absence of homoisocitrate lyase and synthase in, 104
  ATP malate lyase activity in, 101, 123
  citrate synthase in, 136
  cytochromes and electron transport, 223*t*
  elucidation of serine pathway in, 101, 102*t*, 111*t*, 112*t*
  glycerate kinase of, 118
  growth inhibition by itaconate, 114
  isocitrate lyase in, 103, 112*t*, 125, 129, 130
  malate thiokinase of, 121, 122
  malyl-CoA lyase of, 123, 124
  N-methylglutamate dehydrogenase of, 212*t*, 213, 214*t*
  N-methylglutamate synthase of, 210, 211
  oxidation of formaldehyde by serine pathway in, 135, 193
  PEP carboxylase of, 119–121, 134–136

Pseudomonas MA (*continued*)
  regulation of serine pathway in, 129, 130, 134–136
  succinate thiokinase of, 121
  synthesis of 5-hydroxy-*N*-methylpyroglutamate by, 211
  TCA cycle in, 139*t*
Pseudomonas MS, 30*t*
  ATP malate lyase activity in, 103
  cytochromes and electron transport, 223*t*, 224
  elucidation of serine pathway in, 99*t*, 111*t*, 112*t*
  formaldehyde oxidation in, 189*t*
  isocitrate lyase in, 103, 112*t*
  metabolism of trimethylsulphonium salts by, 151
  mutants of, 108*t*, 113
  oxidation of methylamine by, 210, 211, 214*t*
  PEP carboxylase in, 119
Pseudomonas M27, 28*t*
  methanol dehydrogenase of, 167–172, 175
  methylamine dehydrogenase in, 214*t*
  oxidation of alcohols by whole cells, 166
  serine pathway in, 110*t*, 112*t*
Pseudomonas PCTN, 28*t*
  serine pathway in, 110*t*, 112*t*
Pseudomonas PP, 28*t*
  methanol dehydrogenase of, 168*t*
Pseudomonas PRL-W4, 28*t*
  elucidation of serine pathway in, 98, 99*t*
Pseudomonas RJ1, 28*t*
  dye-linked formaldehyde dehydrogenase of, 188*t*, 189*t*, 190
  methanol dehydrogenase of, 168*t*, 170*t*
Pseudomonas RJ3, 25*t*, 26*t*
  methylamine dehydrogenase of, 204
  oxidation of methylated amines, in, 216*t*
Pseudomonas S25, 30*t*
  methanol dehydrogenase of, 168*t*, 170*t*
Pseudomonas TP1, 28*t*
  methanol dehydrogenase of, 168*t*, 170*t*, 172, 179
  serine pathway in, 110*t*

Pseudomonas W1, 25*t*, 26*t*
  incomplete TCA cycle in, 138*t*, 142
  methanol dehydrogenase of, 168*t*
    170*t*, 172, 175, 176
  oxidative routes for methylated
    amines in, 216*t*
  RuMP pathway in, 72*t*
Pseudomonas W6, 25*t*
  cytochromes and electron transport,
    222*t*
  glucose 6-phosphate dehydrogenase
    of, 82*t*
  6-hexulose phosphate synthase of,
    78
  incomplete TCA cycle in, 138*t*
  isocitrate dehydrogenases of, 145
  pyruvate carboxylase, PEP
    carboxylase and pyruvate kinase
    of, 70, 71
  RuMP pathway in, 72*t*, 75*t*
Pseudomonas YR, 28*t*
  serine pathway in, 110*t*, 112*t*
Pseudomonas 1, 28*t*
  formaldehyde oxidation in, 189*t*
  methylamine dehydrogenase in, 214*t*
  serine pathway in, 110*t*
  TCA cycle in, 139*t*
  cytochromes and electron
    transport in, 230
Pseudomonas 8, 34*t*, 48*t*
  TCA cycle in, 140*t*
Pseudomonas 20, 30*t*
  oxidation of methylated amines in,
    215*t*, 217, 218
Pseudomonas 80, 28*t*
  proposed homoisocitrate-glyoxylate
    cycle in, 103*f*, 104
Pseudomonas 135, 28*t*
  formaldehyde oxidation in, 189*t*
  methylamine dehydrogenase in, 214*t*
  serine pathway in, 110*t*
Pseudomonas 2941, 28*t*
  methanol dehydrogenase of, 168*t*,
    170*t*
Pseudomurein, 301–303
Pteridine, *see also* Methanopterin
  possible prosthetic group of
    trimethylamine dehydrogenase,
    198
  similarity to PQQ of methanol
    dehydrogenase, 177, 179

Pyridine, *N*-oxidation by methane
  monooxygenase, 163, 165*t*, 346*t*
Pyrrolo-quinoline quinone (PQQ), 131,
  177–186
  absorption spectrum, 177, 178*f*, 180*t*,
    182
  chemical structure, 177–182
  ESR, ENDOR, NMR and Mass
    spectrometry, 179, 180*t*
  fluorescence spectra, 177, 180*t*, 184
  formation of adducts with water,
    alcohols, aldehydes and ketones,
    179, 180*t*, 181*f*
  in mechanism of methanol
    dehydrogenase, 180*t*, 181*f*,
    184–186
  in mechanism of methylamine
    dehydrogenase, 206, 207*f*
  as prosthetic group of quinoproteins
    other than methanol
    dehydrogenase, 182, 183*t*
  quantitative determination, 180*t*, 182
  X-ray diffraction analysis, 179, 180*t*
Pyruvate
  in assimilation of formaldehyde
    (RuMP pathway), 62, 63*f*, 66*f*,
    67*t*, 70–74, 91–92
  metabolism by methylotrophs,
    147–150
  in oxidation of methylated amines,
    209*f*
Pyruvate carboxylase, in assimilation
  of formaldehyde (RuMP pathway),
  70, 71, 74, 92
Pyruvate dehydrogenase, 146, 147, 149*f*
  absence in *Hyphomicrobia*, 98, 143
  absence in restricted facultative
    methylotrophs, 143
  in bacteria with the RuMP pathway,
    70, 71*f*
  low levels in methanotrophs, 143
  mutants lacking, 106*t*, 143, 146, 147,
    149*f*, 340
  and overproduction of metabolites,
    340
  in oxidation of formaldehyde, 193
Pyruvate kinase
  in assimilation of formaldehyde
    (RuMP pathway), 70, 71, 92
  in assimilation of formaldehyde
    (serine pathway), 98, 135

Pyruvate kinase (continued)
  in the oxidation of formaldehyde, 193
Pyruvate phosphate dikinase, in assimilation of formaldehyde (RuMP pathway), 71f, 92
Pyruvate synthase of methanogens, 305f, 306, 310

Q

Quinol, see also Pyrrolo-quinoline quinone
  intermediate in mechanism of trimethylamine dehydrogenase, 208
Quinoproteins (containing PQQ prosthetic groups), see also Methanol dehydrogenase, Pyrrolo-quinoline quinone, 182–184

R

Rearrangement reactions
  of the DHA cycle in yeasts, 286f, 287, 289f
  of the ribulose bisphosphate pathway, see also individual reactions, 44f, 45f, 46
  of the ribulose monophosphate pathway, see also individual reactions, 62, 67t, 85–88
Reconstitution
  of methylamine dehydrogenases, 207
  of N-methylglutamate synthase, 211
  of pyruvate and 2-oxoglutarate dehydrogenases, 147
  of quinoproteins with PQQ, 179, 180t, 182–184
Red tides, 343
Redox potential
  for growth of methanogens, 298
Redox potentials, see Midpoint redox potentials, Electroenzymology
Reductant supply in co-oxidations, 166, 167, 345
Reductant-limitation of growth yields, see Growth yields, limitation by NAD(P)H
Reductive carboxylic acid cycle, 43
Regulation, see Dihydroxyacetone cycle in yeasts, regulation; Tricarboxylic acid cycle,

Regulation (continued)
  regulation; Ribulose bisphosphate pathway, regulation; Ribulose monophosphate pathway, regulation; Serine pathway, regulation; Yeasts, regulation of methanol metabolism in;
  individual enzymes, regulation
  important role of formaldehyde in, 93, 133, 244, 294
Regulatory enzymes, of the serine pathway, 116, 120, 121, 134–136
Regulatory mutants, 108t, 129, 132f
Resazurin, as electron acceptor for methylamine dehydrogenase, 205
Resting stages of methylotrophs, see also Spores, Lipid cysts, "Azotobacter"-type cysts, 6t, 7t, 9, 10, 12t, 13–15, 39
Restricted facultative methylotrophs, 23, 32t, 33, 35, 139t
  glucose 6-phosphate dehydrogenase of, 83
  metabolic basis, 143
  oxidative routes for methylated amines in, 214t–218
  TCA cycle and, 139t, 141–143
"Reversed electron transport"
  effect on growth yields, 260, 265
  in reduction of NAD$^+$, 50, 263, 265
Rhodomicrobium vannielii, RuBP pathway in, 48t, 53
Rhodopseudomonas (genus), 33, 34t
Rhodopseudomonas acidophila, 34t, 48t, 50
  cell composition and carbon balance of, 50, 263
  formaldehyde dehydrogenase of, 188t, 189t
  formaldehyde oxidation in, 189t
  growth yields on methanol, 50, 263
  methanol dehydrogenase of, 168t, 170t, 172–174, 176, 182
  reversed electron transport in, 50, 263
  RuBP pathway in, 48t, 50
Rhodopseudomonas capsulatus,
  mutant lacking cytochrome c in, 226
Rhodopseudomonas gelatinosa, RuBP pathway in, 48t

Rhodopseudomonas palustris, 34*t*
   RuBP pathway in, 48*t*
Rhodopseudomonas sphaeroides
   ATP malate lyase activity in, 103, 123
   malate thiokinase of, 121, 123
Rhodospirillum rubrum, RuBP carboxylase of, 52
Rhodospirillum sp., RuBP pathway in, 48*t*
Rhodotorula sp., 272*t*, 273
Riboflavin derivatives, production by methylotrophs, 294, 339
Riboflavin kinase in yeast, 294
Riboflavin synthase in yeast, 294
Ribose, growth of yeasts on, 293
Ribose 5-phosphate
   as alternative substrate for DHA synthase, 291
   in assimilation of CO_2 (RuBP pathway), 44*f*, 45*f*, 56, 57
   in assimilation of formaldehyde (DHA cycle), 286*f*
   in assimilation of formaldehyde (RuMP pathway), 63*f*–66*f*, 69*f*, 86, 87
Ribulose 1,5-bisphosphate, in assimilation of CO_2 (RuBP pathway), 44*f*, 45*f*, 51, 57
Ribulose bisphosphate carboxylase, 51–53
   in assimilation of CO_2, 44*f*, 45*f*
   inhibition by 6-phosphogluconate, 53, 59
   of *Methylococcus capsulatus*, 20, 51, 53
   occurrence in methanotrophs, 6*t*, 20, 21, 51, 53
   occurrence in methylotrophs (other than methanotrophs), 48*t*
   in *Rhodospirillum rubrum*, *Chlorobium thiosulphatophilum*, *Thiobacillus intermedius*, *Hydrogenomonas eutropha*, *Ectothiorhodospira*, *Halophila*, *Chromatium*, *Euglena* and higher plants, 52
Ribulose bisphosphate pathway, 42–59
   description, 43–46
   distribution, 46–48*t*
   evidence for, 47–51

Ribulose bisphosphate pathway (*continued*)
   and growth yields, 252*t*, 255, 256*t*, 260–263
   introduction and definitions, 42
   reactions of, see also individual enzymes, 44*f*, 45*f*, 51–58
   regulation of, 58
   relevance to electron transport, 267*t*
   summary equation, 44, 252*t*
Ribulose monophosphate pathway, 60–94
   carbohydrate biosynthesis by, 68–69*f*
   description, 62–74
   dissimilatory pathway for formaldehyde oxidation, see also Dissimilatory (RuMP) cycle of formaldehyde oxidation, 68–70, 76, 89, 93, 188*t*, 192, 193
   distribution of, 23*t*, 72*t*, 74–77
   evidence for, 60, 72*t*, 74–76
   FBP aldolase/SBPase variant, 62, 64*f*, 67*f*, 70, 74–76, 93
   FBP aldolase/transaldolase variant, 62, 65*f*, 67*t*, 70, 74–76, 93
   and growth yields, 252*t*, 255–262, 264–266, 332
   historical introduction, 60
   KDPG aldolase/SBPase variant, 62, 66*f*, 67*t*, 74–76, 93
   KDPG aldolase/transaldolase variant, 62, 63*f*, 67*t*, 70, 74–76
   reactions of, see also individual enzymes, 63*f*–66*f*, 77–92
   regulation of, 78, 89, 93, 94
   relevance to electron transport, 267*t*
   relevance to SCP production, 332, 333
   summary equations, 67*t*, 69, 252*t*
Ribulose 5-phosphate
   in dissimilatory cycles of formaldehyde oxidation, 69*f*, 77, 89, 287*f*
Ribulose 5-phosphate
   in assimilation of CO_2 (RuBP pathway), 44*f*, 45*f*, 57
   in assimilation of formaldehyde (DHA cycle), 286*f*
   in assimilation of formaldehyde (RuMP pathway), 62, 63*f*–66*f*, 68*f*, 69*f*, 77–79*t*, 86, 87

Ribulose 5-phosphate (*continued*)
  in oxidation of formaldehyde
    (DHA cycle), 287*f*
RNA of methanogens, 304
Rosette formation, 7*t*, 9, 19, 35
Rotenone, inhibitor of electron
  transport, 235*f*, 237*f*, 239*f*
RPG effect in methanogenesis, 315*t*,
  322, 323, 326
Ruminants, methanogenesis in, 298,
  299*f*

S

Saccharomyces sp., 271*t*, 273
SBP, *see* Sedoheptulose bisphosphate
SCP, *see* Single cell protein
Seasoning agents, production by
  methylotrophs, 341
Secondary alcohols
  commercial oxidation of, 344
  oxidation by methanol
    dehydrogenase, 170*t*, 172
Secondary alcohol dehydrogenase,
  344, 345
Secondary amines
  oxidation by dimethylamine
    dehydrogenase, 203
  oxidation by trimethylamine
    dehydrogenase, 198
  oxidation by trimethylamine
    monooxygenase, 200
Secondary amine dehydrogenase, *see*
  Dimethylamine dehydrogenase
Secondary amine monooxygenase, *see*
  Dimethylamine monooxygenase
Sedoheptulose 1,7-bisphosphate
  in assimilation of CO_2 (RuBP
    pathway), 44*f*, 45*f*, 55
  in assimilation of formaldehyde
    (RuMP pathway), 64*f*, 66*f*, 88
Sedoheptulose bisphosphatase
  in assimilation of CO_2, 44*f*, 46, 55
  in assimilation of formaldehyde,
    62, 64*f*, 65*f*, 67*t*, 88
  possible regulatory enzyme, 93
Sedoheptulose bisphosphatase variants
  of the RuBP pathway, 44*f*, 46, 55
Sedoheptulose bisphosphatase variants
  of the RuMP pathway, *see*
    ribulose monophosphate pathway,
    variants

Sedoheptulose 7-phosphate
  in assimilation of CO_2 (RuBP
    pathway), 44*f*, 55, 56
  in assimilation of formaldehyde
    (DHA cycle), 286*f*
  in assimilation of formaldehyde
    (RuMP pathway), 63*f*–66*f*, 69*f*,
    86–88
Selenium, in formate dehydrogenase
  of methanogens, 327
Semicarbazide, as inhibitor of
  methylamine dehydrogenase, 205
Semiquinone, intermediate in
  mechanism of trimethylamine
  dehydrogenase ETF, 200
Serine
  in assimilation of formaldehyde
    (serine pathway), 97*f*, 98–100,
    115, 116
  overproduction by methylotrophs,
    341
Serine-glyoxylate aminotransferase
  in assimilation of formaldehyde
    (serine pathway), 97*f*, 101, 102*t*,
    116
  mutants lacking, 106*t*, 108*t*, 132*f*
  regulation of synthesis, 126, 128*t*,
    132*f*
Serine pathway, 95–136
  description, 95–98
  distribution, 109–114
  evidence from enzyme studies,
    101–105
  evidence from mutant studies, 100,
    105–109, 148–150
  evidence from radioisotope studies,
    98–101, 104
  and growth yields, 252*t*, 256*t*, 258,
    259, 260–262, 264
  icl$^+$ variant, 103, 112*t*–114
  icl$^-$ variant, 103–105, 112*t*–114
  oxidation of acetyl-CoA to
    glyoxylate, 96, 97*f*, 103–105, 124,
    125
  in oxidation of formaldehyde, 136,
    193
  reactions of, *see also* individual
    enzymes, 115–125
  regulation, 102*t*, 120, 121, 126–136
  regulation at the genetic level, 101,
    102*t*, 126–132

Serine pathway (*continued*)
  regulation at the level of enzyme activity, 132–136
  relevance to electron transport, 266
  relevance to overproduction of metabolites, 340–342
  summary equations, 96, 252t
  synthesis of carbohydrate, 97
  synthesis of oxaloacetate and succinate, 97, 98f
Serine transhydroxymethylase
  in assimilation of formaldehyde (serine pathway), 97f, 102t, 115, 116
  mutants lacking, 106t, 108t, 132f
  in overproduction of serine, 341
  regulation of activity of, 115, 134
  regulation of synthesis, 115, 127, 128t, 132f
Serological relationships between methanol dehydrogenases, 176
Single cell protein (SCP)
  introduction and summary, 328–331
  from methylotrophic bacteria, 329t, 333–338
  from methylotrophic yeasts, 329t, 331–333
  reviews of literature, 247t, 329t
Sorbitol, growth of yeasts on, 293
"S organism", 300
Spectrum (absorption)
  of alcohol oxidase (yeast), 277t, 278
  of aldehyde dehydrogenase, 190
  of Coenzyme F_{420}, 309
  of component B of methyl-Coenzyme M reductase, 314
  of compounds F_A and F_C, 314
  of dimethylamine dehydrogenase, 203
  of dimethylamine monooxygenase, 202
  of Factor F_{430}, 312
  of methanol dehydrogenase, 177, 178f, 184
  of methanopterin, 311
  of methylamine dehydrogenase, 205
  of N-methylglutamate dehydrogenase, 213
  of pyrrolo-quinoline quinone, 178f, 180t, 182
  of trimethylamine dehydrogenase, 198, 199

Spermidine, oxidation by methylamine dehydrogenase, 205
Spirulina platensis, RuBP pathway in, 48t
Spores of methylotrophs, *see also* Lipid cysts, "Azotobacter"-type cysts, 6t, 7t, 10, 12t, 13–15, 39
Sporobolomyces roseus, 272t
Squalene
  in methanogens, 303f, 304
  in methanotrophs, 10
Stalked bacteria, *see Hyphomicrobium* (genus)
Sterols, in methanotrophs, 10
Strain S50, 31t
  methanol dehydrogenase of, 168t 170t, 175
Streptomyces griseus, Coenzyme F_{420} in, 310
Streptomyces 239, 32t, 33
  RuMP pathway in, 73t
  serine pathway in, 110t, 112t
Styrene, oxidation by methane monooxygenase, 165t, 346t, 348
Substituted methane derivatives, oxidation by methane monooxygenase, 163–167, 345
Substrate specificity, *see* individual enzymes
Succinate
  assimilation equation for, 252t
  in assimilation of formaldehyde (serine pathway), 97f, 98f, 100, 125
  catabolite repressor of serine pathway, 127–130
  growth yields on, 235
  oxidation by methylotrophs, 230, 234
  synthesis in methanogens, 305f, 306
Succinate dehydrogenase, *see also* Flavoprotein dehydrogenases, and growth yields
  in assimilation of formaldehyde (serine pathway), 97f
  iron-sulphur centre of, 199
  low levels in *Hyphomicrobia* and obligate methanol utilisers, 138t, 139t, 141
  proportion of electron transport from, 266
Succinate thiokinase, 121, 122

Sulphanilamide as inhibitor of methylotrophic growth, 114
Sulphate-reducing bacteria in oxidation of methane, 38, 167
Sulphur, acid-labile, *see* Acid-labile sulphide, Iron-sulphur centres
Summary equation
of DHA cycles in yeasts, 252*t*, 285, 287*f*
ribulose bisphosphate pathway, 44, 252*t*
of RuMP pathways, 67*t*, 69, 252*t*
of serine pathways, 96, 252*t*
Supplementary metabolism, 166
Swarmer cells of *Hyphomicrobia*, 35, 36*f*, 37*f*

T

Taxonomy, 2–4, 22
heretical, 4
of methanogens, 300–302*t*
of methanotrophs, 4–22
of methylotrophs unable to use methane, 22–38
of yeasts, 269–273
TCA cycle, *see* Tricarboxylic acid cycle
Temperature optima for growth
of methanotrophs, 6*t*, 8*t*, 9, 10, 11*t*, 21, 39
of other methylotrophs, 26*t*, 37, 273
in production of SCP, 330, 333
Tertiary amines
oxidation by trimethylamine monooxygenase, 200
substrates for trimethylamine dehydrogenase, 197
Tetradecane, oxidation by methane monooxygenase, 346*t*
Tetrahydrofolate
as "protective carrier" of formaldehyde, 115
in regulation of enzyme activity, 132, 133
Tetrahydrofolate formylase, in assimilation of formaldehyde (serine pathway), 102*t*
Tetramethylammonium monooxygenase, 196*f*, 197
Tetramethylammonium salts
oxidation to trimethylamine, 196*f*, 197
utilisation by yeasts, 273

Tetramethylenephenylenediamine (TMPD), reaction with cytochrome *o*, *see also* ascorbate/TMPD, oxidation of, 237
Tetrapyrrole, nickel-containing, *see* Factor F_{430}
Theoretical assimilation equations, *see* Assimilation equations (theoretical)
Thermoacidophiles, 300
Thermodynamic constants, *see also* Midpoint redox potentials, for reactions in oxidation of C_1 compounds, 154*t*
Thermodynamics of methanogenesis, 298, 306–308, 317, 320*f*, 323, 325–327
Thermodynamics of methanol oxidation
relevance to "methanol oxidase", 244
relevance to methane hydroxylation, 265
Thermophilic methanogens, *see also* *Methanobacterium thermoautotrophicum*, 298, 321
Thermophilic methylotrophs, 10, 11*t*, 21
Thermotolerant yeast, 274
Thiamine requirement of yeast, 273
Thioacetamide, as inhibitor of monooxygenase, 161
Thiobacillus A2, 34*t*, 48*t*
methylamine dehydrogenase in, 183*t*, 215*t*
RuBP pathway in, 49*t*
Thiobacillus intermedius, RuBP of, 53
Thiobacillus novellus, 34*t*, 48*t*
RuBP pathway in, 48*t*
Thioglycollate, as stabilising agent for methane monooxygenase, 157
Thiosemicarbazide, as inhibitor of methane monooxygenase, 161
Thiourea, as inhibitor of methane monooxygenase, 161
Toluene, oxidation by methane monooxygenase, 165*t*, 347*t*
Torulopsis sp., 272*t*, 273
evidence for DHA cycle in, 289*t*
peroxisomes and crystalloids in, 282*t*
rearrangement enzymes of, 289*t*

Index 429

Transacylase (E2 component of oxoacid dehydrogenases), 146
Transaldolase
 in CO_2 assimilation, 45f, 46, 58
 in formaldehyde assimilation (DHA cycle), 286f
 in formaldehyde assimilation (RuMP pathway), 62, 63f, 65f, 67t, 68, 69f, 87
Transaldolase variant of the RuBP pathway, 45f, 46, 55, 58
Transaldolase variants of the RuMP pathway, see Ribulose monophosphate pathway, variants
Transamination, see Aminotransferases; Serine-glyoxylate aminotransferase
Transcriptional control, proposed lack of, in obligate methylotrophs, 142
Transformation mapping, in *Methylobacterium organophilum*, 131, 132
Transketolase
 in assimilation of CO_2, 44f, 45f, 56
 "classical" enzyme in methylotrophic yeast, 291
 in formaldehyde assimilation (DHA cycle), 286f, 291
 in formaldehyde assimilation (RuMP pathway), 63f–65f, 68, 69f, 85, 86
Transpeptidation, in methanogens, 301
"Transvestite" enzymes, 117
Tricarboxylic acid cycle
 and basis of obligate methylotrophy, 142, 143
 distribution, 6t, 20, 138t, 139t
 enzymes of, in serine pathway, 97f, 103
 its function in methanotrophs, 6t, 20
 individual enzymes, 144–147
 operation and evidence for, 137–142
 and overproduction of metabolites, 340
 regulation of, 144–147
Trichloromethane, oxidation by methane monooxygenase, 164t
Trichoderma lignorum, 270t

Trimethylamine
 anaerobic growth of *Hyphomicrobium* on, 196f, 203, 217
 assimilation equations for, 253t
 enzymic assay of, 343
 growth of methanogens on, 326
 growth yields on, 257t, 258, 259
 inhibitor of dimethylamine dehydrogenase, 203
 as intermediate in oxidation of other methylated amines, 196f
 oxidation of, 196f–201, 216t–218
 utilisation by yeasts, 273
Trimethylamine dehydrogenase, 196f, 197–200, 216t, 217
 in assay of trimethylamine, 343
 limitation of growth yields and, 253t, 257t, 258, 259
 proportion of electron transport from, 266, 268
 prosthetic groups, 198–200
Trimethylamine monooxygenase, 196f, 200, 217
 limitations of growth yields and, 257t, 258, 259
Trimethylamine N-oxide
 as constituent of fish, 40
 as intermediate in oxidation of other methylated amines, 196f
Trimethylamine N-oxide aldolase, see Trimethylamine N-oxide demethylase
Trimethylamine N-oxide demethylase (aldolase), 196f, 200
Trimethylsulphonium salts, metabolism by *Pseudomonas* MS, 151
Triokinase (dihydroxyacetone kinase)
 in formaldehyde assimilation (DHA cycle), 286f, 289t, 290, 292
 in formaldehyde oxidation (DHA cycle), 287f
 mutants lacking, 290
Triose phosphate, see Dihydroxyacetone phosphate, Glyceraldehyde 3-phosphate
Triose phosphate dehydrogenase, see Glyceraldehyde phosphate dehydrogenase
Triose phosphate isomerase
 in CO_2 assimilation (RuBP pathway), 44f, 45f, 54

Triose phosphate isomerase (*continued*)
  in formaldehyde assimilation (DHA cycle), 286*f*
  in formaldehyde assimilation (RuMP pathway), 64*f*, 65*f*, 91
Trypsin-sensitive component of formaldehyde dehydrogenase, 188*t*, 190
L-Tryptophan, overproduction by methylotrophs, 341
Tungsten, in formate dehydrogenase of methanogens, 327
Type I methanotrophs, membranes of, 15, 16*f*, 17*f*, 22, 177
Type II methanotrophs, membranes of, 15, 18*f*, 22, 177
L-Tyrosine, overproduction by methylotrophs, 341

U

Ultrastructure of methanogens, 297*t*, 301
Uncoupling agents, effect on methanogenesis, 316

V

Valine
  accumulation by pyruvate dehydrogenase mutant, 340
  overproduction by methylotrophs, 341
Valine hydroxamate, in overproduction of metabolites, 341
Valinomycin, 316
Variants of carbon assimilation pathways, *see* individual pathways
Vibrio extorquens, *see* *Pseudomonas extorquens*
Viologen dyes, in biofuel cells, 349
Vitamin B_{12}
  bioassay of, 342
  production by methylotrophs, 339
  similarity of spectrum to factor F_{430}, 312
Vitamin B_{12} derivatives in methanogenesis, 312, 319
Vitamin requirements of methylotrophs, 12*t*, 26*t*, 31*t*, 37, 273

W

Whittenbury, methanotrophs isolated by, 4–8
Wurster's blue
  as electron acceptor for dimethylamine dehydrogenase, 203
  as electron acceptor from methanol dehydrogenase, 175
  as electron acceptor for methylamine dehydrogenase, 205
  as electron acceptor for *N*-methylglutamate dehydrogenase, 212*t*

X

Xylose, growth of yeasts on, 293
Xylulose monophosphate pathway, *see* Dihydroxyacetone cycle
Xylulose 5-phosphate
  in assimilation of CO_2 (RuBP pathway), 44*f*, 45*f*, 56, 57
  in assimilation of formaldehyde (DHA cycle), 286*f*, 291
  in assimilation of formaldehyde (RuMP pathway), 63*f*–66*f*, 69*f*, 85, 86
  intermediate in oxidation of formaldehyde (DHA cycle), 287*f*

Y

Y_{ATP}
  assumed value for prediction of growth yields, 250
  relationship to P/O ratio, 245–248, 255
  introduction to concept, 245–248
Yeasts
  alcohol oxidase of, *see also* Alcohol oxidase (yeasts), 274–279
  alkane utilising (growth yields of), 258, 281
  amine oxidase of, 208
  assimilation of formaldehyde in, 252*t*, 285–292
  ATP-limitation during growth of, 255
  catalase in methanol oxidation, *see also* Catalase in methylotrophic yeasts, 276*f*, 279–281
  a commercial source of flavins, 339
  culture collections, 270*t*–273

Yeasts (*continued*)
   growth on methane, 272t, 273, 283
   growth yields, *see also*
      Dihydroxyacetone cycle in yeasts, and growth yields, 247t, 329t, 332
      growth yields and SCP production, 256t, 331–333
   isolation on methanol, 270t, 273
   methylotrophic, *see also* individual yeasts, 269–274
   methylotrophic metabolism, 269–295
   oxidative metabolism in, 274–285, 287, 288

Yeasts (*continued*)
   peroxisomes in, 258, 274, 276f, 277f, 281–283, 344
   regulation of methanol metabolism in, 289t, 291–295
   a source of single cell protein (SCP), 247t, 329t, 331–333, 337t
   utilisation of amines as nitrogen source, 273
YFC factor, *see* Methanopterin
Yields, *see* Growth yields